# SYMPOSIUM ON APPROACHES TO EARTH SURVEY PROBLEMS THROUGH USE OF SPACE TECHNIQUES

*Symposium Organized by:*

The Committee on Space Research — COSPAR
and the "Deutsche Forschungsgemeinschaft"
of the Federal Republic of Germany
in collaboration with and with the participation of:

International Union of Biological Sciences — IUBS
International Union of Geodesy and Geophysics — IUGG
International Union of Geological Sciences — IUGS
International Geographical Union — IGU
International Union of Radio Sciences — URSI
Scientific Committee on Antarctic Research — SCAR
Scientific Committee on Problems of the Environment — SCOPE
Joint ICSU/WMO Organizing Committee for GARP — JOC

*Symposium Organizing Committee:*

M. Tepper — Chairman
F. W. G. Baker
Z. Niemirowicz
S. Ruttenberg

COSPAR

# APPROACHES TO EARTH SURVEY PROBLEMS THROUGH USE OF SPACE TECHNIQUES

Proceedings of the Symposium held in

CONSTANCE, F.R.G.
23 — 25 May 1973

Edited by:
P. Bock

With the assistance of:

F. W. G. Baker
and
S. Ruttenberg

AKADEMIE-VERLAG · BERLIN
1974

Executive Editor:
A. C. Stickland

*Library of Congress Catalog Card Number 73-94252*


202 · 100/474/74
Gesamtherstellung: VEB Druckhaus „Maxim Gorki", Altenburg
Bestellnummer: 761 927 3 (6201) · LSV 1405
Printed in GDR

# PREFACE

This Symposium on Approaches to Earth Survey Problems Through the Use of Space Techniques has a certain unique aspect: while the speakers are identified, as is usual in COSPAR, as representatives of countries, they are additionally identified by affiliation with another organization associated with the International Council of Scientific Unions (ICSU). For example, there is a speaker from the International Union of Biological Sciences (IUBS), five from the International Union of Geodesy and Geophysics (IUGG), three from the International Union of Geological Sciences (IUGS), one from the International Geographical Union (IGU), and two from the International Union of Radio Science (URSI). In addition, there are five speakers from the Scientific Committee on Antarctic Research (SCAR), one from the Scientific Committee on Problems of the Environment (SCOPE), and two from the Scientific Committee on Ocean Research (SCOR), indicating that three special scientific committees of ICSU are being represented, and finally, there is a speaker from the Joint Organizing Committee (JOC) for the Global Atmospheric Research Program (GARP) which is a committee organized jointly by ICSU and the World Meteorological Organization (WMO) for the planning and overview of the Global Atmospheric Research Program.

In 1973 the World Meteorological Organization completes 100 years of devoted service to meteorology. In congratulating the WMO on its achievements I should like to express the hope that cooperation between COSPAR and WMO will continue to develop in the future as effectively as in the past.

Eight papers describe the preliminary scientific results from the US Earth Resources Technology Satellite (ERTS) and closing the Symposium six additional contributed papers also deal primarily with ERTS results.

In our audience, there were representatives from such international organizations as our parent organization, ICSU, World Health Organization, European Space Research Organization, World Meteorological Organization and the Pan-American Institute of Geology and History. We also had official participants from some of the ICSU Unions who did not give papers but who joined in the discussions.

To explain why we have identified most of the speakers with their union or special scientific committee affiliation, I would like to recall to you the mission of COSPAR Working Group 6 which, in a sense, is the host of this Symposium. Working Group 6 has for its full title the "Application of Space Research to

Meteorology and Earth Surveys". The key word in that title is "application"; and this application is primarily to scientific problems. Thus, we wish to stress that the emphasis of our Working Group activity and its primary concerns is not the development of space technology and techniques as an end in itself; but rather its application to problems that are of major concern to earth scientists.

As an illustration, I would like to narrate briefly our experience in Working Group 6 with the Global Atmospheric Research Program. This program, growing out of UN resolutions in the early 1960's, was undertaken jointly by ICSU and WMO who established a Joint Organizing Committee for GARP. This Committee approached COSPAR, and specifically COSPAR Working Group 6, for assistance in developing the space related aspects of the observing systems required for GARP. Working Group 6 met many times with the atmospheric scientists and the meteorologists concerned with the problems of GARP and through a series of iterations of requirements and space capabilities has finally produced a viable observing system which now represents an important basis for continuing with the global program.

We in Working Group 6 feel that this mode of operation can be and should be extended to other scientific disciplines beyond meteorology, and it is our hope that this Symposium represents a first step in that direction.

We feel that it is important to bring familiarity with, and knowledge of space technology capabilities into the discussions and plans of the scientists in the various Unions; and reciprocally we feel that the concerns of the scientists and the problems with which they deal should be brought to the attention of the space experts to the mutual benefit of both. That is why we have approached the scientific unions and scientific committees and asked *them* to nominate *their* speakers and representatives to come here to this Symposium in Constance and to represent *their* disciplines. We have asked them to prepare their papers so as to bring out three main aspects:

First, to state and explain clearly the scientific problem or problems which they are addressing;

Second, to indicate by illustration, if at all possible, how remote sensing, the major tool for viewing the earth from space, has been helpful, has contributed, to the solution of the problem;

And finally, and here we in COSPAR would consider this third aspect most important, the opinion of these discipline specialists on:

The prospects for the future?

What are the requirements?

What are the possible remote sensing solutions that they visualize, that could guide the consideration, planning and development activity of the space technologists and space scientists in the future.

With this exchange of conversations between the problem oriented scientists and the space technologists and with an iteration of these conversations as often as possible, it is our hope that we will enrich the contribution of each to the mutual benefit of each and finally to the benefit of man on earth.

M. Tepper
Chairman, Working Group 6

## SYMPOSIUM A

### Contents List

*ERTS 1 Results*



*Contributed Papers*



*National Reports on Activities in Eearth Surveys*



*Latest Results in Meteorology and Earth Surveys*

# General Considerations

Earth Survey Problems — Akademie-Verlag, Berlin 1974

# EROS PROGRAM AND ERTS 1 SATELLITE APPLICATIONS TO GEOPHYSICAL PROBLEMS

W. D. Carter

EROS Program, US Geological Survey, Washington, D. C., USA

The US Department of the Interior Earth Resources Observation Systems (EROS) Program is a multi-disciplinary Department-wide effort to evaluate data from remote sensors borne by aircraft and spacecraft as they apply to operational problems and responsibilities. These problems and responsibilities range from managing large tracts of public range and forest lands, Indian reservations, wildlife refuges, national parks and recreation areas, to geologic mapping for mineral and water resource inventories and studies of the environmental impact of proposed public works projects. The EROS Program began in 1966 as a result of earlier feasibility studies conducted in cooperation with the National Aeronautics and Space Administration (NASA). It has grown in both scope and area so that it now has cooperative endeavors with many Federal, state and local agencies, universities, foreign nations and international organizations.

The Earth Resources Technology Satellite, ERTS 1, successfully launched on 23 July 1972, was designed to meet the needs of many earth resource scientists. Specifications for near-orthographic multiband images and a data relay capability were developed from recommendations provided to NASA by the Departments of Interior and Agriculture. Over 320 principal investigators are now evaluating ERTS 1 data from many regions of the globe, and of these more than 70 are geologists. The broad synoptic view provided by small-scale ERTS 1 images, under uniform lighting conditions, has enabled geologists to map lineaments and other structural features over large areas. Near-infrared images which portray vegetation in light tones and soil moisture in dark tones have reduced the problem of distinguishing structural features in some heavily vegetated regions. A limited number of repetitive satellite images acquired after leaves or snow have fallen have suggested new advantages to geological mapping. Yet to be realized is the high probability of identifying areas where anomalous snow-melt patterns may be of importance in prospecting for new sources of geothermal energy.

Experiments from aircraft employing multiband camera and optical-mechanical scanner systems have shown that reflectance minima occur both in the visible and near-infrared spectra in rocks rich in iron-bearing minerals. These minima are due to optical absorption caused by electronic transitions in the ferric and ferrous ions. Increases in near-infrared reflectance and temperature have been noted in coniferous vegetation overlying soils enriched by copper and molybdenum. Such observations have not yet been made at satellite altitudes. However, bi-band ratioing techniques using two bands (3.5—5.5 μm and 10—12.5 μm) of thermal infrared data from the Nimbus 3 satellite have enabled geologists to distinguish several rock and soil types and improve on small-scale maps of desert areas in Saudi Arabia and in Muscat and Oman. Geologic materials have been discriminated from Nimbus 3 and 4 reflectance and thermal data by computing the day-night temperature difference and maximizing the effects of thermal inertia.

## 1. Introduction

Investigations of the geologic applications of remote sensors from air and space-borne platforms by the US Geological Survey began in 1964 as a small effort funded by the National Aeronautics and Space Administration (NASA). Initial results were sufficiently encouraging so that research was expanded the following year into the fields of cartography, geography and hydrology. By the end of 1966 it was determined that many applications could also be developed within other bureaus of the Department of the Interior and to this end, the Earth Resources Observations System (EROS) Program was formed. The EROS Program is a multi-discipline effort in which seven agencies of the Department have vigorously sought, obtained and evaluated photographic and line-scan data from aircraft and spacecraft as they apply to resource inventory and management responsibilities. Nimbus and TIROS weather satellite data and photography from Mercury, Gemini and Apollo manned spacecraft missions provided early experience of viewing large areas of the earth from afar. This experience helped NASA and its sister agencies design the first Earth Resources Technology Satellite (ERTS 1).

ERTS 1, launched on 23 July 1972, flies at 915 km above the earth's surface, in a circular, near-polar orbit to provide synoptic views which cover 34225 $km^2$. Each scene is orthographic or distortion free. A mid-morning sun angle provides shadowing of surface features of a few hundred meters in relief. Two imaging systems are employed. One is a 3-camera, high resolution (4500 line) return beam vidicon (RBV) system in which each camera records a different spectral response. Unfortunately this system operated for only a few weeks before a power failure incapacitated its tape recorder. The RBV system is temporarily shut down but may be reactivated in the future if the second system fails or appears to degrade.

The second imaging system is the multi-spectral scanning system (MSS), consisting of four detectors that operate simultaneously in the green (500 to 600 $\mu$m), red (600—700 $\mu$m) and two near-infrared (700--800 and 800—1000 $\mu$m) bands. All the data, in the form of radio signals, are relayed directly to the nearest NASA reception stations in Alaska, California or Maryland or stored on tape recorders until the stations are in view. Canada also has a station for direct transmission and others are being constructed or considered in Brazil, Europe and Africa.

A third system on the satellite is an S-band data relay system for monitoring ground-based data collection platforms in North and Central America. The ground-based instruments include stream and tide gages, water quality meters, snow measuring devices for hydrologic experiments; tiltmeters and seismic event counters are used in conjunction with volcano monitoring experiments. EROS Program experimenters have a total of 116 such instruments located on a large number of test sites which include major river basins and estuaries, lakes and reservoirs as well as recently active volcanoes. The data relay system has been proven to be effective for a distance of 6500 km: from Iceland where a thermistor array monitored volcanic temperature and relayed this information via ERTS 1 to the Goddard Data Reception Center at Greenbelt, Maryland. Further details of the ERTS 1 system and its products can be obtained by requesting a copy of the *ERTS Data Users' Handbook*, available on sale for $ 10.00 from the General Electric Company, Beltsville, Md.

The ERTS 1 imaging systems operate in the visible and near-infrared spectrum and sense reflected radiation from the sun. Surface reflectance varies greatly from place to place depending on the composition, reflectance and absorption characteristics of the surface materials and also on the local composition of the atmosphere and its scattering characteristics. For example, images of forested areas of the tropics and temperate zones have less contrast than arid regions because of scattering properties of atmosphere in these regions. The green and red bands appear to be equally suitable in desert or arctic regions, while the near-infrared bands provide optimum information in tropical and temperate regions because atmospheric scattering is minimized and vegetative reflectance is greatest at the longer wavelengths. In the near-infrared bands surface water absorbs light and appears black except where silt or other impurities create a scattering effect. Moisture in soils appears as dark tones of gray.

It is (i) the synoptic view of large areas under uniform illumination, (ii) the absorption and reflectance characteristics of the earth surface, and (iii) repetitive observation that make ERTS 1 an excellent addition to our geophysical tools. Let us now consider three examples of geologic and geophysical problems that have presented difficulties to earth scientists for many years.

## 2. Problem 1: Regional and Continental Tectonic Analysis

One of the more difficult and complex geophysical problems that geoscientists have had in the past is that of regional and continental tectonic analysis. Such studies have had to rely on the synthesis of detailed surface mapping completed on a quadrangle by quadrangle basis and pieced together subjectively by scientists interested in a broad overview. These syntheses are then correlated with geophysical data obtained by ground or airborne surveys at different map scales. While many of these analyses are excellent in areas where surface mapping and geophysical information are abundant, the compilations are subjective and often valuable information is discarded. There are many areas of the world where mapping is inadequate or non-existent and, therefore, must be studied by other methods. Views of large areas as recorded by ERTS 1 place geologic structures in their true perspective and relationships. They are, therefore, ideal for regional and continental tectonic analysis. Adequate tectonic maps are fundamental to mineral resource development because they provide a base on which to develop ideas on metallogenesis and metallogenetic provinces which, in turn, are basic keys to exploration and development.

Synoptic, multi-spectral views of the earth provided by satellite imaging systems have already provided significant new information on the geologic structure of Alaska (Fig. 1) and a revision of metallogenic hypotheses in that rugged, less well-known state [1, 2]. Fig. 2a shows the distribution of metallic ore deposits of Alaska according to concepts based on geologic mapping and synthesis prior to ERTS 1. Fig. 2b shows the revision of metallogenic provinces due to new lineament information provided by Nimbus 4, and refined with ERTS 1 data. While it will be several years before mapping and exploration in Alaska can confirm or refute these new concepts, it is important to note that the new hypothesis has narrowed or confined the targets of interest to smaller areas and has resulted in a more rational distribution pattern for some ores, e.g. mercury, considered by many to be fault-associated. This interpretation has identified areas where new work

Fig. 1. A clear day in Alaska from Nimbus 4.

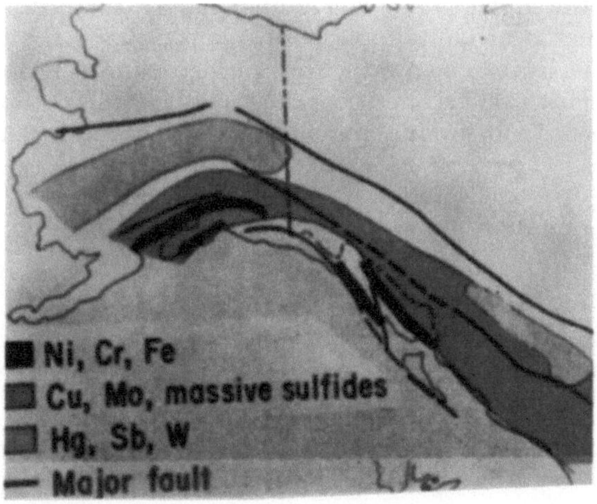


Fig. 2a. Metallogenic provinces of Alaska based on geologic and geophysical mapping.

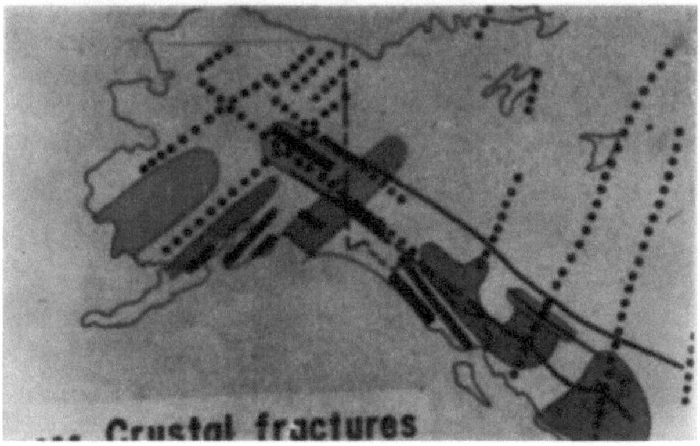


Fig. 2b. Revised metallogenic provinces of Alaska based on geologic and geophysical information combined with new information provided by interpretation of Nimbus 4 and ERTS 1 images.

should be done and should therefore reduce the time and costs of future mapping and exploration.

New views of the earth from ERTS 1 have also provided geologists with the opportunity to "see" large structures in well mapped areas that have not heretofore been defined. For example, Fig. 3 is a mosaic of ERTS 1 images of the State of Nevada where a major lineament is revealed that trends northeastward across the northern part of the state, transecting the northern end of the north-trending Basin and Range province [3]. Projection of this trend (Fig. 4) to the southwest indicates that it also marks the break between the Sierra Nevada batholith on the south and the Cascade volcanic chain on the north. If projected further to the west, it passes through the entrance (mouth) of San Francisco Bay. Here there is a suggestion that valleys, fault extensions and rock formations may be offset to the east along the north side of this projection. To the northeast, in southern Idaho, small segments of faults have been mapped along the trend. Along the trend is the Snake River downwarp, the southern limit of the Idaho volcanic field, the Island Park caldera and the geothermal areas of Yellowstone National Park. The Fromberg fault in Montana not far south of Billings may also be part

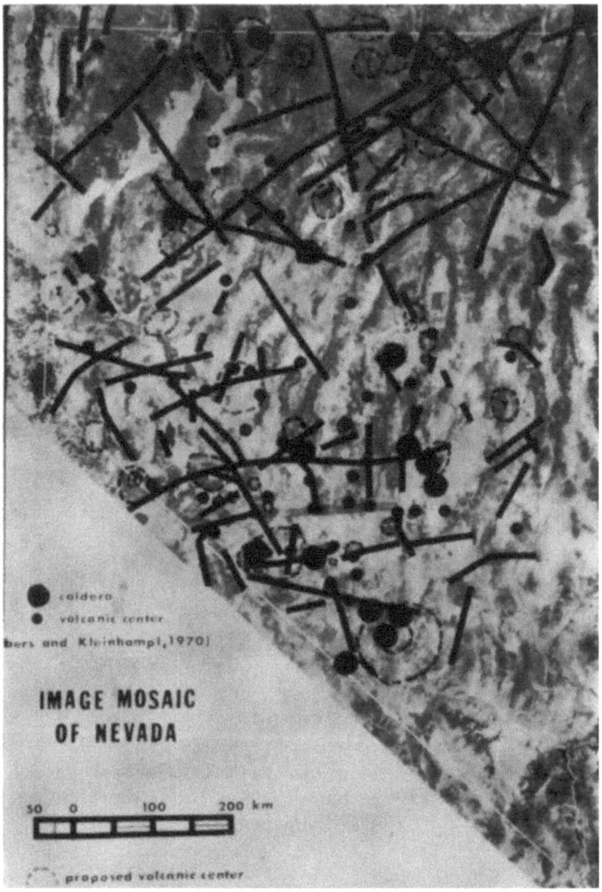


Fig. 3. ERTS 1 image mosaic of the State of Nevada.

of this trend. There is a coincidence of magnetic patterns with this lineament which suggests a tectonic and petrologic relationship.

A third aspect of tectonic mapping problem has to do with exploration for energy resources such as new geothermal and petroleum reserves. The repetitive nature of ERTS 1 permits us, for the first time, to collect data throughout the year; to map snowfall and snowmelt patterns with sufficient resolution to determine where geothermal heat sources may affect the surface. Snowmelt patterns have been recognized in areas of near-surface geothermal anomalies of Yellowstone

Park where snowfall thickness and melting rates have been used as a calorimeter by White [4].

In southern Mississippi arcuate stream drainages and vegetation patterns have been identified on ERTS 1 images that are related to known salt domes that are the sources of petroleum and possible geothermal energy in the state (Fig. 5).

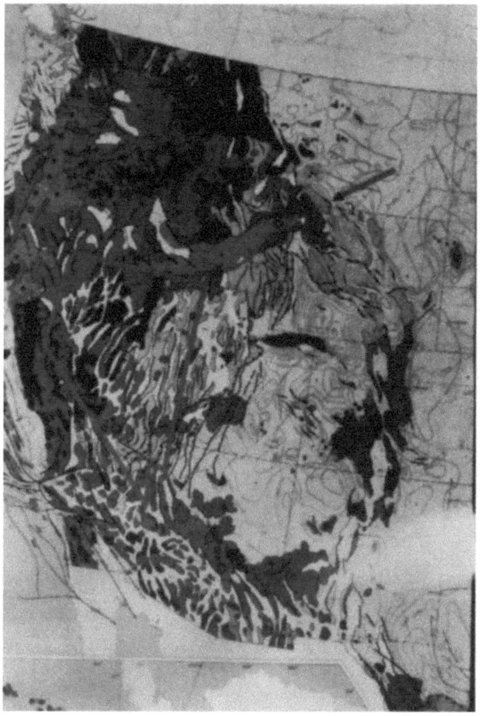

Fig. 4. Generalized tectonic map of the United States showing indicated projection of the northern Nevada lineament.

Similar arcuate patterns have been identified in ERTS images south of the known salt dome region suggesting additional favorable structures for exploration. While most of the domal structures are less than 10 km in diameter, one has been defined that is 70 km in diameter. Although the significance of this feature has not yet been determined, it certainly defines an area that should be studied in greater detail by ground and airborne geophysical methods.

## 3. Problem 2: Volcano and Earthquake Monitoring

Active geologic phenomena such as volcanoes and earthquakes, especially those located in populated areas, present very special problems to man as he attempts to develop monitoring and predictive capabilities. On-site laboratories such as the Hawaiian Volcano Observatory have contributed much to man's understanding of volcanism. New instruments such as tiltmeters and seismic event counters have been developed and tested at such sites to increase our ability to monitor the pulse of the earth.

With the flight of ERTS 1 the US Geological Survey began an experiment to monitor a large number of volcanoes that constitute part of the circum-Pacific "Ring of Fire". Sixteen volcanoes in a zone extending from Alaska to Nicaragua in Central America have been instrumented with tiltmeters, seismic event counters and thermocouples for monitoring changes in land surface, seismicity and temperature. These instruments are coupled to radio relay systems that transmit

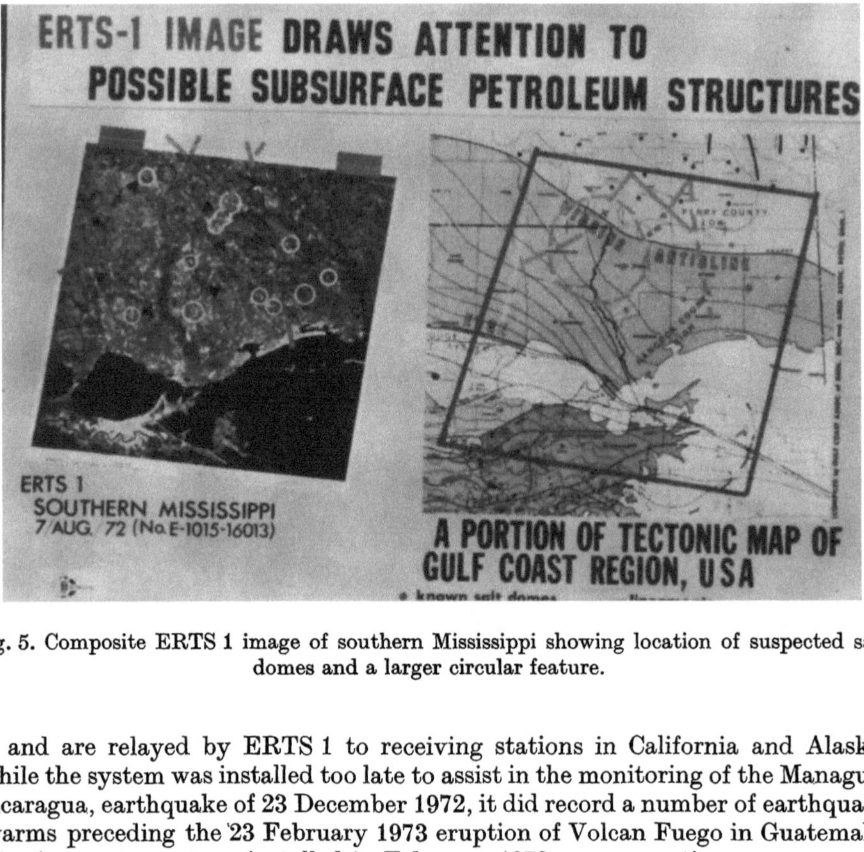


**Fig. 5.** Composite ERTS 1 image of southern Mississippi showing location of suspected salt domes and a larger circular feature.

to and are relayed by ERTS 1 to receiving stations in California and Alaska. While the system was installed too late to assist in the monitoring of the Managua, Nicaragua, earthquake of 23 December 1972, it did record a number of earthquake swarms preceding the 23 February 1973 eruption of Volcan Fuego in Guatemala. Seismic event counters, installed in February 1973, were counting an average of five events a day until the number of events increased to 80 or more. Five days later Volcan Fuego erupted [5].

ERTS 1 data were used on a limited basis due to heavy cloud cover and low sun-angle in a study of the eruption of Kirkjufell on the island of Heimaey, Iceland [6]. However, the NOAA 2 weather satellite with its more frequent repeat cycle, poorer resolution and thermal imagery proved more useful in mapping the volcanic smoke plume and making quantitative measurements of radiant emission. More recent work by Williams in analysis of ERTS 1 imagery of Iceland has revealed several new sub-glacial calderas within the Vatnajökull (ice cap).

## 4. Problem 3: Discrimination of Rock and Soil Types

Remote detection and discrimination of rock and soil types has been attempted from air and spacecraft during the past five years. An early attempt was made by Smedes and others [7, 8] in Yellowstone Park using the airborne multiband scanning system developed by the Willow Run Laboratories of the University of Michigan. Approximately nine generalized terrain types (rock rubble, kame terraces, etc.) were successfully mapped using four optimum bands from the twelve collected that were trained on small test plots (Fig. 6). Accuracies of over

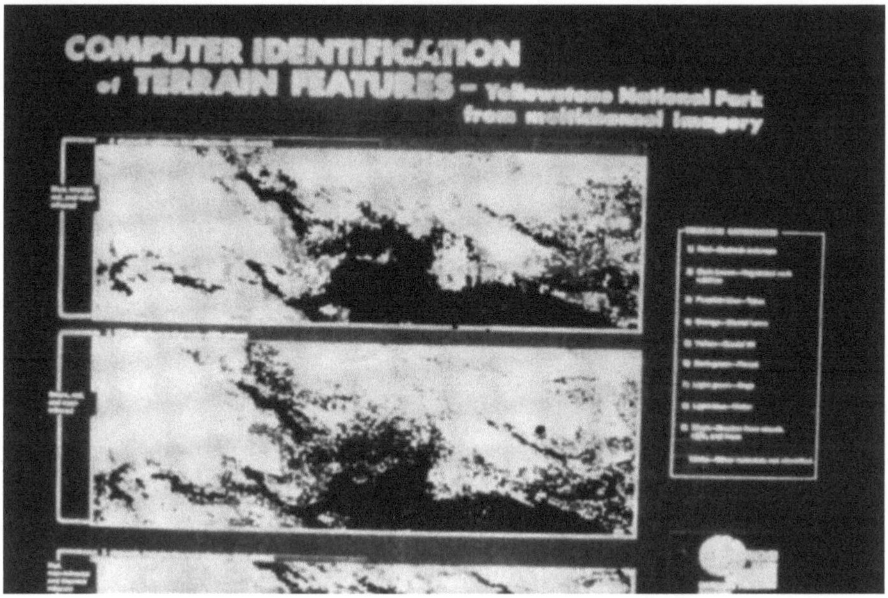


Fig. 6. Multiband scanner image of surface terrain characteristics acquired from aircraft, Yellowstone National Park, Wyoming.

85% were obtained. Using the same bands that had been selected for the ERTS 1 satellite system the accuracy dropped a few per cent but still exceeded 80%.

Testing of a Reconofax IV thermal scanner (8—14 μm) [9] and, later, the same multiband scanning system using visible reflected infrared and thermal-infrared bands in the Mill Creek, Oklahoma, area [10] enabled a clear distinction to be made between limestone and dolomite, silica sandstone and granite. "Such discrimination of rock type was possible because of a combination of the inherently different albedo and thermal inertia properties of the limestones and dolomites" [9]. Pohn and others [11] have also used ratioing techniques in which day and night temperature differences in the thermal bands from Nimbus 3 and 4 satellite imaging systems are weighed against each other. They were successful in discriminating eleven rock and soil types in the Muscat and Oman area of the Arabian peninsula. Estimates of rock and soil density can be calculated by this approach. The clarity of atmosphere and lack of vegetation in this region enabled the investigators to conduct their experiment without interference from such factors. Several changes in the existing generalized geologic maps of the area were suggested by this study.

Fig. 7. Multiband camera image of oxidized breccia pipe in the Homestake Mine area, Montana. 1, blue; 2, green; 3, red; 4, near infrared.

A four-band multi-spectral aerial camera was tested from aircraft using bands and filters similar to ERTS 1 to study the effects of iron absorption in the near-infrared region [12]. An oxidized breccia pipe containing pyrite and copper ores in the Homestake Mine area in the New World mining district of the Beartooth Mountain Region in Montana was flown at moderate altitudes. The oxidized zone appeared dark on the near-infrared band while there was no marked anomaly in the visible region (Fig. 7). A similar response was recorded in the near-infrared band in data acquired over a nearby, chrome-bearing, ultramafic body (Fig. 8).

While these deposits were relatively small in areal extent and below the resolution of the ERTS 1 system, it appears likely that larger but similar deposits can be recognized from ERTS 1. In fact, it appears that several new ultramafic bodies have been identified in color composites of ERTS 1 images in the foothills of the Sierra Nevada of California by Lathram (1973, written communication). These and

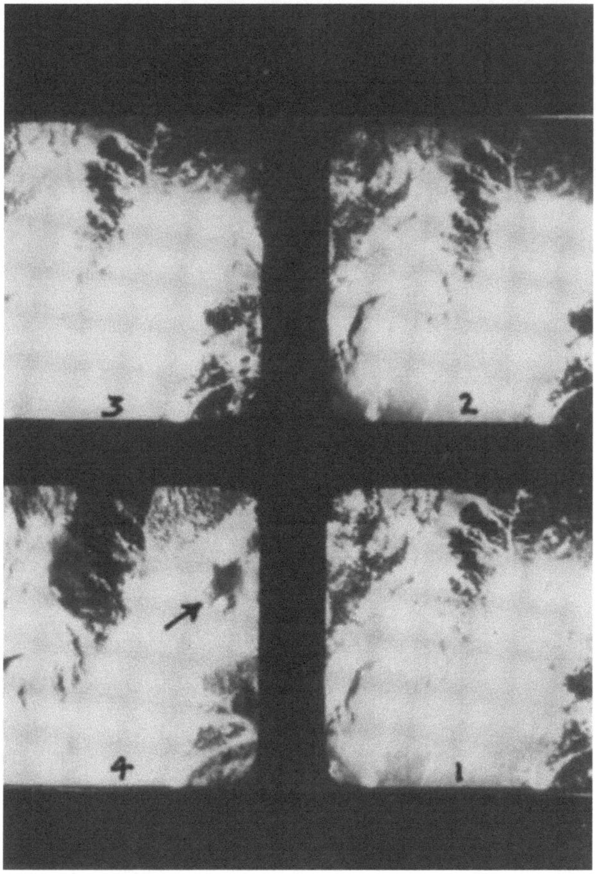


Fig. 8. Multiband camera image of chrome-bearing, ultramafic body near Homestake Mine area, Montana. 1, blue; 2, green; 3, red; 4, near infrared.

other suspected examples of iron absorption are now being field-checked. Thus far, these results are consistent with laboratory spectra for iron-rich and iron-poor minerals measured by Hunt and Salisbury [13] as shown in Fig. 9, and for selected felsic and mafic rocks as in Fig. 10 as determined by Ross, Adler and Hunt [14]. The visible and near-infrared reflectance spectra for limonite, Fig. 11, show diagnostic absorption at about 0.9 μm and 0.65 μm and a reflectance maximum at about 0.80 μm. The positions of the iron-absorption bands depend on the oxidation state and coordination of the iron bands.

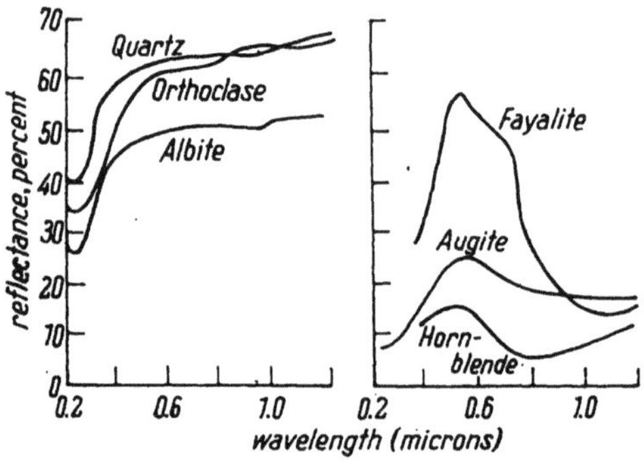


Fig. 9. Laboratory spectra of iron-rich and iron-poor minerals in the 0.2—1.5 μm region.

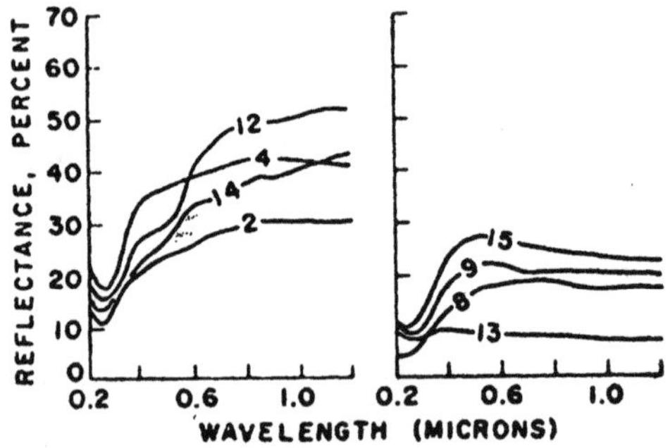


Fig. 10. Laboratory spectra of felsic and mafic rocks in the 0.2—1.5 μm region.

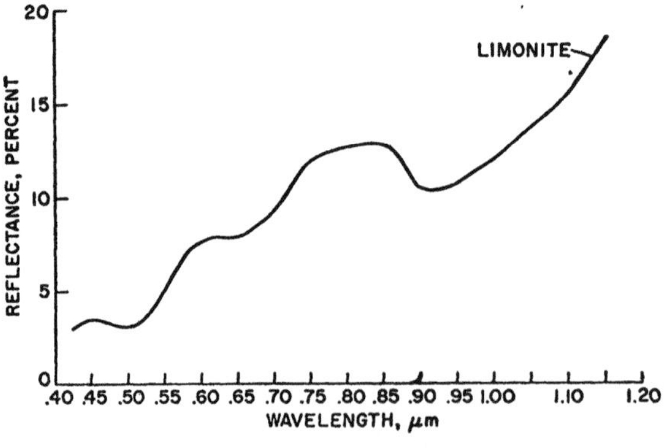


Fig. 11. Visible and near-infrared reflectance spectrum for limonite in the 0.4—1.2 μm region.

## 5. Conclusions

Three major geophysical problems being addressed by studies of ERTS 1 data are: (i) regional and continental tectonic analysis; (ii) volcano and earthquake monitoring, and (iii) rock and soil type discrimination. At this time it can be stated that ERTS 1 imagery and data relay capability can contribute effectively to tectonic analysis and volcano monitoring on an operational basis. Modifications and additional instrumentation will be required before an earthquake monitoring system can reach the same stage and considerably more knowledge must be acquired before an operational predictive capability can be developed.

In the area of rock and soil type discrimination much more work is needed. ERTS 1 provides some of the basic ingredients required by virtue of its multispectral capability and current work is very encouraging. However, one or more thermal bands in the 10.5—12.5 μm range will be required to make such a mapping system more workable, especially if operated in both daytime and pre-dawn orbits. The multiband concept in which several visible, reflected infrared and thermal infrared bands will be tested over special sites during the current Skylab experiment will add to our knowledge. This experience should enable us to refine our requirements for future operational satellite systems.

## Acknowledgment

Publication authorized by Director, US Geological Survey.

## References

[1] E. H. Lathram, Science **175**, 1423 (1972).
[2] E. H. Lathram, I. L. Tailleur, W. W. Patton, Jr. and W. A. Fischer, Proc. NASA Symp. on Significant Results obtained from ERTS 1, Abstracts: GSFC, Greenbelt, Md. 1973 (p. 30).
[3] L. C. Rowan and P. H. Wetlaufer, Proc. NASA Symp. on Significant Results obtained from ERTS 1, Abstracts: GSFC, Greenbelt, Md. 1973 (p. 46).
[4] D. E. White, Geol. Soc. Amer. Prog. with Abs., 1968 Ann. Meeting, Mexico City 1968 (p. 318).
[5] P. L. Ward, J. P. Eaton, E. Endo, D. Harlow, D. Marquez and R. Allen, Proc. NASA Symp. on Significant Results obtained from ERTS 1, Abstracts: GSFC, Greenbelt, Md. 1973 (p. 35).
[6] R. S. Williams, Jr. and others, Proc. Symp. on Significant Results obtained from ERTS 1, NASA GSFC, Greenbelt, Md. 1973 (vol. **1**, Sec. A, p. 317).
[7] H. W. Smedes, N. M. Spencer and F. J. Thompson, Proc. 7th Int. Symp. on Remote Sensing of Environment, Univ. of Michigan 1971 (vol. **3**, p. 2073).
[8] H. W. Smedes, H. J. Linnerud, L. B. Woolaver, M. Y. Su and R. R. Jayroe, NASA 4th Ann. Earth Res. Prog. Rev. **3**, p. 61-1 (1972).
[9] L. C. Rowan, T. W. Offield, K. Watson, R. D. Watson and P. J. Cannon, Bull. Geol. Soc. Amer. **81**, 3549 (1970).
[10] K. Watson, R. C. Rowan and T. W. Offield, Proc. 7th Int. Symp. on Remote Sensing of Environment, Univ. of Michigan 1971 (vol. **3**, p. 2017).
[11] H. A. Pohn, T. W. Offield and K. Watson, NASA 4th Ann. Earth Res. Prog. Rev., **3**, p. 59-1 (1972).
[12] L. C. Rowan, NASA 4th Ann. Earth Res. Prog. Rev. **3**, p. 60-1 (1972).
[13] G. R. Hunt and J. W. Salisbury, Modern Geol. **1**, 283 (1970).
[14] H. P. Ross, J. E. M. Adler and G. R. Hunt, Icarus **11**, 46 (1969).

Earth Survey Problems — Akademie-Verlag, Berlin 1974

# RADIO TECHNIQUES APPLIED TO OCEANOGRAPHY AND EARTH SCIENCE: OCEANIC WIND MEASUREMENT AND OVERLAND IMAGING AS EXAMPLES

R. K. Moore

Director, Remote Sensing Laboratory, The University of Kansas, Lawrence, Kansas, USA

## Abstract

Radio measurements from space will be extremely valuable in future earth surveys. Experiments and scattering theory indicate that microwave backscatter from the ocean is proportional to wind speed, so a satellite scatterometer can provide global wind data. Cloud attenuation of the scattered signal can be measured by use of a microwave radiometer coincident with the scatterometer, thus correcting the wind data. Several questions remain to be answered with both theory and experiment to improve the effectiveness of a wind sensor.

Images are the most important data form for most earth scientists working over land, yet clouds obscure many areas at critical times. Imaging radars on spacecraft can overcome this difficulty, and provide additional information not available except in the radio wavelengths. This has already been proven for geology and vegetation mapping from aircraft in South America and elsewhere; yet only single-wavelength systems have been used. Future systems will use multi-spectral radar images, just as multi-spectral scanners and cameras are used for shorter wavelengths. Much basic research needs to be performed to demonstrate the microwave spectral response of the earth's surface, and the theory for describing this response is in its infancy. Recent URSI meetings have demonstrated the growing interest in this subject by radio scientists.

## 1. Introduction

Radio techniques will be extremely important to future earth surveys from both aircraft and spacecraft. Much progress has been made in the nine years since NASA started research into the applications of radar for earth survey, yet much more research must be done even though operational aircraft imaging radar systems are in use today and spacecraft techniques are being tested. The International Union of Radio Science (URSI), through its Commission II, remains a leading forum for exchange of research data on the microwave responses of the earth and atmosphere. Much additional research is needed, and the fundamental work will remain the special interest of URSI.

Radio waves can penetrate clouds, and usually can penetrate rain. Thus, microwave sensors are most important when the user cannot wait for a cloudfree day to obtain visible and IR images. Furthermore, the microwave response of the earth does not depend on the angle of the sun, and measurements are the same day and night, so constraints on microwave sensing are much less severe than on visible-IR sensing.

Even when clouds are not present, microwave responses are different from optical-IR responses, so combining them permits better identification of the character of the ground sensed. Thus, in clear weather, microwave and optical-IR sensors complement each other.

Radio waves penetrate some vegetation and soil, so the radar provides information about things that are invisible to cameras and IR sensors. Short radar wavelengths penetrate less than longer wavelengths, so the short wavelengths give more information about the plants and the long wavelengths give more about the soil.

Applications of these sensors are already practical in certain cases, particularly where geometric context is the most important tool of the interpreter. Thus, radar images may often be interpreted by photointerpreters with little knowledge of the electromagnetic phenomena. Other applications depending on knowledge of the relative response of different objects and materials, however, take longer to develop. We know little about the mechanism for radar scatter and microwave emission from soil, vegetation, buildings, and other objects found on the land, so both theoretical and experimental studies are needed. Furthermore, the spectral response at microwavelengths is little known, and many experiments are needed.

In this paper we discuss only two examples of radio sensors: a poor-resolution radiometer-scatterometer for oceanography and meteorology over the oceans, and fine-resolution imaging radars for land observations. The theoretical and experimental basis for the first system is well established and it is now time to test performance of such a system—as will be done on Skylab. Research is still required to answer some remaining questions with this system, however. The fine-resolution imaging radars are being used operationally over the land, but the need for research to support new applications of such radars is much greater than for the oceanographic radiometer-scatterometer.

## 2. Oceanic Wind Measurement—An Example of a System with Well-Developed Basis

Ocean wave forecasting computer programs use wind field measurements as their input data. In major shipping lanes, such as those from North America to Europe and Japan, many ships transmit wind observations by radio to forecasting centers, so forecasts for these regions are possible today. In most of the world's oceans, especially in the southern hemisphere, too few ships are present to permit adequate forecasts to be made. Wave forecasts allow routing ships for minimum transit time, and they permit ships to avoid storms that can cause damage to ships and their cargoes. With the many large tankers now voyaging through the southern oceans, this is particularly important.

Weather forecasting at sea is closely related to wave forecasting, but is also important because storms generated at sea control weather over land areas. Thus, oceanic weather forecasting information would be important even if no ships travelled the seas. Rain falling at sea has an important effect on the heat and moisture balance of the world's atmosphere. Here, too, this sensor can provide valuable information not available by any other means. The dimensions of the problem are illustrated in Fig. 1 [1]. At the top a pressure field analysis has been made for the northern Pacific Ocean based on 236 ships available there to provide reports. Below, the number of ships available in the southern Pacific, only 17, has

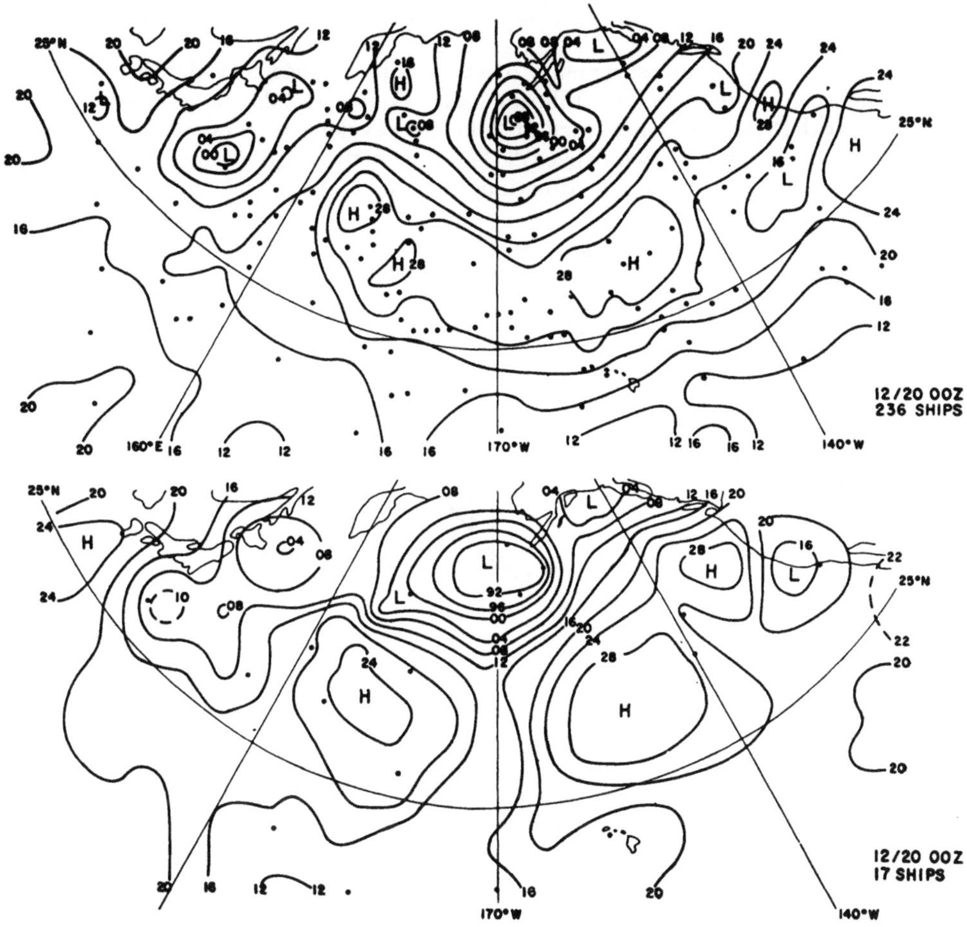


Fig. 1. Comparison of simulated pressure field analysis using RADSCAT (bottom) with 236-ship analysis (top).

been used along with simulated satellite RADSCAT (radiometer-scatterometer) data. Clearly 17 ships would be totally inadequate for a suitable analysis, but addition of the RADSCAT data has made the analysis possible.

Radar scatter from the ocean has been studied extensively, with the first important theoretical work in the early 1950's [2—4]. The last few years have seen a refining of the theory of scatter along with that of the wave spectrum so that today there is good agreement between theory and experiment. Scatter from the rough surface between air and ocean is a classical two-medium random-interface boundary value problem. The most successful theories combine elements from large-scale roughness scatter and from a perturbation technique strictly applicable only to very small-scale roughness [5, 6]. The small scale effect is predominant, and is caused by the radar response to capillary waves (surface tension waves) on the ocean. Pierson has shown the variation of the capillary waves with wind speed [7], and the results compare favorably (Fig. 2) with the wave tank radar measurements of Wright [8].

Our analysis [9] of measurements of the wind response of radar backscatter from the ocean made by NASA/Johnson Space Center at 2.25 cm wavelength and by US Naval Research Laboratory at 3.35 cm wavelength shows that the radar signal is proportional to (windspeed)$^{\alpha}$. The exponent $\alpha$ differs for different incident angles, polarizations, and directions relative to the wind. An example for

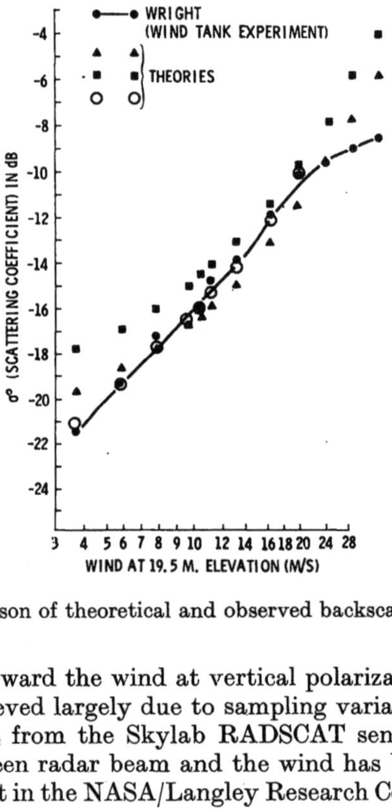


Fig. 2. Comparison of theoretical and observed backscattering values.

observations looking toward the wind at vertical polarization is shown in Fig. 3. The uncertainty is believed largely due to sampling variability, but proof of this hypothesis awaits data from the Skylab RADSCAT sensor. Variation of radar signal with angle between radar beam and the wind has been observed for some time, and an experiment in the NASA/Langley Research Center AAFE RADSCAT

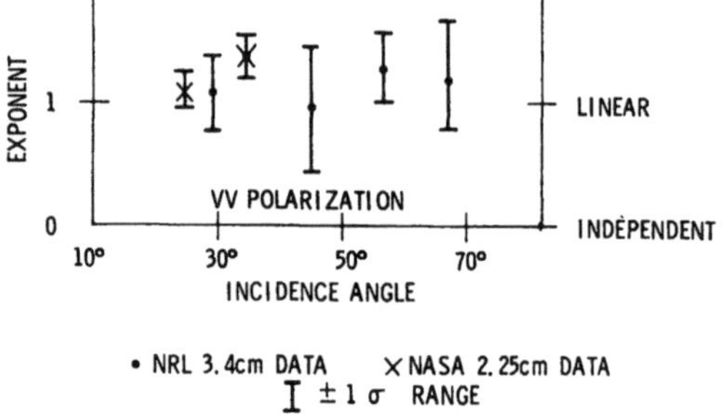


Fig. 3. Wind response of scatterometer (upwind).

program gave the best information to date on the subject [10]. Clearly the wind direction must be determined independently from other meteorological information to complete the measurement with a satellite RADSCAT. Methods are being devised by Pierson for doing this, extending Druyan's work [1].

The source of the radiation observed by the microwave radiometer is much more complex than that for the radar, as shown in Fig. 4. Besides emission from the ocean, emission also comes from clouds, atmospheric gases and extra-terrestrial

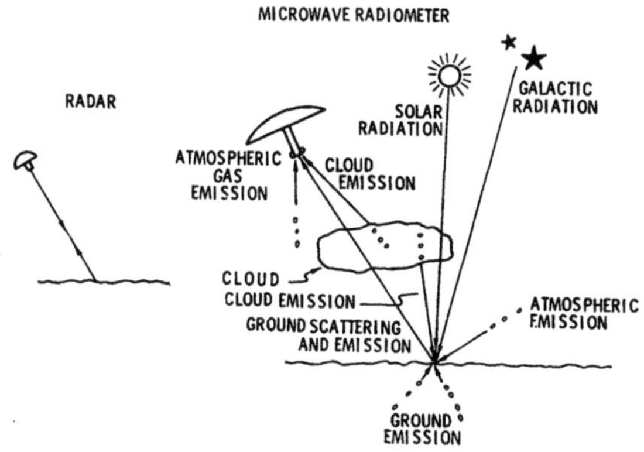


Fig. 4. Comparison of sources of radar and radiometer signals.

sources. When clouds are not present, however, most of the radiation is from the sea. This signal is also governed by the capillary waves, so it is also proportional to wind speed, as indicated in measurements by Hollinger [11] and Nordberg [12] (Fig. 5). The effect of clouds and rain is much greater for the radiometer than for the scatterometer. Wu's calculation of the increase in brightness temperature for

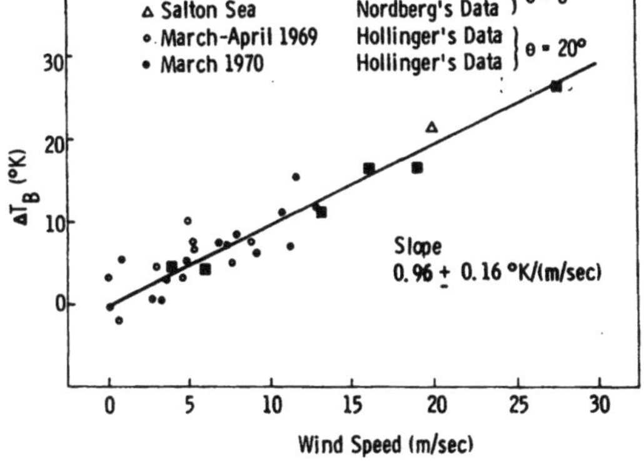


Fig. 5. Radiometer windspeed dependence.

small attenuations is shown in Fig. 6, which indicates large brightness temperature fluctuation for little changes in attenuation [13]. A comparison of the cloud effects on radiometer and scatterometer is shown in Fig. 7. The RADSCAT anemometer takes advantage of this by using the scatterometer as the primary wind sensor and the radiometer as an attenuation sensor. The attenuation measured by the radiometer corrects the radar signal, and thus the wind measurement [14]. The proposed operational RADSCAT system would be on a polar orbiting satellite. An electronically scanned array would scan the beam from side to side, collecting

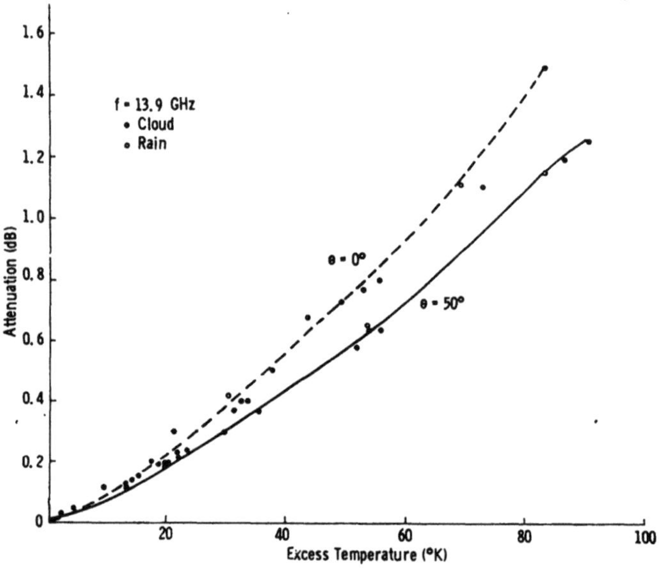


Fig. 6. Effect of cloud attenuation on downward-looking radiometer.

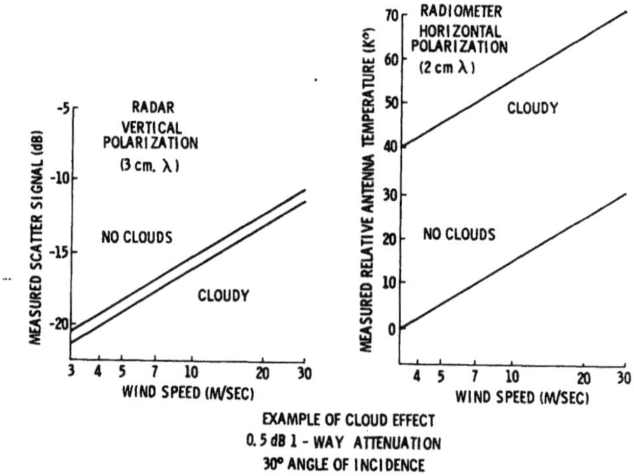


Fig. 7. Example of cloud effect on radar and radiometer.

information from points about 100 km apart (Fig. 8). The axis of the antenna would be about 30° ahead to achieve appropriate incident angles.

Further research on this system will involve confirming the universality of the aircraft data and reducing its variance, further measurement of directional effects, and experimental confirmation of calculated effects of atmospheric attenuation on the radiometer.

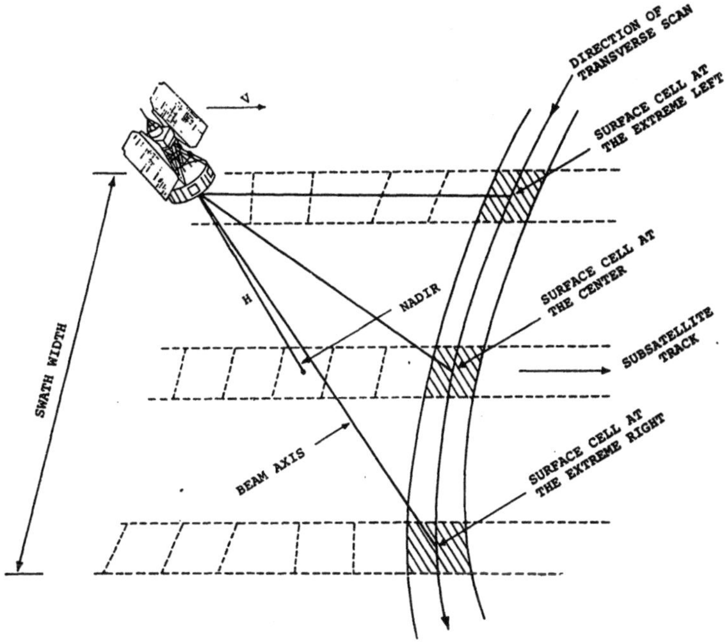


Fig. 8. Proposed operational RADSCAT.

## 3. Imaging Radars over Land—Currently Useful, but Much Research Needed

Imaging radars using the synthetic aperture technique can achieve resolutions from orbit as good as those achieved from aircraft. Resolutions better than 20 m appear quite feasible, although tradeoff studies indicate poorer resolution may be appropriate for some applications because of smaller power requirements and reduced data transmission needs [15]. An example of a radar image with this kind of resolution is shown in Fig. 9. Because such fine resolution is possible with imaging radars using no more power than comparable television systems, many applications requiring images (pictures) at specific times when clouds might obscure the earth for visible-IR sensors seem particularly well suited to imaging radars. Examples where cloud penetration is necessary to obtain timely information include agriculture, soil moisture, lake and sea ice monitoring, natural vegetation study, snow survey, flood and flood-damage assessment, and monitoring of sea traffic and fishing fleets. The special geometry of radar, with illumination always occurring from behind the receiver, makes it useful for some purposes. Applications for which this has been proven include structural geology, construction material surveys, and oil spill monitoring. Evidence indicates that ocean wave patterns may also be monitored for this reason.

Since radar signals penetrate vegetation and soil they are especially useful for soil moisture determination, lithologic mapping, forestry, and in general location of patterns obscured by vegetation. Even if these special cloud-penetrating, geometric, and surface-penetrating properties for radio waves did not exist, radar systems would still be valuable because they operate at different wavelengths from visible-IR sensors. The additional spectral information available from the two to three decades of spectrum available for radar can aid in identification of earth properties in many areas with visible-IR and radar sensors complementing

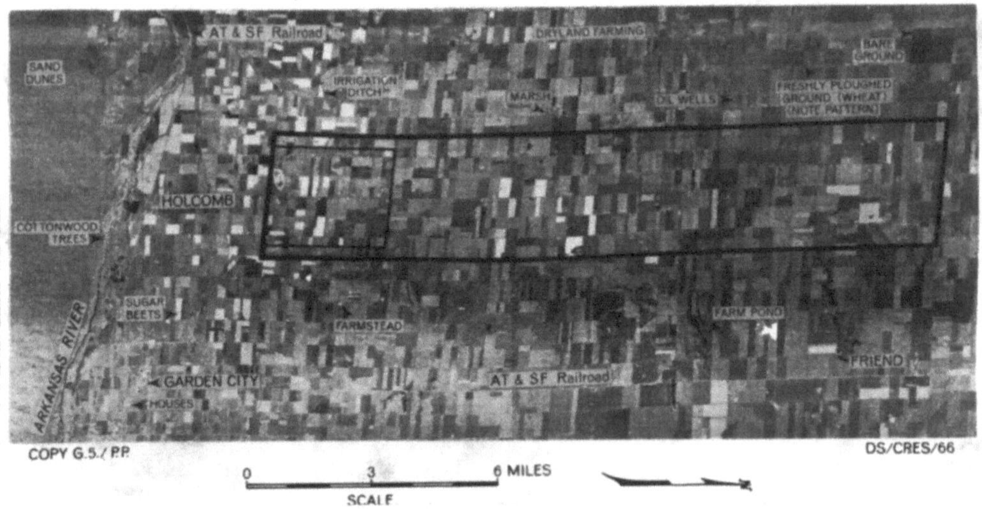


Fig. 9.

each other. Project RADAM in Brazil, mapping about 1500000 km$^2$ from a synthetic-aperture radar flown on a Caravelle, is the most extensive use of imaging radars [16]. The first major project was in Panama, where a real-aperture radar flying for eight hours mapped an area where clouds had prevented adequate air photos for decades [17].

Some current uses of radar images are for geology, vegetation mapping, oil spill location, and sea ice mapping. Most commercial users are geologists searching for minerals. Both the Panama and Brazilian projects have produced vegetation maps, as well as geologic and land-use maps. Oil spill detection, first observed by de Loor [18], has been extensively studied for the US Coast Guard by the Naval Research Laboratory [19] and operational feasibility was shown. Use of radar to observe sea ice has been studied extensively in the USA [20, 21], USSR [22] and Sweden [23]. During the passage of the tanker Manhattan north of North America, ice images such as the one shown in Fig. 10 assisted the ship captains in navigation through the ice pack [24]. New applications of imaging radar await further research, both theoretical and experimental, both fundamental and application-oriented. Research on the complementary role of radar and visible-IR sensors has not yet started.

Two-medium rough-surface scatter theory is well developed, but applies only to the ocean. Land surfaces are inhomogeneous, with subsurface layers and objects complicating the problem. This research has begun, but much more is needed

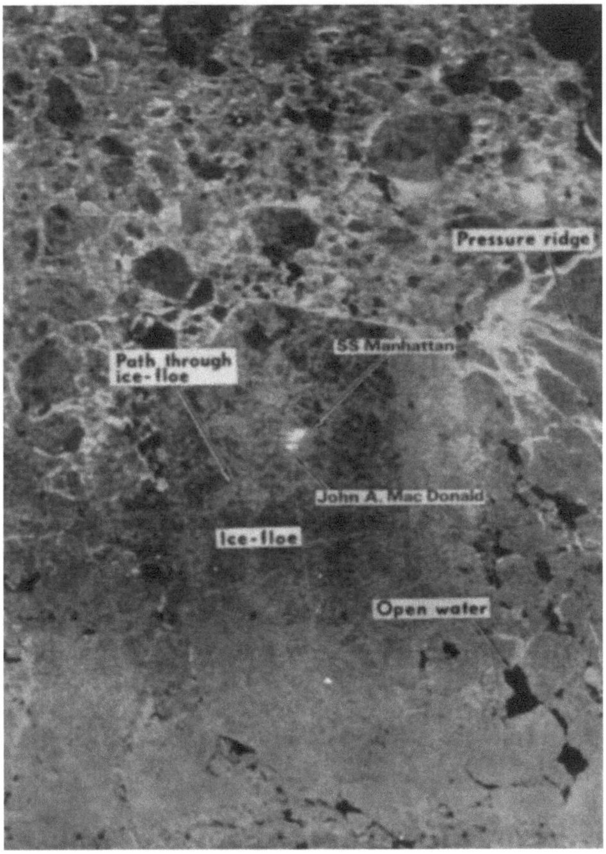


Fig. 10. Radar imagery of ice fields taken at the time of SS Manhattan passage (courtesy Philco Ford).

[25, 26]. The only theory for scatter from vegetation is for a simple cylindrical model that really applies only to grass [27]. More realistic theories must include roots as well as the visible parts of more complex plants.

Although scatter from aircraft and spacecraft can be described theoretically [28] almost no work has been done to treat buildings, fences and other land objects made by man. Quantitative experiments describing radar return from the earth are rare. The data that have been collected are mostly for single wavelengths, or for groups of isolated wavelengths [29]. The kinds of continuous spectral responses available in the visible-IR spectrum are almost nonexistent in the microwave spectrum. At Kansas there are now two summers of octave-bandwidth radar responses in the 3.75—7.5 cm regime [30]. A preliminary result is shown in Fig. 11; we believe use of multiple wavelengths in color, or polypanchromatic, radar imagers will be important in future earth observations [31]; Fig. 12 shows an example of the radar colors produced from the octave bandwidth data. This summer measurement will be over the 1.67—15 cm decade, with both active and passive systems. We believe this is most important research, and we hope

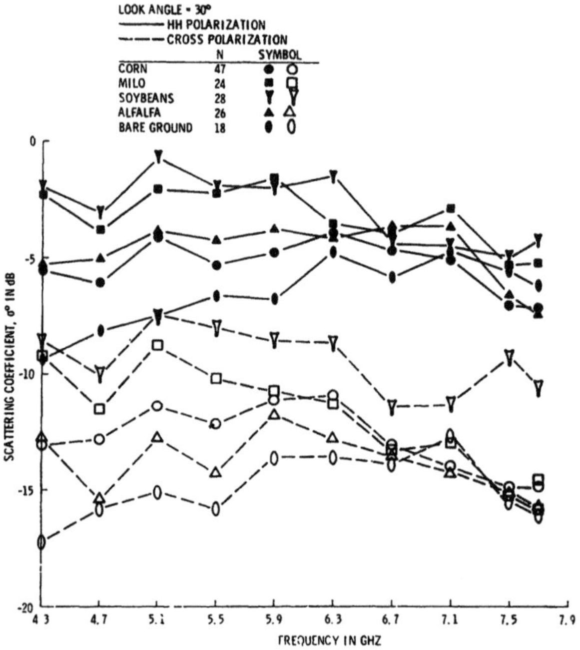


Fig. 11. Scattering coefficient as a function of frequency at 30° look angle for HH and cross polarizations.

other researchers around the world will join us in making such measurements. These ground-based experiments must be complemented by controlled aircraft experiments. Quantitative measurements with *calibrated* polypanchromatic imaging radars will be important in establishing usefulness of such systems. The technology is available now.

## 4. Conclusions

Radar applications to oceanography and land mapping have been demonstrated, with the latter in operational use today. Other important uses depend on future research, much of which is of interest to the URSI scientific community. Applications research falls more in the field of other international scientific unions, but close cooperation between these studies and those of URSI researchers will be most important if the full potential of microwaves is to be realized.

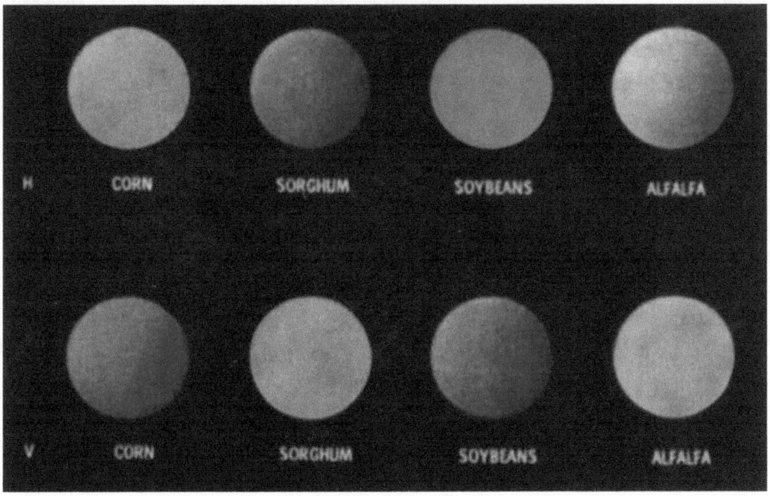


Fig. 12. Example of microwave color.

## References

[1] L. M. Druyan, J. Appl. Meteor. **11**, 413 (1972).
[2] S. O. Rice, Commun. Pure Appl. Math. **4**, 351 (1951).
[3] H. Davies, Proc. IEE, Pt. IV, **101**, 209 (1954).
[4] L. M. Brekovskikh, Zh. Eksper. i Teor. Fiz. **23**, 275 (1952).
[5] B. Semyonov, Radio Engng and Electron. Phys. **11**, 1179 (1966).
[6] A. K. Fung and H. L. Chan, AGARD Conf. Proc. No. 90, CP-90-71 (1971).
[7] W. J. Pierson, Jr. and R. A. Stacy, NASA Langley Research Center Contractor Report (in press, 1973).
[8] J. W. Wright, personal communication to W. J. Pierson, Jr.
[9] J. P. Claassen, H. S. Fung, R. K. Moore and W. J. Pierson, Jr., IEEE Ocean '72 Conf. Record 72 CHO 660-1 OCC, 1972 (p. 180).
[10] W. L. Jones, L. R. Schultz and N. D. Akey, Proc. IEEE/G-AP and USNC/URSI Symp., December 9172.
[11] J. P. Hollinger, IEEE Trans. GE-9, 165 (1971).
[12] W. Nordberg et al., J. Atmos. Sci. **28**, 429 (1971).
[13] R. K. Moore et al., AAFE Prin. Investigators' Review, NASA Langley Research Center 1971 (p. 115).
[14] R. K. Moore and W. J. Pierson, Jr., Proc. 12th Int. Tech. Scient. Mtg on Space, Rome 1972.
[15] R. K. Moore and G. C. Thomann, IEEE Trans. GE-9, 155 (1971).
[16] J. M. de Moura, Proc. 8th Int. Symp. Rem. Sensing of Environment, Univ. of Michigan 1972.
[17] A. Viksne, T. C. Liston and C. D. Sapp, Photo. Engng **36**, 253 (1970).
[18] G. P. de Loor, personal communication.
[19] N. W. Guinard, AGARD Conf. Proc. No. 90, CP-90-71 (1971).
[20] V. H. Anderson, Proc. 4th Int. Symp. Rem. Sensing of Environment, Univ. of Michigan 1966 (p. 845).
[21] J. W. Rouse, Jr., Proc. IEEE **57**, 605 (1969).
[22] V. S. Loshchilov and V. A. Voyevodin, Problemy Artiki i Antarktiki **40**, 23 (1972).

[23] A. Blomquist, Försvarets forskningsanstalt FOA 3 Rapport C 3633-60, Stockholm 1970.
[24] D. J. Johnson and L. D. Farmer, J. Geophys. Res. **76**, 2138 (1971).
[25] W. D. Boles, Ph. D. Dissertation, University of Kansas (1969).
[26] A. Leovaris and A. K. Fung, J. Acoust. Soc. Amer. (March 1973).
[27] W. H. Peake, R. L. Cosgriff and R. C. Taylor, Terrain Handbook, Ohio St. Univ. Engng Exp. Station 1960 (p. 16).
[28] R. E. Kell and R. A. Ross, Radar Handbook, McGraw-Hill Book Co., New York 1970 (Chap. 27).
[29] R. K. Moore, Radar Handbook, McGraw-Hill Book Co., New York 1970 (Chap. 25).
[30] F. T. Ulaby and R. K. Moore, presented at IEEE/G-AP and USNC/URSI Symp., Williamsburg, Va., December 1972.
[31] R. K. Moore, W. P. Waite and J. W. Rouse, Jr., Proc. IEEE **57**, 590 (1969).

Earth Survey Problems — Akademie-Verlag, Berlin 1974

# ON RADIO SCIENCE TECHNIQUES FOR REMOTE SENSING

J. W. Rouse, Jr.

Texas A & M University, College Station, Texas, USA

This paper presents examples of radio techniques for remote sensing which have evolved from activities within URSI. The application areas discussed include remote sensing of polar regions and remote determination of soil characteristics, specifically moisture content. The electromagnetic backscatter and depolarization by ice and terrain are reviewed, and the problem areas needing additional effort are identified. Remote sensing by both active (radar) and passive (microwave and thermal infrared) sensors are considered, and the role of radio science techniques in the future development of orbital remote sensing is described with specific reference to the Skylab and ERTS-B missions.

## 1. General

Fundamental to all earth survey remote sensing problems is the interpretation of the interaction of electromagnetic energy with the environment. Because these interaction mechanisms are extremely complex, analyses of visible and near-infrared sensor data, such as ERTS 1 measurements, have proceeded from a base of empirical results. This approach is often adequate for those disciplines whose earth survey objectives are amenable to photographic interpretation techniques, e.g. geography and geology. However, a more fundamental approach is needed when extracting visible and near-infrared region information beyond the obvious geometrical relationships, in particular, for example, monitoring vegetation stress, where an electromagnetic interaction model of the canopy is important. Such an approach is especially important in preparing for analysis of radar, passive microwave, and thermal infrared data to be obtained by Skylab, Nimbus, and ERTS-B, for which the history of empirical results is far less extensive. Because of this situation, the research activities within URSI on radio techniques for remote sensing are particularly significant.

## 2. Microwave Remote Sensing

The future course of satellite remote sensing of the earth is still too uncertain for accurate prediction despite the fact that ERTS and Skylab are already configured. But for those who have seriously examined the issue, one thing is clear: when eventually synoptic sensing of the earth's surface is done routinely, the primary sensors used will operate in the microwave region of the spectrum. The reason is obvious. The atmosphere poses so serious an impediment to effective satellite sensing with visible and infrared sensors that timely, reliable measurements of the surface cannot be obtained in a systematic manner. Microwave sensors offer the only reasonable solution.

It is equally clear that the proven capability of microwave sensors is inadequate for the task of obtaining the wide variety of remote sensing information now in demand. This is believed to be due to the relatively small number of experimental programs which have been conducted, rather than to any fundamental deficiency in microwave sensor capabilities. Multiple frequency and multiple polarization experiments now in progress are beginning to define more clearly the information content potential of microwave sensors, including the exciting concept of "microwave color" [1]. Basic to this entire effort is the development of improved descriptions of the electromagnetic interaction mechanisms, an activity that has progressed significantly in recent years due, to a great extent, to the work reported through URSI Commission II.

## 3. Recent Advances in Rough Surface Scatter Theories

The electromagnetic interaction mechanisms central to almost all remote sensing activities, have been most vigorously addressed by those examining the phenomena of rough surface scattering. This problem has been studied for many years employing several different approaches in an attempt adequately to describe the incidence angle, frequency, and polarization dependences of experimental measurements. In recent years the work has been accelerated rapidly because of the successes demonstrated by each of two, supposedly different, approaches. The physical optics, or Kirchhoff approach, stemming from the Helmholtz integral has been prominent in literature throughout the world [2—4]. However, early formulations of this approach showed it to have definite shortcomings, especially relative to describing depolarization [5]. The small perturbation method, or Rayleigh-Rice approach [6, 7], has recently risen in prominence among many noted researchers in this field because it yields an incoherent solution which very closely matches actual radar backscatter measurements.

Recently it has been argued that the Rayleigh-Rice approach is actually only a special case of the properly formulated Kirchhoff approach [8] and, hence, that disagreements between the two proponent groups were groundless. Needless to say, this contention has not been unanimously accepted, but some agree that the argument is true if the incident energy is horizontally polarized [9], while others accept it for circular polarization [10]. Even before this recent claim that there are significant areas of equivalence between the two approaches, several attempts had been made to formulate a "composite surface" approach which combines both approaches into a single model [11]. The point is that these diverse groups are converging on an approach to describing electromagnetic interaction with rough surfaces that has considerable promise as a valid theoretical base for interpretation of remote sensing data, especially radar and lidar measurements.

This lively drama has produced numerous side issues shedding new insight on scattering mechanisms. One of these of particular importance to a later issue in this paper is the contribution of subsurface volume scatter to electromagnetic scatter measurements. Independent measurements by Leader [5] and Renau et al.

[12] of rough surface backscatter of laser light suggest that a volume scatter mechanism exists which for some media can dominate the depolarization process. Leader's data also support the physical optics result that there is no appreciable depolarization by the surface scatter mechanism. Rouse [13] and Leader and Dalton [14] have developed expressions for the subsurface contributions which provide rationale for interpretation of the unusual behavior of radar measurements of certain terrains, notably Arctic sea ice.

Interpretation of radiometer measurements of the earth surface requires determination of contributions due to the atmosphere and the terrain. The atmospheric contributions at microwave frequencies are relatively minor, hence the principal concern is development of satisfactory models for the terrain component. Much of this effort follows the work of Peake [15], who determined the terrain emissivity by assuming an equivalence between emissivity and absorptivity, and calculating absorption from the surface reflectance. This requires an assumption of thermal equilibrium between the terrain and the environment, but it permits utilization of the full range of the well-developed scattering models, which can be applied to determine the surface reflectance.

This approach has proven useful for modeling bare surfaces, such as sea ice, water, and bare soils; however, it does not lend itself well to the problem of interpreting passive microwave measurements of covered surfaces, such as vegetated terrain or foamy seas. In an unpublished paper by Johnson [16] an attempt has been made to remove the thermal equilibrium assumption by describing the transfer of the incoherent electromagnetic energy across the surface boundary from within the medium. This approach does provide a means for describing the effect of vegetation over-story, and some preliminary work has been done is this area [17].

Johnson also suggested that surface roughness differences are far less influential on radiometric measurements than on radar measurements, hence that incorporation of complex rough surface models, such as is common in applying Peake's approach, is unnecessary. Johnson introduced a weighting factor proportional to the effective radiating surface area per unit plane area to account for surface roughness effects. Preliminary indications are that this factor does not vary appreciably from unity, and that a weighted smooth surface model may be reasonable for many natural surfaces.

## 4. Radio Remote Sensing of Arctic Ice

The dynamic floating ice pack in the Arctic Ocean has attracted considerable interest among the international scientific community. The recent interest has been motivated, in part, by the discovery of vast mineral and energy resources in the Arctic, and the renewed awareness that the energy exchange in this region significantly affects the meteorological processes in the northern hemisphere.

Major programs by several countries have been conducted in an effort to collect measurements to enable better definition of the physical processes at work in the Arctic Ocean. These have included both airborne and surface investigations. However, because of the vast areas of the Arctic and the severe weather conditions, the existing data provide only a sketchy picture of this region of the globe.

In recent years there have been some attempts to apply airborne remote sensing techniques to the problem of collecting descriptive information about Arctic sea ice. Laser profilometers, thermal infrared radiometers, radar, and passive microwave sensors have been tested. Although the information content of the visible and thermal infrared sensor data is impressive [18], the adverse effects of the weather upon their utility limits their potential as operational devices. The microwave sensors appear to offer the greatest potential for systematic sea ice observations.

Although several experiments have been conducted to study the effects of the Arctic sea ice cover on near-surface electromagnetic wave propagation in the microwave region, little quantitative data exist on remote sensing of sea ice. Consequently, much emphasis has been placed on a series of four aircraft Arctic missions conducted by the National Aeronautics and Space Administration (United States) beginning in 1967. These missions included 2.25 cm wavelength radar data reported by Rouse [19], and multifrequency microwave radiometer data reported by Wilheit et al. [20].

As a result of the limited work in this area, several misconceptions exist as to the nature of the electromagnetic interaction with sea ice. One of these is the belief that microwave signals penetrate the sea ice and reflect from the ice—water interface. Hoestra [21] has shown that attenuation of microwave energy in sea ice is sufficiently high that obtaining recordable signals from the ice—water boundary is unlikely except for very thin new ice. However, it has been found that the near-surface volume has an appreciable effect upon the microwave signal, especially in the case of multi-year ice. For this ice type, the subsurface volume scatter mechanism provides a backscatter component comparable with that produced by the surface scatter process.

Passive microwave measurements of apparent brightness temperature indicate 20—30°K differences between first-year and multi-year ice types at 1—2 cm wavelength [20]. This represents an apparent 20—25% difference in emissivity. Open water contrasts sharply with either ice type, appearing much "colder" than sea ice. The surface structure of the ice types is different, but this difference in roughness factors cannot account for the wide range of measured emissivities. The difference is surely associated with the composition of the subsurface volume. The aging process causes the suspended brine concentrations in sea ice to gather in "pockets", forming scattering centers. Associated with this change is a change in the homogeneity of the near-surface ice with a decrease in the permittivity value near the surface.

Radar measurements of sea ice at 2.25 cm wavelength characterize the aging process of the ice. The dependence of these data on ice age is shown in Fig. 1. The formation stage when thin ice forms over open water leads is indicated by a sharp decrease in the value of the radar scattering coefficient for incidence angles near the vertical. The causes are the change of the relative permittivity from about 55.0 for water to about 6.0 for sea ice, and a change in the surface roughness factor.

The radar scattering coefficient declines and becomes less angle dependent as the ice ages during the first months after formation. This change can be associated directly with changes in surface topography. However, after several months of growth the surface roughness becomes less influential than the subsurface volume, which changes significantly due to settling of the suspended brine. The effect of this change is to increase the radar return. The volume scatter mechanism is especially influential for incidence angles greater than 30 degrees, where scattering coefficient increases of 10 dB between first-year ice and multi-year ice are common.

It has been shown that the 2.25 cm radar measurements of sea ice are sufficiently distinct that remote ice type identification is feasible. This potential is also supported by passive microwave and thermal infrared data at wavelengths in the 1—3 cm and 8—14 nm range. Airborne radar measurements at 0.75 m indicate that the volume scatter mechanism is much weaker for long wavelength radiation,

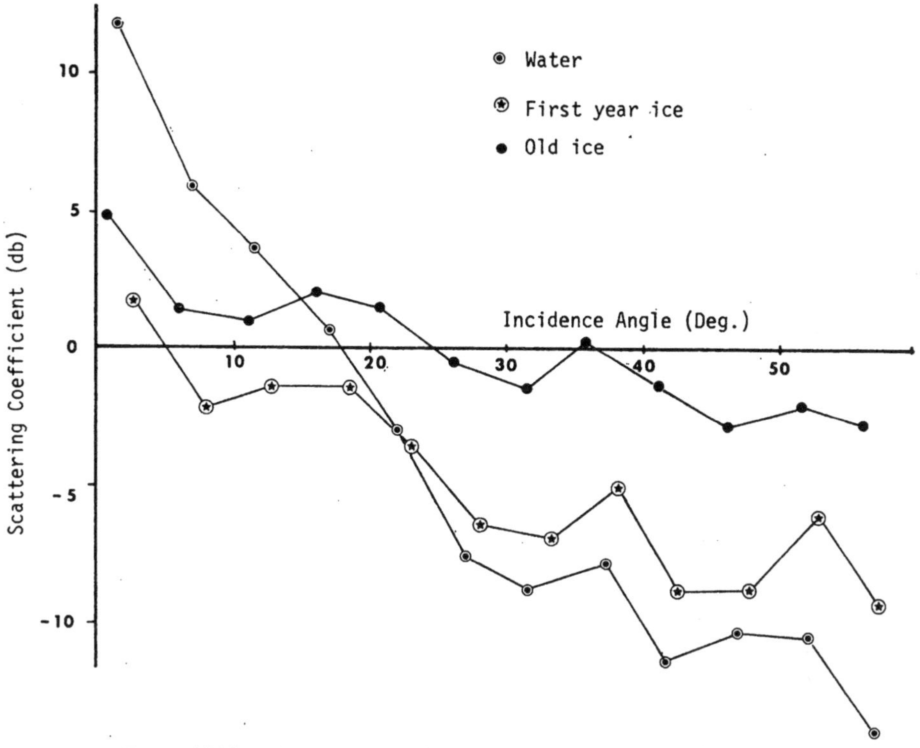


Fig. 1. NASA airborne radar (2.25 cm) measurements of Arctic ice.

and consequently, the multi-year ice is not easily differentiated from younger thinner types.

The implications of these studies is that airborne and satellite microwave remote sensing of sea ice is possible using either imaging or point sensors. Thus, day-night, near all-weather monitoring of the Arctic over broad regions can be accomplished, and data on ice distribution and movement, and energy exchange estimates can be obtained using state-of-the-art microwave devices.

## 5. Radio Remote Sensing of Soil Moisture Content

Hydrologists and agriculturalists have need of synoptic data on soil moisture content to support run-off predictions and agricultural management decisions such as irrigation and crop yield forecasting. Several attempts have been made to obtain this information remotely using visible and thermal infrared sensors [22], but the results have not been sufficiently quantitative to satisfy the needs. The established dependence of microwave measurements upon the near subsurface volume characteristics is being examined by several investigators in an attempt to improve the data source for these applications [23].

Of fundamental importance in this work is the behavior of the electromagnetic properties of soils as a function of moisture content. Fig. 2 shows the variation of the complex permittivity of a sandy soil as a function of moisture content at

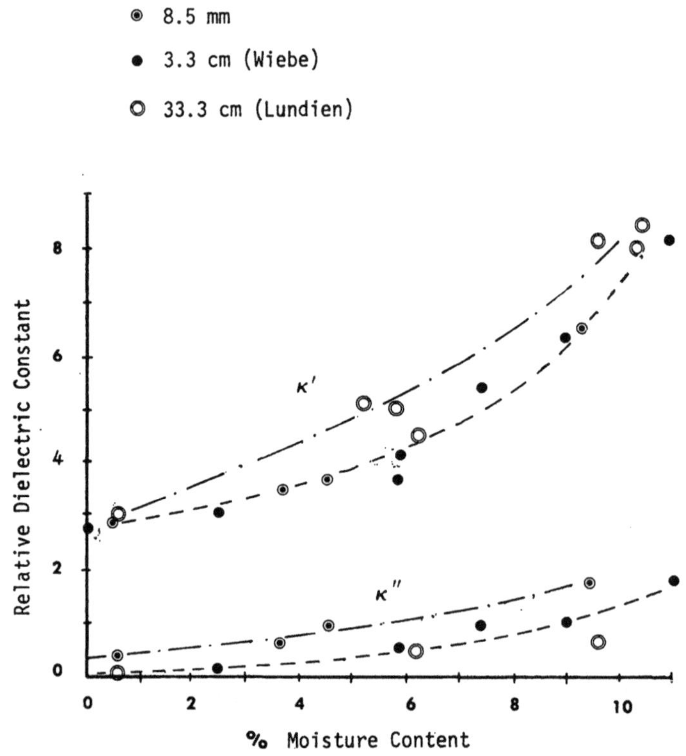


Fig. 2. Variation of relative dielectric constant of sandy soil with moisture.

three different measurement frequencies. Different type soils have different moisture dependence, and the range between sandy soils and clays is appreciable at moisture content levels above about 20%, i.e. permittivity of clay is as much as 25% less at 20% moisture content.

Unpublished work by Jean et al. [24] indicates a strong correlation between microwave radiometric measurements and soil moisture content at several frequencies between 1.6 and 30.7 GHz for both sandy and clayey bare soils. This general result is also supported by several other authors [23]. Fig. 3 shows the relationship of measured microwave brightness temperatures and soil moisture content of clay soils. These airborne data were obtained by NASA over several large bare fields in Texas in 1971. These measurements indicate that airborne passive microwave sensors can record soil moisture content of bare soil to an accuracy of about 5% moisture content, assuming the general soil type is known.

The effect of vegetation cover over the soil on the ability of the sensor to determine soil moisture content is appreciable. The vegetation serves both to attenuate the subsurface radiation and to generate an independent electromagnetic component. The problem has been addressed in limited empirical and theoretical investigations [17] with the preliminary result that vegetation effects can completely mask the soil-moisture-dependent component in the radiometer data. It has not been determined just how dense the vegetation must be before the soil moisture information is destroyed, nor has the potential of longer wavelength sensors been examined relative to minimizing the vegetation problem.

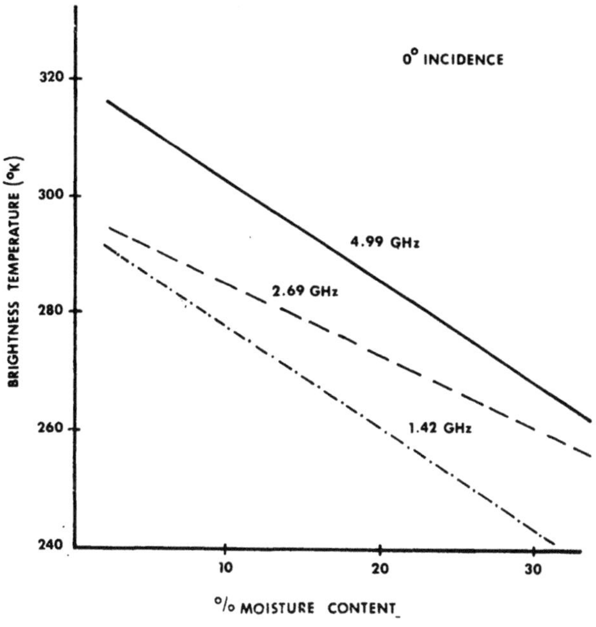


Fig. 3. NASA airborne radiometer measurement trends for bare fields of similar soil type. Abscissa shows soil moisture content.

The Skylab and Nimbus E satellites carry multi-frequency microwave radiometers for earth observations. The Nimbus sensors are designed primarily for atmospheric measurements, but both systems will supply terrain-dependent data, although the resolution cells for each are several kilometers. Similar data were obtained by a Soviet satellite in 1968. Based upon the present understanding of the passive microwave sensing problem, it is doubtful that any satellite data for vegetated terrains of agricultural significance will be interpretable in terms of soil moisture content.

Radar sensors also have potential for measurement of soil moisture content of unvegetated terrain. Recent experiments with airborne radar sensors confirm a strong dependence of radar backscatter on soil moisture [25]; however, the results available at this time are qualitative, and verification of the preliminary findings has not been forthcoming. Theoretical calculations using the Kirchhoff approach suggest that the radar scattering coefficient becomes insensitive to soil moisture content for bare rough soils at levels above 20% [26]; however, this contention has not been confirmed experimentally.

## 6. Discussion and Conclusion

The existing empirical base of visible-region remote sensing data is presently being profitably exploited to yield valuable information about the earth. This is best exemplified by the results being obtained from the ERTS 1 visible and near-infrared data. However, this information base is inadequate to support effective analysis of the thermal infrared and microwave data to be obtained in the near future by Skylab, Nimbus, and ERTS-B.

The steady development of improved descriptions of electromagnetic energy interaction with natural surfaces, such as sea water, sea ice, and soils, is encouraging. However, effective use of satellite sensing techniques will require resolution of several issues in this regard. Although scattering models are now available that appear to predict electromagnetic interaction for several surfaces with reasonable accuracy, several serious restrictions remain. A better understanding of the commonalities in the Kirchhoff approach and the Rayleigh-Rice approach, and the suitability of the composite surface approach should be forthcoming shortly from the ongoing work within URSI. The significance of the volume-dependent depolarization mechanism relative to the surface-dependent multiple scatter depolarization process must also be determined. Finally, the troublesome problem of vegetation cover must be given immediate attention.

The Earth Observations field was created by a scientific community eager for improved data acquisition techniques in the environmental and earth sciences. The tremendous potential of sensor technology, able both to provide new methods for investigating environmental problems and to capitalize on the existing capabilities in space travel, created widespread excitement almost from the outset. For the many geoscience disciplines which depend, in essence, on trained hands and the spectrum range of the human eye, the new dimensions possible with mobile synoptic sensors raised high hopes indeed. The enthusiasm continues to grow but it is being tempered by the realization that full utilization of the available remote sensing technology will require further advances in the understanding of basic electromagnetic interaction mechanisms involved.

## References

[1] R. K. Moore, W. P. Waite and J. W. Rouse, Proc. IEEE **57**, 590 (1969).
[2] P. Beckmann and A. Spizzichino The Scattering of Electromagnetic Waves from Rough Surfaces, Macmillan, New York 1963.
[3] A. K. Fung, Planet. Space Sci. **15**, 1337 (1967).
[4] L. M. Brekovskikh, Zh. Eksper. i Teor. Fiz. **23**, 275 (1952).
[5] J. C. Leader, J. Appl. Phys. **42**, 4808 (1971).
[6] S. O. Rice, Comm. Pure Appl. Math. **4**, 351 (1951).
[7] In. P. Lysanov, Dokl. Akad. Nauk SSSR **87**, 719 (1952).
[8] J. C. Leader, IEEE Trans. Ant. Prop. **AP-19**, 786 (1971).
[9] J. W. Wright, IEEE Trans. Ant. Prop. **AP-14**, 749 (1966).
[10] M. L. Burrows, Can. J. Phys. **45**, 1729 (1969).

[11] J. W. Wright, IEEE Trans. Ant. Prop. **AP-16** (1968).
[12] J. Renau et al., J. Opt. Soc. Am. **57**, 459 (1967).
[13] J. W. Rouse, Radio Sci. **7**, 889 (1972).
[14] J. C. Leader and W. A. J. Dalton, J. Appl. Phys. **43**, 3080 (1972).
[15] W. Peake, IRE Trans. Ant. Prop. **AP-7** (1959).
[16] L. F. Johnson, Texas A & M Tech. Rep. RSC-37 (1972).
[17] C. L. Kroll, T. G. Sibley and J. W. Rouse, Texas A & M Tech. Memo. RSC-69 (1973).
[18] V. E. Noble et al., Proc. IEEE, **57**, 594 (1969).
[19] J. W. Rouse, Proc. IEEE, **57**, 605 (1969).
[20] T. Wilheit, et al., Remote Sensing Environ. **2**, 129 (1972).
[21] P. Hoestra, U.S. Army Corp. of Engr. CRREL Rep. (1969).
[22] H. D. Werner and F. A. Schmer, South Dakota St. Tech. Rep. RSI-72-16 (1972).
[23] A. Edgerton et al., Space Gen. Corp. Tech. Rep. SGD 829-6 (1968).
[24] B. R. Jean et al., Texas A & M Tech. Rep. RSC-32 (1972).
[25] R. K. Moore et al. (unpublished report, 1973).
[26] J. W. Rouse, Texas A & M Tech. Memo. RSC-52 (1972).

# Biology Related Problems

Earth Survey Problems — Akademie-Verlag, Berlin 1974

# BIOLOGY AND REMOTE SENSING FROM SPACECRAFT

J. C. Zadoks and H. D. Frinking

Laboratory of Phytopathology, Agricultural University, Wageningen, the Netherlands

Biological objects studied from spacecraft are necessarily large units, such as are studied in ecology. Types and sizes of the biological units to be studied from space-borne platforms depend on inherent biological properties like homogeneity and continuity in space and time, and hence on the spectral, spatial, and temporal resolution of the satellite's equipment. Marginal conditions of life, short-lived events, and sudden disturbances of living communities are important items. Good temporal resolution is as valuable as spatial resolution. Surveys are carried out to produce vegetation maps, forest inventories, agricultural inventories, etc. Monitoring with adjusted time intervals can be used to record crop acreages, crop vigour, and the occurrence of pests and diseases. Early detection of pests, diseases or forest fires, and yield prediction can have economic value, and may be instrumental in a fair and effective distribution of food and other agricultural commodities among the world's population. Efficient use of satellite-generated imagery by biologists is hampered by poor training in radiation physics. Lack of understanding of the interaction between biological matter and radiation leads to poor interpretation of imagery, poor understanding of the new possibilities offered by remote sensing, and low selectivity in the demand for information. The first remedy should be better training, tailored to the biologist's needs. A plea is made for precision in the formulation of research objectives.

## 1. Surveying and Monitoring

Satellites are platforms at long distance from the earth, from which imagery can be obtained depicting the earth or its parts. The imagery can be used for the purpose of biological survey, i.e. a general view to examine the condition of a subject and to collect by measurement all facts needed for determining the boundaries, size, position, shape, contour, or other attributes of that subject [1]. The biologist can experience great difficulties in handling satellite imagery, because of the small scale, the distortion of the pictures due to the curvature of the earth, the sometimes unconventional signals collected, and the shielding of the subject by cloud cover. Recurrent surveys, made by means of a sun-synchronous satellite which takes up a predetermined position at regular intervals, or incidental surveys through dirigible spacecraft can be used for monitoring. Monitoring can be described as "watching and warning". A monitoring system records change. In the biological field, changes of interest are the "green waves" of vegetational growth proceeding from south to north when winter has passed away and spring sets in, and also droughts, floods, fires, night frosts, or outbursts of pollution.

## 2. The Biologist's Attitude towards Remote Sensing

The biologist is deeply interested in small-scale global surveys, because he feels to some extent responsible for the earth's wealth in natural resources, on land and in the sea. However, the biologist is ill-prepared to utilize the information contained in global surveys. There is a large gap to bridge between those who provide satellite-made imagery and those who can use it profitably.

Traditionally, biologists have been trained to see the trees rather than the forest. The "ecology drive", and more specifically the IBP (International Biological Program), have thoroughly changed this traditional attitude. IBP [2] was initiated by IUBS (International Union of Biological Sciences), and promoted by ICSU (International Council of Scientific Unions). Biologists have learned to study large units of complex nature, often called ecosystems. An ecosystem can be described as a community of interacting species, well defined in space and time, with typical characteristics differentiating it from other communities, and with an inherent stability, or resistance against disruption. Ecosystems can be small and of short duration, like the ephemeral community of micro-organisms on a leaf surface, medium sized and reasonably permanent, like a pond or a forest, or large and nearly eternal, like the tropical rain forest or the arctic tundra. Such large-sized ecosystems are called "biomes". IBP has generated many biome studies, but remote sensing (RS) techniques have seldom been used [3]. There are two reasons: first, the RS techniques were not yet ready for use when the biome studies were planned; second, the biologists were not yet ready to utilize RS techniques. Here lies an educational problem. Biological interpretation of RS data depends on a thorough understanding of the interaction between radiation and matter. The phenomena of reflection and emission in relation to the underlying biological processes are but poorly understood.

## 3. Signals of Biological Interest

The signals of biological interest are well scattered over the electromagnetic spectrum. Besides the wavelengths of the visible light, used in the most classical type of RS, ordinary photography, the adjoining bands of UV and near-infrared are important. UV ($\lambda = 0.3-0.4$ µm) can be of use in pollution studies ($SO_2$ peak absorption at 310 nm and $NO_2$ at 440 nm). Within the near-infrared ($\lambda = 0.76-3$ µm) [4], the photographic infrared ($\lambda = 0.76-0.95$ µm) is of extreme importance in plant studies, because healthy green plants have a very high reflectance, whereas bare soil and dead plants have a low reflectance [5]. Plants or crops injured by insect pests, diseases, drought, or otherwise show an intermediate reflectance [6—8]. Gradually, understanding is growing of the interaction between leaves, foliage or canopy, and radiation [9—11]. This will have great impact on the refinement of near-infrared image interpretation. The importance of near-infrared is such that it has been selected as one of the four bands in the multi-spectral scanning (MSS) system of the ERTS 1 satellite (0.8—1.1 µm).

The middle ($\lambda = 3-6$ µm) and far ($\lambda = 6-15$ µm) infrared, with the main thermal bands at ca. 3 and ca. 10 µm, again provide valuable information. They permit cloud and snow cover identification, monitoring of thermal pollution, the study of soil and crop condition, and the study of animal life on land or in the sea. Healthy, vigorous crops with high transpiration rates are cool in day-time and warm at night-time, relative to the ambient air or the adjacent bare soil. During the day, the difference between the temperature of a leaf surface and the surrounding air can be up to 10 °C [12]. Diseased crops can show an abnormal temperature response, being relatively warm during the day [6, 12, 14], due to lack of cooling by evapotranspiration, and they can be recognized accordingly. At night, large terrestrial animals can easily be differentiated against their background in large-scale imagery. At sea, fish shoals can sometimes be located by means of slight variations in water surface temperature, due to the movements of individual fishes that mix water from different strata and of different temperatures. A typical application of thermal infrared is the monitoring and locating of forest fires, now routinely done by the USDA Forest Service [15], a system that could be made satellite-borne.

Radar shows great promise for surveying and monitoring of vegetation and crops [16, 17]. ESRO has ordered a pre-feasibility study on a satellite-borne side-looking radar system (SARSAT), supposed to be operational by 1980. Finally, mention should be made of specific signals. One is the system of satellite-interrogation. The satellite passing over a location can emit a signal that provokes a response from an instrument at that location. The response can be the transmission of measurements on snow depth or height of the waves at sea, or the location and temperature of an animal. In that way, the behaviour of animals that move large distances, such as whales, reindeer or polar bears, could be studied.

## 4. Visible Light and Near-Infrared

The colour and false-colour photographs from the various Apollo missions have generated great admiration. Actually, they are colour-composites based on multi-band black-and-white photography [18]. These pseudo-photographs have shown their usefulness in vegetation and crop studies. The spatial resolution goes down to about 100 m (Apollo 9 Infrared colour film) [19], and, in military satellites like Big Bird, to about 30 cm. In principle, crop identification is possible on false-colour photographs [5, 20]. Recognition of injuries, pests and diseases seems possible in principle, with the restriction that the affected area is large relative to the spatial resolution. Tidal vegetation [21] and algal growth in the oceans can be seen. The effects of air and water pollution on vegetation can be identified.

However, photography is of little use in routine satellite work. Therefore photography is replaced by line scanning techniques. In MSS the signals are selected from 3 to 24 narrow wavelength bands. The information carrier is no longer the photographic plate but a multitrack magnetic tape. The information can be transmitted over long distances, and used in various ways, one of them being to produce images that look like photographs. These pseudo-photographs in black-and-white, colour, or false-colour, can be handled like ordinary photographs by biologists trained in photo-interpretation.

MSS from aircraft is being tested thoroughly at present for crop monitoring and disease surveying. Leading institutes are the Environmental Research Institute of Michigan, Ann Arbor, and LARS, Laboratory for the Application of Remote Sensing, Purdue. A typical agricultural project is the Corn Blight Watch, corn blight being a disease of maize that in the sudden epidemic of 1970 destroyed 15 million dollars worth of crop in the USA [22].

## 5. Biological Use of Satellite-Generated Imagery

The biological use of satellite-generated imagery is quite diverse. It can be classified according to field of interest, wavelength band, or type of satellite. A classification according to field of interest is, e.g.:

Ecology: Natural vegetation, Oceanography, Pollution.
Natural resources: Agriculture, Range lands, Forestry, Fisheries.

Wavelength band and type of satellite are interrelated, and can be treated together here. The weather satellites have provided interesting information in various ways. First, the weather itself affects daily life profoundly. Second, the movements of weather fronts, cyclones or wind fields can be read from cloud images. Knowledge of such movements leads to understanding, and eventually predicting, of specific biological phenomena like locust attack or plant disease epidemic. The swarms of migratory locusts follow a path determined by air currents along the Inter Tropical Convergence Zone [23], which can be recognized on weather satellite imagery. Cloud patterns are used to locate rain areas, where the locusts can breed. Crops can be threatened by plant diseases. Sometimes, adequate disease control is possible after timely warning. In India, wheat is grown in the Ganges plain; a disease, the black stem rust (*Puccinia graminis*), comes in from the south in the form of wind-borne spores. Suitable high-altitude winds occur, for instance, when a depression formed in the Bay of Bengal travels over Southern India. Weather systems providing these conditions at the suitable time (the month of November) could be identified on ESSA 9 cloud images. Forecasting of wheat rust epidemics by means of satellite imagery is now under study [24].

Weather satellite imagery of low spatial resolution has been used to locate the transition zone between two major vegetation types (biomes: forest and tundra) in the poorly accessible terrain of sub-arctic Canada. The transition zone or ecotone, viewed in continental perspective, is linear. A seasonally recurrent linear east—west signature has been found on TIROS imagery; this was tentatively called the "tree line". The feature studied has a distinct seasonal variation, which makes frequent coverage necessary. High temporal resolution also helps to differentiate cloud cover from snow cover, the first being less permanent than the latter. ESSA (AVCS) and Nimbus (HRIR) imagery was compared with ground truth. The so-called "tree line" is a tonal-textural change on several types of imagery, resulting from some change at the earth's surface. The primary cause of that change is sought in vegetation differences, more precisely the difference between areas with more than 30% woodland (= open coniferous forest) and areas with less than 30% woodland [25].

The ERTS 1 project combines much of the experience and expectation recorded above. There is a great number and variety of evaluation projects. One list of ERTS 1 projects gives the following information:

| Type of project | number | | % | |
|---|---|---|---|---|
| Agriculture and forestry | 38 | | 12 | |
| Ecology and zoology | 32 | | 10 | |
| Total biology | | 70 | | 22 |
| Others | | 241 | | 78 |
| Grand total | | 311 | | 100 |

To return to crop protection and the desert locust, the Centre for Overseas Pest Research planned to use ERTS 1 imagery to detect areas in the desert recently wetted by rain, or covered by young vegetation. Here, the desert locusts find breeding grounds, with food and shelter for the young [26]. The agricultural project of the authors, a contribution of the Netherlands to the ERTS 1 undertaking, failed due to the short functional period of the satellite-borne tape-recorder. It is a pity that real-time satellite to ground transmission is not possible in Europe. The European Space Research Organisation (ESRO) has missed an opportunity.

## 6. Problems of the Near Future

Thinking of the near future of space techniques in relation to biology, we come to the following tentative conclusions:

a) In the technical area of data collection, transmission and storage, there is little to be desired, except everyone's desire for better resolution, spatial, spectral, temporal, and otherwise (polarization, coherence, phase, etc.).

b) In the scientific area there is much to be desired, especially with regard to the understanding of the interaction between living matter and radiation. A breakthrough can be made only when much more labour is invested in the border area between radiation physics and crop or vegetation physiology, an area which we can only indicate by the word biophysics.

c) In the teaching area new approaches will be needed. Radiation physics has a minor position in the biology curriculum; photogrammetry and photo interpretation are taught but rarely. Real progress will only be made when training in the application of remote sensing is offered, and tailored to the biologist's need.

d) In the functional area, the area which covers the relation between objectives and means, little has been demanded, and very wrongly so.

The people working in the functional area, here redefined as the area of the relations between science and society, seem to suffer from undernourished inspiration. There is no committee or government to be blamed. There is only the group of biologists, who, as a group, have little grip on the subject matter. In this perspective, there is sympathy for the present empirical approach of "let's generate information and see what happens" [27].

A great problem is the nearly infinite amount of information that will become available. This can easily lead to such lack of scientific self-discipline, that an economically sound relation between objectives and means is lost. The physical techniques as applied to the study of life on earth are quite unspecific in relation to the biological information required. Information can be obtained about nearly everything, but high selectivity is required to avoid drowning in the information flood. High selectivity is required for surveying and monitoring projects not only with respect to spectral and spatial resolution, frequency of coverage, etc., but also with a view to objectives, all in the light of a proper relation between present investment of means and future returns in human welfare and wellbeing. High selectivity is also required in the information needed from each project. When the objectives of a project are well defined, the amount of information to be transmitted from satellite via ground station to user can be greatly reduced, with resulting large economies.

Techniques of data reduction have been applied to MSS studies [28]. Experimental data were collected using equipment with, e.g., nine narrow spectral bands. Any object is characterized by its position in the multi-dimensional space formed by nine vectors. By statistical calculation a set of three new vectors can be found which contains over 90% of the information on the chosen object in the original nine-vector space. When the object is well defined, the data flow can be reduced by 80%. The goal is, of course, to perform the data reduction by spaceborne computer before transmission. The data reduction programme must be capable of being reset by instructions from earth.

This then is the biologist's problem: How to match technical feasibility with social demand into a happy and fertile marriage. It is a universal problem, and undoubtedly, a biological problem.

## Acknowledgments

The expert advice of Mr. S. A. Hempenius (I. T. C., Enschede) and the assistance of Mr. J. Holle (NIWARS, Delft) are gratefully acknowledged.

## References

[1] J. A. Howard, Aerial Photo-Ecology, Faber and Faber, London 1970.
[2] SCIBP, International Biological Programme, London 1972.
[3] L. D. Miller and R. L. Pearson, 7th Symp. on Remote Sensing of Environment, Ann Arbor, Michigan 1971.
[4] R. D. Hudson, Infrared System Engineering, Wiley, New York 1969.
[5] J. R. Shay (ed.), Remote Sensing, with special references to Agriculture and Forestry, National Academy of Sciences, Washington 1970.
[6] W. M. Ciesla et al., Photogramm. Engng **37**, 867 (1971).
[7] B. S. Ausmus and J. W. Hilty, Phytopath. **62**, 1070 (1972).
[8] V. R. Wallen and L. E. Philpotts, Photogramm. Engng **37**, 443 (1971).
[9] H. W. Gausman and R. Cardenas, Bot. Gaz. **130**, 158 (1969).
[10] W. A. Allen and A. J. Richardson, J. Opt. Soc. Am. **58**, 1023 (1968).
[11] W. A. Allen et al., J. Opt. Soc. Am. **60**, 372 (1970).
[12] P. Stoutjesdijk, Acta Bot. Neerl. **19**, 373 (1970).
[13] W. G. Rohde and C. E. Olson, Photogramm. Engng **36**, 561 (1970).
[14] P. A. Murtha, Bi-monthly Res. Notes **26**, 28 (1970).
[15] S. N. Hirsch et al., USDA Forest Serv. Res. Paper int-49 (1968).
[16] G. P. de Loor, Geoforum **2**, 9 (1970).
[17] R. M. Haralick et al., Rem. Sensing of Environment **1**, 131 (1969—1970).
[18] E. Yost and S. Wenderoth, Photogramm. Engng **37**, 590 (1971).
[19] R. C. Aldrich, Photogramm. Engng **37**, 389 (1971).
[20] H. D. Frinking, Neth. J. Agric. Sci. **20**, 261 (1972).
[21] R. L. Vadas and F. E. Manzer, Proc. 3rd Biennal Workshop "Color Aerial Photography in the Plant Sciences", 255 (1971).
[22] R. Mittauer and J. McDavid, NASA release No. 71—129 (1971).
[23] Johnson, C. G. Migration and Dispersal of Insects by Flight, Methuen, London 1969.
[24] S. Nagarajan and H. Singh, Proc. SABRO Meetings, New Delhi 1973.
[25] S. A. Aldrich et al., Rem. Sensing of Environment **2**, 9 (1971).
[26] D. E. Pedgley, ERTS 1 Rep. No. E73-10098 (1973).
[27] Space Science Board. Priorities for space research 1971—1980. Nat. Acad. Sci., Washington, D. C. (1971).
[28] D. Landgrebe and Staff, 4th Ann. Earth Resources Program Review II: Univ. Programs NASA. Houston. 51.1 (1972).

Earth Survey Problems — Akademie-Verlag, Berlin 1974

# USE OF REMOTE SENSING IN THE STUDY OF ANTARCTIC MARINE RESOURCES

S. Z. El-Sayed and K. A. Green

Department of Oceanography, Texas A & M University, College Station, Texas, USA

Several components of the Antarctic marine ecosystems are biologically or economically important, but poorly understood or measured due to logistic difficulties in working from conventional platforms (ships). Because of the great significance of the krill (*Euphausia superba*) in the Antarctic marine ecosystem, and since krill will eventually be the subject of commercial exploitation, it is imperative that accurate data on the density, dimensions, frequency, and distributions of its swarms be obtained. The distribution of krill with respect to variable ice and water is of great interest. The extent, color, location and other features of krill swarms may render them susceptible to remote sensing for determination of their spatial and temporal distribution. The Antarctic physical/chemical/biological milieu, amenable to remote sensing, will be described.

Remote sensing over large expanses of the Southern Ocean rather than point source information gathered by ship will certainly provide data needed for intelligent and wise management of valuable Antarctic marine resources.

## 1. Introduction

It has been said that one of the biggest advantages of remote sensing is that large areas of the earth's surface can be surveyed in short periods of time, thus providing near synoptic observations [1]. In no other region of the World Ocean is this statement more pertinent and more applicable than in the Southern Ocean, the areal extent of which rivals that of the Pacific Ocean. Since the waters surrounding Antarctica are sparsely covered by commercial ships or research vessels, a strong case can be made for the need of remote sensing in the study of the vast expanses of the Southern Ocean. While there is an increasing interest among the physical oceanographers to enlist the assistance of remote sensing in the study of the distribution of pack ice, the hydrographic phenomena associated with the Polar Front, etc., we are mainly concerned here with the use of remote sensing in the study of the living marine resources of the Antarctic. The use of remote sensing in the exploitation of the seals and whales is discussed in [2], and we will limit our discussion here to the krill resource. Although other resources can be included, e.g. antarctic squids and fish, the krill is the most valuable marine resource in the Southern Ocean.

Our interest in the study of the Antarctic krill is neither casual nor academic, but is desperately needed in a protein-hungry world. With the demise of the stocks of the baleen whales in the past 40 years, attention is now being shifted to the krill, the potential catch of which has been estimated by Moiseev [3] to be well in excess of the total harvest of the millions of tons of fish from the ocean. The standing stocks of krill have recently been estimated to be about 950000000 tons [4]. However, a more reasonable estimate of 150000000 tons has been made by Allen [5]. If the krill resource is to be exploited wisely, it is imperative that we obtain reliable data on the distribution, density, dimensions, and frequency of its swarms. We can ill afford to decimate the krill population as other valuable marine resources have been wiped out. Since, sooner or later, the krill will be the subject of commercial exploitation, we should enlist the assistance of remote sensing techniques in obtaining the kind of data needed to make judicious decisions regarding the rational exploitation of this marine resource.

Before we discuss the possibilities of remote sensing the krill, it would be well to discuss the physical/chemical milieu in which these crustaceans live. We will then discuss their biology and ecology with special emphasis on those biological characteristics which are amenable to remote sensing.

## 2. The Physical/Chemical Milieu

The physical/chemical settings of the Antarctic seas are unique in the World Ocean. Their unique features are represented by (a) the presence of pack ice around the Antarctic continent and the seasonal waxing and waning of the pack ice zone; (b) the variability of the light regime which alternates between complete perpetual darkness in winter to continuous daylight during summer; (c) the nearly homeothermic sea temperature which varies very little with depth or season; (d) the extensive cloud cover; and (e) the circumpolarity of its surface waters and the dominant influence of the circumpolar fronts. Each of these features will be discussed with special reference to their remote sensing properties.

### 2.1. Distribution of Pack Ice

The area surrounding Antarctica is covered with pack ice. This area represents about 7% of the earth's surface, or 12% of the sea surface during the winter [6]. The Ice Atlas of the Antarctic [7] is the only published atlas which gives a circum-Antarctic picture of the ice distribution. In recent years, satellite studies of the ice have provided an excellent means of solving the large scale problems of Antarctic sea ice distribution. For instance, Nimbus 1 and 2 data obtained during the winter months of September 1964 and August 1966 show that large, stationary polynyas are both numerous and more widely distributed during the winter seasons than had previously been recognized [8].

Antarctica came to be known as the "pulsating continent" because of the waxing and waning of its pack ice zone. The area covered with pack ice is reduced from $23.8 \times 10^6$ km² in September to $18 \times 10^6$ km² in February [9]. The greatest change in the ice extent occurs during December, the month in which solar radiation is maximum (Fig. 1). Throughout its cycle of expansion and contraction, the circum-Antarctic ice belt is generally moving. The movement is least during the winter and greatest during the summer and is generally from east to west except for the clockwise circulation of ice in the Weddell Sea, which, in the northern part of that sea, produces a west to east drift. This movement, superimposed on a similar wind drift of water, is a function of the general atmospheric pressure and wind regime of the Antarctic [10].

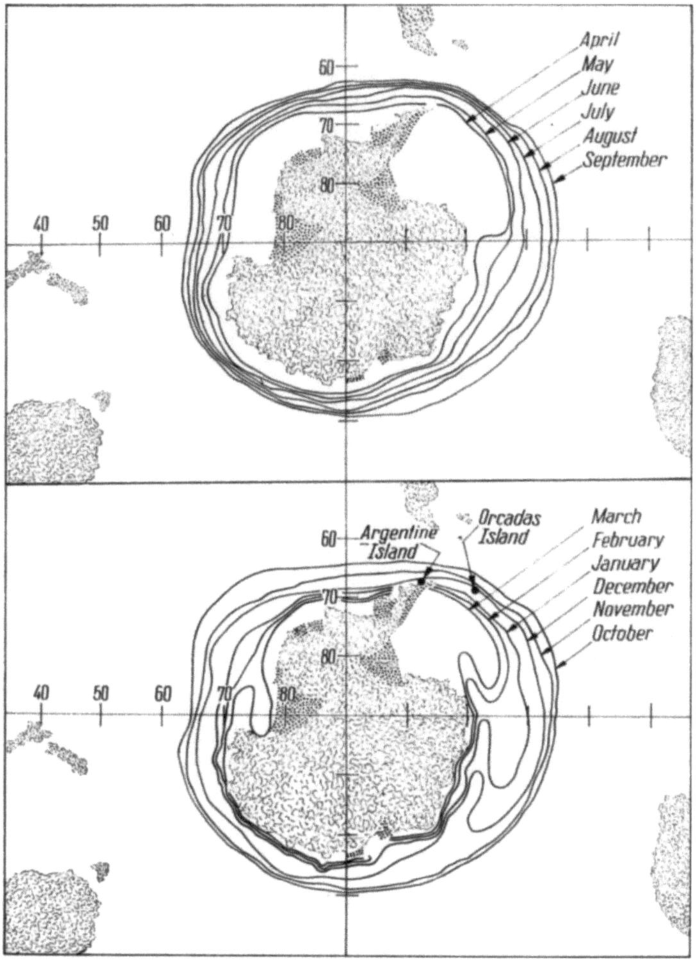


Fig. 1. Monthly variations in the boundary of the pack ice (after J. O. Fletcher, adapted from *Atlas of Antarctica*, 1960).

Our interest in remote sensing of pack ice stems from the fact that several investigators have noted a close association between the distribution of pack ice and the swarming of the krill. Furthermore, net solar radiation in the water mass was also found to be a function of ice distribution. Since primary productivity is a function of net solar radiation, and since krill distribution is closely related to the productivity of the water, the distribution of sea ice and pack ice should be of great concern to those interested in remote sensing of that organism.

## 2.2. Light

Variations in incoming solar radiation between summer and winter in the Antarctic are extreme. These variations have pronounced effect on plant and animal life in the Antarctic. This effect is illustrated in Fig. 2 in which the mean monthly values of radiant energy (with average cloudiness) measured at the Norwegian Station "Maudheim" (71°03′ S 10°56′ W) are plotted against primary production and standing crop of phytoplankton collected in antarctic waters. It is clear from this figure that the rise and fall of the chlorophyll *a* and $C^{14}$ uptake are in direct response to the amount of energy received.

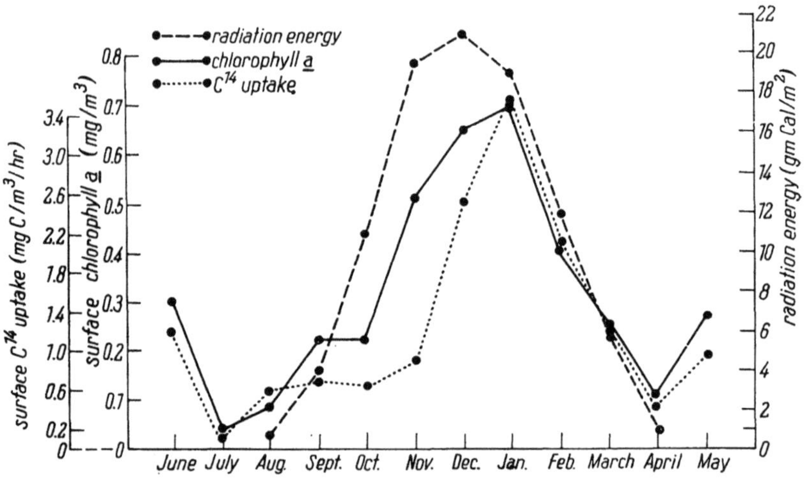


Fig. 2. Monthly variation of radiation energy (with average cloudiness) at "Maudheim" station (71°03′ S 10°56′ W) compared with monthly changes in surface chlorophyll *a* and $C^{14}$ uptake in Antarctic waters (After El-Sayed, 1971).

## 2.3. Cloud Cover

Measurements of cloud cover from ground or sea stations in high latitudes are sparse. On the Antarctic continent, coastal areas are more cloudy than at the interior stations; breaks in the cloud cover are common along the coastline. Cloud cover is greater still over the surrounding Southern Ocean area, varying from about 50% to 90% [11]. Study of ESSA and ITOS satellite infrared photographs of Antarctica, taken almost daily during 1967—1969, has greatly enhanced our knowledge of the extent of cloud cover over the Antarctic Continent and its surrounding waters. Again, as with the case of pack ice, the extent of cloud cover was found to have a significant effect on the primary productivity of the water masses. For instance during *Eltanin* Cruise 51 in the Ross Sea in February 1972, Franceschini (personal communication) noted that incident radiation increased with nearness to the continental shelf as a result of a decrease in cloud amount. Increased reflection from sea ice as well as from the Ross Ice Shelf contributed to this enhancement. Primary productivity measurements made in the Ross Sea did indeed show significant increase.

## 2.4. Physical Oceanography

The Southern Ocean (designated by some as the region between 40° S and Antarctica), can be divided into several well-defined circumpolar frontal zones of which the Antarctic Convergence is the most stationary. The Antarctic Convergence forms one of the major boundary zones of the world's ocean, dividing the southern waters into the antarctic and subantarctic regions.

The main transport of water between 40° and 60° S is from west to east in what is known as the Circumpolar Current. Close to the continent easterly or southeasterly winds prevail that cause transport in a westerly direction. Because

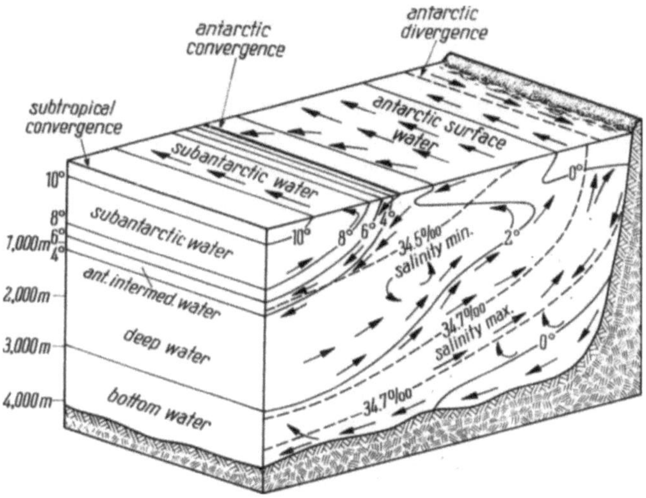


Fig. 3. Schematic representation of the currents and water masses of the Southern Ocean and the distribution of temperature and salinity (modified from Sverdrup, Johnson and Fleming, 1942) (After El-Sayed, 1971).

of the earth's rotation, the westerlies and easterlies produce, respectively, northerly and southerly transports in the upper layer. These transports result in a surface divergence (the Antarctic Divergence), and consequently upwelling along the boundary of these two wind belts (Fig. 3). Superimposed upon this eastward circumpolar movement are north—south components. At its northern limit the *Antarctic Surface Water* sinks beneath the less dense south-flowing *Subantarctic Water*. The region in which this phenomenon occurs is called the *Antarctic Convergence* or *Antarctic Polar Front*. The principal property by which its location can be mapped is the steep temperature gradient at the sea surface. Across the convergence, temperature change in summer is from 3° to 5.5 °C and in winter from 0.5° to 3 °C [12]. One of the most remarkable features of the Antarctic Convergence is that its position changes very little from season to season and year to year. As the *Antarctic Surface Water* sinks below the surface at the convergence, and continues northward, mixing with the water above and below, it forms the *Antarctic Intermediate Water*, which is characterized by a salinity minimum and a relatively oxygen maximum. At the edge of the continent (the Weddell Sea in particular), very cold water sinks and moves northward along the shelf and into the deep, forming the *Antarctic Bottom Water*. This water is found all around the Antarctic continent and even beyond the equator in the Atlantic Ocean but is of more limited extent in the Pacific and Indian Oceans.

Immediately above Antarctic Bottom Water is an exceptionally thick, warm water mass, the *Deep Water*, that is formed in the North Atlantic and perhaps in the northern Indian Ocean by the cooling and sinking of high-salinity water. This Deep Water moves southward and ultimately spreads into all the world's oceans. The upward movement of the Deep Water creates a zone of upwelling which supplies to the surface water vast quantities of nutrients which contribute to the luxuriant phytoplankton growth close to the antarctic coasts.

We are primarily interested in the physical oceanography of the waters surrounding Antarctica because of the close association of the distribution of the krill with oceanic circulation, because of the effect of upwelling on enhancing primary productivity, and finally because of its effect on the distribution of pack ice. As will be discussed later, remote sensing by satellites has been successfully used in studying areas of upwelling on the northwest coast of Africa, and in the Benguala Current [13, 14]. One would hope, before too long, to add the Antarctic Ocean to this list.

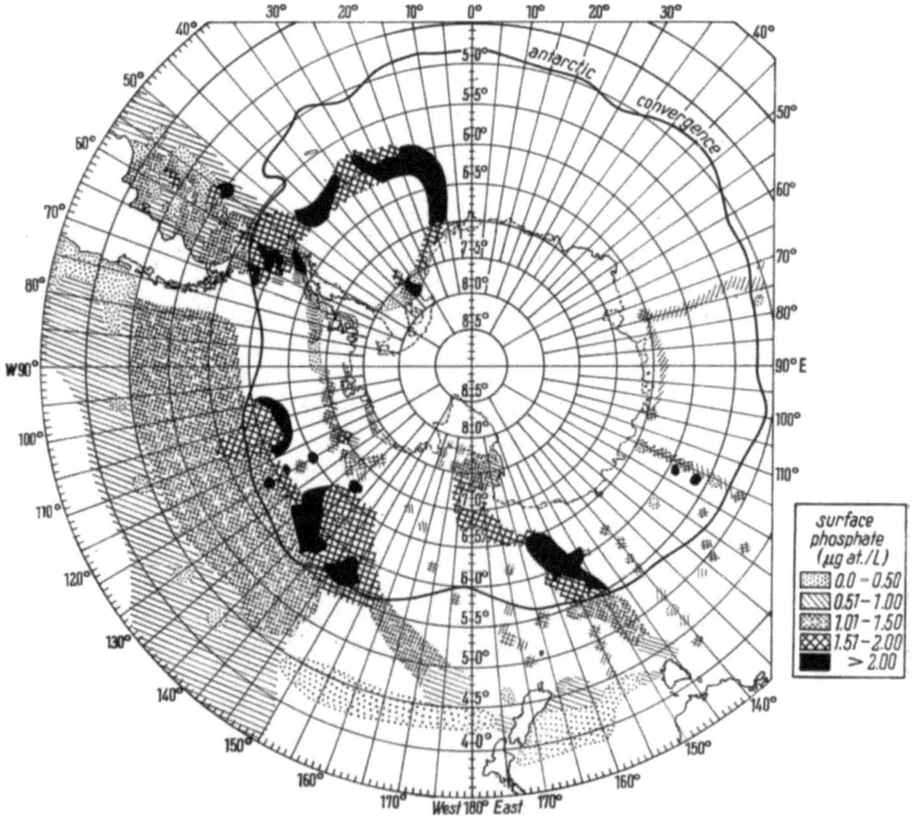


Fig. 4. Distribution of surface phosphates in the Southern Ocean (After El-Sayed, 1970).

## 2.5. Nutrient Salts

As a result of upwelling, the rich nutrient salts in the antarctic waters do not appear to be limiting factors in phytoplankton growth. The distribution of phosphate, silicate, nitrate, and nitrite concentrations in the Southern Ocean are discussed by El-Sayed [15] and are shown here in Figs. 4—7. Despite marked variations in the distribution of the inorganic salts, their pattern of distribution is fairly similar in the Atlantic, Pacific and Indian sectors. By and large, low concentrations of these nutrients are found along the Argentine coast, in the northern Drake Passage, in the Bellingshausen Sea, in the Pacific and Indian waters north of the Antarctic Convergence, and in the Tasman Sea. Moderately high values are found in the southern Drake Passage, west of the Antarctic Peninsula, and south of Tasmania. Very high values of nutrient concentrations are located in the Bransfield Strait, the Weddell Sea, and to the north of the Ross Sea. When comparing the nutrient concentrations in both sectors of the Antarctic, it was found that the distribution of surface phosphates, nitrates, and nitrites showed, in

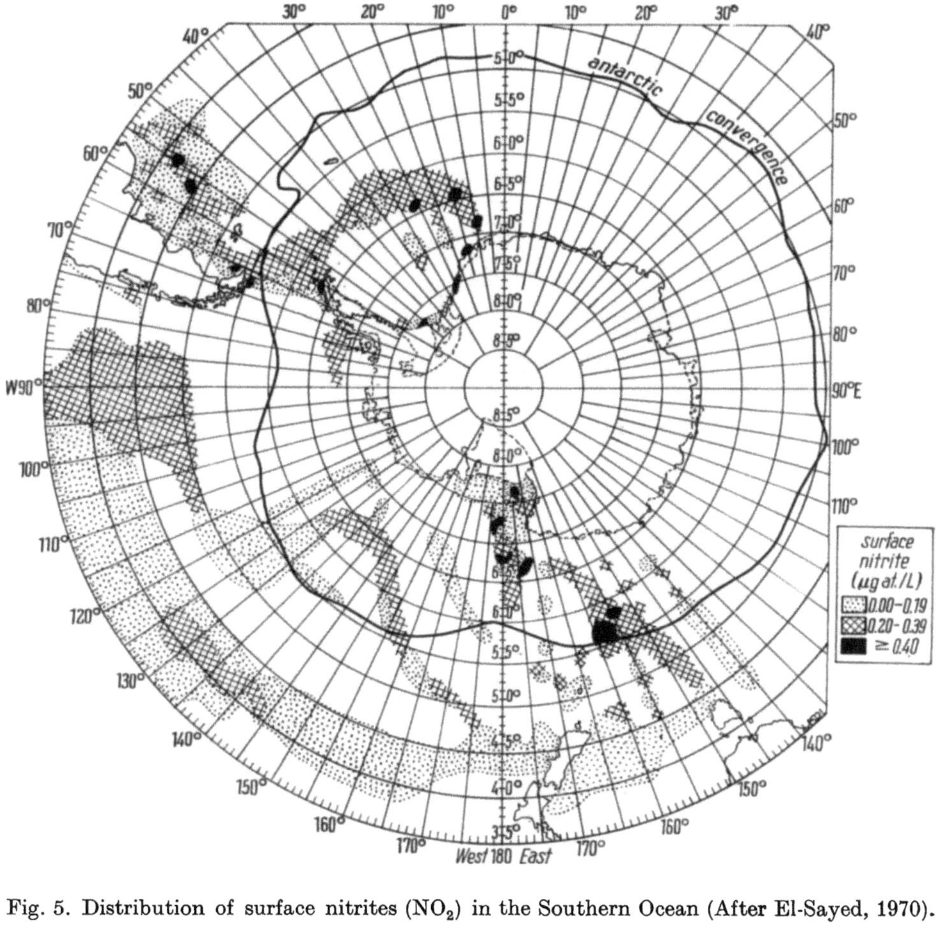


Fig. 5. Distribution of surface nitrites ($NO_2$) in the Southern Ocean (After El-Sayed, 1970).

general, similar values. Silicate concentration, on the other hand, was more than doubled in the Atlantic stations as compared with the Pacific ones.

We have at present no method for remotely measuring the concentration of the nutrient salts in the Antarctic Ocean; it is possible however that we can use the concentration of chlorophyll as indicator of the distribution and concentration of these biogenic salts.

## 3. The Biological Setting

### 3.1. Phytoplankton Population

The distribution of the phytoplankton standing crop (in terms of chlorophyll *a*) in surface water samples (Fig. 8) shows discernible geographic variations in the Atlantic and Pacific sectors [15]. Small phytoplankton standing crops were found in the region of the Antarctic Convergence, the Drake Passage, the Bellingshausen Sea, and in the Pacific sector of the Antarctic, except in isolated pockets in the latter where moderate chlorophyll *a* concentrations (ranging between 0.50 and

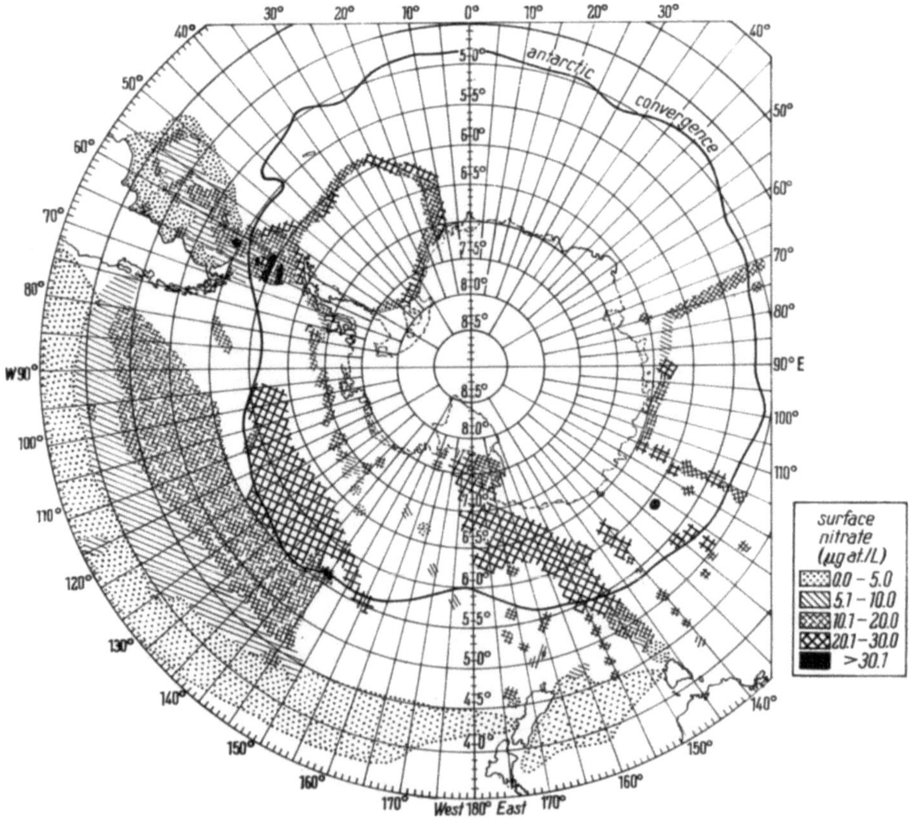


Fig. 6. Distribution of surface nitrates ($NO_3$) in the Southern Ocean (After El-Sayed, 1970).

0.99 mg $m^{-3}$) were encountered. Large phytoplankton standing crops were found in the Scotia sea, in the southwestern Weddell Sea, west of the Antarctic Peninsula, and in the Ross Sea.

Primary production of surface water samples in both the Atlantic and Pacific sectors (Fig. 9) showed a picture more or less similar to that of the distribution of the phytoplankton standing crop. A highly significant correlation was found between surface chlorophyll *a* and primary production in the water samples taken aboard the *Eltanin* Cruises 19—28 in the Pacific sector of the Antarctic and Subantarctic [15].

### 3.2. The Zooplankton Populations

The investigations carried out by the *Discovery* and the *Eltanin*, made it clear that although most, if not all, the zooplankton species are circumpolar in distribution, they frequently show centers of concentration at different depths or in different latitudes (Fig. 10). As a result, there are both quantitative and qualitative differences. The dominant organism in antarctic zooplankton is the krill, *Euphausia superba*, which in the past has supported enormous stocks of whales.

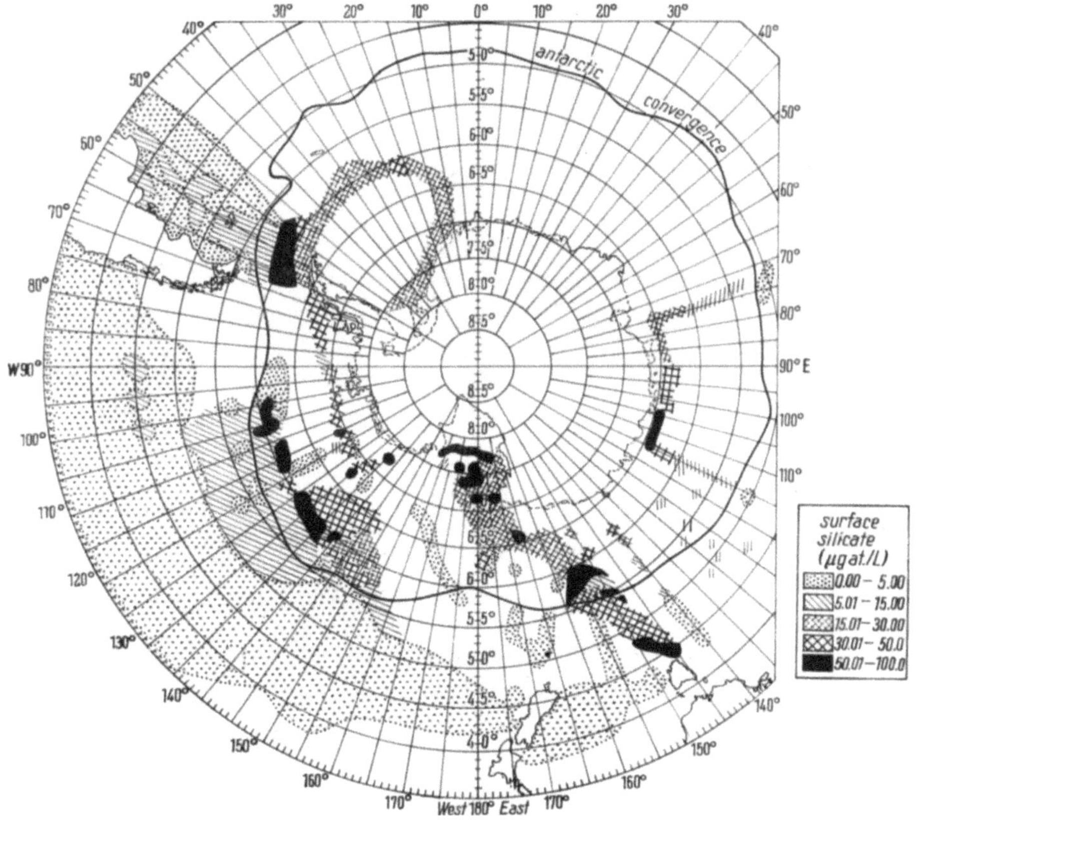


Fig. 7. Distribution of surface silicates ($SiO_3$) in the Southern Ocean (After El-Sayed, 1970).

## 4. The Krill Population (*Euphausia superba*)

### 4.1. Biology

The Antarctic krill, mainly *Euphausia superba*, are small planktonic crustaceans (5—9 cm in length) belonging to the order Euphausiaceae. Eggs hatch in the late spring apparently in deep water and larval development occupies the first summer and fall of the life cycle (November—March). During their second summer, adolescent krill attain an average length of 38 mm with average weight of 800 mg. Krill become sexually mature during their third summer with an average length of 48 mm and average weight of 1200 mg. They die soon after breeding; their average life span is $2^1/_2$—3 years [16].

### 4.2. Distribution

Krill are found in large swarms of a single age class. Spawning probably occurs in the Bransfield Strait and eggs are carried north and east by the cold bottom water. Krill larvae mass in the Weddell Drift. Swarms of adolescents and adults

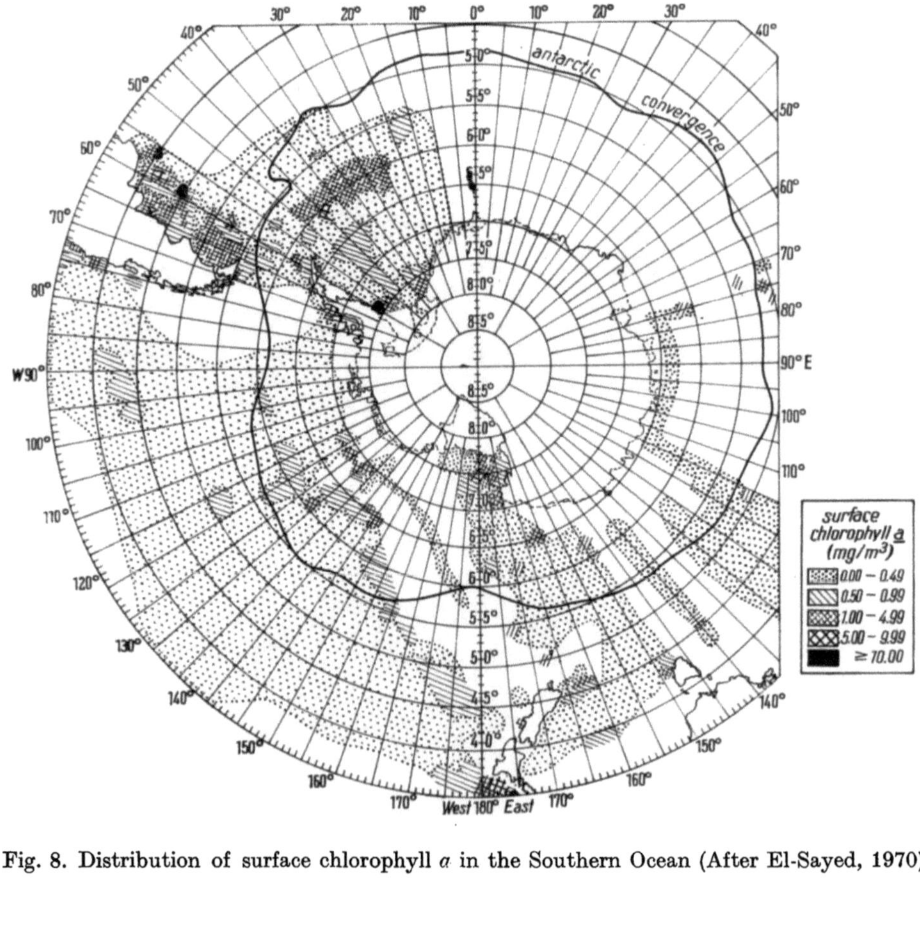


Fig. 8. Distribution of surface chlorophyll *a* in the Southern Ocean (After El-Sayed, 1970).

are distributed separately due to behavioral variation which places them in different water masses. Adolescents are found south of the Scotia Sea during the summer; adults are found in the Weddell Sea or the West Drift. Krill overwinter at the edge of or underneath pack ice. Distribution of the total population is circumpolar [17]. The largest concentration of the krill, according to Marr's maps (Fig. 11) are in the Atlantic sector; however, they are very unevenly distributed.

Recent Soviet investigations show that not all the generations of the krill inhabit this belt, and that there seems to be a very clear spatial separation of the young and the mature *E. superba*. Makarov has shown that finwhale eat mature krill, whereas blue whale feed on the young. The investigations of Fraser and others revealed that the young *E. superba* stay at the ice edges and in polynyas whereas the mature crustaceans are usually found to the north of the ice-free waters. Soviet investigators, however, showed that there is no relationship between the distribution of the Antarctic krill and ice. A large number of young krill were found in clear water, far from the ice, and were also found at the ice edge. In spring and winter, when the ice edge is farther north, mature krill can be found near it, also.

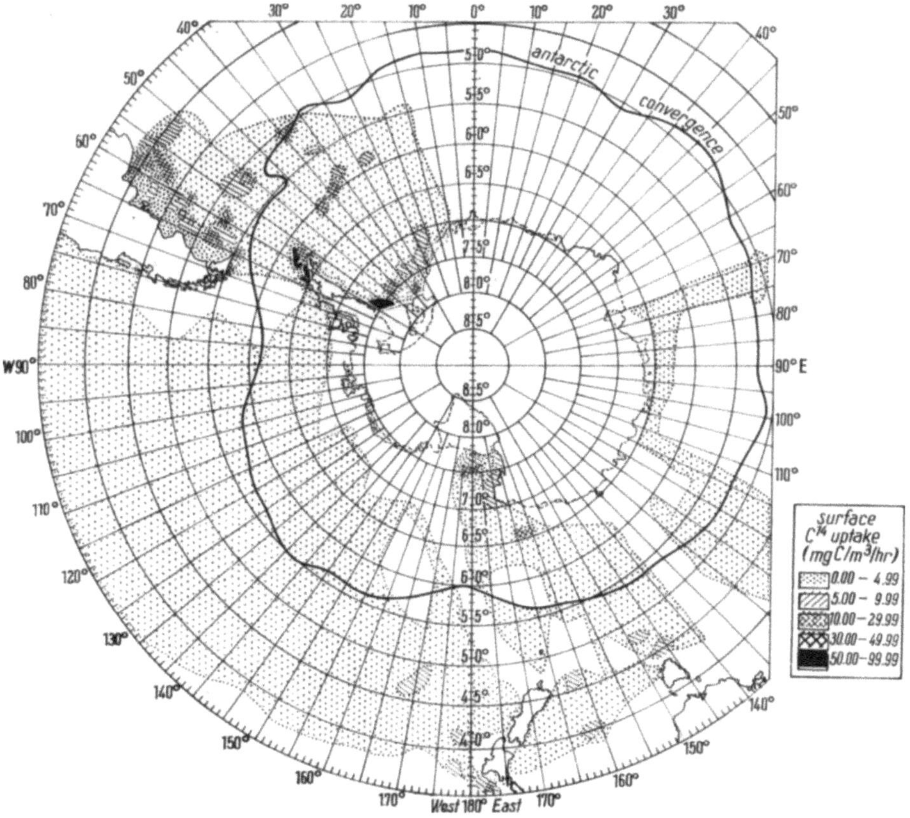


Fig. 9. Carbon-14 uptake (primary productivity) distribution in the Southern Ocean (After El-Sayed, 1970).

Soviet scientists aboard *Akademik Knipovich* showed that the mature and yong crustaceans are rarely found together; mature crustaceans are found, as a rule, to the north and the young to the south. By young, they meant individuals 1 to 1.5 years old, which comprise most of the crustaceans. Soviet scientists attribute the separation of the young and mature krill to the difference in seasonal migration

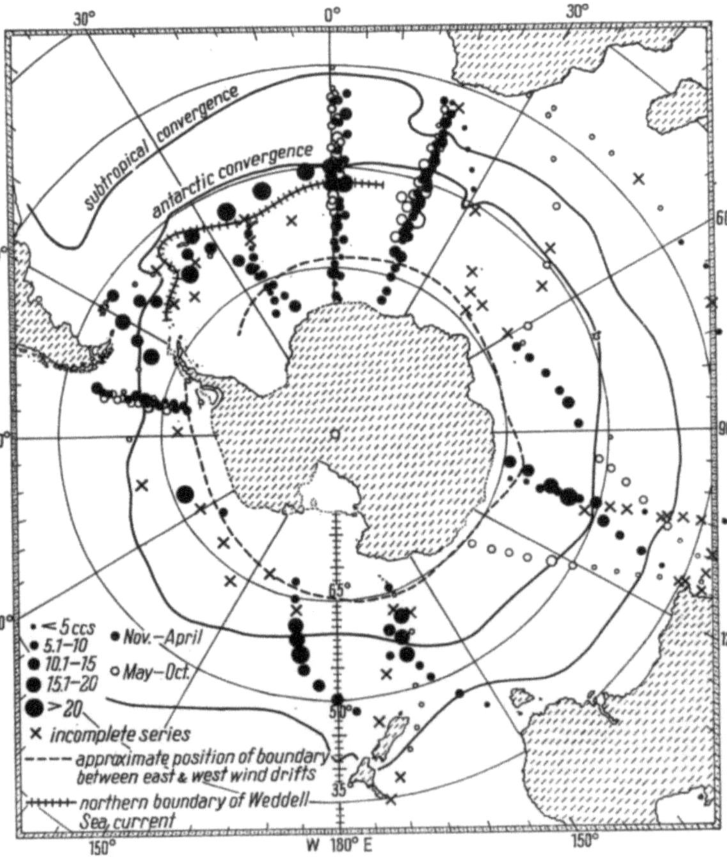


Fig. 10. The total volume of the zooplankton (1000—0 m) at each station plotted according to the position of the station. A few observations in 0°, 20° E and 80° W have had to be omitted for clarity (After Foxton, Discovery Rept. 28, 191, 1956).

which leads to a difference in horizontal movement. The young are carried south as a result of their deep migrations, while mature crustaceans that do not perform deep vertical migrations are found farther north.

The vertical distribution of adult krill (over 20 mm long) has been discussed extensively by Marr [17]. More than 90% of the population is found between the surface and 100 m, with the bulk of that population present above 10 m. Diurnal vertical migrations occur; daytime concentrations are found between 10 and 40 m, while pronounced night-time concentrations are found from the surface to 10 m.

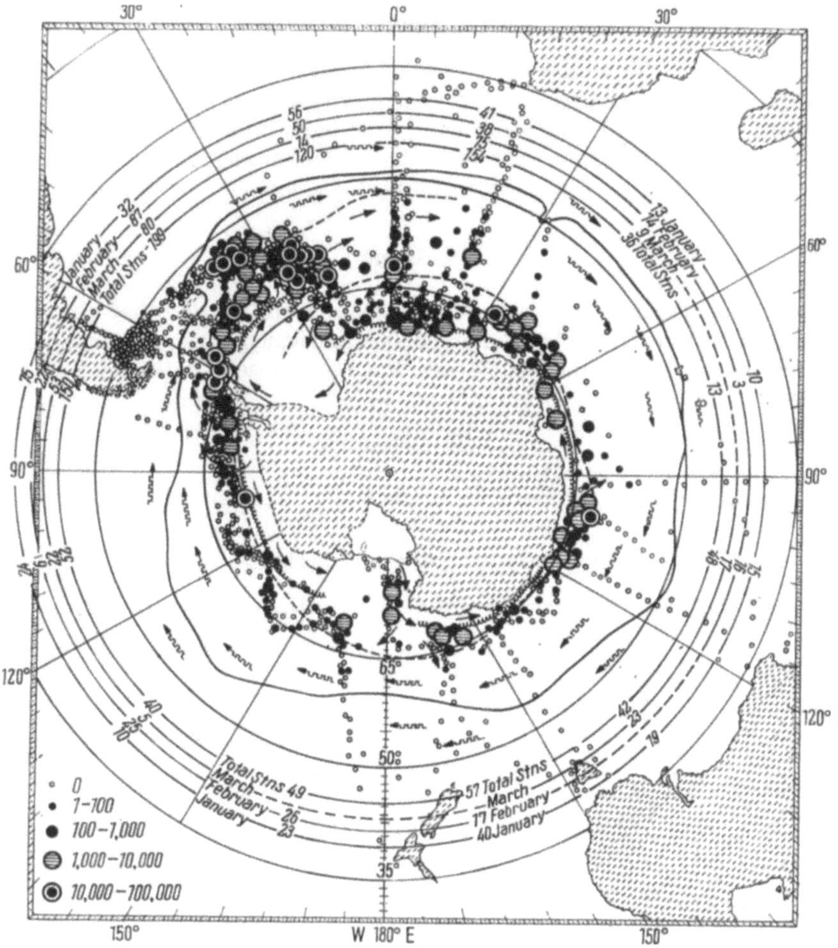


Fig. 11. Distribution of staple whale food in summer (redrawn from J. W. S. Marr [17], p. 394).

## 4.3. Swarms

Krill tend to congregate in large shoals or swarms of a single age class. The swarming habit greatly facilitates feeding by whales on these animals. Swarms average 40 m × 60 m in size; maximum dimension recorded for a swarm is 600 m. An area of 150 km² may be covered by a group of krill patches a quarter mile apart [17].

## 4.4. Color

Krill vary in color from a brillant red to rust, or a pale straw yellow. Swarms are dense enough to discolor sea water and hence can be easily spotted from the deck of a ship [17].

### 4.5. Bioluminescence

Euphausiids are bioluminescent; their spectral distribution peaks around 470—500 nm [18]. Luminescing swarms of *Euphausia pacifica* have been photographed off the coast of California using low light intensifying lenses [19]. Sir Alister Hardy [20] has likened the intensity of bioluminescence of the Antarctic krill swarms to that of a well-lit restaurant—a convenient target for hungry whales!

### 4.6. Sound

Although euphausiids do not have any swim bladders or other gas pockets, they contain oil during the late summer and fall which may have enough differential density to cause sound scattering. They are present in the deep scattering layer [18].

### 4.7. Temperature

Measurements of water temperature in the midst of krill swarms are not available.

### 4.8. Summary of Characteristics

The following is a summary of the characteristics of krill swarms:

| | |
|---|---|
| Size | Average 40 m × 60 m, maximum dimension 600 m |
| Extent | Up to 150 mi$^2$ in patches |
| Depth | 10—40 m (day) |
| | 0—5 m (night) |
| Color | Bright red to rust, straw yellow |
| Luminescence | Low intensity, peak at 470—500 nm |
| Sound | Part of deep scattering layer |

## 5. Prospects of Remote Sensing of the Antarctic Krill

So far we have discussed the broad physical, chemical and biological aspects of the Southern Ocean which are pertinent or amenable to remote sensing. It remains for us now to explore the bearing these aspects have on remote sensing of the krill. What is needed is a simple method to locate these large swarms of krill. At present, we do not have the technological know-how for remotely sensing these swarms. Until such time when we have an easy-to-operate instrument to locate the krill, we will be well-advised to capitalize on: (a) what we already know about the association between the krill and the phytoplankton; and (b) the latest technological advances in remote sensing of chlorophyll *a* and in detecting areas of upwelling.

In recent years there have been some excellent pioneering studies on remote sensing of chlorophyll *a* using low-flying aircraft. Clarke, Ewing and Lorenzen [21] used a specially constructed spectro-radiometer mounted on an aircraft to obtain spectra of upwelling solar radiation from bodies of water having various concentrations of chlorophyll. The spectra were shown to be related to chlorophyll content, with higher chlorophyll resulting in a decrease in intensity in the blue region and a corresponding increase in the green.

More recently the NASA, Ames Research Center, Moffett Field, California, has developed an instrument that gives a real-time indication of chlorophyll along the flight track. The instrument is relatively insensitive to scattering within the water body and the atmosphere [22]. The primary instrumentation used by the NASA scientists includes a differential radiometer (capable of detecting small changes in spectral radiance between two or more wavelength regions) and an infrared filter radiometer. The instrumentation was installed aboard the Cessna 401 aircraft (Fig. 12), and a series of flights were performed to evaluate the capability

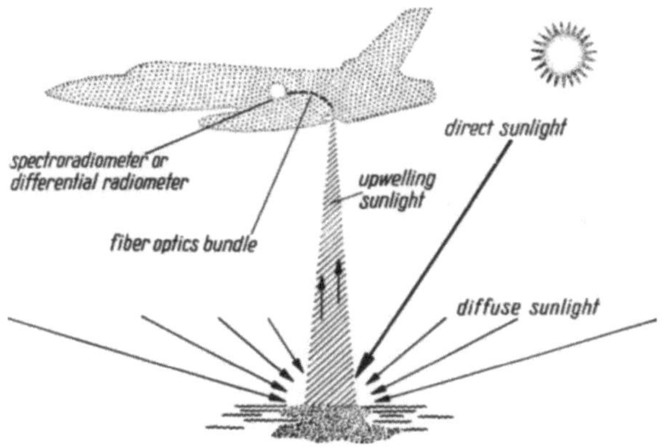


Fig. 12. Airborne remote sensing of subsurface biological characteristics (After Arvesen et al. [22]).

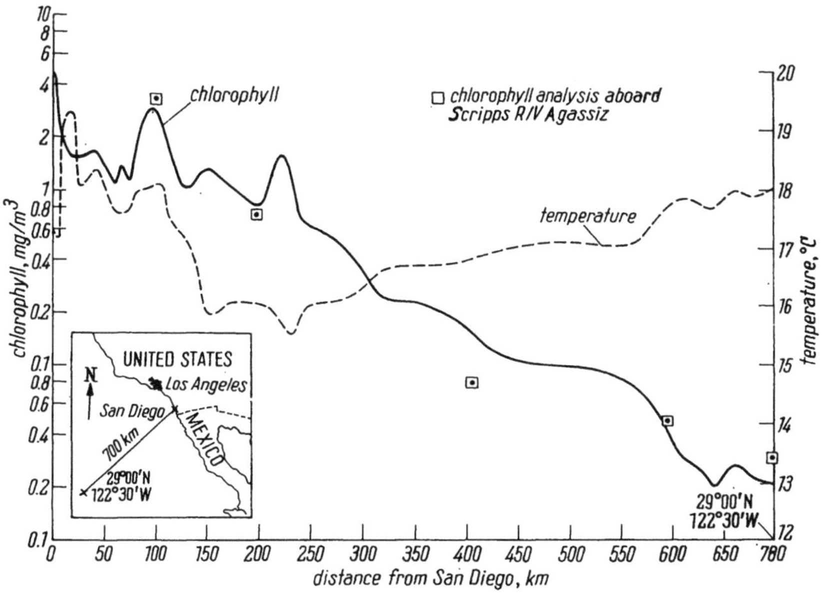


Fig. 13. Chlorophyll and temperature variations along a 700-km track southwest from San Diego on 20 June 1971 (After Arvesen et al. [22]).

of the chlorophyll correlation radiometer to measure chlorophyll content of the water masses.

The results of one of their flights are shown in Fig. 13. In this figure we note that the chlorophyll correlation radiometer recorded an increase in chlorophyll concentration of about two orders of magnitude along a 700-km flight path into San Diego from the deep-offshore waters of the Pacific Ocean. The chlorophyll

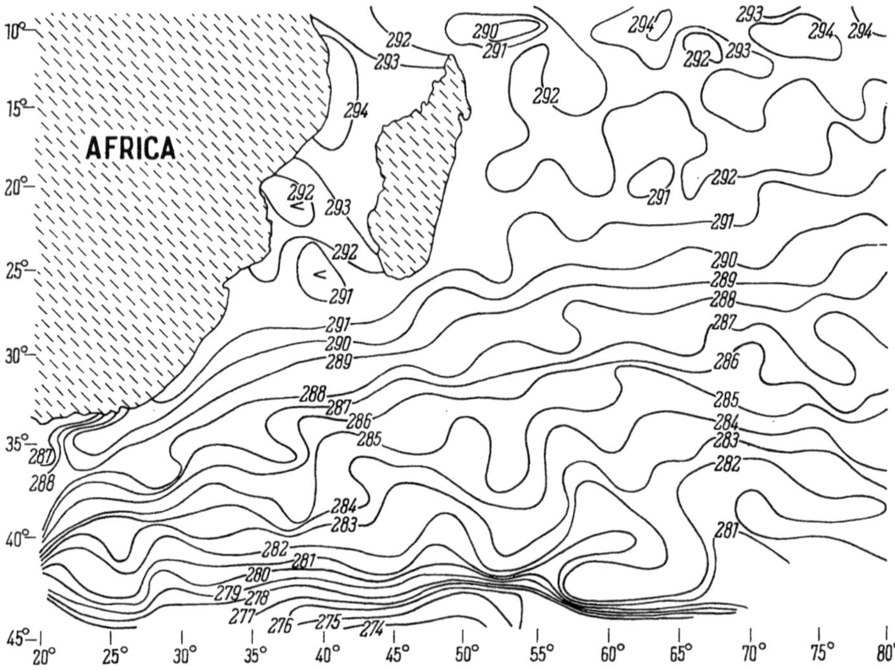


Fig. 14. Temperature distribution along the southern coast of Africa (After Szekielda, 1971).

variation along the flight tracks, and the extremely low chlorophyll concentration found in this offshore area were confirmed by measurements from the R/V *Agassiz* of the Scripps Institute of Oceanography, which took samples at various locations along the track. The same figure clearly indicates that there is a strong inverse correlation between water temperature and chlorophyll from near shore to deep ocean.

Remote sensing of sea surface temperature via satellite has recently shown great promise in delineating areas of upwelling. Scientists at the Goddard Space Flight Center have recently developed a method which detects cloud-free conditions and uses the radiation for the determination of sea surface temperature [14]. This method, together with radiometer recordings analysis, was used to study areas in the oceans where upwelling takes place. Fig. 14, shows a satellite study of the Southern Indian Ocean in which the multi-spectral method was used. The sharp temperature gradient at 40° S is indeed striking; it possibly indicates the Subtropical Convergence between the cold water from the Antarctic and the warmer water masses [14].

The above examples clearly demonstrate that we now have the technological know-how for the simultaneous sensing of chlorophyll and temperature for studies in marine primary production and for detecting such physical processes as upwelling. Texas A & M University investigators are at present collaborating with the research scientists from the NASA Ames Research Center in California to extend the airborne instrumentation and techniques to satellite application in the Southern Ocean. Ground truth data will be furnished by the USNS *Eltanin*. We are also working closely on developing a technique for detecting krill swarms, taking advantage of the physical and biological characteristics of these swarms discussed previously. If successful, this should afford us an excellent opportunity to make reliable estimates of the magnitudes of these swarms that have hitherto been impossible from aboard slow-moving research vessels in the vast expanses of the Southern Ocean.

## Acknowledgment

This work was supported by grants from the National Science Foundation (GV-36215 X).

## References

[1] W. Pearcy, in: Proc. Symp. on Remote Sensing in Marine Biology and Fishery Resources, Texas A&M University, College Station, Texas 1971 (p. 158).
[2] D. B. Siniff and V. B. Kuechle, this volume, p. 65.
[3] P. A. Moiseev, Abstracts of Symp. on Antarctic Ecology, Cambridge, England, 1968.
[4] H. McQuillan, West-Fish **63** (1962).
[5] K. R. Allen, J. Fish. Res. Bd. Canada **28**, 1573 (1971).
[6] W. F. Weeks, in: Symp. on Antarctic Oceanography, Scott Polar Res. Inst., Cambridge, 1968 (p. 173).
[7] N. A. Mackintosh and H. F. P. Herdman, Discovery Rept. **19**, 285 (1940).
[8] W. W. Knapp, Antarctic J. of US **4**, 222 (1969).
[9] N. A. Mackintosh and S. G. Brown, Norsk Hvalfangstiid **45**, 469 (1956).
[10] J. A. Heap, in: Antarctica, Methuen and Co., Ltd, London 1965 (p. 187).
[11] M. J. Rubin and W. S. Weyant, in: Antarctica, Methuen and Co. Ltd., London, 1965 (p. 375).
[12] T. J. Hart, Discovery Rept. **8**, 1 (1934).
[13] K. H. Szekielda, Goddard Space Flight Center Preprint X-651-70-419, 1970.
[14] K. H. Szekielda, Deutsch. Hydrogr. Z. **24**, 26 (1971).
[15] S. Z. El-Sayed, in: Antartic Ecology, Academic Press, New York 1970 (p. 119).
[16] R. R. Makarov and V. V. Shevtsov, Lzdatelstvo, "Nauka", Moscow 1971 (p. 81).
[17] J. W. S. Marr, Discovery Rept. **32**, 33 (1962).
[18] J. Mauchline and L. R. Fisher, Adv. Mar. Biol. Vol. 7, Academic Press, New York 1969.
[19] C. M. Roithmayr and F. P. Wittmann, Mar. Tech. Soc. 8th Ann. Conf. Reprints, 1971 (p. 277).
[20] A. C. Hardy, The Open Sea: Its Natural History, Houghton Mifflin, Boston 1965.
[21] G. L. Clarke, G. C. Ewing and C. J. Lorenzen, Science **167**, 1119 (1970).
[22] J. C. Arvesen, E. C. Weaver and J. P. Millard, Joint Conf. on Sensing of Environmental Pollutants, AIAA paper No. 71-1097 (1971).

Earth Survey Problems — Akademie-Verlag, Berlin 1974

# REMOTE SENSING OF ANTARCTIC BIOLOGICAL RESOURCES: VERTEBRATES

D. B. SINIFF and V. B. KUECHLE

University of Minnesota, Department of Ecology and Behavioral Biology, St. Paul, Minn., USA

The application of remote sensing methodology to the assessment of Antarctic vertebrate populations is reviewed. The ERTS missions are discussed and their application to the assessment of Antarctic vertebrates considered. Multi-spectral scanners are discussed and considered useful for the evaluation of pack ice conditions and to establish relationships among vertebrate densities and environmental conditions. High resolution is required to identify individual seals and/or small groups of penguins, and it would appear that the return-beam-vidicon offers characteristics which may be useful in this respect. However, the prospects for obtaining such resolution in the near future are not considered good. The telemetering of information from remote platforms to satellites passing overhead would provide data on movement patterns and other biological parameters. Also remote sensors placed in pack ice regions could provide considerable information by using hydrophones in a passive listening context. Clearly, space technology could aid greatly in the evaluation of the Antarctic biological resources.

## 1. Introduction

It is well documented that the Antarctic pack ice region is biologically rich. Ocean currents converge in a manner which produces upwelling, making these waters rich in nutrients and hence this richness is reflected through the total food chain. Because the region is rather inhospitable, the magnitude and extent of the biological resources in this region are poorly known. There currently exists much interest in the Antarctic seal resources [1] and concern for other vertebrates such as the whales and penguins has brought a public focus in this region. Thus biologists are looking more and more to remote sensing to aid in the assessment of these resources and also to aid in our understanding of the Antarctic pack ice ecosystem. This paper is directed toward suggestions as to how this technology can aid in furthering our understanding of the vertebrate populations of this region. The major emphasis of this paper will be on the application of remote sensing to the bird and mammal species because they are periodically visible. The problems in the evaluation of the fish species is more closely related to the assessment of the invertebrates and will not be treated here.

The paper will be divided into two major areas. First, we will consider the application of aerial or satellite sensing in the assessment of the visible vertebrate populations. In this discussion, we will discuss current hardware and resolutions which are required to identify various animal concentrations. Second, we will try to furnish guide-lines as to the type of instrumentation needed to assess different types of animal concentrations. Floating platforms or remote land stations which can communicate to satellites passing overhead will be mentioned and their applications discussed.

## 2. The Antarctic Pack Ice Region

In order to make suggestions concerning the application of remote sensing in the Antarctic region, it is important first to consider some of the physical characteristics of this area. In this region, nearly all the biota is supported by the sea. The oceans surrounding the Antarctic continent are covered by annual sea ice which extends north in winter and retreats to the south by melting in summer. This region is referred to as the area of pack ice. Large icebergs originating from the ice sheet covering most of the Antarctic continent are often seen floating among the pack ice, drifting about as current and wind conditions dictate. Large populations of seals, whales, penguins and flying birds can be sighted in this pack ice region. Naturally, life is more visible through the austral summer when the region is exposed to continuous daylight. During this time changes occur quickly; phytoplankton multiply rapidly and most of the vertebrate populations go through their annual reproductive cycle. By late March the pack ice has receded to its southernmost limit and the annual increment to the vertebrate populations has occurred. Thus at this time, the visible vertebrate densities are highest because the pack ice area is smallest and because the highest population levels in the yearly cycle now exist.

Each species has characteristics which make it unique when considering remote sensing applications. For example, the penguins congregate on rookeries to breed and nest. During this nesting time, the number of individuals per unit area is extremely high and colony sizes vary from a few thousand animals to over 100000. The seal species show different behavior patterns during this time. Although detailed information is poorly known, it appears that the crabeater, Ross and leopard seals have pups in the floating pack ice and do not congregate to any great degree during this phase of their life cycle (Fig. 1). On the other hand, however, the Weddell seal which occupies heavy pack ice regions, generally close to the Antarctic continent congregates along annual tide cracks to have pups and breed (Fig. 2). Clearly the problems in remote sensing to determine the magnitude of these different species varies depending on the variations in life style. As will be discussed, current hardware probably possesses sufficient resolution to document the presence and status of penguin rookeries; however, the same instrumentation will not detect lone seals or penguins on ice floes.

Besides the problems of determining densities, there are other parameters of the vertebrate resources which are of importance. As previously mentioned, there is currently interest in utilization of the Antarctic seal resources. As a consequence guide-lines were recently formulated with the purpose of assuring that these species will not become endangered because of man's actions. The implementation of such guide-lines could be aided greatly by the application of space technology to assist in the collection of biological data on the life history processes of these species. For example, it is important that we understand something of the discreteness of individual populations so that the influence of localized actions can be evaluated. Thus, devices for monitoring the movement of individuals are important considerations. There are many other parameters which could be measured remotely which would shed considerable light on the life history of the species. These include such areas of endeavor as the measurement of physical factors which are correlated with species abundance and distribution; and physiological mea-

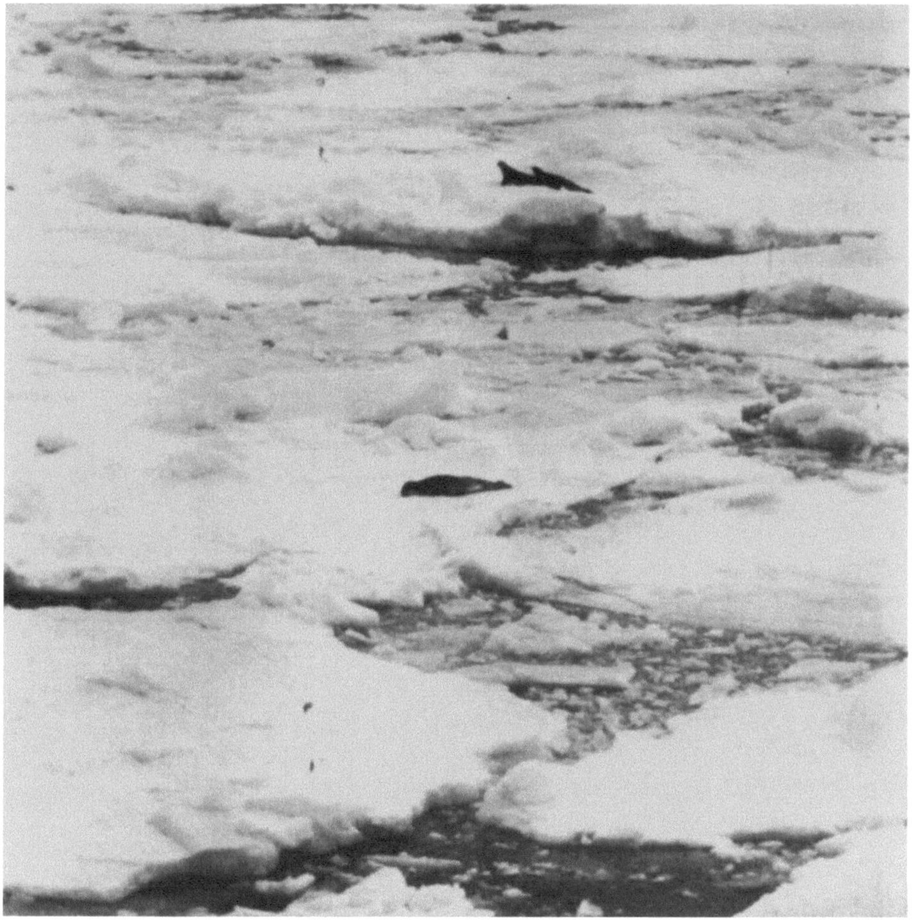

Fig. 1. Seals in pack ice.

surements which reflect the general condition of individuals in the population. There no doubt are other measurements to be considered but these represent worthy areas for remote sensing endeavors.

## 3. Current Hardware

The first useful data for the Antarctic Ocean regions which originated from a satellite came from the Nimbus series. From these photos, notions of the extent and movement of the pack ice are available; however, very little can be said about ice density or uniformity. This instrumentation is not adequate for habitat evaluation or the location of populations which have not already been known from other sources.

Some improvement has occurred from the present ERTS-A mission and further improvement will come with the upcoming ERTS-B. Although a great many test

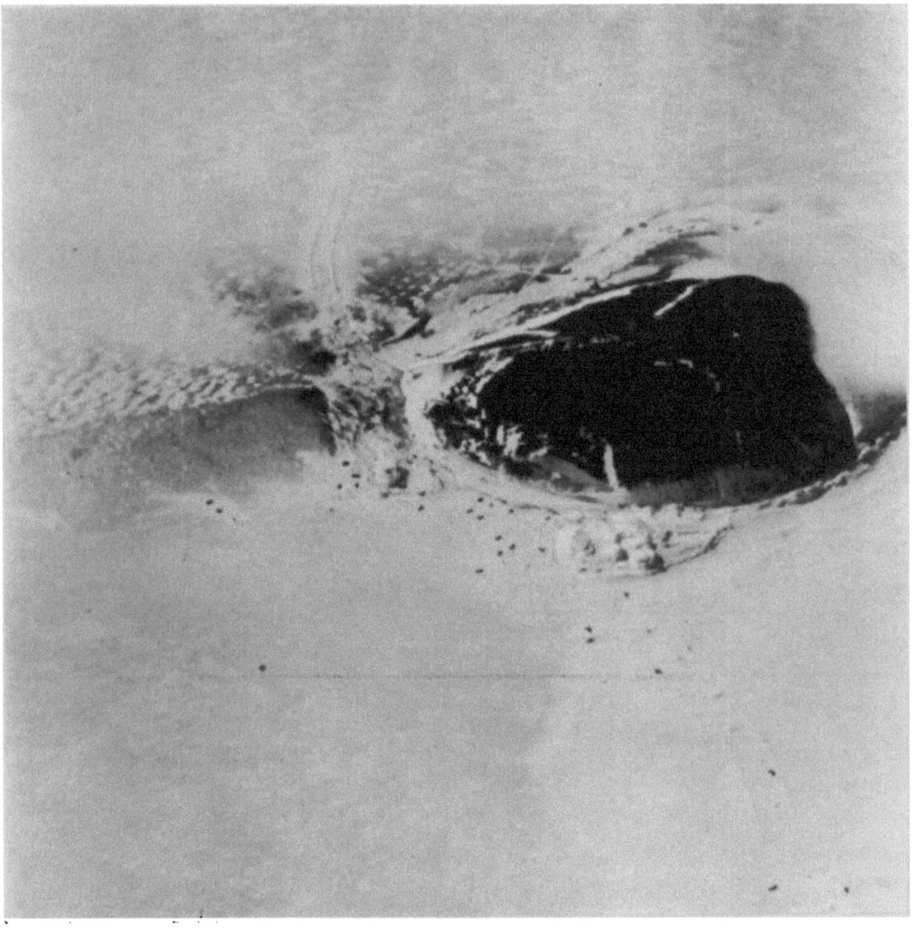

Fig. 2. Weddell seals at Turtle Rock, McMurdo Sound, Antarctica.

data are available on the ERTS-A system and its performance in the laboratory, its performance in the field is only now being evaluated [2] and its ultimate performance will probably not be determined for several years. Unfortunately there are no charts or graphs that would allow a biologist to specify what resolutions would be expected from the ERTS mission with a given set of conditions. We already know of course that these missions will not identify objects the size of penguins or seals in the pack ice regions.

The following is a discussion of possible contributions to the studies of the Antarctic vertebrates which might be envisioned by application of ERTS imagery. The guide-lines for this discussion were obtained from [3]. We have available the photos from the Return Beam Vidicon (RBV) camera which takes three separate video pictures of the same area in different spectral bands. Each of these pictures views an area of side approximately 185 km (100 n mi) with coverage of any given area repeated every 18 days. Since the satellite is in polar orbit and is designed for minimum overlap at the equator, there is increasing overlap as the poles are approached. For the region of interest, approximately 65° to 75° south, there is considerable overlap in coverage (Table 1). This coverage could help to alleviate

Table 1
Overlap of Adjacent ERTS Coverage

| Latitude (°) | 0 | 10 | 20 | 30 | 40 | 50 | 60 | 70 | 80 |
|---|---|---|---|---|---|---|---|---|---|
| Image Overlap (%) | 14.0 | 15.4 | 19.1 | 25.6 | 34.1 | 44.8 | 57.0 | 70.6 | 85.0 |

Data Users Handbook, pages 1–3.

the cloud cover problem, although since the orbit period is 103 min, little actual improvement between passes could be expected. Also available on ERTS is the Multi-spectral Scanner (MSS) which images the surface in several bands simultaneously. For ERTS-A there are four bands in the solar reflected region from 0.5 to 1.1 μm. ERTS-B will have the same four bands plus an emissive spectral region from 10.4 to 12.6 μm and will cover the same 185 km width as does the RBV, but it will cover a continuous band in the forward direction versus the discrete photo format of the RBV. In addition to these two modes ERTS-A has a data collection system to collect data from remote stations and relay them to a data collection point.

The primary limitation of the ERTS mission for our purposes is, of course, resolution. From the ERTS specifications a ground resolution of 90 to 180 meters is expected. The final resolution is dependent on several factors, the most important of which is space contrast. Space contrast is used to describe a target's radiance ratio which, when placed in terms of a seal on an ice floe, is the apparent radiance (usually measured in mW $cm^{-2}$) in the optical band of the seal contrasted to the higher radiance of the seal—background combination. We can get some idea of the contrast/resolution relationship from Fig. 3, where resolving ability versus contrast is given for a standard test pattern. In the Antarctic we would expect contrast to be high in the pack ice regions. Unfortunately the prediction of resolution is highly complex. However, we can say that the best we might expect is a resolution of 40 m/bar for high contrast scenes which is still much less than that needed to detect individual animals under optimum conditions. This roughly means that an object would have to be at least 40 m wide for it to be detectable. We know that in most of the scenes the objects of interest will be smaller than this resolving ability. How many seals would be needed in a given area to make it look different from an area without seals is purely speculative. Some theoretical work in this area has been done by Nalepka et al. [4]. For Weddell seals in fast ice areas the requirements are not as severe as for the species that inhabit the pack ice where open water would reduce the total radiance, making it difficult to determine whether we were looking at open water and ice or seals and ice.

Thus, in order to detect populations of seals either directly or via environmental correlates, a great deal of work needs to be done before any conclusions can be drawn. However, our best estimate indicates that the outlook for direct sensing is not particularly promising and perhaps the most useful information may come from environmental correlates [5]. For example, since the distribution of seals is correlated with ice class [6], ice surveys and classification of ice types is

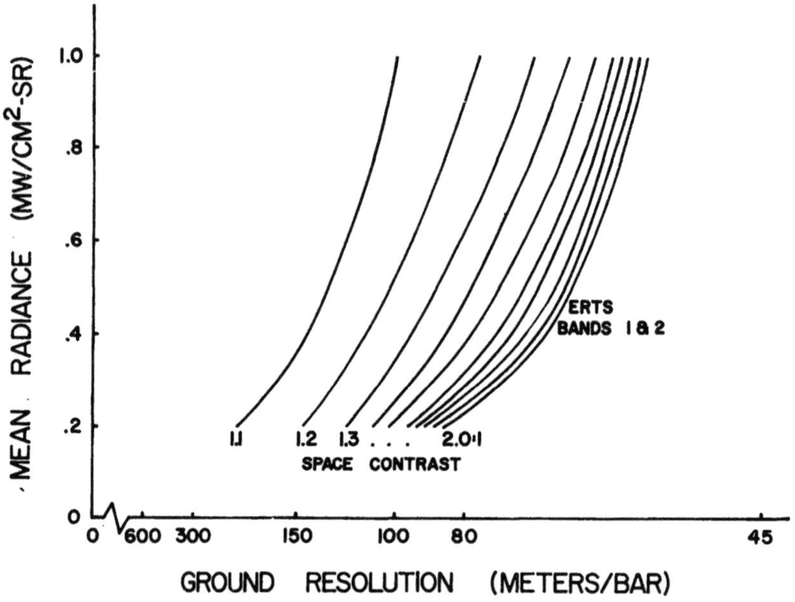


Fig. 3. Estimated ERTS vidicon resolution.

of great interest. With ice surveys from ERTS and ground surveys from icebreakers much more accurate methods for population assessment could be developed.

The assessment of penguin populations particularly during congregation at rookeries, shows much more promise. Penguins tend to congregate in one area year after year. These areas are on outcrops of land and because of the yearly deposit of guano they could be expected to have a different radiance from that of the surrounding area (Fig. 4). Due to the characteristics of these areas we should be able to observe population variations from year to year. In addition some of the rookeries used in the past but now vacant should be visible.

## 4. Needs in Remote Sensing in the Antarctic

If we are going to resolve individual seals we need to be able to detect an object 0.5 m by 2 m. This would allow a sampling technique to be developed with a view toward enumeration of populations. Such an evaluation would allow no species classification, but these data are available from ground surveys. This resolution is obtainable from aircraft surveys; however, with the current aircraft available

Fig. 4. Penguin rookery showing differential radiance between rookery and surrounding area.

surveys of the pack ice region are a difficult task. Such enumeration also seems to be within the capability of return-beam-vidicons. The practical way to increase the resolution would be to increase the picture size; cameras with double the 25 mm square vidicon format used in ERTS-A and B have already been built [7]. This would mean that the volume of data would be much higher and the exposure time would probably have to be shorter or orbital characteristics would have to be changed. Studies of all these factors are needed but current needs in this area seem to be feasible with today's hardware. Whether these specifications can be economically achieved remains to be considered. Experimental and theoretical work needs to be done to determine what can be done in terms of locating concentrations of animals as opposed to surveys of individuals.

It is desirable to know what happens during the periods of darkness. This could be done by infrared scanning in the range of 10.4—12.6 μm as planned for ERTS-B. The accuracies available with current technology are probably sufficient to survey ice type and extent; however, at the present time it does not seem that the resolutions needed to detect individual animals by emissive radiation are within the realm of possibility. Whether concentrations of seals or penguins could be detected by emissive radiation is doubtful because of the small relative difference in emissivity of the larger vertebrates and ice and the lower resolution of the infrared systems.

### 4.1. Telemetry from Remote Sensors

Another area where space technology is most important to Antarctic biology is the capability to collect data from remote earth or ocean based sensors and to retransmit information to a ground-based data processing system (Fig. 5). This capability is especially valuable in the Antarctic where accessibility during the summer season is difficult and during the winter months nearly impossible. Radio frequency tags have been used to locate seals when they were out of the water,

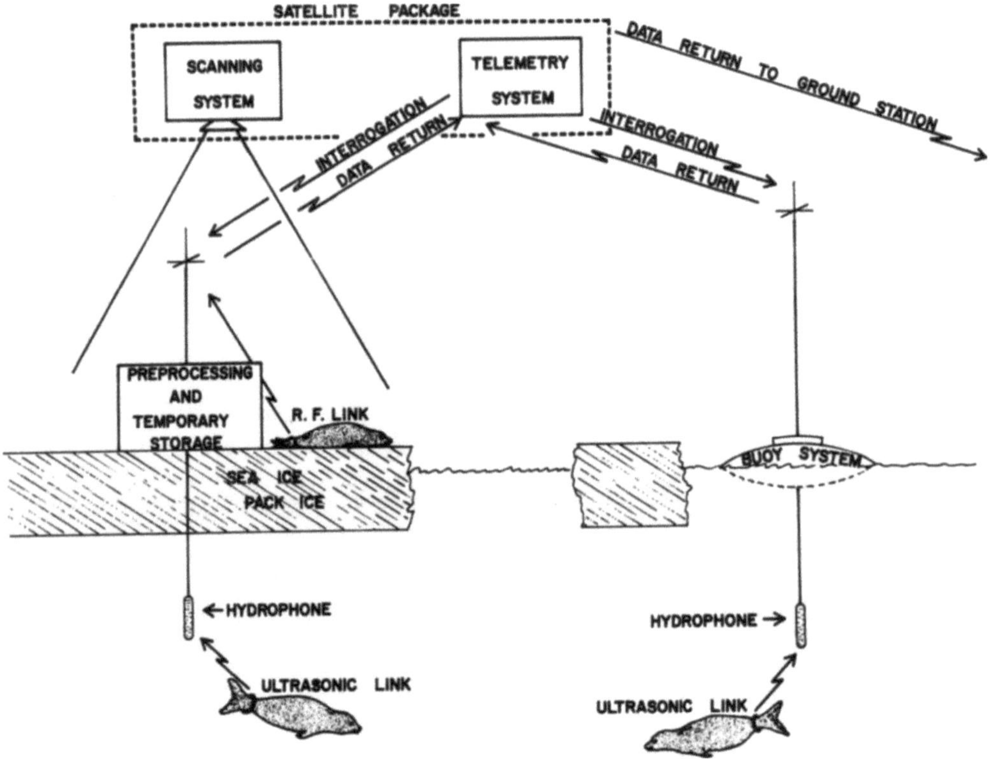


Fig. 5. Remote data collection system.

and current work is underway on a system locating seals underwater using ultrasonic tags [8, 9]. This system has been used in the field for one season and although a completed evaluation is not available, the evidence indicates that the system is rather flexible and holds promise for future work.

Remote platforms could be located on land in the coastal areas with links to hydrophones in the water. Such a system would probably serve rather well for Weddell seals. Toward this end we have proposed a system in the McMurdo area to use collection stations in several areas and transmit the data by radio link to a central data logging point. By placing sensors or "listening posts" throughout the area we could determine the seasonal movement patterns of these animals. This system would work without a spacecraft link; however there are very few coastal stations that are manned year-round, and if further coverage is to be obtained a spacecraft link would be needed.

The most attractive feature of remote data collection is its capability in pack ice regions where remote stations could be put in place by icebreakers during summer cruises or air dropped with the required instrumentation suspended in the water below. These stations would have to be designed to withstand the pressure of the pack ice but this should not be difficult because the packages would be

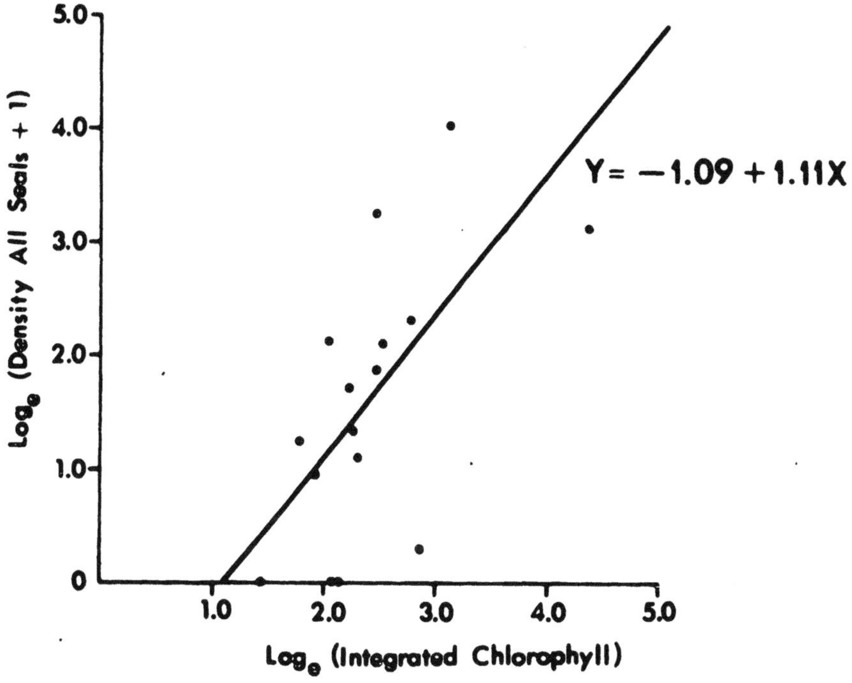


Fig. 6. Relationship found between crabeater seal densities and chlorophyll found in the Weddell Sea in 1968: linear regression of $y = \ln$ (density total seals $+1$) versus $x = \ln$ (integrated chlorophyll).

quite small, and could easily be built to withstand such pressures. In addition to the active packages placed on the seals a great deal of information could be gained by passive listening using vocalizations as an indicator [10].

Such remote stations would also be valuable in simultaneously measuring biological productivity. This would relate productivity through the entire food web. Although we are not certain what factors need to be measured; it has been documented [11] that a correlation between productivity and seal densities exists (Fig. 6). Simultaneous to the monitoring of the biological properties, physical environmental parameters could also be monitored.

Although the parameters we want to measure using a relay are new, most of the hardware already exists. Most of the meteorological/environmental hardware has already been developed for Nimbus F as part of the Tropical Wind, Energy

Conversion, and Reference Level Experiment (TWERLE) and most of the other experiments could be readily interfaced with this package. The costs of the packages are reasonable, making it possible to establish a high number of stations and at the same time consider the packages expendable at the end of the experiment. Although we know of no application involving space relay of biological productivity, there is a great deal of effort currently taking place in remote monitoring; thus it would appear that refinement in existing sensors should not be difficult in order to integrate sensors into a data collection platform.

## References

[1] Polar Record **16**, 271 (1972).
[2] S. C. Freden, in: Earth Resources Technology Satellite 1, Symp. Proc., Goddard Space Flight Center, Greenbelt, Md, 1972 (p. 1).
[3] Data Users Handbook, General Electric Document, No. 71SD4249, Goddard Space Flight Center, Greenbelt, Md, Nov. 1972.
[4] R. F. Nalepka et al., in: 4th Annual Earth Resources Program Review, **32**, 1 (1972).
[5] G. C. Ewing, in: Symp. on Significant Results obtained from ERTS 1, Goddard Space Flight Center, Greenbelt, Md, March 1973, vol. **2**, p. 146.
[6] A. W. Erickson et al., in: SCAR Symp. on Antarctic Ice and Water Masses, 1971, (p. 55).
[7] O. H. Slade, Sr., in: International Workshop on Earth Resources Survey Systems, NASA publication SP-283, Vol. 1, 245 (May 1971).
[8] D. B. Siniff and R. Reichle, in: Symp. on the Biology of the Seal, Guelph, Ontario, Canada 1973 in press.
[9] D. B. Siniff et al., The Antarctic Journal of the US **8** (4), 1973.
[10] L. Midturn, Proc. Remote Sensing in Marine Biology and Fishery Resources, College Station, Texas (1972).
[11] D. B. Siniff et al., in: Antarctic Ecology, Academic Press, New York 1970 (p. 377).

Earth Survey Problems — Akademie-Verlag, Berlin 1974

# A REVIEW OF SOME POSSIBLE USES OF REMOTE SENSING TECHNIQUES IN FISHERY RESEARCH AND COMMERCIAL FISHERIES

J. Joseph and M. R. Stevenson

Inter-American Tropical Tuna Commission, Scripps Institution of Oceanography, La Jolla, Calif., USA

Fishery research is the study of the biology, environment, abundance, availability and exploitation of fish or other aquatic organisms for the purpose of facilitating their utilization by mankind. Efficiently to exploit such resources it is necessary to assess their abundance prior to exploitation, to monitor exploited resources by predicting changes in their abundance and distribution, and to provide management information such as weather and fishing conditions, and vessel activities and position, to industry and government. A discussion of conventional approaches (non-remote sensing) utilized to attain the objectives of fishery research is given. It is concluded that such conventional approaches do not provide all the necessary information to attain the goals of fishery management. In many instances expanded data systems which provide coverage on a synoptic as well as on a real-time basis are needed. Remote sensing techniques appear to provide a potential in this regard and their current and future use in fishery research is discussed. Particular attention is called to the problems of sampling and data storage, retrieval and reduction.

## 1. Introduction

The renewable resources of the sea have provided man with considerable revenue throughout recorded history. More important, these resources have provided him with a significant and increasing source of protein. As the world population has increased, the harvest of living marine resources has likewise increased. In fact, during the last two decades fish production has grown at a much more rapid rate than has the human population. There is, however, concern over future production. Indeed, if the production of fish tapers off, which it appears to have done during the last three years, and population growth continues, then the already significant animal protein gap will widen.

We therefore are faced with the need to realize the full potential from the living resources of the sea. Two major things are necessary in order to accomplish this. On the one hand there is a need to evaluate the latent resources of the sea and to develop methods to harvest them. On the other hand there is required an understanding of the effect of man and of the environment on the presently exploited stocks of living marine resources and to provide management information or advice based upon this understanding to government and industry.

To accomplish these major tasks requires a broad spectrum of research on the fish and fisheries of the world. Such research must encompass studies of the biology, abundance and availability of living marine resources and the effects of exploitation and the environment upon these factors. Fishery scientists have been unable in many cases to obtain sufficient understanding of the effects of the fishery and the environment on fish populations because of their limited ability to collect and examine synoptic data on the populations of fish and on the large-scale features of ocean processes. With the advent of remote sensing techniques it has been suggested that a tool is now available to collect such synoptic data. It has been often stated that through the use of presently available remote sensing equipment, and equipment which will be developed in the near future, it is or will be possible to assess directly fish abundance and distribution. Notably popular has been, among others, the suggested use of visible spectrum cameras, spectrometric techniques to detect ocean productivity and natural oil effluents from concentrations of fish or smaller organisms, and infrared (IR) imagery.

It is the purpose of this paper to examine ways in which remote sensing equipment and techniques have been used in seeking solutions to the problems posed in fishery research and ways in which such techniques might be employed in the future. In the section which follows, a discussion of fishery research, exploitation, and management will be presented as background information for further discussions of remote sensing.

## 2. Fishery Research and Exploitation — Aims and Objectives

### 2.1. The Structure of a Fishery

The basic components of a fishery are presented in the form of a flow chart in Fig. 1. The population biomass shown in the large rectangle represents the most important component in the system, for obviously without it there could be no

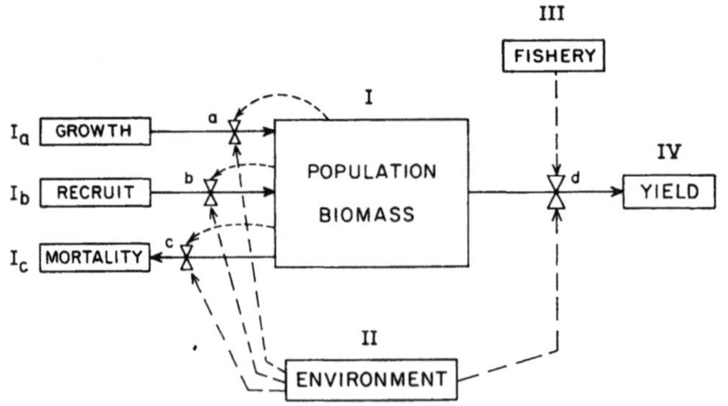


Fig. 1. Flow diagram of the structure of a fishery.

fishery. This biomass is added to by growth of individuals within the population and by the recruitment of juvenile fish to the population. It is decreased by deaths due to natural factors such as predation and disease. All three of these factors, growth, recruitment and mortality, are probably density-dependent, their rates being affected by the size of the population at any point in time. In addition to density-dependent effects, density-independent effects of the environment also influence the three factors. In the natural state such a population is maintained at its average maximum size, which is determined by the ability of the environment to support it. As conditions in the ocean fluctuate the population fluctuates about this average maximum size, as does its spatio-temporal orientation.

Man's primary interest in these populations has been, and probably will be for some time, concerned with their exploitation. This is represented in the flow diagram as yield, the flow of which is determined by the population biomass and man's intervention into the system, in this case represented as the fishery.

As we have already noted, fishery research is concerned primarily with two major problems represented within this system. These are prediction of yields over various time—space surfaces when a fishery is operating in the system, and estimation of potential yields prior to the introduction of a fishery.

## 2.2. Exploited Resources

Considering first the situation wherein a fishery already operates we can examine the classical appro aches which have been employed to predict yields from a fishery. These approaches can be conveniently grouped into three categories. The first category for assessing trends in the abundance of an exploited population of fish in the sea involves sampling the system shown in Fig. 1 at points III and IV. At point IV estimates of the catch by strata of time and area are collected. The amount of fishing effort, in terms of number of boat days fishing, boat days absence from port, vessels making landings per month, or some other measure is recorded at point III. Catch and effort data used for this type of modelling is generally collected by personal interview with vessel masters in order to obtain records of amount, date and location of catch. These data are used with the so-called general production models to present yields from the resource. Such models are not concerned with estimation of the elemental rates shown in $I_a$, $I_b$ and $I_c$, but rather with their combined effect as a single-valued function of population biomass. They are described deterministically by a general differential equation, where changes in the size of the population are represented merely by the difference between net additions to the population by growth and recruitment less natural mortality, and removals by fishing. Such models can account for a significant proportion, say 60—70%, of the observed variability in population biomass and catch. The variability not accounted for by the model is primarily the influence of anomalies attributable to the effect of the environment on I and III and it is assumed that these are random and their effects are cancelled out over the long term.

The second category or level of investigation for monitoring exploited resources involves the use of models in which more variables are examined. At this level the models consist of equations which express the yield as a function of the component rates of growth, mortality and recruitment. Once again, however, environmental effects on the abundance of the fish are not considered at this level and estimates of population growth and potential yield are for average conditions. The data necessary for assessing abundance and associated potential yield at this level of investigation include, in addition to the statistics of catch and effort discussed above, sampling the yield (IV) additionally to determine the age structure of the fish in the catch. Assuming that the age structure of the fish in the catch is representative of the age structure of the fish of catchable size in the population, such data in conjunction with data on catch and effort provide useful estimates of growth, recruitment and mortality. The estimation of such population parameters permits the formulation of models on a higher level of sophistication, which often provide more precise estimates of yield. In addition to sampling at points III and IV, techniques of sampling at point I, in order to estimate yields, are employed. These include the use of direct assessment methods such as sonar surveys wherein concentrations of fish are measured as well as actual counts of such animals as salmon, seals or whales. Indirect assessments such as mark-recapture experiments and egg and larva surveys are also used.

With models formulated at the two levels of research mentioned above it is possible to provide predictions of yield over time horizons extending beyond one year. As already noted, such predictions are accurate only to the extent that they explain man's effect on the populations; they do not take into account the effect of natural factors. To increase the precision of the estimates and to make short-term (generally less than one year) predictions of yield requires an understanding of the effect of environmental parameters on the behavior and subsequent distribution of the fish in time and space; this forms the third category of investigation.

Fish and other marine organisms spend most or all stages of their life in the sea, and therefore are strongly influenced by ocean conditions. They are not randomly distributed throughout their range, but concentrate in areas where the conditions for their survival are best. In order to maintain themselves in such optimal conditions they undertake migrations of various magnitudes. They are probably not responding to a single variable, but to the interaction of a number of variables, even though their distribution has on occasion been correlated with the distribution of a single ocean variable such as temperature, salinity or depth of the thermocline. Such a relationship may be the result of the covariability of the single observed variable with the other variables determining the optimal conditions for the organism. One can make the analogy of a multi-dimensional response surface over which the dependent variable, in this case the fish, is searching for optimal conditions. Naturally this optimum is not fixed in time and space but is a 3-dimensional continuum in the ocean.

Such changes in the environment affect not only the distribution of the fish, but also their behavior and abundance. These effects show up in two ways at the man/fish interface. Referring to Fig. 1, an alteration in the environment can affect the vital rates $I_a$, $I_b$ and $I_c$, thereby affecting biomass and hence flow through Valve d. This is a change in real abundance which affects the catch. Also, the environment may affect the behavior of the fish in such a way as to make them more or less vulnerable to the fishery even though not changing the size of the biomass. This will affect the yield, of course.

Fishery scientists are therefore vitally interested in the effects of the environment on the abundance, distribution and behavior of the fish, for it is only through the understanding of such relationships that the aims of fishery research and management can be fully realized.

### 2.3. Latent Resources

In addition to predicting yields of exploited populations, fisheries science is concerned with the determination of the potential yields from latent resources. A number of techniques are employed to make such determinations. The most direct and accurate are survey cruises in which exploratory fishing is conducted in a systematic fashion. Sonar cruises in which sound is transmitted through the water and reflections from fish are monitored, are also routinely used to assess resource potential. Additionally, surveys of egg and larva abundance are used to make inferences of the parent population size. A less direct technique involves the concept of energy flow through various trophic levels in the food web. For example, if the object is to estimate the abundance of a population of apex predators, then by estimating the production at any lower level and knowing the transfer rates between levels the abundance of the apex predators can theoretically be estimated. Many trophic-dynamics studies used to assess resource potential use the amount of carbon fixed per time—area stratum as a base.

### 2.4. Factors affecting the Harvesting Sector

In addition to factors related to the resource itself, fishery science is concerned with the study of oceanic and atmospheric conditions which affect the ability, either positively or negatively, of the harvesting sector to capture fish. Modern fleets of fishing vessels range from hundreds to thousands of miles from shore in search of fish, remaining on the high seas for periods of weeks or months. They of course must know where they are and what weather conditions to expect. Such conditions as wind direction and force, sea swell, height and direction, and current set and speed are all important factors in the determination of fishable conditions. To operate at a profit such high-seas vessels, which are capital-intensive, must be able to fish a large number of days out of the year. Therefore knowledge covering oceanic and atmospheric conditions is essential. Likewise, in order to reduce the time spent in searching for fish, environmental data which can be used to identify optimal conditions for the object species must be available to the fleets on a real-time basis.

As more and more stocks of sea animals come under exploitation and the fishing fleets increase in size, controls on the levels of harvest are needed to limit the catches to levels corresponding to the ability of the stocks to replenish themselves. Such controls most often take the form of an open season of fishing within specific areas. As some of these areas are rather large, say 17000000 km$^2$, the logistics of monitoring vessel activities and locations become staggering. The position of as many as 300 to 400 vessels from a number of nations will need to be monitored on a daily basis. Such positions will often need to be accurate to within 5—10 km. Though such monitoring is not the province of fishery science, the problems of developing methods to monitor the activities of the vessels fall to fisheries agencies and hence indirectly to fishery scientists.

## 3. The Adequacy of Conventional Approaches to Fishery Research

Having discussed the techniques used to assess sea animals and the forms of data necessary to employ these techniques, it is next important to consider the question of whether these techniques have been adequate to reasonably carry forward the task assigned to fishery science.

Fishery scientists have been moderately successful in their efforts to assess the abundance of some species of exploited sea animals based on the examination of trends in the catches and the amounts of effort generated to make those catches. However, these assessments apply to average conditions, and deviations from such averages may be great. For example, in the eastern Pacific Ocean removals from the yellowfin tuna stock by the fishery account for about 65—75% of the subsequent change observed in the abundance of the stocks. The remaining 25—35% of the observed variability can be attributed in large part to anomalies in average abundance which are caused by the influence of the environment on the stock. For other species of animals such as Penaeid shrimps and many species of fin fishes, no, or at most very poor, relationships between removals by the fishery and subsequent abundance are detected, the major share of the observed variability in abundance in such cases being caused by environmental factors.

With respect to the assessment of latent resources, the direct assessment by exploratory fishing has been the most successful technique employed, but such surveys are limited in scope and extent by expense and the ability of the platforms to sample adequately within required time—area strata. Sonar surveys in most cases are limited by the ability to identify the target species, and target identification must rely on multi-stage sampling techniques. Egg and larva surveys pose extremely difficult sampling problems, as well as the additional difficulty of relating eggs and larvae to parent stock size. The most difficult approach to estimating the potential of latent animal resources is the examination of trophic dynamics. In such cases estimates of transfer rates of energy from one level to the next are necessary, as well as the knowledge of the constraints imposed by limiting factors at the nutrient level. Not only are estimates of biomass and transfer rates necessary, but flux of energy within each level, that is rates of productivity, are needed too.

It therefore appears that fishery science is capable of utilizing "standard" techniques of assessing the abundance of resources in some cases, but if we are to improve the efficiency of such estimates as well as directly estimate abundance, it will be necessary to collect environmental data on a synoptic and nearly real-time basis. Additionally, there is a need to supplement both direct and indirect assessment of latent resources. It appears that remote sensing techniques may possibly offer a capability for such analysis. The greatest advantage that airborne remote sensors provide us with is that they can collect data much more rapidly and over broader areas than can be accomplished by conventional techniques. Indeed satellites provide us with an advantage heretofore unattainable by man, that is, they enable us to see the entire world ocean, minus that area covered by clouds, in a single look.

## 4. Remote Sensing and Fisheries Research

In this section of the report it is our intention to review what is currently being done with remote sensing relative to fisheries research and exploitation and what might be possible in the near future.

### 4.1. Direct Assessment

The most straightforward approach to assessing and monitoring the abundance of sea animals is to count individually each one or a constant sub-sample of the population. Unfortunately this is relatively impossible to do from sea-level vantage points, with the exception of some stocks of salmon and sea mammals. A number of remote-sensing devices might however, provide such a capability.

### 4.1.1. Visual-Range Cameras

To be able to detect and identify sea animals such as fish, schools of fish, or mammals and groups of mammals they must be at or near the surface and cameras with a resolution great enough to detect the object species must be available for use. To identify individual fish would require a resolution of from a few centimeters to about 0.5 m. If however schools of fish are the targets to be photographed then a lower level of resolution is permissible; in this case resolution from say 10 m for small schools to about 1 km for larger ones.

Of course there are numerous other measures in addition to ground resolution which affect picture quality. The color of the object being photographed and its contrast with the surroundings are extremely important. For example it may be entirely possible to detect in photographs taken from 25000 m altitude a white paper plate floating on the surface of the ocean, but a blue green plate of the same dimensions would probably be indistinguishable from its surrounding medium. Also important would be the albedo or reflectivity of the object, the angle of incident solar radiation, the differential in height between the object and its surrounding medium, and the optical opacity to incident solar radiation. Unfortunately the clearest ocean water is also usually the least productive. It is important to remember when considering photographic detection of sea animals that they have evolved protective coloration and hence generally have a very low level of contrast with their surrounding medium, and are below the surface of the water most of the time. The presence of intervening atmospheric haze or clouds may seriously degrade image quality or make the photographs useless for analysis.

Recent work at the Tiburon Laboratory of the National Marine Fisheries Service (NMFS) of The National Oceanic and Atmospheric Administration (NOAA) of the USA has been conducted on the evaluation of remote-sensing cameras to detect, identify, and count California gray whales [1]. Overflights were made with NASA aircraft of the area off Central California in the regions through which gray whales migrate. Synoptic coverage was made of the test area with a variety of sensor systems. Photographs taken in the visual range with a number of camera systems were interpreted for gray whale identification. The only system for which photographs of gray whales were clearly detected were from visual range color film transparencies from the RC 8 camera. Individual whales were detectable in photos taken from altitudes of up to 2000 m. However, species identification was possible to only 1200 m, and then only under ideal atmospheric conditions.

Cameras with long focal lengths appear to offer the best prospects for identification of individual fish or schools of fish at the present time. Because spatial resolution of a camera system is a function of the focal length, the film emulsion used or the electro-scanning beam resolution, and the distance of the object, it is worth considering the use of specially designed cameras from aircraft as well as satellites for fishery research. Reasonably compact cameras with folded optical systems that possess focal lengths of up to 610 cm were proposed several years ago [2] for use in civilian spacecraft and may be in limited use today for military reconnaissance. With such a camera and fine grain film it would be theoretically possible to resolve a 30 cm object at an altitude of 180 km.

Such a camera system carried aloft by an airplane flying at an altitude of 18 km would produce a 0.01 mm image corresponding to a 30 cm fish. The 87 lines/mm required for this image is within the capability of present films. Resolution of a fish 15 cm in length would require a film capable of handling 174 lines/mm, which though technically feasible is near the upper limit. Better resolution requires that the airborne sensors be flown at lower elevation or that the focal length be further increased. Flights at lower altitude restrict the areal extent of coverage although at an altitude of 9 km the camera's resolution would be approximately 8 cm. It appears then that if other factors are favorable, such as contrast and atmospheric transparency, it would be theoretically possible to observe individual fish or schools of fish from a folded optical system aboard an aircraft at high altitude.

If the same camera system were used aboard a satellite such as NOAA-2 (1450 km altitude), the ground resolution would be about 122 cm, an unacceptably large number for identification of individual fish. A platform such as ERTS-A (910 km altitude) would improve the resolution to about 75 cm which is still not adequate for individual fish, but is for schools. An optical system of 1465 cm focal length placed aboard an ERTS-A satellite would be needed to improve the resolution to about 30 cm which is still minimal for detection of some individual fish. The use of camera systems aboard satellites to identify individual fish does not appear technically feasible although the maximum resolutions referred to above would allow some fish schools to be resolved assuming the schools to be several times the effective ground resolution of the camera system. The feasibility of monitoring schools of fish by species is, of course, dependent upon their exhibiting a species-specific schooling pattern identifiable from a vantage point above the school.

If such very high resolution "wet" camera systems were used aboard an ERTS platform to monitor fish schools, further technical problems of data retrieval are raised. The film can be either developed aboard the satellite and photo-electronically scanned or it can be released from the satellite and dropped back to earth. While automated film processing is widely used, the film scanning equipment to read these film strips aboard the spacecraft would need to be of high precision to achieve 170 lines/mm. Automated photo-electronic scanning equipment presently on the market routinely scans filmstrips with a 50-$\mu$m raster (20 lines/mm) at high speed. To increase the scanning resolution by one order of magnitude, however, increases the data handling requirements by 100 times. The alternate procedure would be to drop the undeveloped filmstrips in a canister periodically from the satellite platform in the same manner as is now done with some reconnaissance satellites. In any case the task of handling these data is formidable. Digitization of the film at 30 cm resolution would produce $1.3 \times 10^{11}$ bits of information from an area of only 108 km $\times$ 108 km. To transmit such a data set to the earth in 10 minutes would require a frequency of 210 MHz. By comparison, ERTS-A telemetry for the MSS camera system is only 15 MHz. Since the memory cores of large size computers routinely handle $10^6-10^7$ bits of information, the task of handling even one such image is formidable. The answer is obviously not to attempt such a high resolution digitization of even 1° squares but rather to use multi-stage screening techniques to relieve the computer of handling such a large quantity of data.

Further development of "dry" cameras such as the Return Beam Vidicon or Multi-spectral Scanners aboard ERTS-A to improve the 100 m spatial resolution may be expected with time and effort, although the extremely high resolution requirements indicated above may well require the combined benefits of a folded optical system and say, the Multi-spectral Scanner System (MSS). The underlying basis for successful operation of folded camera systems from aircraft or satellites for fishery research lies in recognizing the targets as species rather than merely pelagic fishes. The authors are not aware of any plans at present to place such a high resolution camera system aboard earth resources satellites in the next 5—10 years. The highest resolution provided by cameras aboard Skylab launched into a 460 km orbit this month (May 1973) is estimated to be 50—100 m.

There appears to be no capability at the present time to photograph individual fish or mammals with either "wet" or "dry" camera systems aboard satellites. Likewise it does not seem probable that such a capability will be available in the foreseeable future. If such systems were available they would most probably be of limited use.

It appears to be theoretically possible to detect schools of fish and/or mammals in photographs taken from spacecraft, but to our knowledge this has not been done. For purposes of stock assessment it is necessary that the schools of animals being detected be identified as to species. To do this would first require that the conformity of a fish school is species-specific and that a school-signature for each species of fish be available. Assuming school conformity is species-specific, the development of signatures would pose a formidable task. Therefore the use of remote-sensing visual range cameras for directly assessing the abundance and distribution of sea animals appears to hold only limited promise in the future.

### 4.1.2. Infrared Imagery

Direct assessment of individual sea animals or schools of sea animals from thermal measurements represents another possible method which has been considered for resource evaluation. The present thermal sensors aboard satellites were designed for meteorological requirements and possess a dynamic range in the neighborhood of 130 °K and cover a range from about 45 °C to —85 °C. Effective thermal resolution is determined by factors such as the sensor's dynamic response (volts/°K) and the kind of A/D converter used to process the sensor signal. Sensors aboard current NOAA and NASA satellites provide relative thermal imagery that is good to within approximately 0.5—1.0 °K. Spatial resolution of the Direct Readout Infrared Radiometer (DRIR), a scanning radiometer, is approximately 8 km at nadir subpoint; by comparison Direct Readout Visual Imagery (DRVIS) resolution is 4 km. The Very High Resolution Radiometer (VHRR) system aboard the NOAA-2 satellite improves the spatial resolution to 1 km for IR and VIS. Plans have been made for the development of a satellite to carry sensors of particular interest to the oceanographic community (*ca.* 1975) but the developments have been delayed by fiscal limitations. The technology exists however to increase the thermal sensitivity of available scanners to 0.05—0.1 °C by decreasing the total dynamic range of the thermal sensors to a range of greater interest to Marine Science such as —2 °C to 30 °C. The same fundamental questions remain, however, as to exactly what is being measured by the sensor, i.e. what part of the infrared radiation received by the radiometer is due to back-radiance from the sea surface and what proportion is due to atmospheric noise.

All marine mammals and many of the pelagic species of tuna and tuna-like fishes maintain internal body temperatures which are more than negligibly higher than the temperature of the surrounding water. For the fishes, this difference between internal body temperature and ambient varies from 2—10 °C, and for the mammals may be in some situations near 35 °C. It has been assumed that this differential in temperature should provide enough contrast to differentiate the organisms from their surrounding media. However experiments with IR sensors capable of detecting relative differences in temperature of about 0.5 °C failed to differentiate warm-blooded organisms from their surrounding media. While the internal temperatures of mammals are 35—38 °C and well above ambient water temperatures, the insulating capabilities in the skin and subcutaneous layer are so effective that the thermal gradient between skin and water is probably less than 0.1 °C under normal swimming conditions. It thus appears that even if remote thermal sensors were to penetrate to the depth of the tunas or mammals, the thermal gradients produced between the skin and water would be too small for detection. Consequently the use of IR systems to directly monitor abundance holds little or no promise unless sensors with the thermal resolution indicated above become operational.

The presence of fish and mammals at or near the sea surface may in theory also be detected by the thermal wake or "scar" produced in the surface water as the animal swims along and disturbs this thin layer. During times of high insolation the surface water layer is typically warmer than the underlying water with the reverse condition true for the night-time. When the animal is swimming at the surface, the surface layer is stirred with the cooler or warmer water below and a horizontal thermal wake is produced in the direction of motion. Examples of this thermal wake have been observed by IR scanner imagery at night from aircraft when tracking such targets as whales and porpoises.

If temperature differences of 0.5 °C are produced by stirring of surface water and the extremely high spatial resolution (e.g., 2—10 m) previously referred to is assumed, detection of fish schools swimming at or near the surface would appear to be technically feasible. Though such a technique for monitoring fish schools might be practical from a low flying aircraft with present technological capability, it is our opinion that this methodology will not be implemented using resource satellites until the high resolution sensors already discussed above become available.

Although only in the early experimental stage, recent development in passive microwave detection also offers a capability for detecting such temperature discontinuities. The advantage of passive microwave over IR imaging is that the former will penetrate cloud cover and therefore will allow monitoring on overcast days—of which there are a significant number in many areas of high productivity.

As was indicated in the previous discussion of visual range cameras, given a capability to identify warm-blooded animals with either IR imagery and/or passive microwave, species identification would be required before the results would be very useful. Species-specific signatures or multi-state sampling techniques would be required. If the signature were based on temperature differences due to the temperature of the animal, then the number of species-specific signatures would be much less than in the case of the visual range cameras because only a subset of the entire set of sea animals are "warm-blooded". If on the other hand the detection of organisms is the result of a temperature discontinuity created by surface agitation, then the array of species-specific signatures becomes extremely broad.

### 4.1.3. Sound Propagation

The use of sound waves to detect the presence of sea animals has been common for a number of years. Sonar gear which involves the transmission of sound waves and monitoring reflected echoes has been used to detect and quantify fish in commercial fishing operations, and assessment cruises. Multi-stage sampling is generally necessary for species identification. Sound emitted by fish, or by their contact with the surrounding water is also used to detect the presence of sea animals, although the use of such information is largely confined to research and military use.

The use of aircraft or satellites appears to offer little benefit in the direct acoustic assessment of fish stocks. The use of aircraft is one step removed in the acoustic assessment process since low flying aircraft would need to drop portable sonar gear to transmit and receive acoustic information over a several mile radius surrounding each sonar transmitter. The information would be relayed back to the aircraft in the same manner as is done with present sonobuoy technology. While considerable spatial coverage could be obtained from the deployment of a large number of these sonar floats, the basic problem of signature recognition still remains. The direct use of acoustic transponders from aircraft is not feasible due to the very poor acoustic impedance match for the air/sea interface.

With respect to the application of remote sensing systems to such techniques, there does not appear to be any unique advantage which might be gained other than the fact that satellites would be useful as relays for telemetering such information to shore stations, where it could be processed, and returned to vessels at sea on a real-time basis.

## 4.2. Indirect Assessment

In the context of our discussions, indirect assessment refers to estimates of the abundance of sea animals which are a step removed from counting them directly. A number of techniques have been suggested for which remote-sensing systems appear to provide a capability. For these discussions, only those for which some considerable feasibility exists and for which developmental work has been completed will be discussed.

### 4.2.1. Oil Slicks

Oil slicks on the surface of the ocean are a common phenomenon which has been observed for centuries. Fishermen on occasion have been known to use such slicks as an indication of the presence of fish. Whether these slicks are a result of material escaping from the fish, either the predator or prey, or from petroleum effluents of some sort is in most instances unknown. However it is possible to distinguish between the two types on the basis of spectral signatures, where such analysis can be conducted. This fact has led to a great deal of interest in the possibility of developing remote-sensing systems which would be able to qualify and quantify fish oil slicks.

Work by the NMFS and Barringer Research Ltd. has shown that studies of oil film in ultraviolet light produced the set of absorption curves for a thickness of oil approximately 0.001 inch seen in Fig. 2 (after Barringer [4]), which shows strong absorption in the ultraviolet for a number of species of fish. Species-specific differences among the fish examined are evident. Further research with airborne spectrometers has demonstrated that such techniques in the ultraviolet range have a low probability of success because of atmospheric interference problems [5]. Further work with spectrometric techniques is not planned.

A newer method theoretically capable of detecting oil films on the sea surface involves the use of passive microwave sensors. Recent studies [6] with microwave

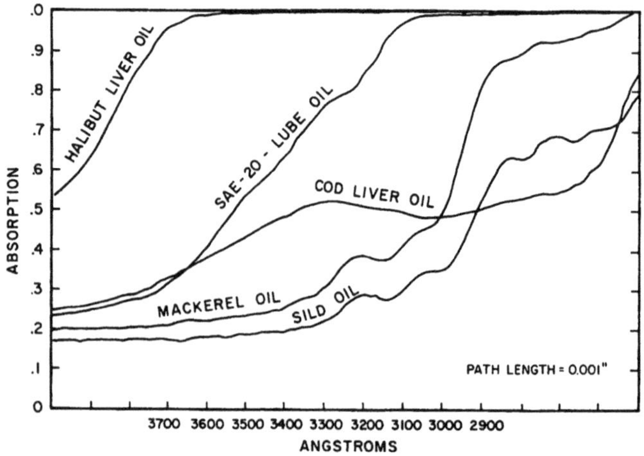


Fig. 2. Ultraviolet absorption characteristics of selected oil slicks (from [4]).

systems have shown that such radiation is sensitive to the presence of surface oil slicks. Further evaluation of microwave sensors for the purpose of detecting different kinds of oils on the sea surface will indicate whether or not the method possesses sufficient sensitivity for use in detecting different fish oils from satellite platforms.

### 4.2.2. Chlorophyll

The sun's energy striking upon the surface of the ocean is converted directly by marine plants into food. The amount of carbon fixed at this level in the food web provides the basis for all animal life in the sea. By estimating the productivity of a given area of the ocean, and assuming certain conditions and transfer rates among trophic levels, it is theoretically possible to estimate biomass at various trophic levels. A measure of the productivity of the sea can be made by quantifying the amount of chlorophyll per unit measure of sea water which is an index of standing crop, and by measuring the rate of carbon fixation by photosynthesis. Such analyses have been routinely made from shipboard by spectrophotometric measurements of chlorophyll content from a unit volume of water sampled, and by incubating samples of water tagged with the radioisotope $^{14}C$. Such experiments are time-consuming and expensive, and only limited sample coverage can be made.

Recent advances with airborne multi-spectral scanners suggest the practicability of measuring chlorophyll with remote sensors and from this to deduce the productivity. As another paper has been presented on this entire subject at this symposium [7] no further discussion will be given here.

### 4.2.3. Image Intensifiers (Bioluminescence)

Many small marine organisms, particularly dinoflagellates, are bioluminescent, i.e. they are capable of emitting biologically generated light. They generally do this when stressed. A common source of such stress or stimulation is provided by the turbulent wake of fish or vessels. Such bioluminescence makes it possible to detect schools of fish at night. This bioluminescence, which takes the shape of the fish or fish school, provides a signature which appears to be species-specific. The Pacific sardine which at one time formed the basis for one of the largest fisheries in the world was caught primarily at night during the dark of the moon with purse seine nets used to surround luminescing schools. In the California wet-fish fishery, spotter planes fly at night and locate schools of fish by the bioluminescent light they emit and direct fishermen to the fish, and assist them in catching the fish. Pilots of these aircraft, who are highly experienced and proficient fish-spotters, are able to identify more than a dozen species of fish from their school shapes and sizes and the patterns of emitted flashes of light [8]. Similar techniques are employed in numerous other fisheries around the world.

Recognizing the potential benefits to stock assessment work that automation of this approach might provide, fisheries scientists of the NMFS at Pascagoula, Mississippi, initiated the development and testing of airborne low light sensors to detect luminescing schools of fish at night. Preliminary work was conducted with the US military Starlight Scope, which uses an image intensifier tube that amplifies reflected starlight more than 40000 times. Tests from research vessels at sea demonstrated the capability of this equipment to detect schools of fish both at the surface of the sea and below the surface to about 3 meters. Second stage experiments in which more advanced equipment was used were conducted from aircraft. In these experiments a sensor which amplified light more than 55000 times was used to sample with a closed circuit television camera and video-tape recorder. Schools of fish were detected and recorded on tape from an altitude of nearly 2000 meters.

NMFS fishery scientists are continuing their research in the development and utilization of the image intensifier. Methods of automated image analysis designed to compute stock density are being investigated. Such analysis concentrates on the determination of surface area of the school and transforms the numerical data to magnetic tape for processing.

In South Africa, where the stocks of pilchards support one of the major fisheries of that nation, image intensifiers are used from aircraft on an almost routine basis. The sampling program is designed to provide estimates on a continuing basis of the total biomass of pilchards. It is anticipated that such estimates will provide some of the information necessary for managing the resources so as to maximize production over the long term [9].

It appears that at the present time there exists a technological capability to detect the presence of marine animals by the bioluminescence emitted as a result of their activity in the surrounding medium. Such bioluminescence has been detected on a routine basis from aircraft at an altitude of 2000 meters. In some cases it has been apparently possible to identify the species of fish being observed by their characteristic patterns of bioluminescence.

Remote sensing image intensifiers are not presently employed in resource satellites, nor does it appear probable that they will be in the very near future as numerous technological and logistic problems must be overcome first. If in fact such a capability were available, the problem of cloud cover would likely create serious difficulty. As is the case with other remote sensing techniques species-specific signatures of bioluminescence patterns would be necessary before the technique will have very wide application and this would involve considerable multi-stage sampling. The use of such techniques will probably remain in the realm of airborne sensing for some time to come.

### 4.3. Environmental Effects on Fish and Man

The importance of the ocean environment on the distribution and abundance of marine animals has already been discussed in an earlier section of this paper. It was indicated that in order to develop accurate predictive models for exploited stocks of marine animals, some understanding of their physiological requirements and their reaction to environmental stimuli (which environmental variables and measured over what scale) would be necessary. Many attempts to relate environmental factors to the distribution of the animals of the sea have failed because such studies have in many instances only been concerned with one or two variables, such as average indices of surface salinity or surface temperature, whereas the animals themselves may be responding to real-time three-dimensional changes in a number of variables. Such studies have been primarily confined to these sorts of intermittent, two-dimensional data because these have been the only data available or obtainable due to the high cost of collecting data at sea. Certainly to examine average annual, or even monthly, surface temperature or salinity from a single shore station in an attempt to predict deviations in abundance of a population of fish distributed over a broad oceanic region will most often give indeterminate results. What is needed is some understanding of the organisms' reaction to environmental stimuli, the rate at which these are changing and the extent of the area in which the changes are taking place and which are in the animals' sphere of detectable stimuli.

It is to be emphasized that the relation between the organism and its environment is not clearly understood. At the present time a great deal of effort is being directed to studies of the physiological requirements of marine organisms. A recognition of the importance of the energetics of marine populations, nutritional requirements, heat budgets, osmotic field preferences, and other factors is emerging. It is becoming clear that many environmental variables are acting upon the spatiotemporal distribution of the organisms and their general levels of natural abundance. It is also becoming clear that sampling of certain factors over short intervals of time, and over broad areas is important. The availability of remote sensing techniques provides an opportunity to collect environmental data on a fine enough time scale and over broad enough areas to permit the type of analysis that might result in definitive conclusions. Although most remote sensors sample in only two dimensions, the capacity to probe the third is developing.

Because of the poor understanding of physiological requirements of marine animal populations and the lack of identification of relationships between the behavior of the animals in the ocean and ocean variables, fishery science must rely heavily upon the search for empirical relationships between the animals and the sea.

For fishery forecast models, and particularly weather and sea state forecasting models, to be useful to the harvesting and processing segment of the industry, information must be available in order to make such forecasts on a real-time basis. Such advice and forecasts would be of no value after the fact.

In the following section a discussion of certain aspects of remote sensing of the environment is given.

### 4.3.1. Sea Surface Temperature

Of the various environmental parameters that have been examined in relation to fish distribution, the most often examined has been sea surface temperature (SST). Past studies have shown that many species of fish apparently respond to the surrounding water temperature and when present are usually found in commercial quantity within relatively narrow temperature ranges. Realizing the important role of surface temperature in the distribution of tunas, agencies such as the NMFS in La Jolla, California, have issued monthly surface temperature charts as part of a fishery advisory service for a number of years. The data base presently used to construct these charts consists primarily of data from commercial and research vessels. Since commercial vessels normally make traverses along the most direct routes, large ocean areas are therefore not sampled. Satellite sensors are in a unique position to fill these data voids and to supplement the other better covered regions as well.

Because tunas are continuously moving in search of food they are likely to respond to other environmental factors that result in greater food abundance, such as the presence of oceanic fronts. Recent tuna oceanography studies conducted at the NMFS Laboratory in La Jolla suggest that in their eastward migration across the N. Pacific, Albacore tuna may orient themselves on a thermal front or boundary hundreds of miles out at sea. The front is found between the subarctic and subtropical confluence in the Central Pacific. While more field observations are needed before this information can be used in a predictive scheme, the data suggest that tunas may migrate in a more spatially coherent manner when the front is strong (1 °C/36-54 km) than when the front is weak and sinuous. This behavior pattern is possibly beneficial to the fish since a convergent front concentrates forage.

Because of the importance of SST and surface fronts in the distribution and behavior of tunas, the Inter-American Tropical Tuna Commission has explored the feasibility of using IR data from NOAA satellites as a basis from which to construct special maps showing the sea surface temperature and the location and character of fronts [10]. An example of a SST "map" is shown in Fig. 3. The land/sea margins are sharply outlined due to the high temperatures associated with Baja California. The data represented in this figure were obtained from the NOAA-1 DRIR sensor and indicate a spatial resolution of approximately 10 km (§ 4.1.2.). Several locations in the Gulf also possess thermal gradients. Though this map provides some information on variability in surface temperature, frontal features are best seen as portrayed in Fig. 4. The central axis of each front is shown as a locus of zero values while the warm and cold sides of the fronts are indicated by + and − values that are found on each side of the front axes.

### 4.3.2. Thermocline Depth

Fishermen and fishery scientists are all aware of a fish's ability to modify its distribution because of temperature changes in the surface water, however it is only recently that the interaction between thermocline depth and the distribution of fishes has been recognized as being of importance. In the eastern Pacific the success of fishing for tropical tunas seems to be related to the depth of the thermocline which is generally shallow and strong (i.e. $\leqq$50 m depth and 1 °C/1-1.5 m).

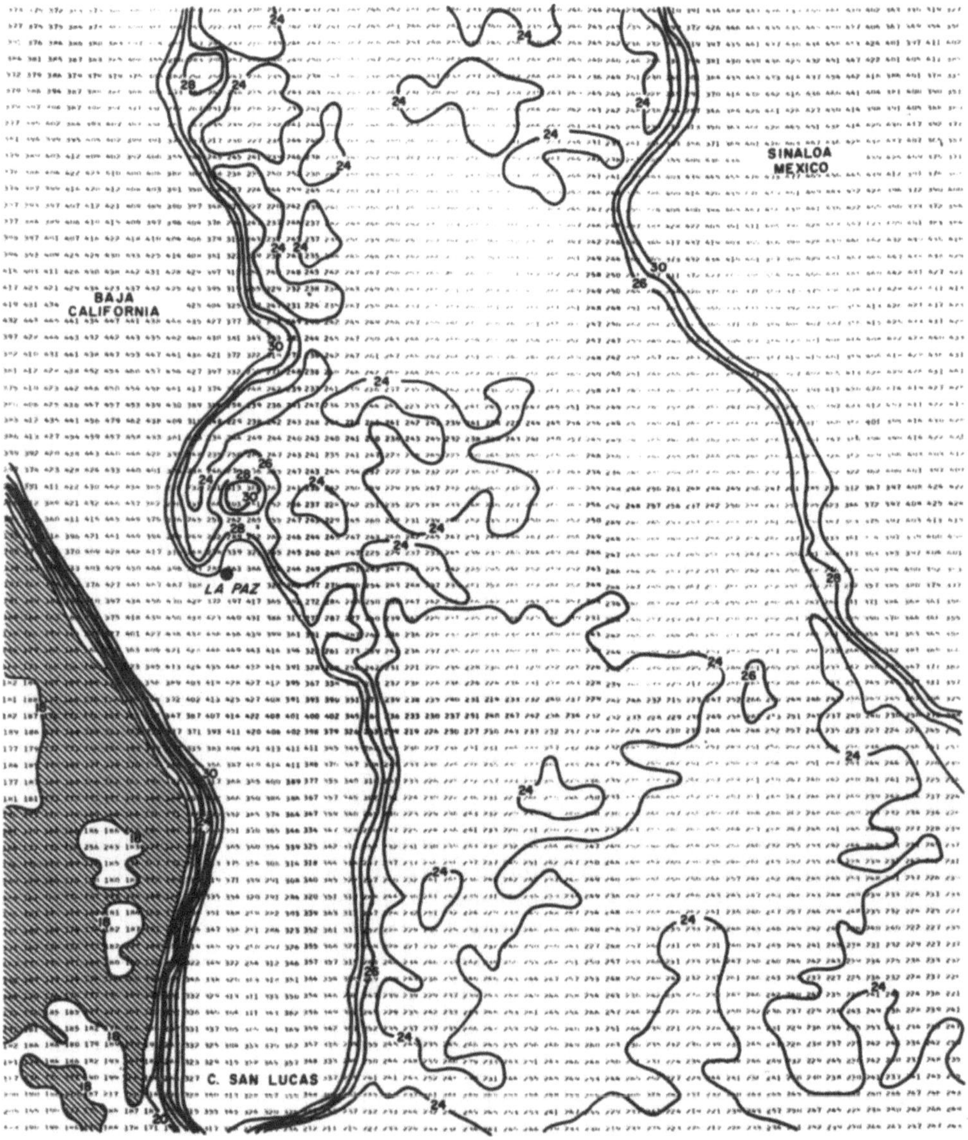


Fig. 3. DRIR computer formated imaging of SST in the southern Gulf of California. Thermal infrared map (in °C) of direct readout (DRIR) data from NOAA-1, orbit number 1835 (2146:53 La Paz time; 6 May 1971) of the Little Window region. Shaded regions are cool and signify surface conditions obscured by clouds.

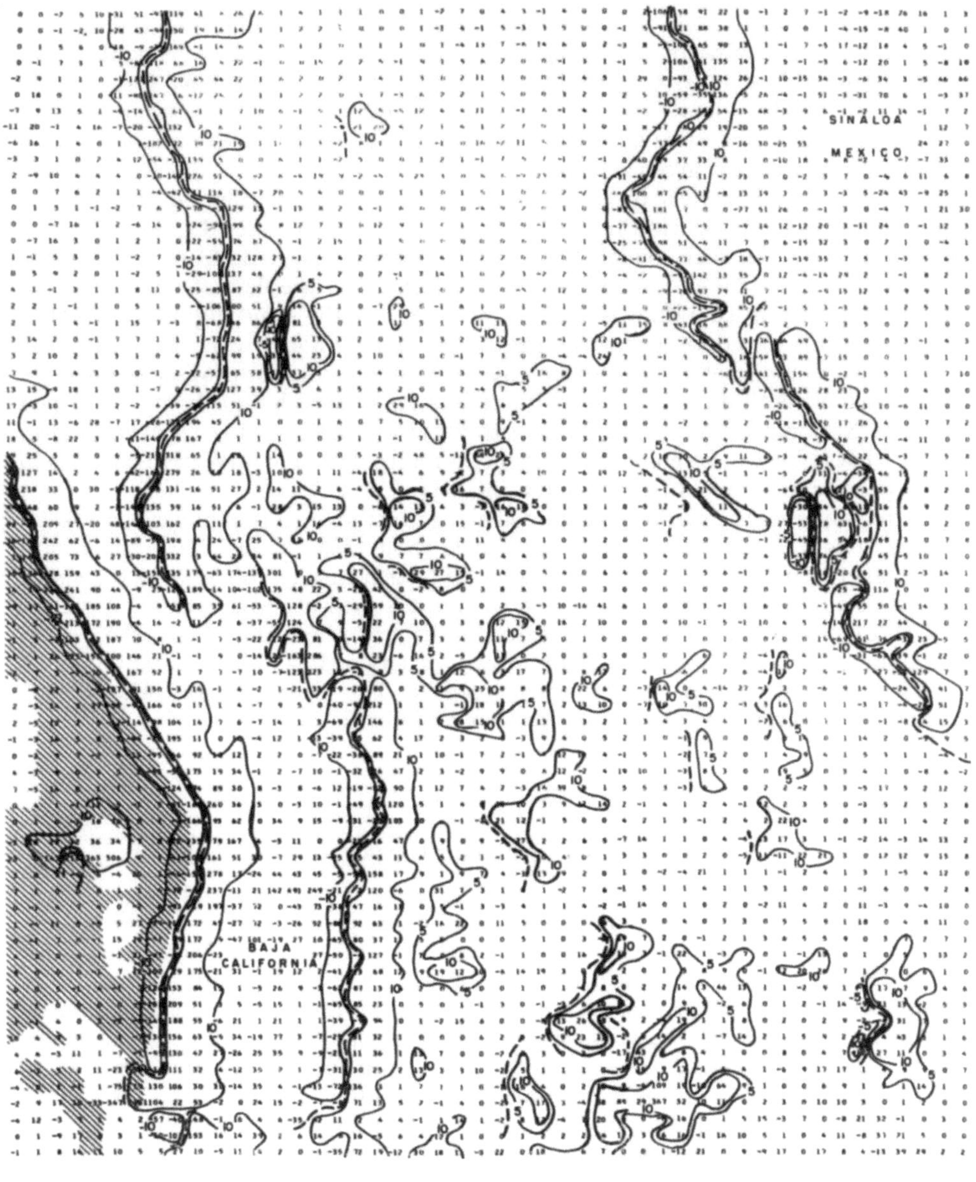


Fig. 4. Frontal enhancement display of horizontal thermal gradients using the same data as in Fig. 3. Gradient · gradient ($-\nabla \cdot |\nabla|$) map (X. 1 °C) for NOAA-1, orbit 1835, showing areas containing thermal fronts. The land/sea margin and axes of frontal features are shown by heavy dashed lines. Positive and negative regions on either side of the frontal axes represent the warm and cold sides of the fronts respectively.

The tuna's preference for swimming above the thermocline may also be related to the presence of a pycnocline (density gradient at the same depth), where the pycnocline causes plankton and other organic matter to collect at the base of this layer in the course of its sinking. Tunas are normally found and caught within the surface mixed layer located above the top of the thermocline. Information about thermocline depth therefore is of considerable use to both fishermen and fishery scientists since this knowledge can appreciably affect catch rates and movement of fish within a given region.

Semi-empirical models presently exist for the determination of thermocline depth as a function of wind intensity, duration and, to a lesser degree, SST. The wind data used in such models come from observations made from ships at sea plus selected coastal and island weather stations. The problem of representative areal sampling of wind data is limited, however, by the same ship traverses that provided the water temperatures previously noted. Some wind data coverage therefore is much better than other coverage. While it is not possible to estimate thermocline depth directly by satellite, a data analysis technique being developed at The National Environmental Satellite Service (NESS) of the USA has shown that wind fields can be constructed from an interpolation of advective changes in cloud patterns obtained from satellite imagery.

After this analysis has been evaluated and put on an operational basis, it should be possible for NESS to produce large scale wind vector fields. These wind data may then be used as input to generate maps of thermocline topography for major parts of the world's ocean, wherever the need is present. Such a product will be of considerable interest to fishermen and fishery scientists. When the GOES satellite is launched in late 1973, improved sensors will be able to obtain 0.5 mile resolution in both visual and thermal imagery. The visual imagery will provide increased accuracy in determining the wind vector fields.

### 4.3.3. Sediment Outfall

In coastal zones of the world ocean, the quantity and manner in which terrigenous sediments are brought into the coastal waters may significantly influence the viability and thereby the sustained population sizes of coastal stocks of fish. Research done by NMFS scientists of the Penaeid shrimp fishery along the coastal waters of Texas and Louisiana, states bordering the Gulf of Mexico, has shown an interaction between the availability of shrimp for harvest and the load of suspended sediment. Sediment outfall can be readily observed from an aircraft although daily flights detailing the amount of outfall over a region 650 $km^2$ or more in area per day is both time consuming and expensive to support. By contrast VHRR imagery costs \$ 2.00 per photograph for more than $6.5 \times 10^4$ $km^2$ of low angle coverage. Imagery from the ERTS-A satellite has already shown that it is feasible to observe the sediment load in some coastal locations. Results from colored satellite imagery for the upper Gulf of California have been used to determine the bathymetry and sediment load [11]. The ability to observe changes in the sediment load in coastal waters via satellite imagery offers another type of use for remote sensors now and in the near future which *may* have some relation to fisheries.

### 4.3.4. Salinity

There have been little data to show that the behavior of pelagic fishes is significantly altered by the small horizontal salinity gradients usually found on the high seas. Where strong salinity gradients occur, as along coastal margins and at a few current boundaries, however, the change across these interfaces may operate to modify the behavior pattern of selected species of coastal and high seas fishes. Although some progress has been made to develop passive microwave systems to measure surface salinity, the resolution of 1 to $3^0/_{00}$ is not yet sufficient to observe the small salinity variations on the high seas and is marginal for detection of larger salinity gradients associated with oceanic fronts (e.g., $1^0/_{00}/100$ km for the Equatorial Front off of northwestern South America). It is presently possible to detect strong salinity gradients which exist in estuaries with currently available microwave sensors. Additional development and evaluation of this microwave sensor will be required before such a system can be considered operational from satellites. It would appear that a usable satellite salinity sensor for high seas work with $0.1-0.5^0/_{00}$ sensitivity and 1 km resolution is at least 5—10 years away.

### 4.3.5. Cloud Cover

There is increasing evidence that the location of major currents and their boundaries may be reflected in the distribution and kind of cloud cover over surface water. Cloud patterns such as the Inter-Tropical Convergence Zone (ITCZ) in the eastern tropical Pacific may, together with the IR satellite data, provide a basis for monitoring those areas through which certain tuna species may pass on their migratory routes. The utility of using cloud cover to indicate the presence of a current was recently substantiated by observations made from the French R/V *Calypso* as it passed through the Peru Current. An Automatic Picture Transmission (APT) receiver aboard the *Calypso* received images that were used to delineate the general boundaries of that current, which were verified from ground-truth.

Currently operational sensor systems are adequate to provide the necessary visual data base for making such studies. However, little effort has been expended to date in this area.

### 4.3.6. Weather Conditions

Information on current meteorological conditions is also useful to the fishing fleet as it affects the operational efficiency and safety of the vessels. Cloud cover data obtained from ESSA-8 and NOAA-2 satellites have been used in a NMFS (La Jolla) fishery advisory program for the past $1^1/_2$ years. Special maps are made daily to indicate those localities where poor weather conditions are either present or are to be expected shortly. These daily products are transmitted via facsimile machine to vessels on the high seas possessing the necessary receiving equipment. Meteorological observations from satellites collected as recently as 2—3 hours before FAX transmissions are incorporated into these charts. Knowledge about present or impending weather conditions also provide fishery scientists with a more accurate index of the intensity of fishing effort since there is little fishing effort expended when high winds and heavy seas are encountered.

### 4.3.7. Management and Enforcement

There has already been mentioned the need to monitor the position of high seas fishing vessels for purposes of enforcing management agreements designed to conserve exploited resources. There are at present many such agreements in effect, and as the demand for and harvest of sea animals continues to increase it is certain that many more agreements will be entered into by the nations of the world. In many cases the conservation regime may apply to a number of species of fish and extend over very large areas, and in fact may be ocean wide. These agreements often take the form of open and closed areas and seasons. To enforce such agreements it is therefore necessary to be able to monitor vessel locations.

At the present time such monitoring is generally done by observers placed aboard vessels, surveillance with aircraft over-flights or patrol vessels, and by triangulation on radio transmissions. All of these techniques are limited in scope by physical and logistic constraints. Satellites, on the other hand, are not limited in the same manner, and by the nature of their synoptic coverage of the earth theoretically could provide a more efficient system for monitoring vessel positions.

Two modes of operation seem possible. The first would be visual monitoring, i.e. taking photographs of specific areas of the ocean surface at specific times. A number of factors would however make this approach difficult. Particularly troublesome would be cloud cover, which would make viewing the surface of the ocean difficult in certain areas during certain seasons, and problems of screening the photographs.

The other approach would be to determine the position of vessels by monitoring radio transmissions from the vessel to the satellite. Apparently the technology presently exists to monitor vessel locations with satellites. Two recent programs have involved the use of automatic radio transmitting systems (passive "black box") to monitor the location of atmospheric balloons and buoys. These were the IRLS (Interrogation, Recording and Location System) of NASA and the Eole (Greek god of wind) experiments of the French. Both of these were short-term projects and are no longer operational.

For monitoring vessel position for purposes of insuring compliance with fisheries agreements a number of factors are important of consideration. The position of as many as 300 to 500 vessels will need to be monitored at a single time. Position will need to be fixed within 5—10 km; areas as large as 15—20000000 km$^2$ will have to be monitored. If a system is developed wherein a human operator on a vessel transmits information to a satellite some type of mechanical check on the veracity of the human operator will be necessary or else a completely automatic monitoring system (passive "black box") will be required. The cost to the vessel owner of the shipboard system or "black box" must be kept reasonable, say below $ 5000, and must be tamper and jam proof.

At the present time we are aware of only one organization, NOAA, outside of the military, which has plans for a continuing operational satellite program and which is concerned with such monitoring systems.

## 5. Problems of Data Acquisition, Reduction and Analysis

Satellites provide a unique service in that they are capable of viewing numerous factors over a very large area of the surface of the globe in a single glance. Though this fact provides science with the capacity to monitor both small- and large-scale features over the entire globe on a nearly real-time basis, the quality and quantity of data generated create formidable problems in data handling and analysis. Some of these problems are discussed in the following sections.

### 5.1. Selection of Data Formats

The use of data from airborne remote sensors, and more particularly from sensors aboard satellites, requires a different approach from that normally used in the acquisition, processing and reduction of conventional oceanographic data. The most conspicuous difference between data collected aboard a research vessel and data transmitted from a spacecraft is the quantity. Marine scientists have been accustomed to considering the data obtained at say, 500 hydrographic stations on a 30-day cruise, as representing a sizable data base in terms of effort and expense. If we assume that the hydrographic casts sample at 14 depths to 500 m, and eight physical and chemical parameters are measured at each level, the resulting total data set consists of 56000 observations. By contrast, to collect the same number of SRIR thermal observations from NOAA-2 may require only 70 seconds for 56 scan lines and a meridional traverse of 217 miles! SRIR values corresponding to a nadir angle greater than 45°, however, are considered of limited value due to the curvature near the horizon. Clearly the semi-automated method of processing much of the conventional oceanographic data is not applicable to data telemetered from spacecraft sensors. In the processing and reduction of satellite data the computer plays an indispensable role. An evaluation of some of the factors which might be encountered by researchers in studies using satellite data in hypothetical analysis, is discussed below.

In reviewing data requirements for such a study, the location, size of the region of interest and the frequency of coverage required are considered first. Frequently, visual images from geosynchronous (ATS) satellites are useful for such purposes because they are taken from the same perspective day after day. The spatial resolution of 10 km (satellite subpoint) often provides an excellent overview of the region of interest and enables a more accurate assessment to be made of data requirements from polar orbiting satellites. Preliminary screening of these higher resolution data from SRIR scanners and Vidicon cameras can normally best be done from photographs or transparencies available from NASA, NESS and the National Satellite Data Center (NSDC). Very little time is required to produce these photographic products so the cost is quite modest. After the available visual and infrared imagery has been inspected to locate the desired time/space strata of coverage for the project, computer-gridded arrays of surface temperatures (effective black-body radiation temperatures) are then ordered. Except for special project needs, it is best to request data that have been calibrated and corrected for atmospheric absorption and limb darkening.

If the project's objectives include intercomparisons of satellite data with other parameters it may also be desirable to order the same data on magnetic tape to facilitate numerical calculations by computer. We have found it very desirable to use small magnetic disks for storage of satellite data that are to be used in numerical comparisons. Each disk holds about 200000 temperature observations. Much of our data processing has been done on an IBM 1800 system.

As an example of the data handling requirements, we recently completed a small case study in which we evaluated numerical arrays corresponding to 15 orbital passes over the same region. These data were placed on two disks with adjacent space allocated on each file for storage of the numerical results. This procedure effectively filled both disks and it was necessary to transfer their contents to magnetic tapes before the disks could be used again for other work. The cost of the two IBM-compatible disks is \$ 200. About 5 minutes of computer time, costing about \$ 10, were required to process each array of data.

The amount of effort and expense entailed in a large-scale analysis would be considerably greater. Assuming that the above spatial resolution requirements would be satisfactory, it is possible to make a rough estimate of the probable cost of a modest fishery—oceanography study. Let us assume that archived SRIR data will be used, since such data can be obtained with relatively little difficulty and expense, and that the survey is to cover an area of 15° lat. by 10° long. in the low latitudes on a daily basis for one month to determine the temperature field and the presence of oceanic fronts.

With a scan line data density of 10 km per data point, and a scan line separation of about 7 km, the resulting array will contain approximately $3 \times 10^4$ observations. To process this quantity of data on a daily basis would fill a small disk to 50% capacity after only about 3 days. The transfer of data from this disk would be required at least twice each week. Each transfer requires about 10 minutes of computer time, or a total of 100 minutes per month. The actual processing of each daily array to produce the temperature and frontal maps would require 5 minutes of computer time. For 30 days of operation the computer time would be approximately 4 hours, costing about \$ 500 exclusive of labor required to prepare the data and to interact with the computer.

If the objective is to use VHRR type data from NOAA or NASA satellites the handling and computing costs would increase significantly since the data density would be ten times greater along each scan line and from line to line. Only a large installation would be adequate for processing these data. For the same geographic region then, each daily array would consist of about $3 \times 10^6$ observations. The computer costs would increase by about two orders of magnitude. To this operation one must add a procedure for screening out cloudy regions.

An admittedly rough estimate of the computer time required for these operations would amount to 100 minutes per day, or 50 hours per month at \$ 6.00/minute totaling \$ 18000 per month. The large quantity of data would require the use of a large computer for increased efficiency in handling the arrays.

Such a project is to be considered "primitive" since no provision has been made for automatically screening out cloudy areas or contouring the data arrays. Furthermore, no provision for determining fish schools or mammals is included. If the spatial resolution of the IR sensor were improved to 50—100 m instead of 1 km and if species-specific signatures had been determined, a multi-stage pattern recognition procedure could be included in the automated data processing. The addition of higher resolution data and the pattern recognition scheme would appreciably increase the computer time requirements and attendant costs. Given the hypothetical possibility that schools could be detected and identified, an additional problem of spatiotemporal sampling remains. Tunas are strong swimmers; tagging experiments have shown tunas may move at mean speeds of 0.5 m/sec or more over large distances. Repetitive satellite coverage of an area can now be obtained once every 48 hours during which a school may have moved a distance of 86 km. To attempt to track even a small number of schools poses the additional requirement of cloud-free conditions for several days at all times (§ 5.3.). Therefore, the problem of multiple counts of the same school would be a troublesome one. In order to obtain a coverage greater than 50% for the above region it would be necessary to combine daily results from a 5-day period.

### 5.2. School Density and Problems of Sampling

The use of remote sensors for assessing the abundance of marine resources will create special sampling problems related to school density and distribution. To examine some of these problems we have chosen two of the world's major fisheries—yellowfin tuna in the eastern Pacific and anchoveta in the coastal waters off Peru.

When at its optimum size the biomass of the Peruvian anchoveta averages about 18000000 tons. This species is found along the entire coast of Peru out to about 40 km, an area of 65000 km². Since individual schools of anchoveta average about 50 tons each, there should be about 360000 schools in the population or approximately one school for each 0.18 km².

The yellowfin tuna on the other hand is distributed over a 22000000 km² area in the eastern Pacific Ocean. The mean population size is about 200000 tons and the average school size is 20 tons. Thus there are about 10000 schools of tuna distributed throughout the area giving a density of about one school for each 2200 km².

These large differences in areal density of fish schools would require differences in sampling the two species from the same satellite. Naturally any estimate of abundance made from sampling in a two-dimensional plane would result in estimates which are biased or at best only relative because the schools of fish actually occur in three dimensions.

If the schools of fish being sampled over these large areas were randomly distributed, sampling problems would be much simplified, but this is not the case as populations of marine organisms are generally distributed in a contagious manner. Since schools of anchoveta and tuna appear to be contagiously distributed also, some finite sampling scheme would need to be devised for sampling both stocks of fish. If we assume that the distribution of schools within an area is the same for both species but that the distances between contagions differs between species in proportion to the difference in schools per unit area, then sampling of yellowfin tuna would have to be more intense than for anchoveta. The differences in sampling levels would be proportional to the differences in areas if the estimates were to have the same statistical reliability. This fact, coupled with the additional fact that an orbiting satellite like ERTS-A passes over the same area only once each 18 days with a swath path of about 180 km at sea level obstructed by bad weather from 60 to 80% of the time, suggests that statistically reliable estimates of abundance of sparsely distributed sea animals may be difficult to obtain.

Sophisticated subsampling procedures will need to be formulated not only because of the statistical problems noted above, but because of the volumes of data which will need to be handled. For example, let us assume that schools of fish can be readily observed and identified from a visual system such as that aboard ERTS-A. Let us further assume that we wish to identify schools of fish which are about 100 m in diameter. Such a school would appear as a 0.1 mm dot on a photo from the ERTS-A camera. In order to enlarge the dot to a size at which the school could perhaps be identified would require a magnification of about 100 times. A photograph taken off the Peruvian coast would have to be expanded, in simple terms, to the equivalent of 10000 photographs to identify anchoveta schools. If one minute is required to scan each photo to establish whether schools are present or not, then 21 working days would be required to examine 10000 photos. Because of the density of schools one would expect to find 0.6 school per photo. However, because of the vertical distribution of the schools, probably only a fraction, say less than one third, would be visible at the surface. If it is assumed that 5 minutes are required to identify and count each school per photo, then about 35 days would be needed to analyze all magnified photos, resulting in a total of 56 working days to analyze a single ERTS photograph. Ten photographs would be required to make one count of all the anchoveta schools off the coast of Peru, which would be the equivalent of 2.3 man-years of labor.

For tuna, the figures are much more formidable. In this case, $7.0 \times 10^6$ magnified photos would have to be scanned. Based on the fact that there would be about one school per 2200 km$^2$, one out of each 550 pictures would show a school. Assuming the same time constants to scan, and to count and identify, as were used for the anchoveta example above, 14140 man-days, or 59 man-years would be necessary to make a single assessment of the number of schools. The study of time trends is a necessity for stock assessment analysis.

It should be recognized that these examples are most likely biased upward, but they serve to exemplify the fact that sophisticated subsampling techniques and automated screening techniques are definite prerequisites to the use of satellite information for stock assessment analysis.

### 5.3. Cloud Cover

Many of the remote sensors discussed in this report operate on wavelengths emitted in the visible spectra. Therefore if the sensors are "looking" at the surface of the ocean, the atmosphere must be relatively clear if they are to "see" it. To evaluate the impact of cloud cover on the ability of sensors to sample the sea surface in the visible range, we have compared the persistence of cloud cover by major geographical areas with the quantities of fish taken in those same areas. To calculate an index of cloud cover we used an atlas of relative cloud cover based on data from meteorological satellites [12]. The data on catches of fish are from the FAO Yearbook of Fishery Statistics for 1971 [13]. For each of the major FAO areas the number of months which, on the average, were cloud free 50% of the time during the period 1967—1970, are shown expressed as a percentage in Table 1. The 1969—1971 average annual catch is shown in the same table for the corresponding areas. It can be seen generally that the areas where the most fish are caught are the same areas which are clouded over most often. A strong relation exists between fish production and cloud cover. This is not surprising since the high production necessary to sustain large numbers of fish is supported by air—sea processes such as wind induced upwelling. This relationship is even stronger if Area 87 is subdivided into an inshore and offshore area. The inshore portion of Area 87 which is clouded over nearly all the time produces 99% of the fish catch in that area.

On the basis of the data presented in Table 1, it appears that nearly 75% of the world catch of marine animals is taken from waters that are clouded over from about 40% to 80% of the time. This fact would suggest that the successful use of visual range sensors from high altitudes would be seriously impeded in areas of high fish production.

Table 1

Fifty-percent Cloud-free Skies by Month expressed as a percentage and average catch of fish in millions of tons, by major areas

| Month | Area | | | | | | | | | | | | | | |
|---|---|---|---|---|---|---|---|---|---|---|---|---|---|---|---|
| | 21 | 27 | 31 | 34 | 37 | 41 | 47 | 51 | 57 | 61 | 67 | 71 | 77 | 81 | 87 |
| Jan. | 10 | 25 | 90 | 95 | 70 | 75 | 80 | 100 | 100 | 25 | 15 | 90 | 50 | 50 | 75 |
| Feb. | 5 | 15 | 85 | 95 | 70 | 75 | 80 | 100 | 100 | 30 | 10 | 90 | 80 | 70 | 80 |
| Mar. | 5 | 5 | 90 | 90 | 80 | 75 | 80 | 95 | 95 | 20 | 5 | 90 | 80 | 70 | 70 |
| Apr. | 100 | 95 | 100 | 100 | 100 | 70 | 60 | 100 | 100 | 95 | 95 | 100 | 100 | 60 | 60 |
| May | 90 | 80 | 100 | 95 | 100 | 95 | 90 | 100 | 100 | 95 | 95 | 100 | 100 | 95 | 95 |
| Jun. | 30 | 40 | 100 | 95 | 100 | 100 | 95 | 80 | 100 | 40 | 65 | 100 | 65 | 90 | 80 |
| Jul. | 25 | 20 | 100 | 90 | 100 | 100 | 90 | 80 | 60 | 10 | 55 | 90 | 95 | 95 | 80 |
| Aug. | 25 | 20 | 95 | 80 | 100 | 80 | 30 | 75 | 75 | 70 | 10 | 85 | 80 | 60 | 40 |
| Sep. | 40 | 20 | 95 | 80 | 100 | 75 | 75 | 75 | 85 | 70 | 10 | 80 | 75 | 80 | 60 |
| Oct. | 35 | 20 | 90 | 95 | 90 | 50 | 35 | 60 | 60 | 50 | 10 | 90 | 90 | 40 | 50 |
| Nov. | 10 | 5 | 95 | 90 | 90 | 80 | 80 | 85 | 85 | 40 | 5 | 95 | 80 | 80 | 75 |
| Dec. | 10 | 30 | 90 | 80 | 70 | 65 | 70 | 80 | 80 | 25 | 15 | 95 | 80 | 65 | 75 |
| Aver. | 32 | 31 | 94 | 90 | 90 | 78 | 72 | 86 | 87 | 48 | 33 | 92 | 82 | 72 | 70 |
| Catch | 4.3 | 10.4 | 1.4 | 2.4 | 1.0 | 0.8 | 2.6 | 1.6 | 1.1 | 13.2 | 2.3 | 3.8 | 0.8 | 2.1 | 11.9 |

Area and corresponding geographical description

21 NW Atlantic
27 NE Atlantic
31 W. Cent. Atlantic
34 E. Cent. Atlantic
37 Mediterr. and Black Sea
41 SW Atlantic
47 SE Atlantic
51 E. Indian
57 W. Indian
61 NW Pacific
67 NE Pacific
71 W. Cent. Pacific
77 E. Cent. Pacific
81 SW Pacific
87 SE Pacific

## 6. Recommendations

In preparing this review of the application of satellites to fishery research a number of problems were touched upon. The following recommendations are made to facilitate solutions to some of these problems.

1. Pamphlets describing the kinds, formats and specifications of data collected with their satellites should be made available to the scientific community by NASA and NESS.

2. Procedures should be instituted to facilitate obtaining in timely fashion satellite data from agencies such as NASA, NOAA and the National Satellite Data Center (NSDC). For many of the results of fishery research to be useful data must be available to the users on a nearly real-time basis.

3. For some satellite systems, the dynamic range of thermal SRIR and VHRR sensors should be modified to $-2$ to 30° and the sensitivity should be extended to 0.05–0.1 °C.

4. The spectral bandwidths for cameras such as the MSS system should be modified, so that chlorophyll and other properties in the marine environment can be surveyed better from satellites. It is realized that the successful completion of Skylab experiments will provide valuable information on the effectiveness of such a technique.

5. More complete calibration data should be provided for the DRIR imagery received by an APT satellite receiver from NOAA satellites. This is an important source of satellite data to scientists in other countries who wish to utilize remote sensing techniques without a heavy commitment in the development of national programs in space research.

6. Similar calibration data for VHRR type imagery on films or prints, obtained from NOAA or NASA, should be provided. A gray scale integrated into each image that is representative of temperature across the thermal range would permit a quantitative estimate to be made of gray tones on these transparencies or prints.

## References

[1] P. SUND, Nat. Mar. Fish. Svc., Tiburon, Calif., unpublished (1973).
[2] T. GREENWOOD, Scient. Amer. **228**, No. 2, 14 (1973).
[3] T. J. WALKER, in: Oceanography from Space, WHOI Rep. No. 65-10, 1965 (p. 321).
[4] A. R. BARRINGER, Ocean Industry **2**, No. 5, 28 (1967).
[5] H. BULLIS, Calif. Mar. Res. Comm. CALCOFI Rep. No. 14, 1970 (p. 57).
[6] M. J. VAN MELLE, H. H. WANG and W. F. HALL, J. Geophys. Res. **78**, 969 (1973).
[7] G. C. EWING, this volume, p. 101.
[8] J. L. SQUIRE JR, Nat. Mar. Fish. Svc, Fish. Bull. **70**, 1005 (1972).
[9] D. CRAM, Fish. News Intl. **12**, No. 3, 35 (1973).
[10] M. R. STEVENSON and F. MILLER, in: 4th Ann. Earth Res. Program Review, **4**, Sect. 100, 1 (1972).
[11] R. W. AUSTIN, in: 4th Ann. Earth Res. Program Review, **4**, Sect. 106, 1 (1972).
[12] US Department of Commerce and US Air Force, Global Atlas of Relative Cloud Cover, 1967–1970, Washington, D. C., September 1971.
[13] Yearbook of Fishery Statistics, Catches and Landings 1971, Vol. 32, FAO, Rome 1972.

Earth Survey Problems — Akademie-Verlag, Berlin 1974

# REMOTE SENSING OF OCEAN COLOR AS AN INDEX OF BIOLOGICAL AND SEDIMENTARY ACTIVITY

G. L. Clarke[a] and G. C. Ewing[b]

[a]Woods Hole Oceanographic Institution, Woods Hole, Mass., USA
and Harvard University, USA
[b]Woods Hole Oceanographic Institution, Woods Hole, Mass., USA

The potential usefulness of satellite-acquired information as a means of monitoring biological productivity of the world oceans, and the accomplishments, problems, and prospects of this procedure are reviewed. Remote sensing has the great advantage over conventional methods that with it great areas of the ocean can be covered more quickly with less expense, and that surveys can be repeated at small time intervals. Examples of the current use of ERTS data, and of our own measurements, are given. It is shown that at low concentrations the optical effect of plankton, biochromes, and silt cannot easily be distinguished from that due to radiance, foam, and glitter. This presents a problem for photo-interpretation. At higher chlorophyll levels, remote sensing can be used as an aid in the wise management of the biological resources of the sea.

## 1. Potential Utility

The long range aim of marine biology as applied to the exploitation of marine biological resources is to understand the dynamics of marine ecosystems. With this basic understanding, we may attempt to adjust the harvest to a rate that can be sustained without depletion, to provide timely warning of reduced yields resulting from natural calamities, and to assure that no controllable activity of man shall reduce the potential of the environment.

As with extraction of other renewable resources, there is no direct cost to the fin-fish industry for raw materials nor to some types of shellfish industry. It follows that all of the cost components in these cases may be allocated to capital investment or labor. This puts special emphasis on forecasting trends in the availability of fish products and on efficiency in harvesting these resources by minimizing time spent locating the most favorable time and place for the harvest. As primitive stocks of fish close at hand become depleted, longer voyages become necessary with international and logistic complications. Thus in recent years, the tuna fleet from Southern California ports has been cruising as far as the coast of West Africa to make its catch. Thermal and chlorophyll mapping is a proven indicator of seasonally favorable tuna water [1], and probably is within the capability of ERTS sensing [2, 3].

In addition to increased costs due to normal economic development, hardship is sometimes caused by natural events beyond human control. A recent example is the drastic curtailment of the anchovy fishery off the coast of Peru caused by a catastrophic influx of warm water on a gigantic scale [4].

Similarly, economic adversity to the Icelandic herring fishery due to natural causes is reported by Noble and Wilkerson [5] as follows:

> ... the National Research Council of Iceland requested that additional SST (sea surface temperature) measurements be made in the Norwegian Sea to provide environmental data in support of the Icelandic herring fisheries. The economic return from the herring fishery (which constitutes a significant portion of the Icelandic gross national product) is heavily dependent on the efficiency of the industry in terms of catch effort and in the time required to return the catch to Iceland for processing. If, as in normal operations, the fish are caught close to Iceland and rapidly returned to Iceland for processing, the herring are marketed as a table product with a correspondingly high economic yield. On the other hand, if the fishing grounds are far removed from Iceland, the quality of the catch degrades in the return transit and must be used for alternate products, such as fish meal, with a relatively low economic yield.
>
> The East Iceland Current brings partially mixed polar water southward in the Norwegian Sea to the east of Iceland, where the polar water mass intersects with the North Atlantic Current (the Gulf Stream), forming an oceanic thermal front. The significant portion of the herring stock fished by Iceland spawns off the Norwegian coast and migrates westward to the region of the thermal front where there is a high biological productivity as a result of the mixing of the polar and North Atlantic water masses. Locating this thermal front during the fishing season, ... is of prime importance since most of the productive herring fishing occurs in the region of this convergence zone.
>
> During the 1967 fishing season, the southward projection of the polar front ... caused the herring stocks to turn northward and head for Spitzbergen rather than to the traditional fishing grounds east of Iceland. The degradation of the quality of the herring catch on the long return transit from Spitzbergen required that much of the catch be processed as alternate fishery products ..., which resulted in a heavy economic loss to the herring fishery.
>
> In this region of the globe, airborne techniques provide the most efficient means of making synoptic surface measurements to augment basic research programs to improve the management and exploitation of the herring fishery.

Lastly, it is cause for concern that activities of man have already damaged the environment in several fairly large areas of the ocean, either through an unexpected modification of the environment or by pollution. As an example of the former, George [6] has shown that the damming of the Nile, Fig. 1, has caused severe changes in the marine biota throughout the Levant. According to Professor George:

> During the past millennia and at an ever increasing rate, the original hydrology of the Nile has been changed by man. The most notable of events today is the construction of the new High Dam at Aswan, which is now near completion and will eventually impound a five-hundred-kilometer-long body of water called Lake Nasser. This lake will hold the annual flood and then release it slowly throughout the year for irrigation and navi-

gation purposes. The floodwaters used to have both a freshening and a nourishing effect on the Levant Basin because of the quantities of fresh water and plant nutrients they carried. ...

Since 1965, Egypt has suffered a great decline in the eastern Mediterranean fisheries. In 1962, the total fish catch was 30600 metric tons (excluding shrimp). The impoundment of the rich floodwaters has all but eliminated the 18000 metric tons of sardines (a loss of nearly 60% of the 1962 total fisheries intake). This eastern Mediterranean decline has caused a destabilization of the Egyptian fishery.

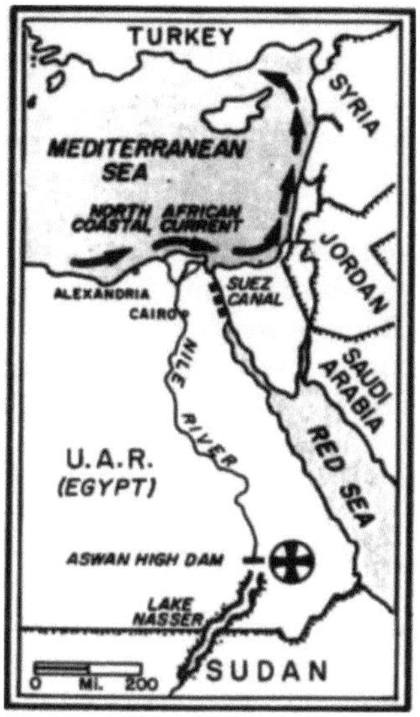


Fig. 1. Route of Nile sediments in the eastern Mediterranean (after George).

As an example of the damage caused by pollution, it has been reported that [7]:

> ... In Japan, 111 cases of mercury poisoning occurred (with 41 deaths) as a result of eating fish of shellfish taken from Minimata Bay. The bay had been contaminated by waste from an acetaldehyde production plant that had used mercuric oxide. Another outbreak occurred in Niigata City with 26 cases (and 5 deaths) due to the same cause. ...
>
> Mercury's toxicity is permanent. In addition, when fish, shellfish, birds, or mammals containing mercury are eaten by other animals, the mercury may be absorbed and accumulated. Vertebrate animals reported as poisoned or killed by mercury, either under natural conditions or in experiments, include various species of fish, chickens, wood pigeons, Japanese storks, pheasant, quail, sheep, cattle, guinea pigs, dogs, rats, pigs, horses, cats, rabbits, and man.

Experience has shown that nothing is more damaging to an animal population than destruction of its environment. The interactions of a fishery with its environment occur at all stages of the life cycle of the fish so that the ultimate yield is frequently controlled by events far from the site of the harvest. Slow deterioration of the environs often escapes detection until the damage has become irreversible. In this context, one might cite as a topic for concern the recovery in the West Indies of DDT which apparently travelled in the easterly air stream across the South Atlantic from Africa [8]. Little is known about the direct effects of DDT fall-out on plankton in the open oceans. However, the laboratory evidence so far accumulated suggests that this may have far reaching and unsuspected impact on marine bio-production [9].

Successfully to monitor the subtle changes that affect oceanic productivity requires a wide and thorough understanding of the whole gamut of processes involved. This will require: intensive studies on a local scale, such as may be performed in laboratories and from small boats; more extended studies in critical areas, best carried out from sea-going vessels; regional studies repeated at frequent intervals from aircraft; and, at the other extreme, repeated extensive surveys of the whole ocean such as can only be accomplished from orbiting or stationary satellites.

In the past, such extensive surveys have had to be pieced together from many discontiguous and intermittent observations. These are time-consuming, expensive, and sometimes misleading. Remote sensing from aircraft and spacecraft provides rapid and continuous measurements of oceanic phenomena. Such surveys can be repeated within short periods of time, thus enabling changes to be followed in detail. We do not today have a thoroughly integrated model of how life processes function over the whole global ocean and how they vary from week to week, season to season, and year to year. Remote sensing will help to fill this need. We also lack information on the intercorrelation of fluctuations in different localities. It is to be hoped that, with the advent of suitably equipped satellites, we can learn economical methods of executing this extensive operation. If successful, remote sensing will improve our understanding of the oceanic system as a whole. It can also enhance the effectiveness and efficiency of ship operations by directing attention to the most critical areas at the most critical times. We cannot now foresee the ultimate limits to the usefulness of such undertakings, either independently or as part of an integrated exploration of the ocean. Nevertheless, the ability to acquire sea surface temperatures and estimates of chlorophyll concentrations from satellites has already proved its feasibility, value, and economy.

It is true that remote sensing over the ocean, as over the land, is limited to the shallow sunlit layers. However, since this is the site of all the significant energy interchanges across the ocean boundaries, and is also the locale of the photosynthetic process, an understanding of events in the shallow layers is fundamental to any understanding of ocean biodynamics as a whole. Life in the sea, as on the land, is fundamentally dependent on surface phenomena. Consequently, the remote sensing problems and methods employed in marine productivity studies are in certain respects similar to those employed in agriculture. In both cases we take account of favorable versus unfavorable environmental factors, such as weather, and make direct assessment of the well-being of the standing crop. In both cases, concentration of chlorophyll and related biochromes is a valuable indicator of vigorous productivity and growth. The outlook for useful biological studies of the oceans from satellites is indeed bright.

## 2. Accomplishments

Although it is true that past accomplishments are modest compared with the potential opportunity, it should be borne in mind not only that the ERTS satellite is the first space device specifically designed to look at agricultural resources, but also that is has been available for less than one year. It remains to be seen what can be accomplished in a longer reach of time, especially with remote sensors specially tailored to marine studies. In the space available here, we can only sketch a few examples to show what is being done and what may be possible in the future.

We start with the search for areas of chlorophyll abundance. Heavy concentrations of this material are typically found in waters near the coast or over shoals such as the Grand Banks or the Dogger Bank. Fig. 2 shows that the fisheries of the world are mostly concentrated in such areas. Fig. 3 shows regions of the oceans where conditions are most conducive to phytoplankton fertility. Over most of the ocean, insolation leads to thermal and density stratification. In this monotonous state, available nutrients are soon exhausted by the plankton and are not replenished. In consequence, most areas of the open ocean are relatively sterile and poor in marine life. Productive areas are found only where dynamics stir the water, or promote up-welling. Evidence that these active areas are indeed disclosed can be deduced from the following:

The color of the light which emerges from the sea surface is controlled by the joint roles of scatterance and absorptance along the optical path (Fig. 4). In the absence of foreign particles or substances, the effect on color is dominated by Rayleigh scattering by water molecules, since these are very small compared with the wavelength of visible light. Absorption, in this case, may be neglected, and the emergent light, having been subjected to scattering inversely proportional to the fourth power of the wavelength, is blue. On the other hand, in the presence of plankton, bubbles, or non-living particles, Mie scattering becomes dominant. For large particles of this sort, the scattering is non-selective and the component added to the emergent light is nearly achromatic. Where the particles are living plankton, they are translucent and contain colored substances, chiefly chlorophyll, which strongly absorbs blue light. In addition, water may contain dissolved biochromes, such as biological degradation products, tannin, etc. Where these factors are active, absorption plays a significant optical role such that the emergent light appears greenish or sometimes slightly red.

To cite some specific examples: (i) in clear water (Crater Lake), Tyler [10] found that the maximum up-welling irradiance at 5 meters below the surface occurred at 416 nm, i.e. deep blue; (ii) in San Vincente Lake, "heavily contaminated with phytoplankton, zooplankton, and plankton decomposition products", he found maximum irradiance at 576 nm, i.e. yellow; (iii) in the surface water of the Sargasso Sea, Jerlov [11] evaluated the color as having a dominant wavelength of 475 nm, with a purity of 85%, i.e. deep blue, whereas off the California coast Tyler found the color to be 519 nm and the purity 8%, i.e. a very pale green.

The effect of chlorophyll-bearing plankton in the water is evidently to desaturate the deep blue of clear ocean water and, depending on the concentration, to shift the dominant wavelength into the green or yellow. In our 1969 results [12] where chlorophyll levels were as high as 3 mg $m^{-3}$, the green shift in hue was

Fig. 2. Distribution of fishing grounds of the world. The yield is suggested by the relative intensity of the shading (after Walford).

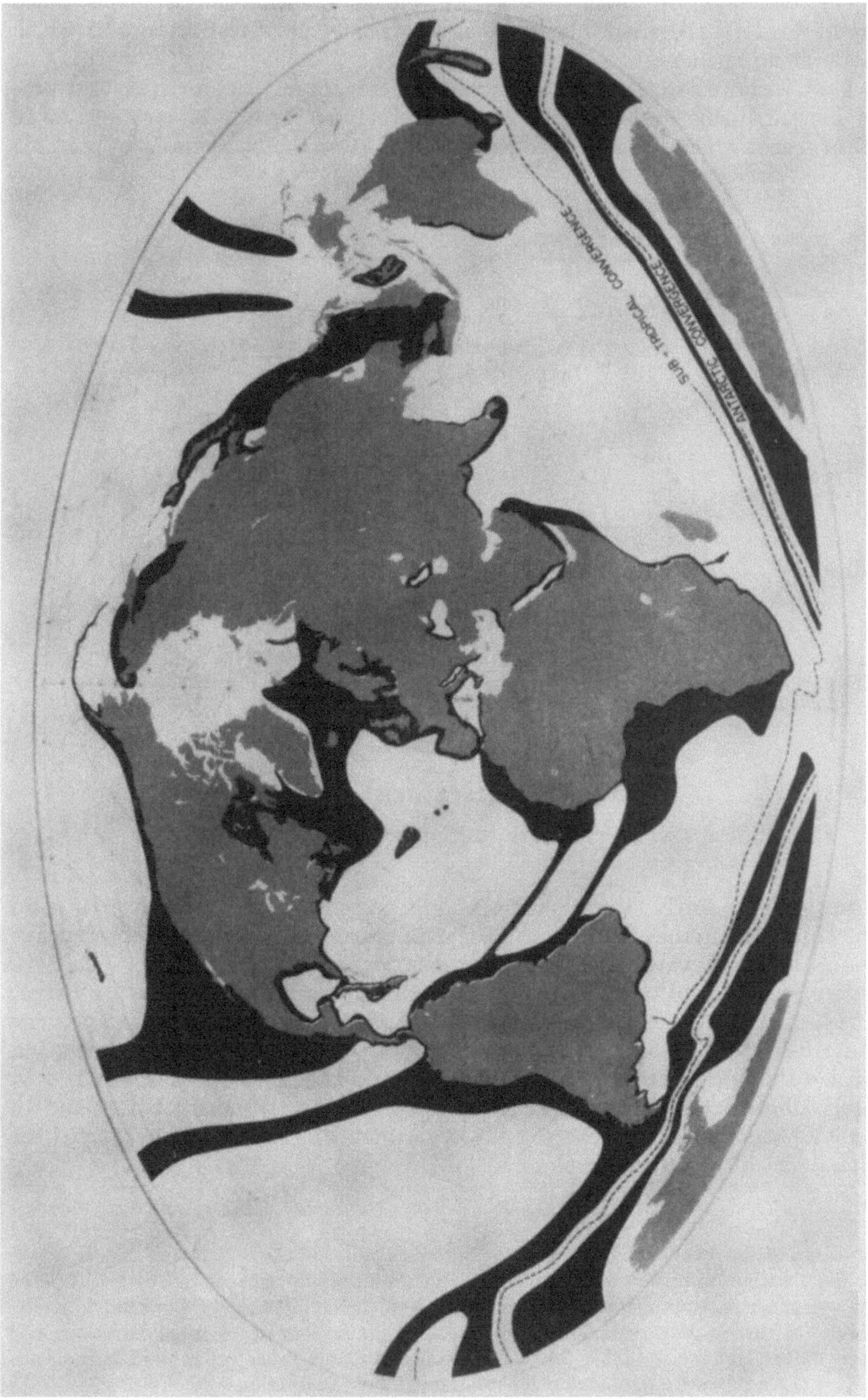


Fig. 3. Fertile areas of the seas where oceanographic processes are conducive to enrichment of surface waters, hence to production of plants and animals (after Walford).

prominent, whereas in 1970, where the measured chlorophyll levels were 0.23 mg $m^{-3}$ or less, the effect of plankton was largely desaturation toward white with little or no hue shift.

This shift is similar to that due to bubbles, glitter, foam, and airlight, and presents a most difficult problem of photo-interpretation. In the biologically more important case where large concentrations of active absorbers are present,

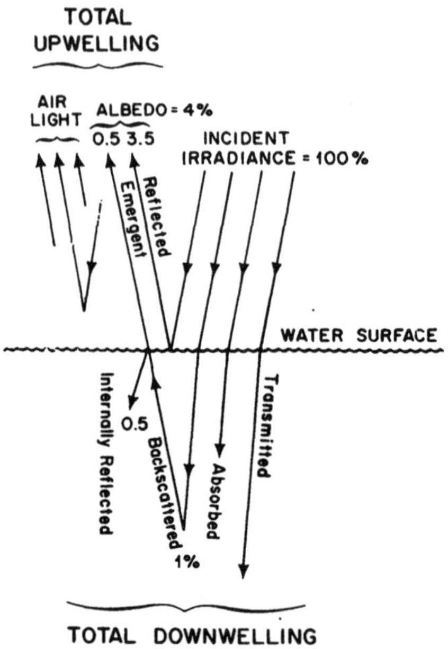


Fig. 4. Schematic light paths and their approximate relative intensities.

plankton produces fairly strong shades of green, yellow, or red. Thus in Fig. 5a [13], the blue water in the central South Atlantic corresponds with the unproductive areas, and the green water with the productive ones.

An interesting exception to this rule is seen in the plume of the Amazon River, where the water is strongly green, but the fertility and productivity are very low. According to Yentsch [14], it exhibits color due to dissolved degradation materials, although it is biologically exhausted and is devoid of chlorophyll.

Appreciation of the relation of sea color to productivity caused Strickland in 1960 to undertake aerial surveys over Georgia Strait in the Canadian Northwest with the object of identifying rich areas [15].

⟶

Fig. 5. A series of charts of the South Atlantic ocean: (a) distribution of color of the sea (after Schott); (b) distribution of phosphate in mg $m^{-3}$ in the upper 50 m layer; (c) distribution of plankton organisms, thousands/liter, in the upper 50 m layer (after Hentschel and Wattenberg 1930); (d) distribution of zooplankton (metazoa), numbers per 4 liters, in the upper 50 m layer (after Hentschel 1933); (e) distribution of organic gross production in summer (g C $m^{-2}$ day); (f) distribution of annual net production (g C $m^{-2}$) (from Steemann-Nielsen [13]).

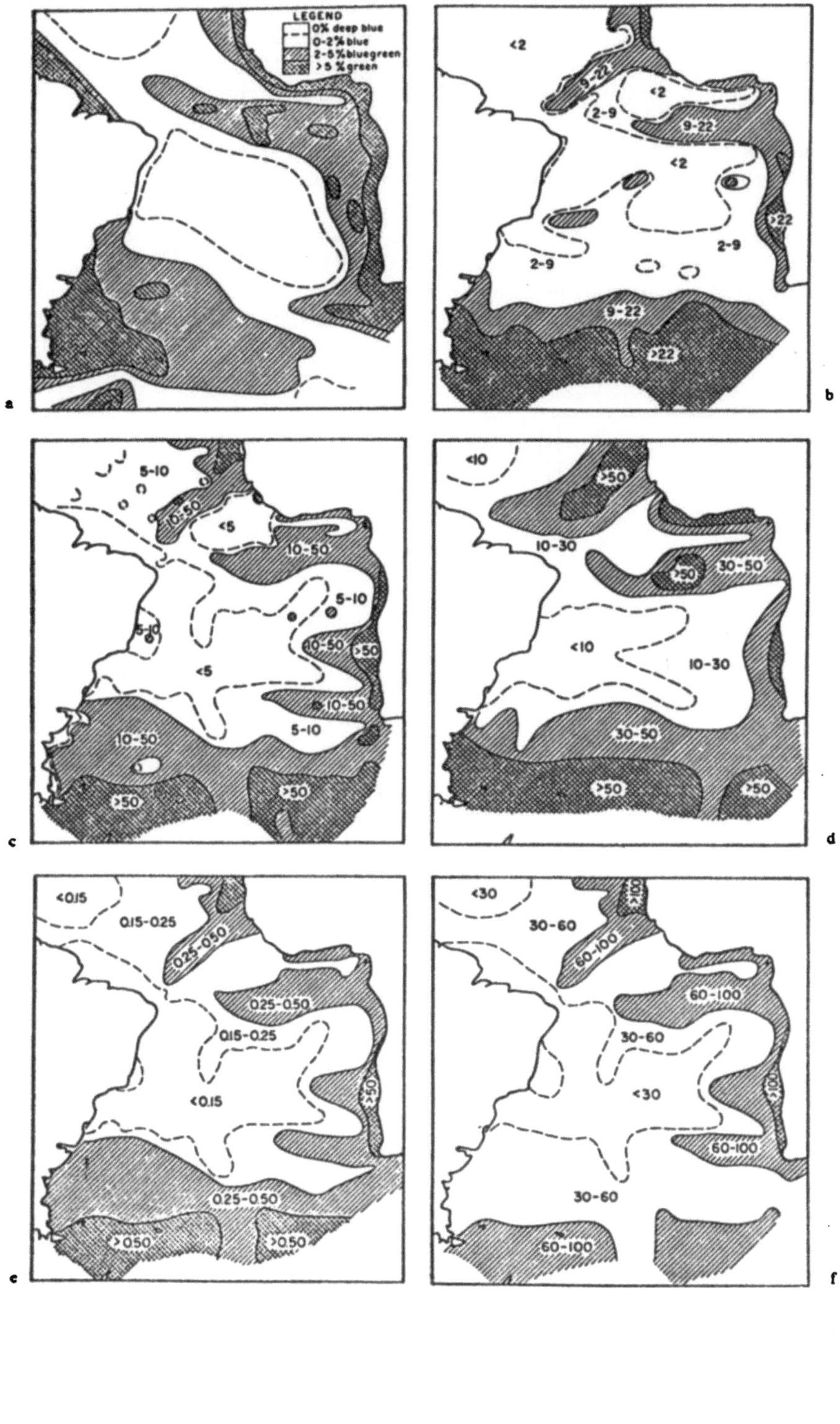
LEGEND
0% deep blue
0-2% blue
2-5% bluegreen
>5 % green
a
<2
9-22
<2
2-9
9-22
<2
>22
2-9
2-9
9-22
>22
b
5-10
10-50
<5
10-50
5-10
5-10
10-50
>50
<5
10-50
5-10
10-50
>50
>50
c
<10
>50
10-30
>50
30-50
<10
10-30
30-50
>50
>50
d
<0.15
0.15-0.25
0.25-0.50
0.25-0.50
0.15-0.25
<0.15
>.50
0.25-0.50
>0.50
>0.50
e
<30
30-60
60-100
60-100
30-60
<30
60-100
30-60
60-100
f

In 1968 and 1969, the authors made aircraft surveys over Vineyard Sound and Georges Bank on the New England coast which showed a definite correlation between observed sea color and the measured concentration of chlorophyll A. A sample of the salient result is shown in Fig. 6. In 1970 the work was extended to a much larger area by means of a 10000 km overflight of the Gulf of Mexico, the Caribbean, and the coasts of Central America. A regional analysis of the data obtained is presented in Fig. 7, in which the shaded areas represent locations where the spectra of the ocean color indicate areas that were relatively rich in biological matter. (None of these, however, approached chlorophyll levels found on Georges Bank in 1969.) Wherever the opportunity was presented, stations were occupied over the R/V *Discoverer* from which quantitative measurements of chlorophyll concentrations were made from water samples. These data are included in the figure. Other discolored areas were found in coastal zones believed to be of high fertility, although no ground truth was available. In particular, the

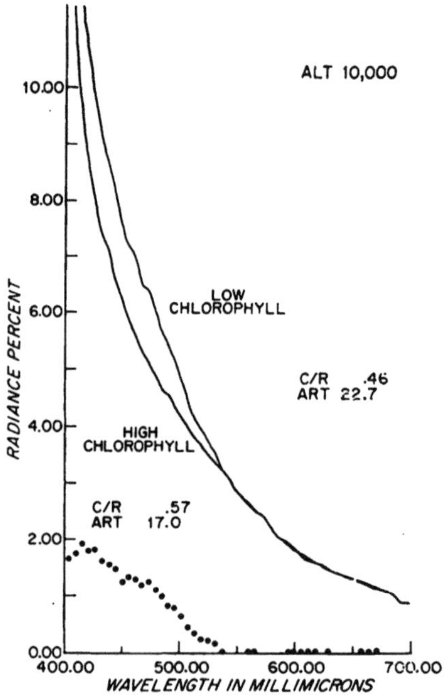


Fig. 6. Spectral radiance from 10000 feet (3048 m) over water containing high and low concentrations of chlorophyll. The dotted curve represents the difference between the two spectra.

ocean color indicates relatively strong bio-activity over the Campeche shrimp banks off Yucatan, known to be a highly productive fisheries area. A photograph of this area taken from Gemini 5 (Fig. 8), shows clouds of mud and chlorophyll streaming from the Laguna de Terminos. It is well established that the shrimp harvest is closely correlated with abundant run-off from land. The association of fertility and sediment is common in shallow water areas along the whole US coast. From this circumstance, it can be inferred that, for our present purposes, discrimination between the optical effects of chlorophyll and silt may not always be necessary.

The foregoing illustrations were obtained from color photographs or spectrophotometers with considerable chromatic sensitivity. However, because of instrumental design of the multi-spectral sensor on the Earth Resources Satellite (ERTS), the imagery obtained over the ocean contains little chromatic information. The MSS is color-blind to blue, and the strongest contrast ratios appear either in Channel 4 (500—600 nm) or in Channel 5 (600—700 nm). ERTS images taken over the ocean show little contrast in Channel 4. Photo-interpretation therefore depends largely on monochromatic data related to size, shape, and location of areas of varying reflectance from optically large particles. Areas of plankton

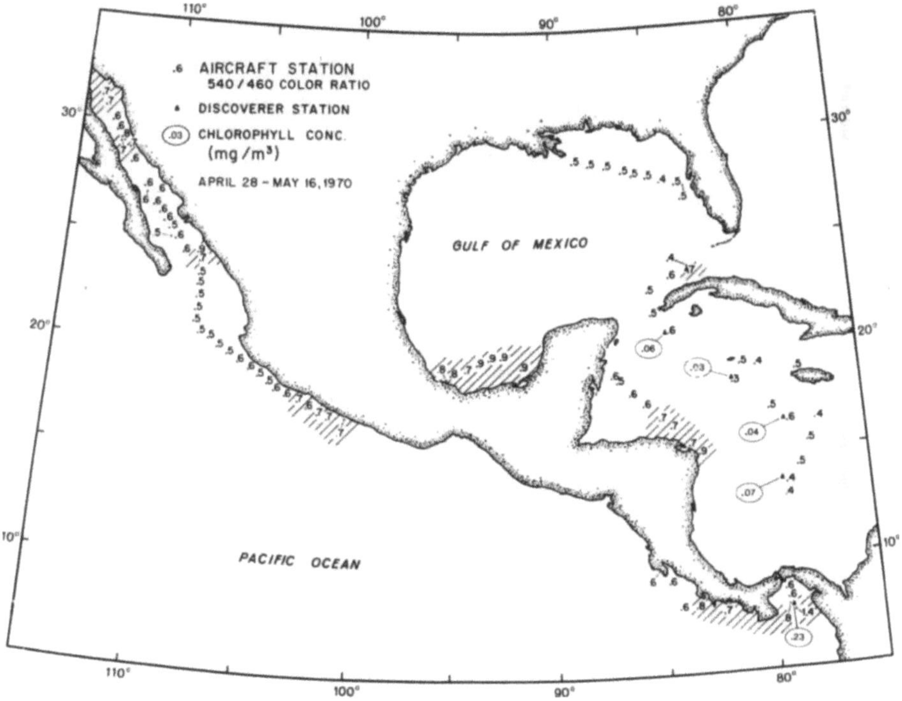


Fig. 7. Map showing aircraft stations (dots) and ship stations (triangles). The circled numbers refer to chlorophyll concentrations (mg m$^{-3}$) measured from a ship. The shaded portions are regions where the spectra indicate high chlorophyll concentrations.

or sediment appear as bright clouds on a dark ground. The effect, shown in Fig. 9, is similar to black and white photographs made through a Wratten 15 filter which has a dominant wavelength of 579.2 nm and an excitation purity of 99% [16].

The association of turbidity and productivity has been used successfully in locating concentrations of menhaden in Mississippi Sound by Maughan and associates [17], as well as by Kemmerer and associates [18] from the National Marine Fisheries Service (NMFS) station at Bay Saint Louis.

Menhaden are the object of the largest volume fishery in the United States. The pattern of turbidity in this region of shallow water is highly complex, as can be seen in Fig. 10, which is derived from ERTS imagery. By relating the distribution of turbidity with the daily catch and with school sightings by spotter aircraft, Kemmerer established that the menhaden congregate on the boundary between the turbid and the clear water (Fig. 10 (a)). This result suggests a strong analogy to the so-called "principle of edges" [19] familiar to wildlife managers on land, namely that the abundance of animals in an area depends on the variety of environments available to the local fauna and, more specifically, on the length of the boundaries between differing environments.

In the form in which images from satellites are first made available, they are not suitable for quantitative analysis because the eye does not assess absolute

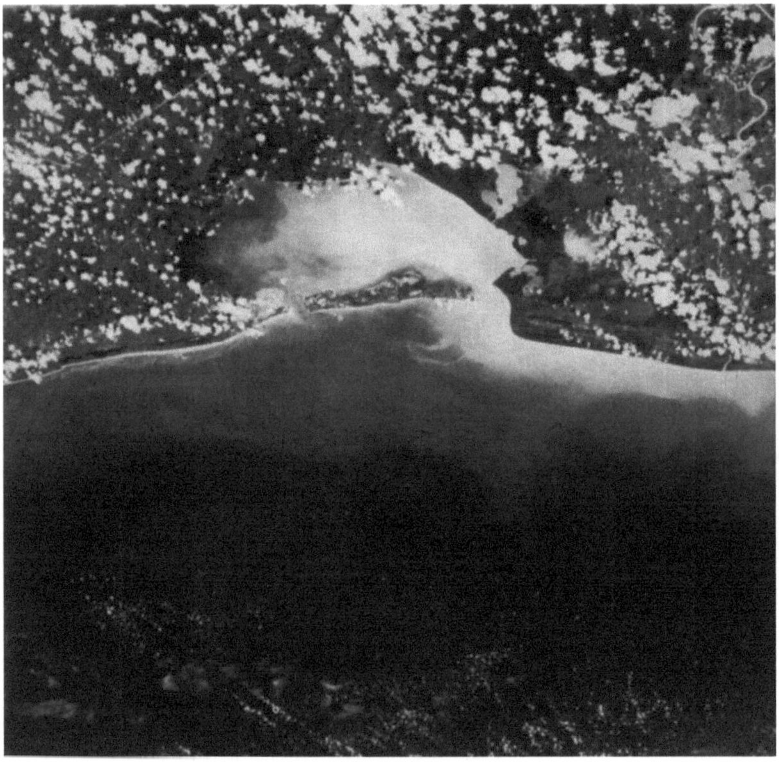

Fig. 8. Satellite photo of Campeche Bay (Gemini 5).

levels of gray in a nebulous scene. This vagueness was largely overcome by resort to photometric density slicing which corresponds to contouring a continuous field. By this means the favorable fishing areas were narrowed down to fairly precise limits. The points at which actual schools were fished or located by spotter aircraft are marked on the drawing in Fig. 10 (a) which relates to a portion of Fig. 10. The location of these areas moved from day to day with the changing

tides and currents, but were located in successive trials when the satellite passed overhead. If this method of locating favorable fishing continues to be fruitful, it may lead to systematic briefing of fishermen and may significantly decrease the time needed to make daily catches.

We turn next to use of thermal images as a guide to locating areas of high productivity associated with renewal of nutrients in surface water by up-welling. Newly up-welled water is cooler than its surroundings and gradually loses temperature contrast with increasing "age" as it comes into thermal equilibrium with the overlying air. Such temperature gradients are readily observable with an accuracy of 2 °C in data from the High Resolution Infrared Radiometer (HRIR) of the Nimbus 2 satellite. Such an area of transient up-welling and resulting high biological fertility is engendered annually off the Somali coast during the southwest monsoon [20]. Satellite imagery, selected by Szekielda, shows the location of the cold water [21]. It is of interest to note that by means of these data, the time of onset and the areal extent of the up-welling were reassessed with improved quantitative accuracy. Compared with conventional survey by ship in these inaccessible waters, the operation was synoptic and economical. More recently Szekielda

Fig. 9. Cloud of plankton in up-welling off coast of Mauritania, south of Cap Blanc [2].

Fig. 10. Turbid water of Mississippi Sound, site of menhaden fishery.

has located ERTS imagery showing strong up-welling of cold water with a heavy concentrate of chlorophyll along the Northwest coast of Africa [2], as shown in Fig. 9.

## 3. Problems and Prospects

Although many more examples of work accomplished have been described in the published literature, the foregoing may be sufficient to encourage some speculation as to future problems and prospects. The most intractable problem of photo-interpretation occurs under conditions of low chlorophyll or silt concentrations. As already explained, it is difficult in this situation to discriminate between those spectral changes in the emergent light which indicate conditions of biological importance and those changes in the up-welling light which are due to glitter and foam on the sea surface. In reality, psychological discrimination

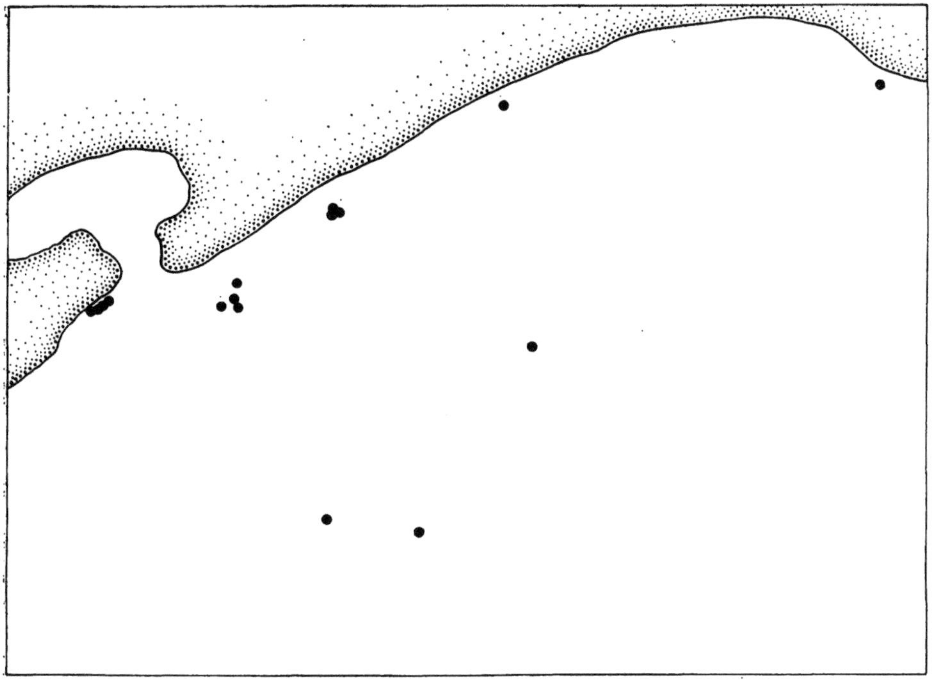

Fig. 10(a). Insert to Fig. 10. Location of menhaden schools in Mississippi Sound [18].

is not as hard to achieve as physical, for reasons that are complex and somewhat difficult to identify. This ability undoubtedly has to do with secondary characteristics of the image and with the intuition of the photo-interpreter. Most experienced viewers can readily distinguish a cloud of plankton from a patch of glittering white water. However, it would be helpful for machine processing if one could express the criteria for discrimination in explicit physical terms.

The human eye perceives discrete areas in a scene by the induced sensations of hue, brightness, and saturation, aided by many additional attributes and modes of appearance. A generally accepted list includes the following [22]:

| | | |
|---|---|---|
| Hue | Saturation | Brightness |
| Lightness | Texture | Location |
| Duration | Gloss (lustre) | Size |
| Shape | Insistence | Pronouncedness |
| Transparency | Fluctuation (glitter, flicker, sparkle) | |

Some attributes of the list such as hue, saturation, and brightness are readily observed by means of realizable instruments. They are related to the psychophysical quantities: dominant wavelength, excitation purity, and luminous reflectance. Others, such as flicker, sparkle, and glitter, would over-tax the capability of presently available sensors or require special instrumentation.

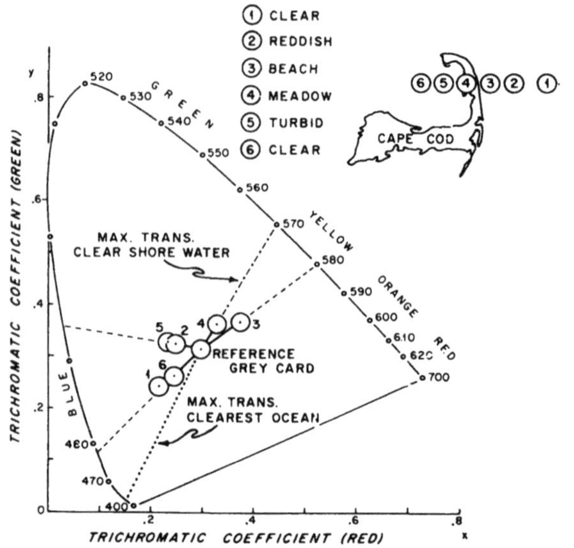


Fig. 11. Chromaticity of various marine scenes.

For many purposes, simple monochromatic discrimination is sufficient to outline areas of major activity. However, the addition of saturation and hue discrimination (which would require correlating the signals in other channels on a point-to-point basis) would serve to classify different regions more critically. The chromaticities of water types observed by us over Cape Cod Bay with various amounts of chlorophyll are shown in Fig. 11. The difficulty of making adequate photo-interpretations at very low chlorophyll levels arises because chlorophyll in the water, although inherently greenish in color, is mostly concentrated in plankton

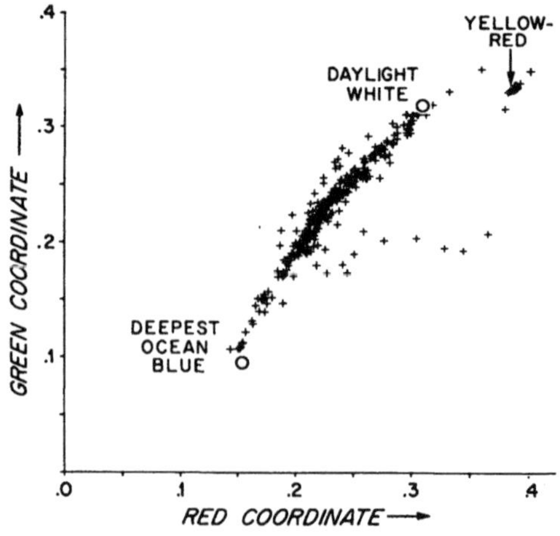


Fig. 12. Chromaticity coordinates of 330 spectra from the Caribbean and Central American waters.

bodies of dimensions large compared with the wavelength of visible light. As a result, the effects of plankton in the water are chromatically similar to those produced by glitter, bubbles, foam, dust, and air light. Shown in Fig. 12 are the chromaticities of 330 spectra taken over the Caribbean and along the Mexican coast, plotted on a standard chromaticity diagram. Many of these spectra were

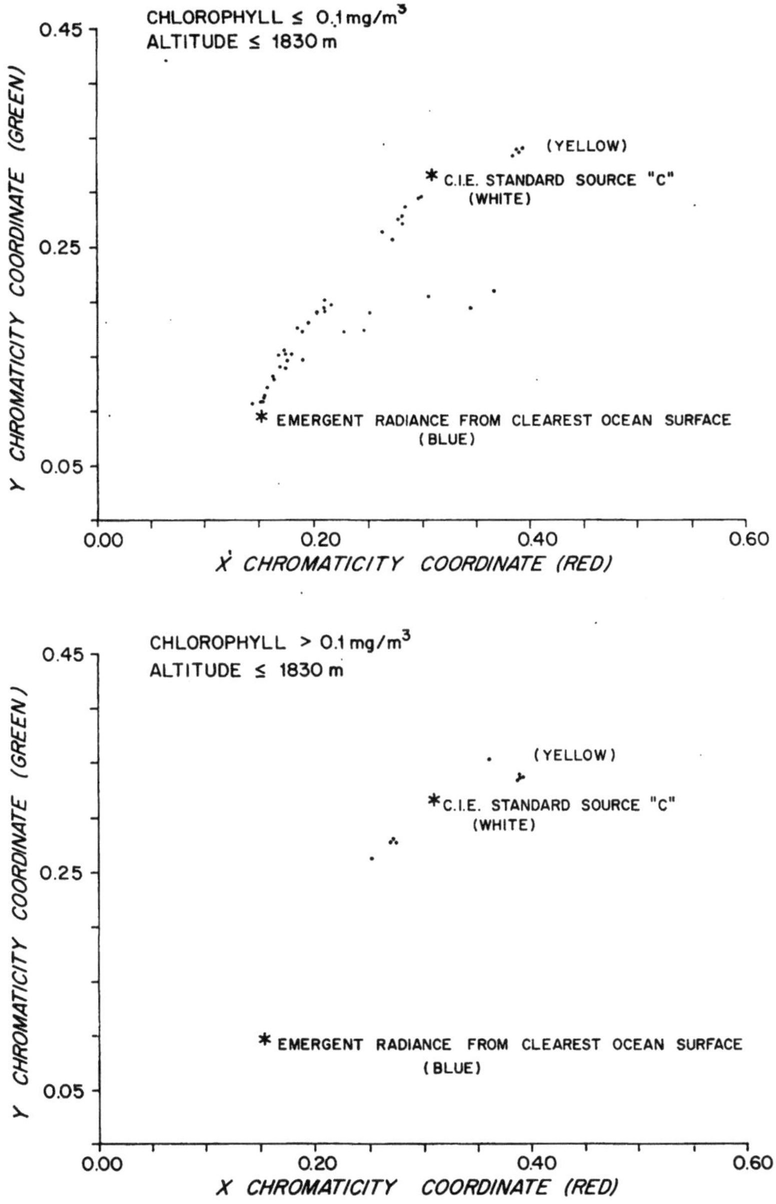


Fig. 13. Chromaticity coordinates of spectra from water containing less and more chlorophyll, measured from lower flight levels.

from water with very low chlorophyll concentrations, while others were from somewhat richer water. The figure shows that the chromaticities of these spectra are arranged along a band of blue with varying degrees of saturation from the deepest blue found over the sterile Sargasso Sea to nearly pure white of reflected

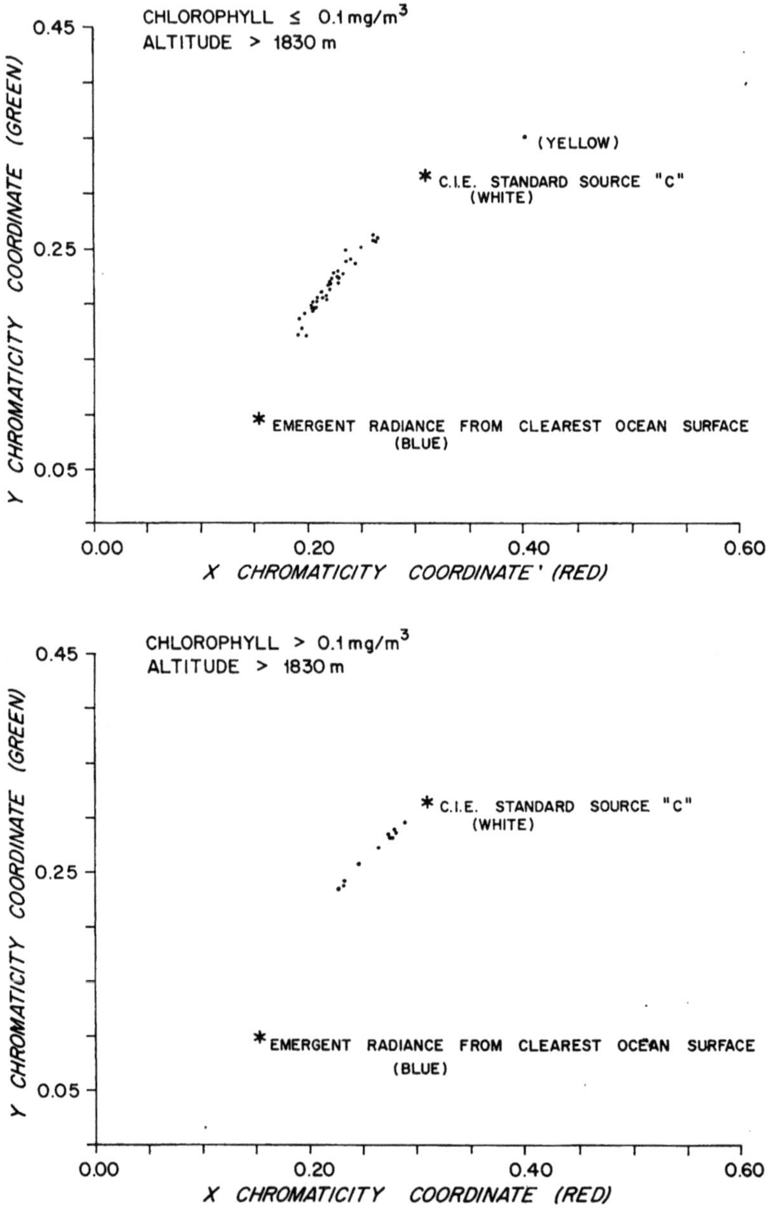


Fig. 14. Chromaticity coordinates of spectra from water containing less and more chlorophyll, measured from higher flight levels.

sunlight. The desaturation of the normally blue water is caused as readily by extraneous effects as by chlorophyll. In Figs. 13 and 14, the chromaticities of spectra where adequate ground truth was available are shown sorted into groups with more and less chlorophyll and from lower and higher altitudes. Unlike our 1969 results, in this study at all the stations where chlorophyll was measured the values were less than 0.25 mg chl $m^{-3}$, and hence a detectable increase in the green region of the spectrum would not be expected. Most of the spectra appear to have indistinguishable hue except for those in the yellow-red region. This latter group, similar in color to the findings of Tyler off the coast of California, is comprised of the spectra over regions known to be relatively rich in chlorophyll plus a few where no ground truth is available. The yellow-red of this cluster is also seen in high altitude color photographs of sediment-laden river plumes [23]. Although the hues or dominant wavelengths are undifferentiated, variations in reflectance and saturation remain. Further discrimination would be provided by correlation with thermal infrared channels.

As for the future, it appears probable that with increased acquisition of satellite imagery over the high seas and with improved rapid distribution of data, combined with refined discrimination and the sensing of surface roughness and temperature, we shall eventually be able to read a more complete story of the distribution and changes in the spectra of emergent light in many regions of the oceans nearly synoptically. From this, we should be able to deduce major differences in biological productivity in the world oceans, as well as to observe daily and seasonal fluctuations. Such an accomplishment will undoubtedly enhance our understanding of global ocean dynamics to a significant degree and will be of fundamental assistance in the wise management of the biological resources of the sea.

## References

[1] M. Blackburn, Pacific Sci. **20,** 36 (1966).
[2] K. H. Szekielda, Distribution Pattern of Temperature and Biomass in the Up-welling Area Along the Northwest Coast of Africa, in preparation.
[3] K. H. Szekielda, V. V. Salomonson and L. J. Allison, NASA Rep. No. X-651-70-416.
[4] New York Times, 28 January 1973, III:58:1.
[5] V. E. Noble and J. C. Wilkerson, Rem. Sens. of Env. **1,** 187 (1970).
[6] C. J. George, The Careless Technology, Natural History Press, New York 1972 (p. 159).
[7] Massachusetts Institute of Technology, Man's Impact on the Global Environment, MIT Press, Cambridge, Mass., 1970 (p. 137).
[8] R. W. Riseborough et al., Science **159,** 1233 (1968).
[9] D. W. Menzel, J. Anderson and A. Randtke, The Susceptibility of Zooplankton to DDT, unpublished. See also [7].
[10] J. E. Tyler and R. C. Smith, Measurements of Spectral Irradiance under Water, Gordon and Breach, New York 1970 (p. 41).
[11] N. G. Jerlov, Optical Oceanography, Elsevier, Amsterdam 1968 (p. 149).
[12] G. L. Clarke, G. C. Ewing and C. J. Lorenzen, Science **167,** 1119 (1970).
[13] E. Steemann-Nielsen, Galathea Report, Vol. I, Copenhagen 1959 (p. 78).
[14] C. S. Yentsch, Proc. Symp. on Remote Sensing in Marine Biology and Fisheries Resources, Texas A. & M., Texas 1971.
[15] J. P. H. Strickland, Fisheries Research Board of Canada, Manuscript Rep. Series 88 (1961).
[16] Kodak, Kodak Filters, Publ. No. 3-B, Rochester 1970.

[17] P. M. MAUGHAN et al., Report of Earth Satellite Corporation, NASA/GFSC Ref. No. SR-258, Washington, D.C. 1973.
[18] A. J. KEMMERER and J. A. BENIGNO, Symp. of Significant Results obtained from ERTS 1, Goddard Space Flight Center, Greenbelt, Md, 1973.
[19] G. L. CLARKE, The Elements of Ecology, Wiley and Sons, New York 1954.
[20] J. G. BRUCE, Deep Sea Res. **15**, 665 (1968).
[21] W. DUING and K. H. SZEKIELDA, J. Geophys. Res. **76**, 1481 (1971).
[22] Committee on Colorimetry, The Science of Color, Opt. Soc. America, New York 1963.
[23] NASA, Earth Photographs from Gemini III, IV, and V, NASA Rep. No. SP-129, S-65-45768, 1967 (p. 178).

# Atmosphere and Climate Related Problems

Earth Survey Problems — Akademie-Verlag, Berlin 1974

# VERTICAL TEMPERATURE PROFILES FROM SATELLITES: RESULTS FROM SECOND GENERATION INSTRUMENTS ABOARD NIMBUS 5

W. L. Smith[a], D. H. Staelin[b] and J. T. Houghton[c]

[a]National Environmental Satellite Service, National Oceanic and Atmospheric Administration, Washington, D.C., USA
[b]Massachusetts Institute of Technology, Boston, Mass., USA
[c]Oxford University, Oxford, UK

The Nimbus 5 spacecraft carries several radiometers for sensing the temperature distribution of the atmosphere. A second generation scanning infrared radiometer observes the temperature distribution of the earth's surface, and the troposphere and lower stratosphere with a spatial resolution of 15 n.mi. An improved selective chopping infrared radiometer developed in the United Kingdom enables temperature profiles to be obtained up to the stratopause. A microwave radiometer has permitted for the first time the observation of the vertical temperature distribution through and below clouds. The unique capabilities of each temperature sounding instrument are discussed and illustrated using results obtained from Nimbus 5. The cooperation of the scientific investigators of these experiments has led to the development of methods for obtaining optimum temperature profiles from the combined set of infrared and microwave radiation measurements. Results show that the detailed thermal structure of the earth's surface and the atmosphere over the entire globe can be obtained solely from the combined set of infrared and microwave radiation observations. Conclusions are made which might impact the development of future operational satellite vertical temperature sounding systems.

## 1. Introduction

It has long been recognized that the only practical means of specifying surface and air temperature on a global scale is from passive radiometric observations conducted from artificial satellites. In recent years numerous radiometric techniques have been proposed for this purpose, several of which have already been tested aboard satellite platforms. Low spatial resolution downward viewing infrared spectrometers first demonstrated the remote air temperature sounding concept aboard the Nimbus 3 and 4 satellites, beginning in 1969. Second generation medium spatial resolution scanning infrared radiometers which were designed to circumvent the cloud problem were successfully implemented aboard the operational NOAA 2 and experimental Nimbus 5 satellites during the latter part of 1972. The capability of temperature sounding through clouds was also demonstrated for the first time from Nimbus 5 by a downward looking microwave radiometer. The next Nimbus satellite, Nimbus F, will carry a third generation multi-spectral infrared device expected to improve the vertical temperature resolution in the middle and lower troposphere, a second generation spatially scanning microwave radiometer to provide complete temperature profile coverage in all weather conditions, a first generation horizon scanning infrared radiometer for obtaining temperature profiles throughout the stratosphere with high vertical resolution, and a pressure modulated $CO_2$ gas cell radiometer for observing the temperature distribution of the upper stratosphere and lower mesosphere. Many of these developments in remote vertical temperature sounding are expected to become operational by the end of this decade.

## 2. Physical Basis

The vertical temperature profile of the atmosphere can be obtained from the spectral profile of outgoing radiance within an absorption band due to a constituent whose concentration is known (e.g. $N_2O$, $CO_2$ or $O_2$). Observations are made in spectral intervals ranging from the center of the band out to the wing of the band. Thermal energy measured near the band center arises from higher altitudes because of atmospheric absorption of energy emitted at lower levels. On the other hand, energy measured in the wing of an absorption band comes from lower altitudes because of the high transparency of the upper atmosphere. Consequently, the variation of outgoing radiance with wavelength can be interpreted in terms of the variation of atmospheric temperature with altitude.

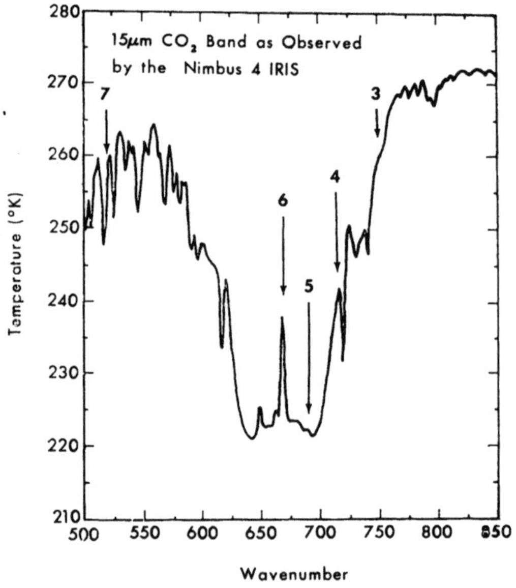


Fig. 1. Spectra of outgoing radiation equivalent blackbody temperature near 15 μm observed by the IRIS on Nimbus 4 (the arrows denote the spectral regions sampled by the ITPR instrument).

Fig. 1 illustrates spectra of outgoing radiation equivalent black-body temperature in the 15 μm $CO_2$ band as measured by the Infrared Radiation Interferometer Spectrometer (IRIS) from the Nimbus 4 spacecraft. We see that the radiation temperature generally decreases as one approaches the center of the absorption band and this corresponds to the decrease of tropospheric temperature with altitude. The minimum near 690 $cm^{-1}$ is associated with the cold tropopause and the subsequent increase to a maximum at the band center is due to the increase of temperature with altitude in the stratosphere. The entire spectrum need not be measured to infer the temperature profile but just several carefully selected spectral intervals (e.g. those, shown by arrows, chosen for the Nimbus 5 ITPR). This is because the energy measured at any wavelength arises from a relatively deep layer (8—12 km) of the atmosphere so that the measurements at adjacent frequencies are highly redundant.

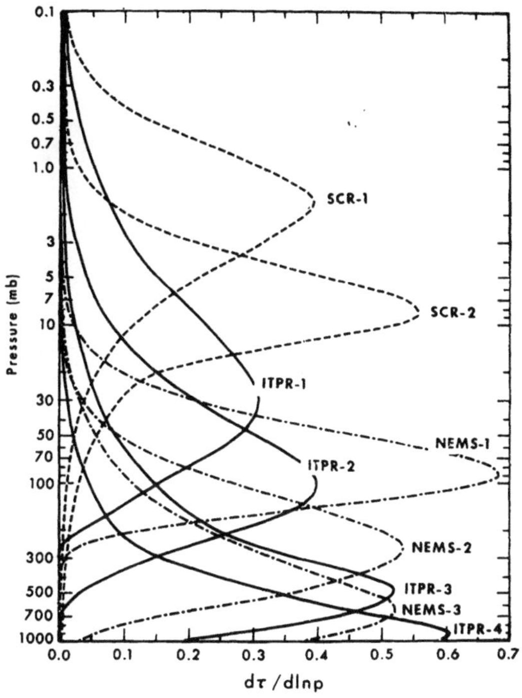


Fig. 2. Atmospheric weighting functions for ITPR, NEMS and SCR spectral channels. Derivative of transmittance with respect to the logarithm of pressure.

Fig. 2 illustrates the vertical layers sensed by various spectral channels of the Selective Chopper Radiometer (SCR), Infrared Temperature Profile Radiometer (ITPR), and the Nimbus-E Microwave Spectrometer (NEMS) instruments flown aboard the Nimbus 5 satellite. The curves, usually referred to as weighting functions, depict the layers most vividly sensed by the particular spectral channels. The SCR weighting functions pertain to high spectral resolution measurements of radiation due to the strong absorption lines at the center of the 15 μm $CO_2$ band. The ITPR weighting functions are due to four spectral intervals of the 15 μm $CO_2$ band ranging from the center to the wing of the band. The NEMS weighting functions are due to a complex of oxygen absorption lines near a wavelength of 0.5 cm. Differences in the weighting function widths and peak values are due to differences in the effective spectral resolutions of the various instruments. As will be discussed below, the vertical temperature profile of the atmosphere may be calculated from measurements by these instruments independently or in combination with each other.

## 3. Nimbus 5 Vertical Temperature Profile Radiometers

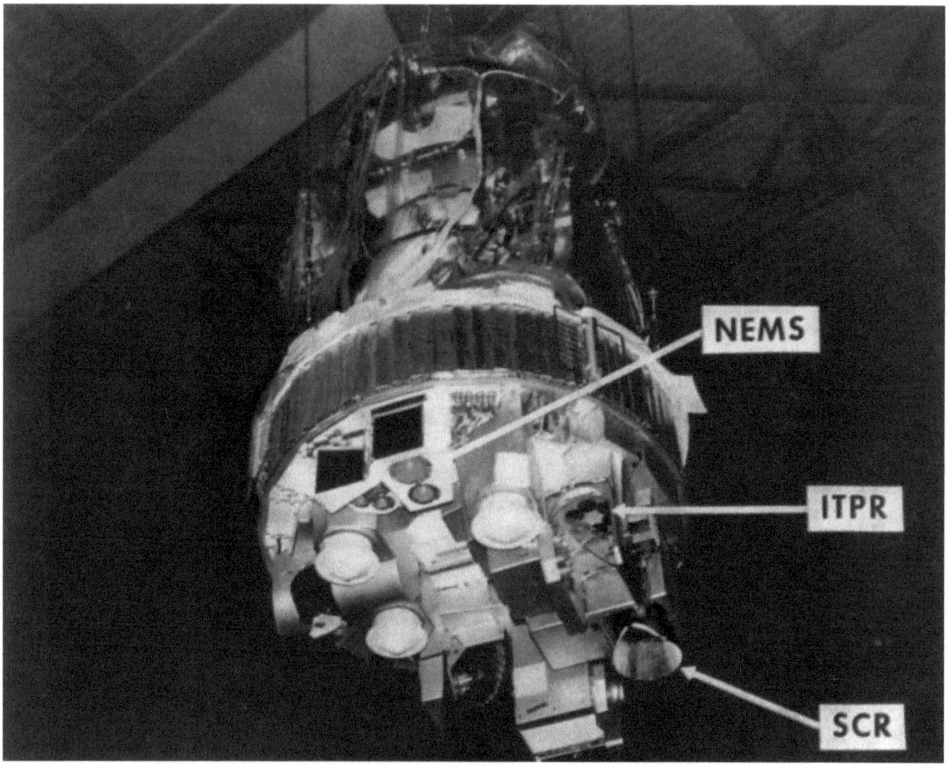


Fig. 3. Photograph of the Nimbus 5 spacecraft showing the ITPR, NEMS and SCR instruments.

Fig. 3 shows the ITPR, NEMS, and SCR instruments aboard the Nimbus 5 spacecraft. The ITPR [1] measures radiation in seven different spectral intervals. Two of the spectral intervals are in the window regions at 3.7 μm and 11 μm for the purpose of detecting clouds and obtaining surface temperatures even when a partial cloud cover exists in the instrument's field-of-view. There are four atmospheric temperature profiling channels in the 15 μm $CO_2$ band (whose weighting functions have been shown in Fig. 2) and a single water vapor channel at 20 μm in the rotational water vapor absorption band. The instrument has a linear spatial resolution of 15 n.mi. and scans spatially in order to circumvent clouds and to obtain uniform earth coverage.

The NEMS instrument [2] is composed of five channels, three of which are in the neighborhood of the 0.5 cm oxygen absorption band and whose measurements are therefore useful for obtaining atmospheric temperature profiles. The instrument has relatively low spatial resolution (100 n.mi.) and does not scan spatially. However, these limitations are offset by its most important characteristic: the radiances it senses are not attenuated significantly by clouds. Thus, this instrument's measurements will yield temperature profiles through and below clouds. Also because it is possible to achieve much higher spectral resolution at microwave frequencies, the weighting functions are relatively sharp, as shown in Fig. 2. This characteristic is particularly helpful in resolving the tropopause.

The SCR [3] is composed of sixteen channels. Four of its channels are in the most strongly absorbing portion of the 15 μm $CO_2$ band and employ $CO_2$ gas cells to separate $CO_2$ absorption line center emissions from contributions from the line wings. When combined, they provide unique measurements, with relatively narrow weighting functions (two of which are shown in Fig. 2), of the temperature of the middle and upper stratosphere.

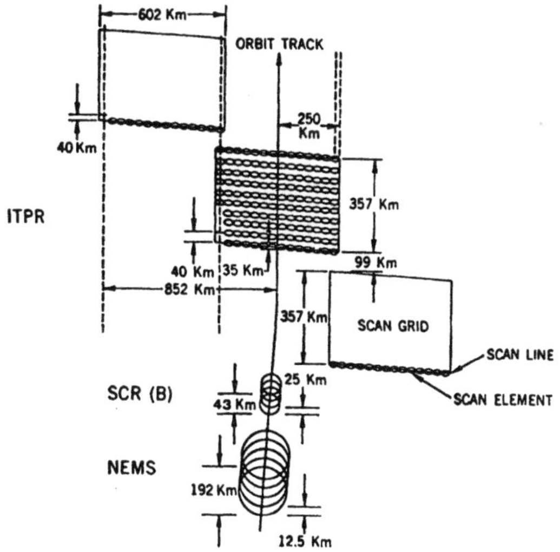


Fig. 4. The earth fields-of-view for the ITPR, NEMS and SCR instruments.

The earth fields-of-view for the ITPR, NEMS and SCR are shown in Fig. 4. As shown, the ITPR cluster samples within three different grids, one to the right of the orbital track, one in the center, and one to the left. NEMS and SCR samples only along the orbital track. The instantaneous resolutions of the ITPR, NEMS and SCR B are 35 km, 192 km, and 43 km, respectively.

### 3.1. The ITPR Experiment

For useful temperature profile determinations from infrared radiation data, it is necessary to detect the existence of any cloud contributions to the observed radiances and to correct for these contributions before attempting to calculate the atmospheric temperature profile. For this purpose, the ITPR experiment employs moderate spatial resolution (15 n.mi.), spatial scanning, and channels for obtaining simultaneous radiance observations in the 3.7 μm and 11 μm windows. Since clouds impose the major obstacle to sounding using infrared measurements, it is worthwhile to explain how the ITPR experiment attempts to alleviate this deficiency.

Since molecular absorption is small in the 3.7 μm and 11 μm window regions, these ITPR window channels sense only the radiation from the earth's surface and any clouds within the instrument's field-of-view. When sensing a uniform and opaque scene (e.g. the earth's surface) they observe the same brightness temperature. However, when sensing a non-uniform or non-opaque scene (e.g. broken clouds within the instrument's field-of-view) they observe different brightness temperatures. This is because of the vastly different radiance dependence upon scene temperature for the two different wavelengths, a phenomenon described by Planck's law. At 11 μm the radiance is approximately proportional to the fifth power of temperature, whereas at 3.7 μm the radiance is approximately proportional to the fifteenth power of temperature. Consequently, when looking at a scene composed of targets of different temperature, the 3.7 μm radiance will have a much larger relative contribution from the higher temperature target than will the 11 μm radiance. As a consequence the 3.7 μm channel senses a higher brightness temperature than the 11 μm channel when the instrument's field-of-view is composed of a non-uniform temperature scene.

Fig. 5 shows false color images created from the ITPR brightness temperature data obtained in two different scan grids. Each picture is generated from observations for the 140 fields-of-view within the instrument's 200 n.mi. × 300 n.mi. scan grid area. The first scan grid (Fig. 5a) was observed over the South Pacific Ocean. The images of both the 11 μm and 3.7 μm window channel data clearly reveal the clouds (colder blue regions) within the scan grid. The cloud features, especially the highest cloud bands, are even more clearly depicted in the pictorial image of the brightness temperature differences for the two window channels. The blue regions correspond to the large negative window channel differences observed in cloudy fields-of-view and the magenta regions to observations of the opaque ocean surface where the window channel brightness temperature difference is near zero. The second grid (Fig. 5b) was observed just two minutes later over the Northeast coast of Australia. In this case both window channels sense the cold (blue) land surface in contrast to the slightly warmer (magenta) ocean surface. It is evident from the window channel difference image that this feature is a surface coastline and not a cloud edge since the window channel differences are small (indicated by the magenta color) and their image is uncorrelated with either of the window channel images. These examples illustrate how simultaneous

→

Fig. 5. Images derived from ITPR window, water vapor, and $CO_2$ channel brightness temperature data for a geographical grid over the South Pacific Ocean (5a) and over the Northeast Coast of Australia (5b) on 12 December 1972.

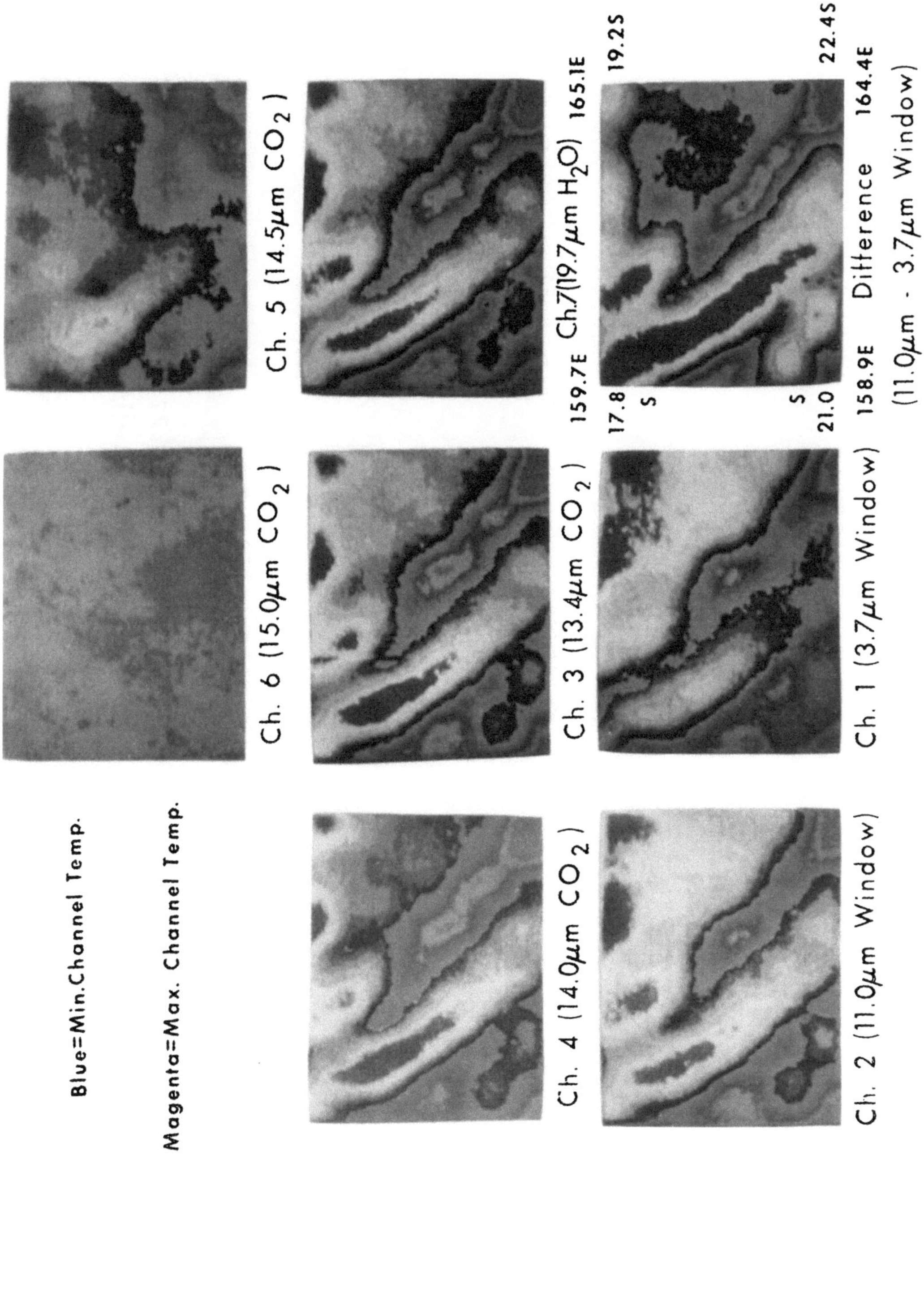
Blue=Min.Channel Temp.
Magenta=Max. Channel Temp.
Ch. 6 (15.0μm CO2)
Ch. 5 (14.5μm CO2)
Ch. 4 (14.0μm CO2)
Ch. 3 (13.4μm CO2)
159.7E Ch.7(19.7μm H2O) 165.1E
17.8 S
19.2S
Ch. 2 (11.0μm Window)
Ch. 1 (3.7μm Window)
21.0 S
22.4S
158.9E Difference 164.4E
(11.0μm - 3.7μm Window)

Fig. 5a

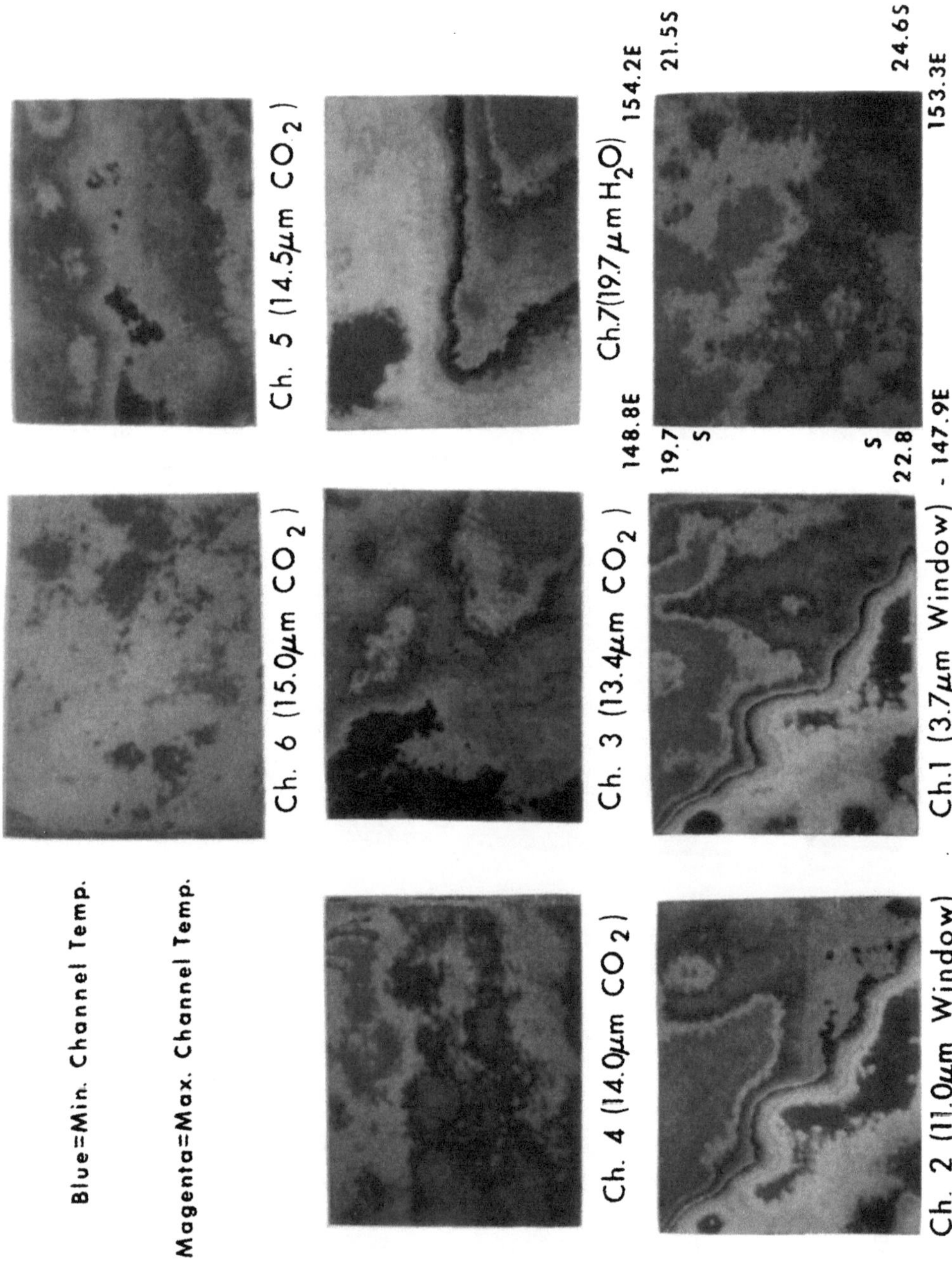
Blue=Min. Channel Temp.
Magenta=Max. Channel Temp.
Ch. 5 (14.5μm $CO_2$)
Ch. 6 (15.0μm $CO_2$)
Ch. 4 (14.0μm $CO_2$)
154.2E
21.5S
Ch.7(19.7μm $H_2O$)
148.8E
19.7 S
24.6S
153.3E
22.8 S
- 147.9E
Ch. 3 (13.4μm $CO_2$)
Ch.1 (3.7μm Window)
Ch. 2 (11.0μm Window)

Fig. 5b.

measurements in the 3.7 μm and 11 μm window channels detect the existence of cloud within the instrument's field-of-view unambiguously from a temperature contrast caused by a change in surface characteristics.

The images shown in Fig. 5 reveal the sensitivity of the ITPR observed radiances to mesoscale features of the surface and atmosphere. Channel 7 water vapor brightness temperature images reveal dry (wet) vertical columns when the temperatures are high (low). Channels 3, 4, 5 and 6 depict variations in the temperature of atmospheric layers centered near the 2-, 6-, 12- and 20-km levels. Of particular interest is the dry zone (magenta colored region) sensed by Channel 7 and shown in Fig. 5b. This is apparently due to strong downward vertical motion associated with an intense high pressure system centered near 24 °S 156 °E. Temperature and water vapor determinations from these radiance data revealed that in the region of sharpest gradient the water vapor concentration tripled in magnitude in a distance of less than 50 miles.

Simultaneous 3.7 μm and 11 μm window observations by the ITPR are used to specify surface temperature even when partial cloudiness exists within the instrument's field-of-view. This is possible using a set of geographically independent observations in which the radiance variation from one element to another is due primarily to clouds rather than surface temperature. Variations in cloud cover produce linear variations of the radiance in one window channel with respect to the other and this characteristic is utilized to extrapolate the observations to their corresponding values for the cloudless sky condition. Once the clear column window radiance is defined, clear column radiance values for the sounding channels are defined from their linear radiance relations to the

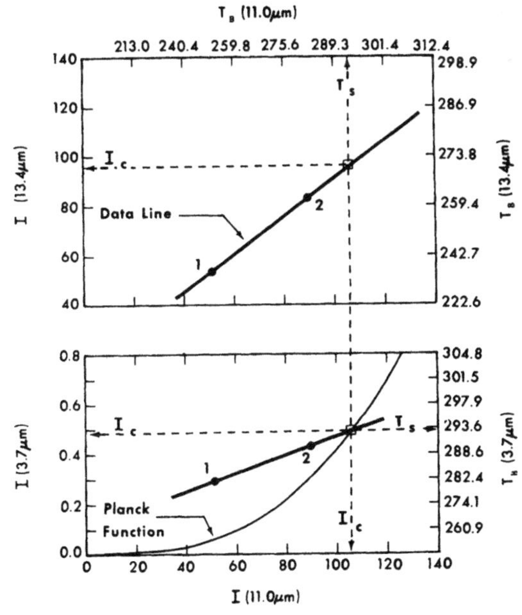


Fig. 6. Plot of ITPR 3.7 μm, 11 μm and 13.4 μm radiance observations in a partly cloudy atmosphere. Also shown is the Planck function which relates the 3.7 μm to the 11 μm radiances for an opaque surface.

window channel radiances. This technique is demonstrated graphically in Fig. 6 which shows a plot of 3.7 μm, 11 μm and 13.4 μm radiance observations in two adjacent scan elements, both of which possess some broken clouds but of different amounts. Also shown is the Planck function which relates the 3.7 μm to the 11 μm radiances for an opaque surface (i.e. a curve of equal 3.7 μm and 11 μm brightness temperature). Linear extrapolations of the observed values to the zero cloud amount condition, which is the intersection of the observation line with the Planck function curve, yields the window radiances which would have been observed if no clouds had existed within the instrument's field-of-view. The brightness temperature corresponding to this radiance is the surface skin temperature (in this case 293.5 °K). Once the clear column radiance for the window regions have been determined, the clear column radiance for any sounding channel (e.g. the 13.4 μm $CO_2$ channel shown) can be specified. In the ITPR data processing, observations from groups of 16 geographically independent and contiguous ITPR fields-of-view are used to define "clear-column" radiances. This enables temperature profiles down to the earth's surface to be obtained under most cloud conditions with a geographical resolution of about 75 n.mi.

The vertical temperature profiles are obtained from Nimbus 5 radiance data using the "Minimum Information" inverse solution of the radiative transfer equation. The bias vector (guess profile) used in this solution is obtained from a regression model whose coefficients were derived from a historical sample of 1200 radiosonde-rocketsonde profiles and corresponding theoretically calculated radiances. Since the routine data processing does not use contemporary conventional data, determinations in remote areas of the world are of the same quality as determinations in conventional data rich areas.

Shown in Fig. 7 is a vertical cross section of the atmosphere's temperature distribution calculated from the ITPR radiance data taken along the orbital track of Nimbus 5. (Also shown are the cloud heights and amounts, total water content, and outgoing long wave flux which are also deduced from the ITPR data.) As demonstrated, the temperature distribution down to the earth's surface could be specified from infrared data even under relatively severe cloud conditions. The latitudinal temperature distribution obtained is typical for this time of the year. The tropical regions possess a relatively warm lower troposphere and cold lower stratosphere. The polar regions have a moderately cold lower troposphere with a nearly isothermal stratosphere. A region of relatively warm air in the lower stratosphere exists in the temperate latitudes of the southern hemisphere.

Fig. 8 shows how the ITPR data can be used to improve the specification of the atmospheric temperature distribution in the vicinity of intense weather systems. Shown are the ITPR derived 500-mb temperatures as the Nimbus 5 passed over an intense low pressure system over the Northern Pacific. (The symbol M denotes missing data due to extensive cloudiness above the 500-mb level.) It is worthy to note that the spatially random error of the deduced 500-mb temperatures appears to be of the order of a few tenths of a degree. As shown, the adjustments required to fit the NMC (National Meteorological Center) operational objective analysis to the satellite data in this area where conventional data are sparse are substantial, especially between 35° and 50 °N. These adjustments result in improvements in the specification of the wind field in this area which could lead to an improved weather forecast.

Fig. 9 shows a similar example, but over Northern Japan where there are numerous radiosonde observations which verify the ITPR measurements. As can be

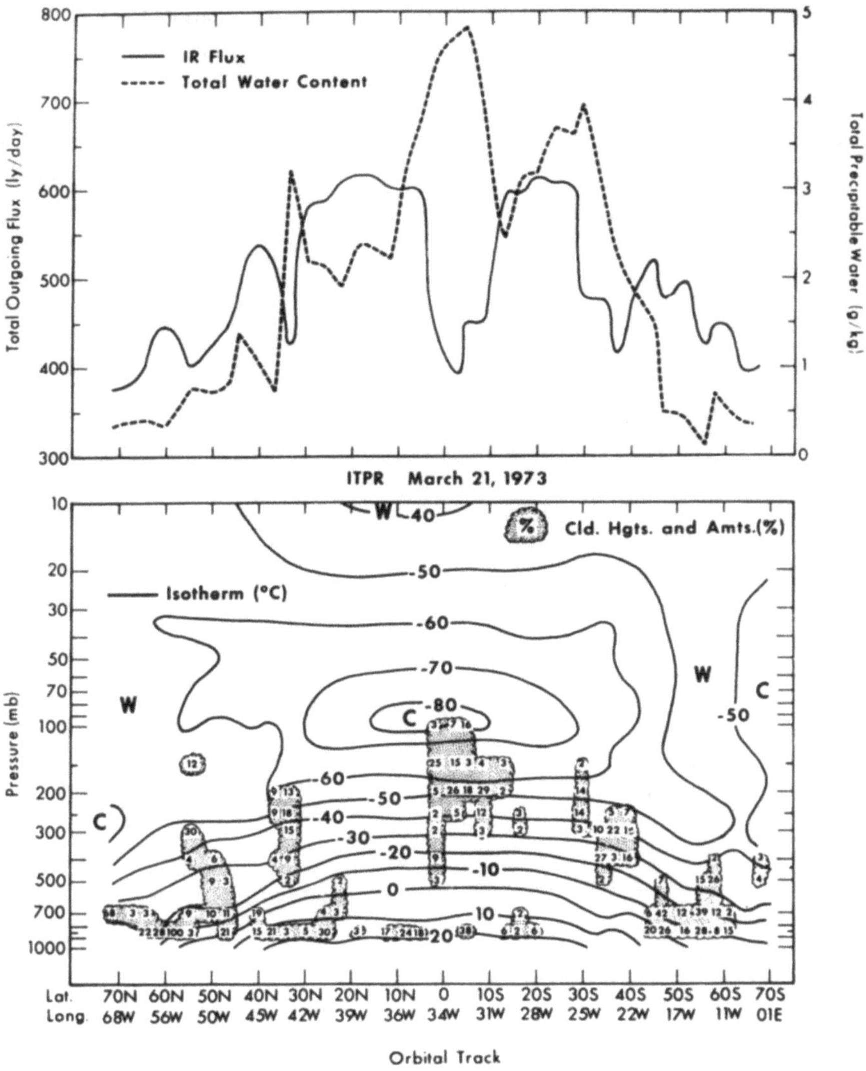


Fig. 7. Vertical cross section of the atmospheric temperature distribution, cloud heights and amounts, total outgoing long wave flux, and vertically integrated water vapor content, deduced from ITPR radiance data along an orbital track on 21 March 1973.

seen, the ITPR deduced temperatures as well as the radiosonde temperatures reveal a much more intense temperature gradient than the NMC objective analysis. In fact, the hand analysis of the ITPR and radiosonde data implies maximum thermal wind speeds (which are proportional to the horizontal temperature gradient) and hence jet core winds which are twice the magnitude indicated by the operational analysis. It is significant to note that these small scale features

**NIMBUS 5 ITPR 500mb Temperature Observations (°C)**
**March 6, 1973 (1311 - 1321 GMT) Over Pacific**

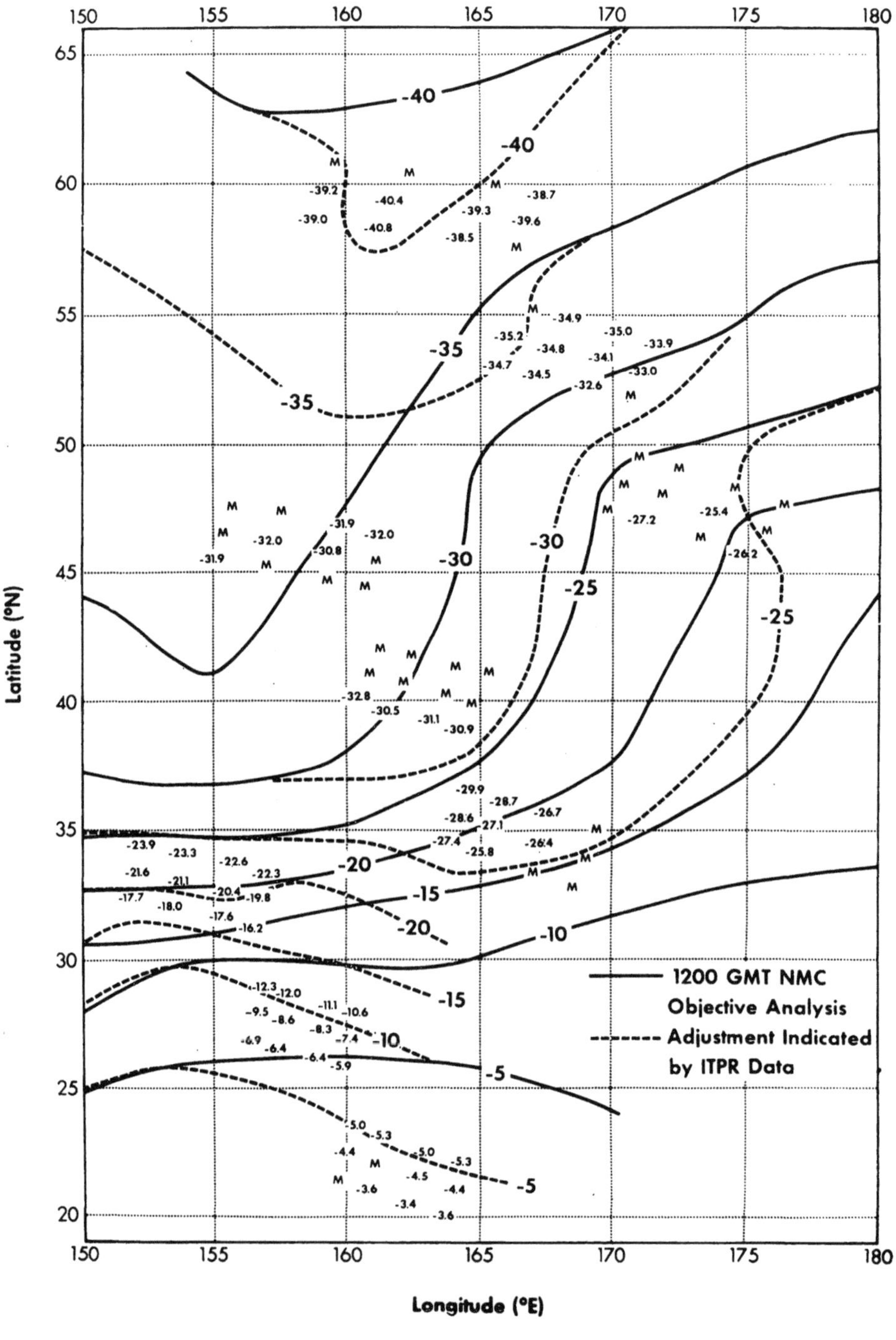

were accurately diagnosed from the ITPR radiance data even though more than half the area was covered by cloud, most of which was above the 500-mb level.

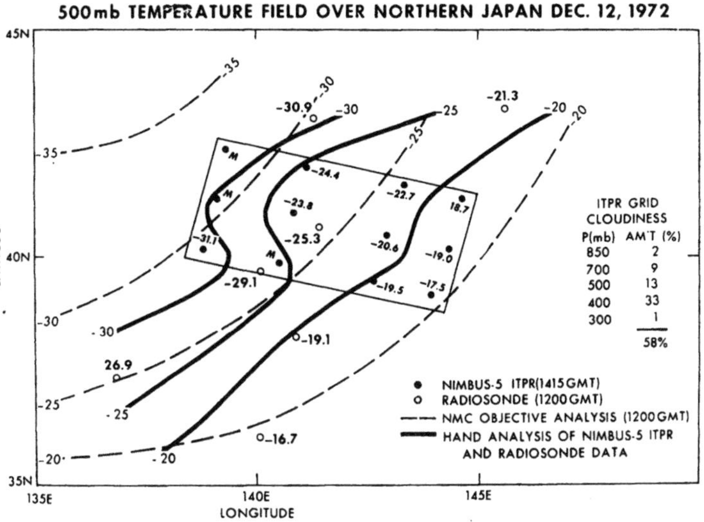


Fig. 9. ITPR derived 500-mb temperatures (solid circle) and radiosonde data (open circle) over Northern Japan on 12 December 1972. The dashed lines are the 1200 GMT NMC objective analysis and the solid lines are an analysis of the ITPR and radiosonde data.

### 3.2. The NEMS Experiment

The NEMS experiment was conducted to show that sounding through clouds can be accomplished using microwave sensors. The absorption by clouds and aerosols in the 0.5 cm oxygen band is known to be orders of magnitude less than that in the infrared. Theoretical calculations show that most ice clouds are completely transparent to 0.5 cm radiation and liquid water clouds are only partially absorbent, the opacity depending on the total liquid water content. Since liquid water clouds are usually confined to the lower atmosphere, the effects of clouds on the microwave soundings would be expected to be negligible, unlike the case with infrared soundings.

Fig. 10 shows several comparisons of soundings obtained from the NEMS microwave radiances with those obtained from ITPR infrared observations and by conventional radiosondes. These cases were selected because the cloudinesss observed (by the ITPR) was extensive, probably of the convective type, and penetrated to high levels, thereby being representative of somewhat complex

←

Fig. 8. ITPR derived 500-mb temperatures (12 per grid unless missing) and analyses over a North Pacific low pressure system on 6 March 1973, 1311—1321 GMT. The solid lines are the 1200 GMT NMC objective analysis and the dashed lines denote the adjustment required to fit the ITPR data.

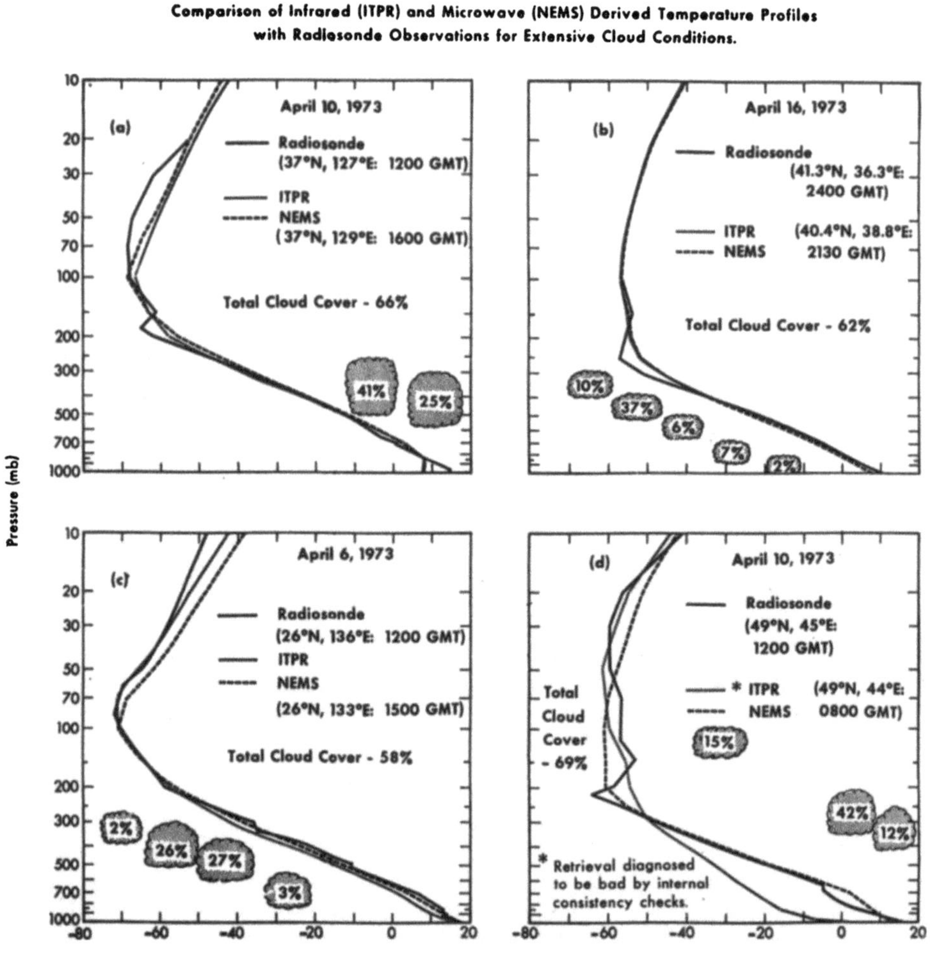


Fig. 10. Comparison between Nimbus 5 temperature soundings derived from NEMS (microwave) and ITPR (infrared) observations with conventional radiosonde observations during April 1973.

cloud conditions. Examples are shown in Figs. 10a and b of the general case where the microwave and infrared soundings both show good correspondence with the conventional radiosonde observation even though extensive cloudiness penetrated into the upper troposphere. The discrepancies between the satellite soundings and the conventional radiosonde near the earth's surface are probably due to the time differences between the two observations, the satellite observations being taken at midnight and the radiosonde observations being taken four hours earlier. Figs. 10c and d show similar cloud situations but in these cases the infrared soundings are erroneous below cloud level whereas the microwave soundings showed good agreement with the conventional radiosonde observations. In fact, the errors incurred in the calculation of the clear column infrared radiances for case d were drastic enough to cause a ridiculous result from the highly error sensitive algorithm for radiance to temperature inversion. In the data processing such results are usually diagnosed to be erroneous from objective and internal checks on temperature lapse rate and the disagreement of the observed and calculated radiances. This, however, is not the case for the more subtle but relatively large (about 3 °C) bias error in the infrared sounding shown in Fig. 10c.

Fig. 11 shows the effect clouds associated with a tropical storm have on the NEMS microwave radiances and the soundings obtained from them. Shown at the top of this figure are the brightness temperatures observed by the NEMS 0.5 cm (50 GHz) oxygen channels and ITPR 15 μm $CO_2$ channels as the Nimbus 5 satellite passed over an intense tropical storm in the South Pacific at 9 °S and 1545 °W. The maximum attenuation due to the storm is only 2.5 °C for the microwave

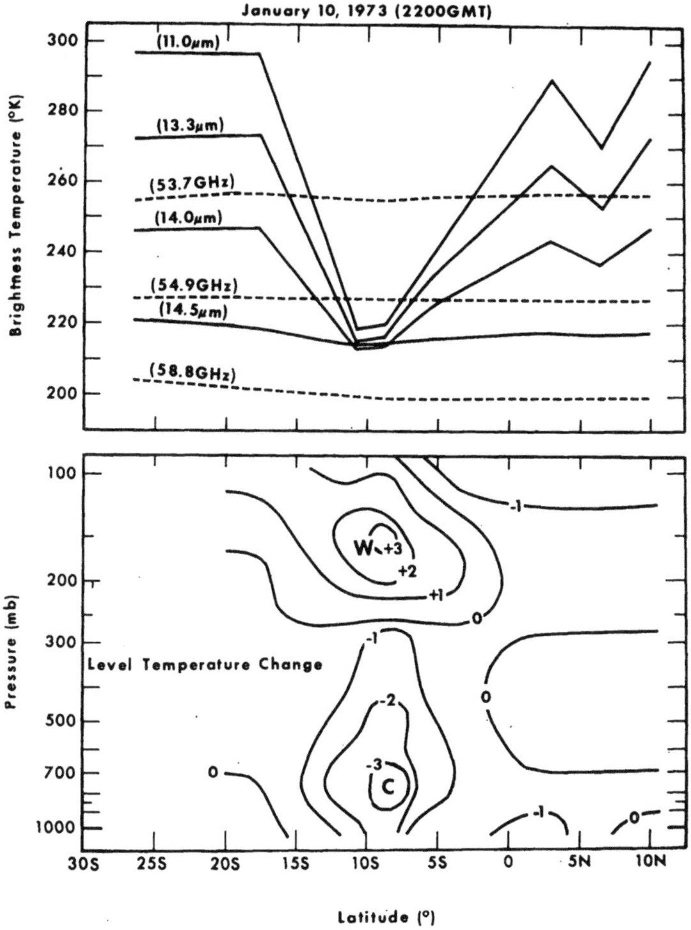


Fig. 11. Brightness temperatures observed by NEMS (microwave) and ITPR (infrared) channels over a tropical storm in the South Pacific on 10 January 1973, 2200 GMT (top); level mean difference between temperatures within the storm and those outside the storm's boundary (bottom).

oxygen radiances whereas it is as much as 55 °C for the infrared carbon dioxide radiances. As a result reasonable temperature profiles could be obtained, from microwave radiances, through the clouds down to the earth's surface, even at the center of the storm. Shown in the bottom portion of Fig. 11 is the difference, at each level, between the atmospheric temperatures obtained within the storm and those obtained in the cloudless air at the storm's boundary. Assuming that no actual variation in free air temperature exists, the largest error incurred by this worst case cloud condition is only 3.5 °C. This latter example indicates that microwave instruments have almost complete cloud penetration capability under less drastic cloud conditions.

### 3.3. The SCR Experiment

As stated earlier, the SCR experiment has four channels at the center of the 15 μm $CO_2$ band which employ gas cells in the optical systems, each cell containing a different amount of $CO_2$. The $CO_2$ in the gas cells absorbs the radiation from the atmosphere near absorption line centers, the degree of which depends upon the amount of $CO_2$ in the cell. The upper layers of the atmosphere are observed by extracting the radiation from near the line centers. This is done by differencing the incoming radiation through a cell containing considerable $CO_2$ with that coming through a cell containing a smaller amount of $CO_2$. By this means unique observations of the temperature of the middle and upper stratosphere are achieved.

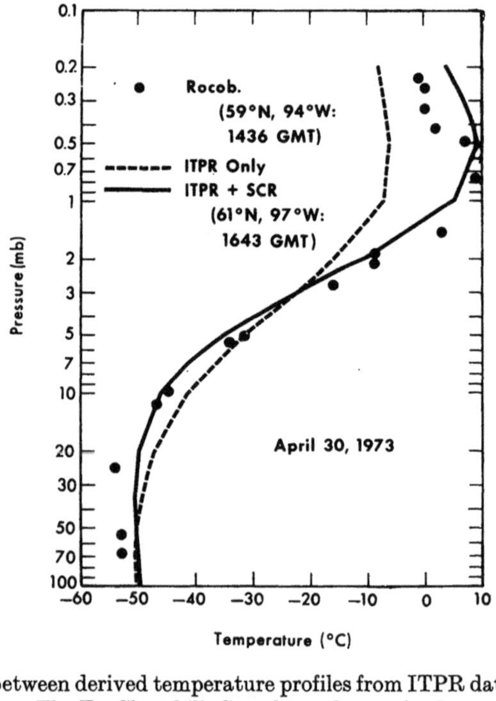


Fig. 12. Comparison between derived temperature profiles from ITPR data alone and combined with SCR radiance data. The Ft. Churchill, Canada, rocketsonde observation is given to verify the profiles.

Fig. 12 shows an example of temperature profiles obtained (i) from SCR upper channel radiance data combined with ITPR radiance data, (ii) using only ITPR radiance data and (iii) by a conventional rocketsonde observation. As illustrated, the SCR upper channel radiance data greatly improve the temperature profile specification in the middle and upper stratosphere.

## 4. The Amalgamation of Nimbus 5 Sounding Data

As can be surmised from the previous discussion, the infrared and microwave radiometers on Nimbus 5 provide highly complementary observations. As a result, one would expect to obtain better temperature profile results using the combined set of infrared and microwave observations than with either set individually. The NESS (National Environmental Satellite Service) with the financial support of NASA and with the technical cooperation of the NEMS and SCR experimenters at the Massachusetts Institute of Technology and Oxford University respectively has developed algorithms for obtaining temperature profiles on a global basis using infrared and microwave radiances individually and in combination. This data processing is done at NESS in order to provide global

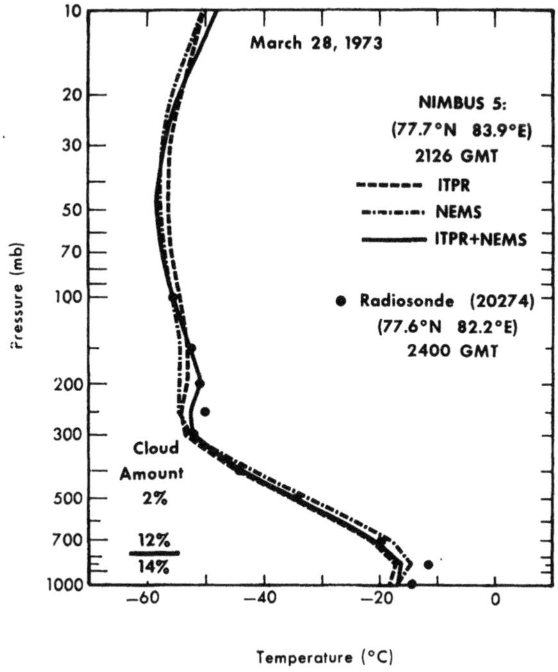


Fig. 13. Comparison between temperature profiles derived from ITPR data alone, NEMS data alone, and ITPR and NEMS data combined. A conventional radiosonde observation is given to verify the results.

meteorological data sets to support various GARP (Global Atmospheric Research Program) and other research programs. Table 1 shows the results of a statistical analysis of ITPR and NEMS brightness temperature data and radiosonde observations of free air temperature in the troposphere and lower stratosphere. As can be seen from the individual channel brightness temperature correlations with radiosonde free air temperature observations, the ITPR and NEMS radiance channels are very complementary. The multiple correlation coefficients and standard errors indicate that significantly better profile results could be achieved from the combined set of measurements than can be achieved by either set used individually. This appears to be most dramatically evident in the tropopause region (i.e. the 100—300 mb region). This characteristic is illustrated in Fig. 13 which shows an example comparison between soundings obtained from ITPR and

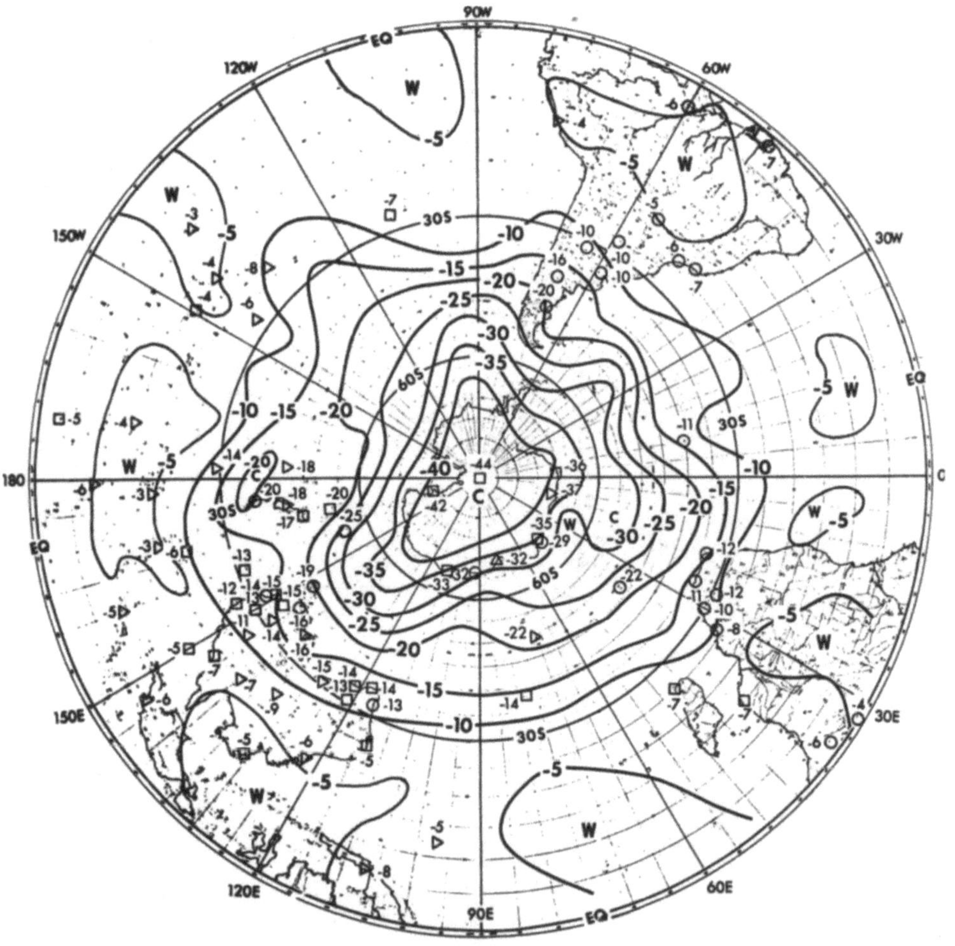


Fig. 14. Southern hemisphere 500-mb temperature analysis based solely on Nimbus 5 radiance data on 5 April 1973. Available radiosonde observations are shown to verify the Nimbus 5 data. ▷, 0000 GMT; ○, 1200 GMT; □, 2400 GMT.

Table 1

Statistical Relations between ITPR and NEMS Brightness Temperature Measurements and Radiosonde Air Temperature Observations (All Latitudes, March—April 1973)

| Channel | Individual Correlation Coefficients | | | | | | | Multiple Correlations ($r$) | | | | Standard Error (SE) | | |
|---|---|---|---|---|---|---|---|---|---|---|---|---|---|---|
| | 13.4 μm | 53.7 GHz | 14.0 μm | 54.9 GHz | 14.5μm | 58.8 GHz | 15 μm | ITPR | | NEMS | | ITPR and NEMS | | |
| Pressure Level (mb) | | | | | | | | $r$ | SE (°K) | $r$ | SE (°K) | $r$ | SE (°K) | SD* (°K) |
| 1000 | 0.99 | 0.95 | 0.94 | 0.34 | −0.47 | −0.86 | 0.63 | 0.99 | 2.3 | 0.98 | 4.0 | 0.996 | 1.9 | 20.2 |
| 850 | 0.94 | 0.98 | 0.96 | 0.45 | −0.55 | −0.91 | 0.51 | 0.98 | 2.5 | 0.98 | 2.2 | 0.984 | 2.1 | 12.1 |
| 700 | 0.93 | 0.98 | 0.97 | 0.48 | −0.54 | −0.90 | 0.50 | 0.99 | 2.1 | 0.99 | 2.1 | 0.991 | 1.6 | 12.0 |
| 500 | 0.91 | 0.98 | 0.97 | 0.55 | −0.52 | −0.89 | 0.50 | 0.98 | 2.2 | 0.98 | 2.3 | 0.987 | 1.9 | 11.7 |
| 400 | 0.86 | 0.97 | 0.96 | 0.62 | −0.50 | −0.87 | 0.47 | 0.98 | 2.3 | 0.97 | 2.5 | 0.983 | 2.0 | 10.9 |
| 300 | 0.71 | 0.86 | 0.87 | 0.79 | −0.36 | −0.78 | 0.42 | 0.95 | 2.5 | 0.96 | 2.3 | 0.965 | 2.1 | 7.9 |
| 250 | 0.08 | 0.28 | 0.32 | 0.77 | −0.02 | −0.27 | −0.02 | 0.81 | 3.8 | 0.88 | 3.1 | 0.908 | 2.7 | 6.4 |
| 200 | −0.59 | −0.46 | −0.42 | 0.30 | +0.39 | +0.41 | −0.31 | 0.83 | 3.3 | 0.85 | 3.1 | 0.903 | 2.6 | 6.0 |
| 150 | −0.84 | −0.85 | −0.84 | −0.22 | +0.54 | +0.85 | −0.45 | 0.89 | 2.8 | 0.91 | 2.5 | 0.925 | 2.3 | 6.0 |
| 100 | −0.83 | −0.91 | −0.90 | −0.42 | +0.57 | +0.93 | −0.43 | 0.95 | 2.8 | 0.94 | 3.0 | 0.967 | 2.3 | 8.9 |
| 70 | −0.81 | −0.90 | −0.87 | −0.44 | +0.61 | +0.93 | −0.38 | 0.95 | 2.5 | 0.94 | 2.6 | 0.957 | 2.3 | 7.8 |
| 50 | −0.78 | −0.83 | −0.79 | −0.27 | +0.74 | +0.93 | −0.26 | 0.93 | 2.0 | 0.93 | 2.0 | 0.940 | 1.9 | 5.5 |
| 30 | −0.15 | −0.10 | −0.04 | +0.26 | +0.57 | +0.30 | +0.27 | 0.67 | 2.8 | 0.51 | 3.2 | 0.686 | 2.7 | 3.7 |

* SD = standard deviation of radiosonde.

NEMS independently and in combination and a conventional radiosonde observation.

Finally, Fig. 14 shows an analysis of the southern hemisphere 500-mb temperature distribution obtained solely from the amalgamated Nimbus 5 radiance data over a 24-hour period on 5 April 1973. Available radiosonde observations (in the southern hemisphere) are given to validate the Nimbus 5 data. The agreement between the satellite derived temperature distribution and the radiosonde observations is generally within 2 °C. The significance of this result (and those for other levels not shown) is that it indicates that the temperature distribution of the atmosphere can be diagnosed accurately from the proper satellite radiation data without the aid of conventional observations.

## 5. Future Prospects

The advances made over the past few years in vertical temperature profiling from satellites have been great. The Nimbus 5 ITPR experiment results have shown that 15-n.mi. resolution, spatial scanning, and simultaneous 3.7 μm and 11 μm window observations enable temperature profiles to be obtained down to the earth's surface with infrared measurements for broken cloud conditions. The NEMS microwave experiment has proven the capability of sounding through even extensive overcast clouds. Since the microwave approach is limited near the earth's surface due to the variability of surface emissivity in this spectral region, optimum results are expected to be obtained from the utilization of measurements in both spectral regions. Advances, to be realized over the next few years, should even further improve our temperature profiling capability. For example, the use of radiation measurements in the 4.3 μm $CO_2$ band, which will be provided for the first time from the Nimbus F satellite next year, is expected to improve greatly our present vertical resolution capability in the middle and lower troposphere. Also, the effect of clouds on 4.3 μm radiation will be much less than it is on 15 μm radiation. Horizon limb scanning techniques, also to be tested aboard Nimbus F, are expected to improve greatly the vertical resolution of the temperature distribution of the stratosphere. After the successful testing of these improvements, we should be able to implement a global remote satellite temperature sounding system which will provide accurate observations with high spatial resolution and with unprecedented vertical and horizontal coverage. Linked with an onboard computer, this system could provide processed temperature data in real-time to local weather forecast facilities as well as to National Meteorological Centers. Such an advanced temperature profile observing system is a realistic prospect for the not too distant future.

## Acknowledgments

The NEMS and SCR experiments were developed under the direction of Professor David Staelin of the Massachusetts Institute of Technology and Dr. John Houghton of Oxford University respectively. We wish to acknowledge the cooperation of these experiment teams with the ITPR team at the NESS in the joint processing of the ITPR, NEMS, and SCR data. Messrs. H. Woolf and M. Chalfant of the NESS developed the computer software for processing the Nimbus 5 sounder data in their combined form. Drs. C. Hayden, P. Abel and N. Grody and Messrs. H. Howell, F. Nagel and J. Yotka have also participated in various aspects of the data reduction. Messrs. P. Pellegrino, L. Mannello, R. Ryan and C. Jacobson assisted in the analyses of the numerous results obtained.

## References

[1] W. L. Smith et al., The Nimbus 5 User's Guide, Goddard Space Flight Center, Greenbelt, Md, 1972 (p. 107).

[2] D. H. Staelin, F. T. Barath, J. C. Blinn III and E. J. Johnston, The Nimbus 5 User's Guide, Goddard Space Flight Center, Greenbelt, Md, 1972 (p. 141).

[3] J. T. Houghton and S. D. Smith, The Nimbus 5 User's Guide, Goddard Space Flight Center, Greenbelt, Md, 1972 (p. 131).

Earth Survey Problems — Akademie-Verlag, Berlin 1974

# BEYOND FGGE

R. W. STEWART

Environment Canada, Victoria, B.C., Canada

The nature of the Global Atmospheric Research Program after the 1977—1978 Global Experiment is of course not yet entirely clear, but some outlines are beginning to emerge. It appears that much of the emphasis will be on prediction of the behaviour of the atmosphere over periods from a season to a year or longer. Such prediction is very unlikely to be of weather. Rather it will be of some statistical properties of weather. It is almost certain that a dominant influence — perhaps the dominant influence — will lie in the upper layers of the ocean. Observing systems will be required capable of describing the temperature structure and perhaps the velocity field of the upper layers, so that predictions can be made of their future evolution.

In addition, the results of the analysis of data obtained during the Global Experiment should have very important consequences on the observational systems used routinely for weather forecasting. For example, if the Global Experiment confirms the present idea that it is essential that many levels of wind be determined in the tropics, then some operational observing system will be needed to provide these data.

The First GARP Global Experiment is now planned to commence in 1977 and to run into 1978. Decisions concerning the data gathering systems and data handling techniques to be used are becoming steadily firmer and it is increasingly possible to sketch in the details of the configuration of the experiment. Is it now time to say "where do we go from there?" To almost anyone except COSPAR one would be inclined to say "let us see what the results of at least the preliminary analyses are before deciding too firmly where we go from there". However, the lead time in the space business is so long that members of COSPAR would not easily accept such an attitude. Probably we have to try to read the future as best we can.

*My* crystal ball is clouded. I may be misinterpreting some of the images which I think I see there. And there are undoubtedly things there which I don't see at all. Nevertheless, I think, I and others like me must make the attempt.

I suppose the first question to be asked is whether after FGGE there will be any GARP. Will there be a second, third, fourth, etc., GARP Global Experiment? Of course nothing is certain in this uncertain world, but neither of the founding bodies for the Joint Organizing Committee for GARP — ICSU and WMO — has ever indicated any desire to see a cut-off date for the program. Further, the notion that much of atmospheric research should be internationally coordinated and that this coordination should include both the academic type bodies affiliated with ICSU and the operational type bodies joined together in WMO makes rational sense. There may, and probably will, be some restructuring and reorganizing but I can't help but believe that something like the Joint Organizing Committee for something like the Global Atmospheric Research Program will be with us for a long time to come.

However, to say that there will still be GARP is not to say that there will be a succession of global experiments. Certainly following FGGE there will be a substantial period of time when a major effort will have to be applied to interpretation of the experiments now underway or planned: AMTEX, GATE, CAENEX, MONEX and the Global Experiment. Much of the energy now being applied to planning will be directed towards interpretation.

Put this way, there seems to be little role identified for the sort of people who make up COSPAR. However, to put it that way is misleadingly incomplete. Out of this analysis will come results, and these results will certainly have most serious implications for members of COSPAR. Let me guess a few possible results. I don't suggest that all or any of them will in fact be reached but I think they do not misrepresent the *kind* of thing we may expect.

(1) Present belief will be vindicated that detailed vertical structure of tropical winds is required to determine tropical dynamics, and that the determination of tropical dynamics will permit importantly useful extensions in the validity and range of forecasts, both tropical and extratropical. This proved true, it will be necessary to discover some *operational* technique for measuring this wind structure. Here is a real challenge to ingenuity which I believe has not yet been met.

(2) Satellite derived temperature structures prove enormously valuable, but there are strong indications that great rewards would result from an improvement in the detail with which this structure is described and an improvement of a factor of four or so in the accuracy of the temperature determination. Here again it seems that new thinking and new ideas would be needed.

(3) Superbly accurate one to two day forecasts may be possible if sufficient data are fed into large enough computers soon enough. This may put great demands on ingenuity for data handling procedures.

(4) Useful forecasts for a week or longer are possible, but depend upon much more accurate knowledge of aerosol distribution and type than we now have in order to get the radiation effects right. Again we will need new ideas and new techniques.

In addition to this sort of thing, I think we can seriously hope that from the analyses of the Global Experiment data it will be much more possible to reach reasonable estimates of the cost benefit figures for various mixes of observational systems. Hopefully such information will importantly influence the future development of the World Weather Watch.

All this is associated with what is now usually referred to as the First GARP Objective—that of improving weather forecasts for periods from one day to several weeks. However, very much of the attention of those concerned with GARP has in the last year or so been turning toward the second GARP Objective— that of attempting to understand climate and climatic fluctuations [1, 2].

COSPAR Working Group 6 was one of the first groups to recognize the increased attention which this problem demands and has already issued an important document [3] directed to it. In some people's minds JOC-GARP has been belated in its response, but is now moving rapidly in this direction and is planning a major study conference for 1974 which will concentrate on climate [2].

Much of climate research has to do with the past, and of course there is a very great deal to be learned from the evidences of past climatic changes. However, with respect to climate, the past of space research is extremely short. We must look to the present and the future. Over periods of tens of millennia, climatic fluctuations are dramatic and unsubtle. Mean global temperature changes by 10 degrees or so and there would be no difficulty whatever in measuring such changes with accuracy. However, a drop in the temperature of that amount resulted in the complete obliteration under ice of the area which is now my country of Canada. Clearly we are importantly concerned with very much smaller changes which can nevertheless produce quite dramatic effects. In the last century or so, during which reasonably reliable measurements have been available, global temperatures seem to have fluctuated by rather less than a degree. Yet there have been important changes in glaciation. Major fisheries have come and gone. Navigation near the borders of ice infested water has had to, or been able to, make important adjustments. To measure such changes, which are presumably those we are likely to face in the next few years, demands a really challenging accuracy from us.

Considering the nature of some of the parameters which go into it, in particular albedo from clouds and from snow cover, I think one has to argue that the earth's climatic history over the last few centuries indicates almost incredible stability. It seems that the earth's global mean annual temperature has fluctuated by a rather small fraction of 1% about its value of two hundred and fifty odd degrees Kelvin. Clearly there are some very strong negative feedback effects acting. We are going to need observational programs to look hard at these negative feedback effects. Climate is global, and nothing can match space-based systems for global observations. However, space-based systems necessarily are going to deal with those small fluctuations we have now, not the large ones of the ten millennia time scale. It seems to me that we are going to need some very accurate measurements of both cloud albedo and surface albedo, and we will have to be able to distinguish between them. We will also have to get a far better handle on what is going on with respect to ozone and its distribution in time and space than we now have. We have to do much better with aerosols than we are now able to do.

To my mind a large fraction of this effort should be aimed at trying to understand those small year-to-year fluctuations that we are actually dealing with, with the hope of elucidating the negative feedback mechanisms and ultimately to see whether and how these may be over-ridden by positive feedback mechanisms important enough to produce an ice age.

In all this I have not mentioned the ocean, although oceanography is my own field, and its importance for climate may be comparable with that of the atmosphere. I think we are going to need global sea surface temperature, and perhaps some other quantities, determined sufficiently frequently that we can make a serious attempt at trying to understand what it is that determines sea surface temperature well enough to predict its behaviour. Global sea surface temperature with an accuracy approaching half a degree is high on my priority list of what one would like to get from space-based systems.

It may seem superfluous to say so, but I will risk the banality of pointing out that this study of the earth's climate is of far more than academic interest. One can see from such studies as SCEP [4] and SMIC [5] that the possibility of human influence upon climate is not taken lightly. It may well be that the results of studies of climate will become important factors in major decisions which will be made in the next years concerning many of the largest scale and most important of man's activities. In addition, recent work [6] indicates that nature herself *can* (although she has not recently) produce very large climatic changes in quite a short time. There is some evidence of dramatic climatic events occurring within the space of around a hundred years. There is also some evidence that a serious deterioration in our climate is perhaps overdue, in that if we regard the present period as an inter-glacial, it has lasted rather longer than average [6]. It could be that the results of climate study, rather than being used to prevent inadvertent man-made changes in climate, may be employed in a desperate attempt to prevent a natural change in climate. More prosaically, and probably nearer to us in time, there is a real hope that our understanding of these climatic processes may enable us to make useful predictions of short term (seasonal) or annual fluctuations in weather statistics. These would be of immense value to many of man's activities.

In sum, I can foresee much that is exciting in the future of those concerned with understanding the global atmosphere, and I can see much that will be demanded of the space community in obtaining the observations necessary to raise this excitement above the level of speculation, to the level where it is important to the affairs of man.

## References

[1] Global Atmospheric Research Program, 1972. Report of the Seventh Session of the Joint Organizing Committee (available from the World Meteorological Organization, Geneva).
[2] Global Atmospheric Research Program, 1973. Report of the Eighth Session of the Joint Organizing Committee (available from the World Meteorological Organization, Geneva).
[3] COSPAR Working Group 6 for the United Nations and Scientific Committee on Problems of the Environment (SCOPE), 1972. Report on the Application of Space Techniques to Some Environmental Problems. Preliminary Observing System Considerations for Monitoring Important Climate Parameters. 87 pp.
[4] Massachusetts Institute of Technology, 1970. Man's Impact on the Global Environment. Assessment and Recommendations for Action. Report of the Study of Critical Environmental Problems (SCEP). 319 pp.
[5] Massachusetts Institute of Technology, 1971. Inadvertent Climate Modification. Report of the Study of Man's Impact on Climate (SMIC). 308 pp.
[6] G. J. Kukla, and R. K. Mathews, Science **178**, 190 (1972).

Earth Survey Problems — Akademie-Verlag, Berlin 1974

# EARTH SATELLITE MEASUREMENTS AS APPLIED TO SEA ICE PROBLEMS

E. P. McClain

National Environmental Satellite Service, NOAA, Washington, D.C., USA

Data from NOAA operational satellites, as well as from NASA research satellites such as Nimbus and the new Earth Resources Technology Satellite (ERTS), are finding increasing operational and research use for sea ice surveillance and study. Methods are being developed to improve the mapping and monitoring of ice pack concentration, character, and condition from satellite observations in the visible, near-infrared, and thermal infrared parts of the spectrum. Recently acquired NOAA and ERTS measurements are higher in spectral and spatial resolution than previously available, and the newest Nimbus carries the first passive microwave imager in space. Examples of some of these newly available data and their applications are presented, and a brief discussion of future sensor systems expected to be of interest to sea ice researchers given.

## 1. Sea Ice Problems, with emphasis on the Antarctic

Although knowledge of arctic pack ice is incomplete, even less is known about the ice cover of the Southern, or Antarctic, Ocean. Arctic sea ice has been far more accessible logistically, but extrapolation to the Antarctic from results of studies obtained in the Arctic is of limited usefulness, if not actually risky, because of the significantly different geophysical contexts of the two areas.

Fletcher [1] has stated that about 10% of the ocean area in the northern hemisphere is covered by ice, whereas in the southern hemisphere the fraction is about 13% of a larger ocean area. At its maximum extent, it has been estimated that the area of the antarctic pack is approximately 1.7 times that of the arctic pack. More importantly, the seasonal variation in the Southern Ocean is much greater than that in the Arctic Ocean: 75—85% of the maximum extent compared with 20—25%. It is known, furthermore, that most antarctic sea ice is first-year ice, which contrasts with the large areas of thicker, multi-year ice in the Arctic. Although large variations in ice extent are known to take place, relatively long-term observations have been taken at only a few isolated locations.

The importance to weather and climate, even global climate, of the behavior of the sea ice cover in the Arctic and the Antarctic has been receiving increasing attention in recent years as technological and scientific progress have made it possible to investigate these difficult problems effectively. Untersteiner and Maykut [2] have stated that the surface albedo and its variations, which are almost totally a function of the presence or absence of ice and the type or condition of the ice, are probably the most important regional factors affecting the heat and mass budgets of the Arctic Basin in summer. Similar statements have been made about the Southern Ocean by Fletcher [1] and others. Ice is the major factor in polar regions that controls the exchange of heat and momentum between ocean and atmosphere in winter. It has been calculated that the heat flux into the atmosphere from a newly formed lead is about two orders of magnitude greater than it is from the surrounding ice [3]. In order to monitor the seasonal patterns of surface heat exchange over the polar oceans, one should systematically observe the location and extent of pack ice, open leads, and polynyas.

Gathering the observations required for a number of fundamentally important studies of the Antarctic Ocean, including its influence on global weather and climate, is totally impractical or impracticable through the use of ships, shore stations, or even aircraft because of the vastness and remoteness of the area and the need for repetitive coverage. Remote sensing from polar orbiting satellites appears to be the only realistic answer, although satellites will probably never obviate certain types of observations made at or below the ocean surface (even here the satellite could provide assistance in the form of data collection and relay).

## 2. Satellites and Sensors suitable for Sea Ice Studies

The usefulness of pictures taken by meteorological satellites for gross ice mapping was quickly recognized after launch of the first TIROS research and development satellite in 1960 [4]. Better photographs and more complete geographical coverage became available from the operational ESSA environmental satellite system in 1966. Global coverage was provided by the Advanced Vidicon Camera System (AVCS) on ESSA 1, 3, ... 9, whereas local coverage was provided by direct readout to simple ground stations in many countries and on ships by the APT (Automatic Picture Transmission) system on ESSA 2, 4, ... 8. AVCS and APT were also carried on board the first two of NASA's Nimbus research satellites, which carried High Resolution Infrared Radiometers (HRIR) as well.

The next generation, called Improved TIROS Operational Satellites (ITOS) (which like the ESSA series were in near-polar sun-synchronous, circular orbits at a nominal altitude of 1500 km) began with ITOS 1 early in 1970. ITOS 1 combined the AVCS and APT functions in a single spacecraft, and it also carried the first dual-channel (visible and thermal infrared) Scanning Radiometer (SR) for global and local direct readout coverage. ITOS 1 was followed by NOAA 1 about a year later, NOAA 2 being placed in orbit in October of 1972 [5]. A significant change in sensor complement took place with NOAA 2: no vidicon cameras were included and their global and local readout functions were assumed by the SR, and a dual-channel Very High Resolution Radiometer (VHRR) was added.

The suite of sensors aboard the Nimbus satellite series also changed somewhat during this period. The AVCS and APT on Nimbus 1 and 2 were replaced by an IDCS (Image Dissector Camera System) on Nimbus 3 and 4. A daytime channel sensitive in the reflective infrared band was added to the HRIR on Nimbus 3, and the HRIR was replaced by the THIR (Temperature-Humidity Infrared Radiometer) on Nimbus 4 and 5. Nimbus 5, launched in December 1972, also included mong its sensors an Electrically Scanning Microwave Radiometer (ESMR) operating at a frequency of 19.35 GHz and a three-channel SCMR (Surface Composition Mapping Radiometer) [6]. Nimbus satellites, which are in near-polar orbits similar to ITOS but at somewhat lower altitude (about 1100 km), have also carried IRLS (Interrogation, Recording, and Location System). IRLS is capable of data relay from, and positioning of, fixed or moving instrument platforms at the surface or in the atmosphere.

To conclude this brief overview of satellites and satellite sensor systems suitable for sea ice studies, one should note the launch of the first Earth Resources Techno-

Table 1

Some Characteristics of Satellite Sensors

| Sensor | Satellite | Nominal Spectral Band (μm) | Nominal Resolution at nadir (km) | Remarks |
|---|---|---|---|---|
| APT | ESSA 2, 4, 0008; ITOS 1, NOAA 1; Nimbus 1, 2, 3 | 0.5—0.7 | 4.6 | Direct readout services for visible and IR provided by SR on NOAA 2, HRIR or THIR on Nimbus 3, 4. |
| AVCS | ESSA 1, 3, ..., 9; ITOS 1, NOAA 1; Nimbus 1, 2 | 0.5—0.7 | 1.9—3.5 | Stored visible provided by SR on NOAA 2, IDCS on Nimbus 3, 4. Global coverage. |
| ESMR | Nimbus 5 | 1.55 cm | 25 | Global coverage. |
| HRIR | Nimbus 1, 2, 3 | 3.4—4.2<br>0.7—1.3 | 8.0 | Thermal IR provided by THIR on Nimbus 3, 4. Global coverage. Near-IR on Nimbus 3 only. |
| IDCS | Nimbus 3, 4 | 0.5—0.7 | 2.2 | Global coverage & local readout. |
| MSS | ERTS 1 | 0.5—0.6<br>0.6—0.7<br>0.7—0.8<br>0.8—1.1 | 100 m | Direct readout only since March 1973 due to tape recorder failure. |
| RBV | ERTS 1 | 0.5—0.6<br>0.6—0.7<br>0.7—0.8 | 100 m | Inoperative since August 1972. |
| SCMR | Nimbus 5 | 0.8—1.1<br>8.3—9.2<br>10.5—11.3 | 0.8 | Inoperative since January 1973. |
| SR | ITOS 1, NOAA 1, 2 | 0.5—0.7<br>10.5—12.5 | 3.5<br>8.0 | Global coverage and local readout. |
| THIR | Nimbus 4, 5 | 6.5—7.5<br>10.5—12.5 | 22.2<br>7.5 | Global coverage and local readout. |
| VHRR | NOAA 2 | 0.6—0.7<br>10.5—12.5 | 1.0 | Mostly direct readout. |

logy Satellite (ERTS) in July 1972 [7]. Although also in a near-polar, sun-synchronous orbit, the ERTS system, unlike the imagers on ITOS and Nimbus, does not provide daily global coverage. It has both direct readout and stored data capability, but the data swath of its four-channel Multi-spectral Scanner (MSS) or its three-channel Return Beam Vidicon (RBV), is 185 km wide, yielding return coverage of most areas in the world only at an 18-day interval. ERTS also is equipped with a Data Collection System (DCS).

A summary of the spectral and spatial resolution characteristics of the satellite sensor systems briefly discussed above is contained in Table 1. All of the sensors listed are primarily imaging systems, but in most cases digitized data can be derived from them as well.

## 3. Application of Earth Satellite Data to Sea Ice Problems

Most use of satellite vidicon photographs for detection and mapping of sea ice in the Arctic or Antarctic has emphasized photo-interpretation techniques to delineate the ice boundaries and features [8, 9]. One of the basic problems encountered in this approach is the discrimination between ice, with or without a snow cover, and clouds, the latter often having reflectances in the same general range as the pack ice. A simple but effective technique takes advantage of the relatively conservative behavior of the ice fields in comparison with the often rapidly changeable cloud masses. Careful study of the area in question for several successive days generally enables differentiation of ice-covered areas from clouds. Also helpful in this regard is the circumstance that ice often has characteristically different patterns, shapes, or texture from clouds; and that clouds cross coastlines, cast shadows, or are partially transparent. The better the spatial resolution of the sensor, the more effectively are these factors brought to bear on the interpretation problem. Spatial resolution also affects the ability to deduce pack ice type and concentration. After comparing aerial reconnaissance and ESSA 7 satellite data, Potocsky [10] concluded that concentrations of three or less oktas are nearly impossible to detect by satellite; gray and gray-white ice appeared as open water in the satellite photographs; and brash ice was less reflective than larger floe sizes in the satellite pictures and thus tended to be underestimated in concentration. Swithinbank [11] came to similar conclusions, but he also stated the opinion that the availability of daily satellite coverage "represents an advance of unprecedented importance towards the safe and expeditious passage of vessels through ice-filled seas. Moreover, since ice conditions, like weather conditions, vary from year to year, there is no doubt of the value of the assembled photographs as a permanent record of the range of variation".

Aber and Vowinckel [12] have recently studied the North Water area between northwest Greenland and Ellesmere Island using APT pictures for two years (March—September). They concluded that "it is possible to obtain a reasonably detailed picture of ice distribution, and its changes in time, from an interpretation of a series of good satellite pictures. Furthermore, motions of ice fields can be observed ..." Strübing [13] has concluded, following a study of ESSA 8 APT pictures in the spring of 1969, that the different gray tones and textures in the satellite photographs essentially represent the manifold forms of ice cover in accordance with the Baltic Ice Code, and that this is applicable to the different stages of development of sea ice as well as to the degree of ice concentration.

Cloud filtering and suppression from visual satellite images can also be achieved by computer manipulation of the satellite brightness values or reflectances after the data have been digitized and mapped [14]. The brightness data are usually normalized with respect to solar zenith angle and corrected for system errors during this process. The minimum brightness at each mapped point during a selected set of successive days, usually five or ten, can then be composited to produce a Composite Minimum Brightness (CMB) chart in the form of an image or computer printout or contoured chart. McClain and Baker [15] have demonstrated that the highly reflective areas in the CMB correspond to close pack or greater concentrations of first-year ice. The longer the compositing period the more effective is the cloud filtering but the more vulnerable the CMB chart becomes to errors due to changes in ice extent. This could be overcome to some extent by generating CMB charts overlapped in time, i.e. at shorter intervals than the length of the compositing period. McClain [16] has developed a procedure for external calibration of the digitized CMB values by normalizing them with respect to natural earth targets whose brightness varies within very narrow limits. Characteristic brightness values were then found to correspond to the ice pack concentration and the presence or absence of extensive snow cover or puddling. Wendler [17] used such a procedure to study ice movements in a small area of the Arctic Ocean off the shore of north Alaska and to derive monthly mean albedo maps. Fletcher [1] has recommended the systematic production of CMB charts "to establish a basic observational picture of time and space variations of Antarctic sea-ice extent".

Thermal infrared imagery and measurements from the HRIR and THIR on Nimbus satellites and the SR and VHRR on the NOAA satellites have begun to be used increasingly in operational and research applications for sea ice surveillance, e.g. the studies by Barnes et al. [18, 19]. These investigations have shown that the 8 km resolution IR imagery provides a means of mapping gross ice boundaries during periods of polar darkness, although it is generally inferior to the 3.5 km resolution imagery that can be used during periods of sufficient solar illumination. During those times when visible and thermal IR can be used jointly, preliminary results indicate that additional ice information, such as ice concentration and possibly thickness, can be inferred. Some ice areas were observed to have higher IR temperatures combined with lower reflectances, which would most probably be a result of lower ice concentrations. Other areas were observed to have higher IR temperatures without significantly lower reflectances, which would indicate a thinner ice cover or possibly ice with less of a snow cover. An analysis of a small sample of daytime and night-time IR data gave evidence that diurnal variations are up to three times greater over close pack or higher concentrations (7/10 or more) than over open and very open pack (6/10 and less). Thermal contrasts between ice and water are not as large as reflectance contrasts, therefore IR imagery generally requires some special treatment, i.e. enhancement, in order to make optimum use of it in ice analysis. Nearly all clouds are opaque to radiation emitted at 10.5—12.5 μm, but many of these same clouds are partially transparent to energy in the wavelengths 0.5—0.7 μm, thus permitting delineation of ice features over broader areas in the visible than in the infrared. Although ice surfaces can sometimes have IR temperatures in the same range as cloud layers in the same area, particularly low stratus clouds, differentiation of ice from clouds can often be made by the same photo-interpretation techniques described earlier for use with visible imagery, except of course that cloud shadow effects are not present in the IR imagery. When the cloud tops are distinctly colder than the underlying ice surface, the IR imagery can aid in the identification of ice even when simultaneous visible imagery is available because the cloud and ice reflectances can be much the same. High latitudes in winter are often characterized by strong low-level temperature inversions, and this can easily result in the tops of low cloud layers being warmer than the underlying pack ice. Therefore, the ice analyst should be aware that clouds can appear either brighter (colder) or darker (warmer) than ice cover in the thermal infrared imagery as it is conventionally displayed.

The newly available VHRR visible and thermal infrared data, whose one kilometer ground resolution yields a fourfold improvement over most previous visible observations and a tenfold improvement over previous IR measurements, represents a substantial gain in ice mapping capability from space. Fig. 1 is a VHRR image (0.6—0.7 μm) showing first-year ice in the Bering Sea; the data were received at the NOAA acquisition station in Virginia via the stored data mode. Cloud discrimination techniques are applied more effectively with the higher resolution images. Furthermore, details of floe, lead, and fracture patterns stand

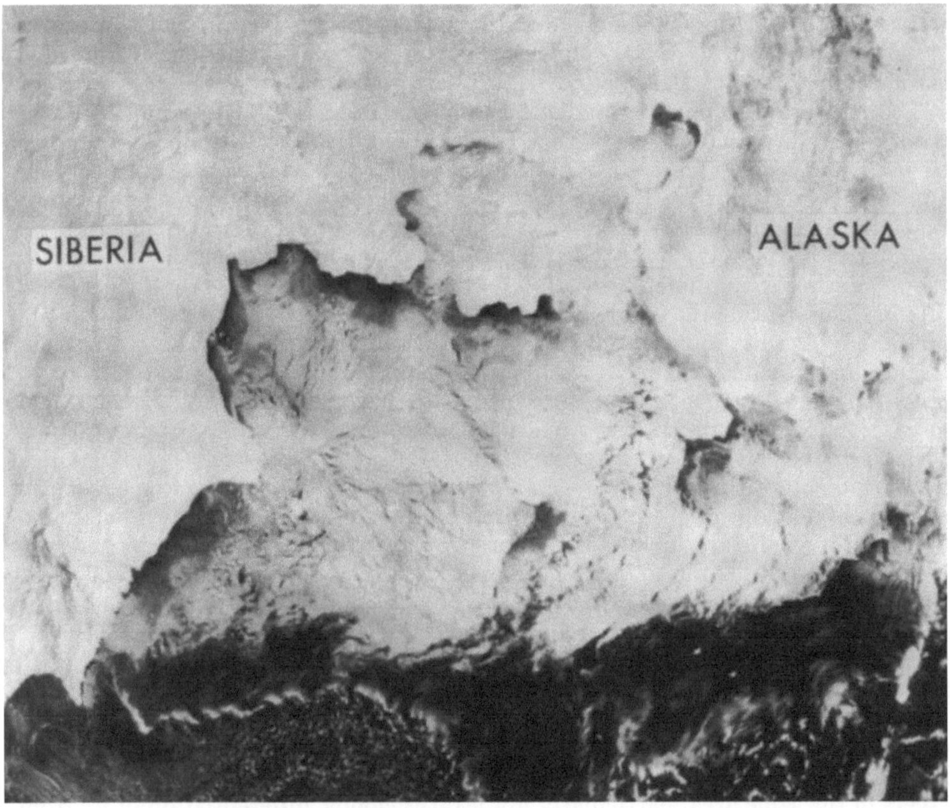


Fig. 1. Visible band image from VHRR on NOAA 2 satellite, 31 March 1973, in Bering Sea area.

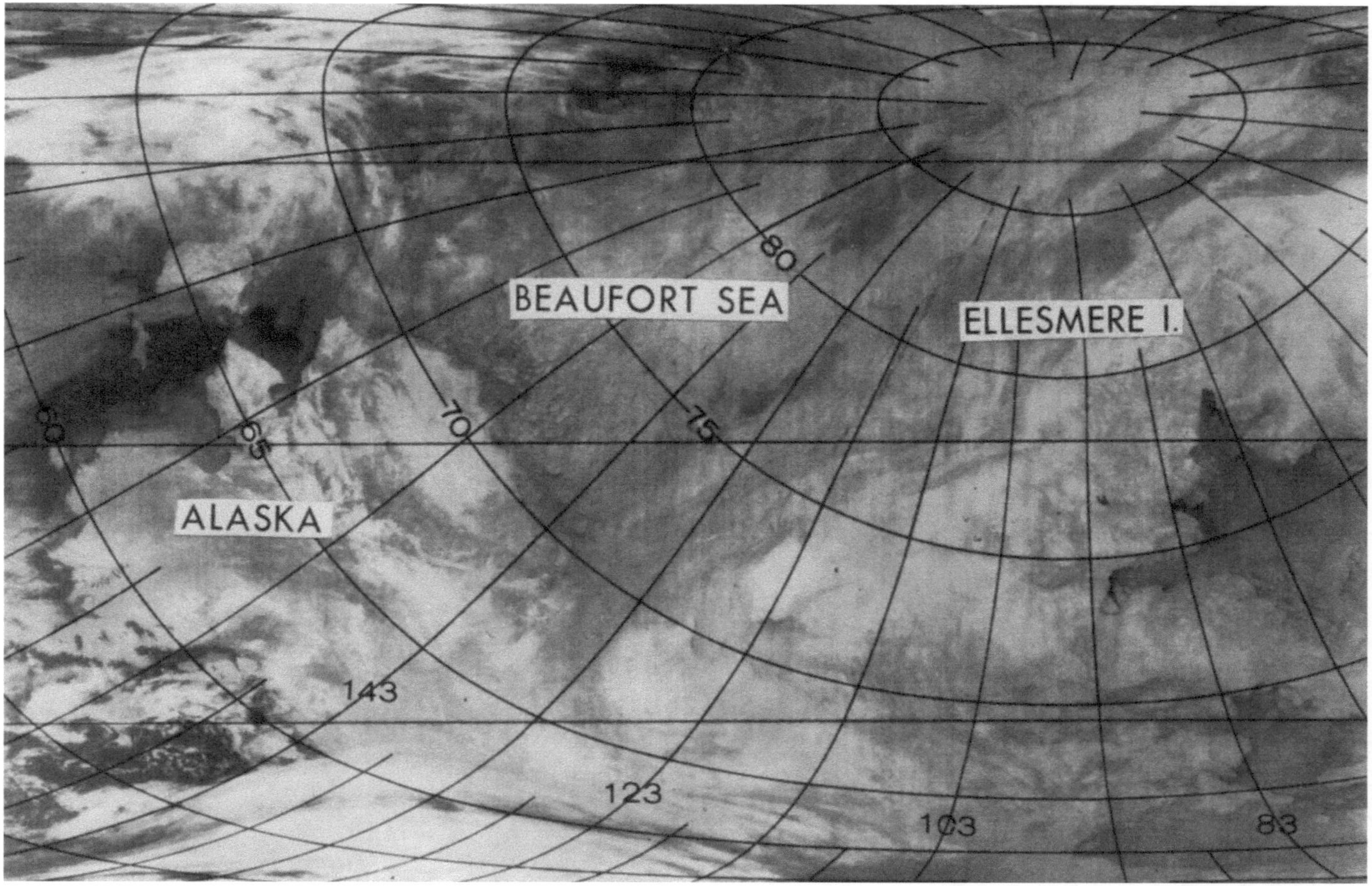


Fig. 2. Thermal infrared image from VHRR on NOAA 2 satellite, 15 December 1972, in North American Arctic. The darker the tone, the warmer the radiating surface.

out clearly in the new data, the improvement in ice information content being especially marked in the infrared imagery. It should be noted, however, that VHRR data coverage is considerably more limited than that of the SR. The VHRR is primarily a direct-readout sensor, although a small portion (about 8%) of several 108-minute orbital passes each day can be sensed and then stored on magnetic tape on board the satellite for subsequent readout to a suitably equipped ground station. The only stations presently equipped to read out stored or direct VHRR data are in Virginia and Alaska; they can acquire direct readout coverage up to a maximum of about 22 minutes of those orbital passes near the readout station. Fairly complete VHRR coverage of the North American side of the Arctic from Greenland and Labrador westward to beyond the Chukchi and Bering Seas was obtained by direct readout during this past winter and spring. Fig. 2 is a VHRR infrared image (10.5—12.5 μm) covering the area from northern Baffin Bay to eastern Siberia; the data were received by direct readout at the Alaska ground station. Ice structure varies from first-year in the Bering Sea, Foxe Basin, and

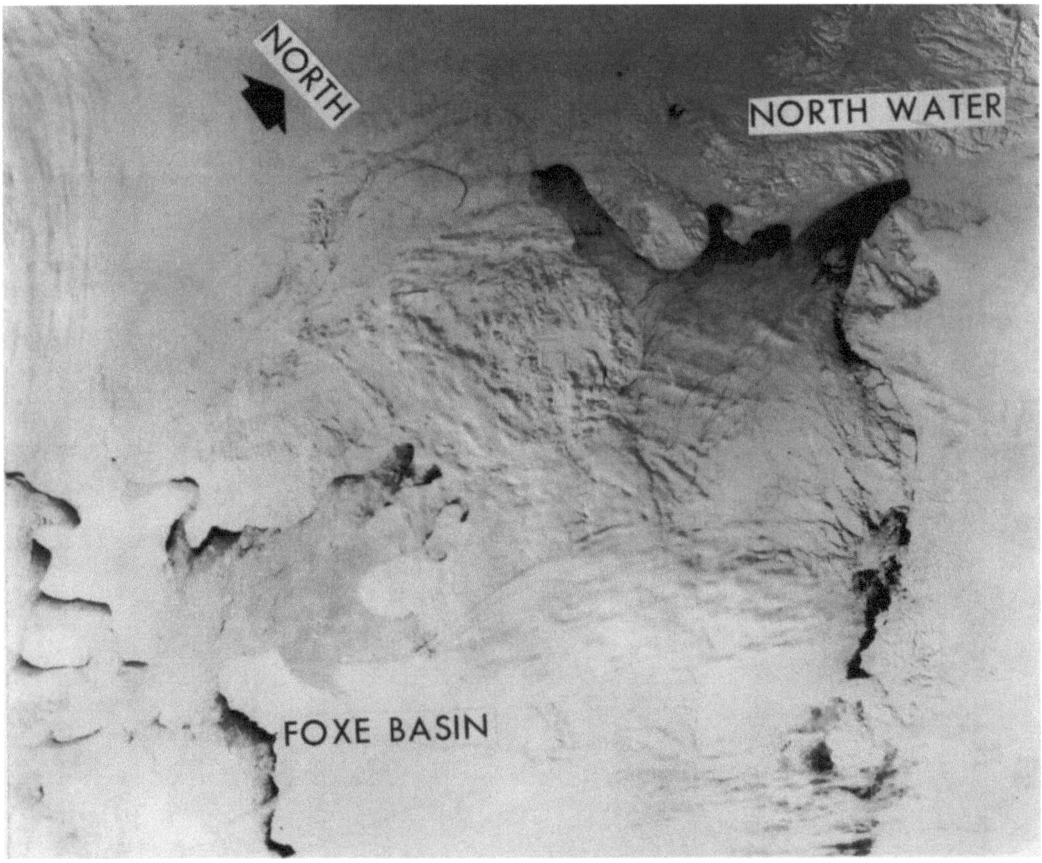


Fig. 3. Visible band image from VHRR on NOAA 2 satellite, 13 April 1973, in eastern part of North American Arctic.

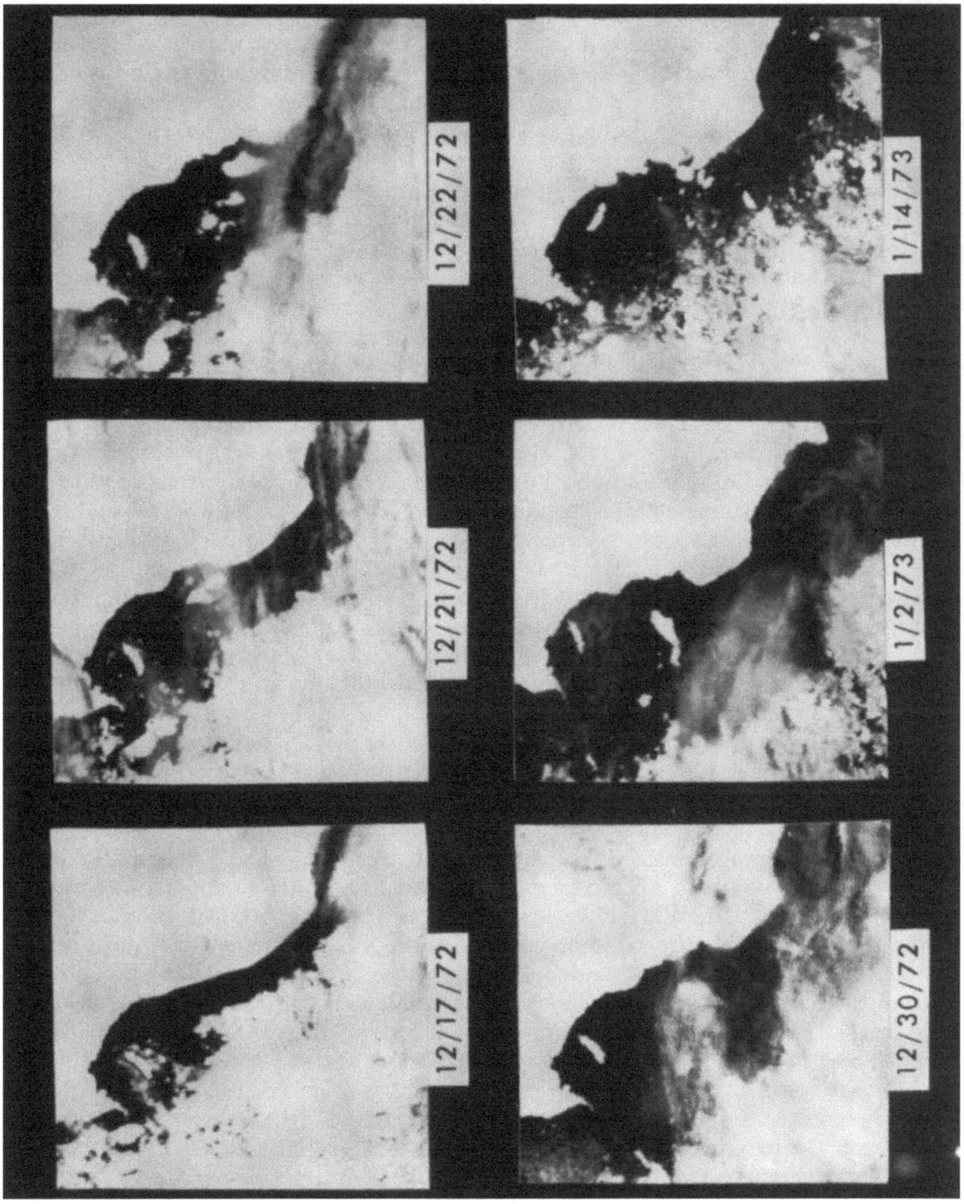


Fig. 4. Visible band images from VHRR on NOAA 2 satellite, 17 December 1972 to 14 January 1973, in Getz Ice Shelf area (75° S, 130° W) of Antarctica.

northern Baffin Bay (including Smith Bay and Lancaster Sound), to multi-year ice with giant floes in the Beaufort Sea and a very long flaw lead along the western perimeter of the Queen Elizabeth Islands near 80° N, 125° W. Special stored VHRR data coverage was obtained of the Ross Sea area of Antarctica from November 1972 through January 1973 to assist US Navy ice forecasters in their support of re-supply missions. This data series has also been used for research purposes within NOAA/NESS, including a study of ice movements and the calving of tabular icebergs in the vicinity of the Ross and Getz Ice Shelves. Fig. 3 is a composite of portions of six VHRR visible-band images over a one-month period in the Getz Ice Shelf area of Antarctica. The first two images in this sequence are just before and after the calving of several large tabular icebergs from the shelf. Several other large bergs, as well as much pack ice, are also visible. Subsequent pictures illustrate the motion (translation and rotation) of these ice features over a three-week period. Special VHRR coverage of several other areas was provided for shorter periods during the spring of 1973: North Water area, Foxe Basin, East Greenland Current, Iceland, and northern Scandinavia. Fig. 4 is a visible channel image from the VHRR on NOAA-2 derived from stored data. It shows first-year and multi-year ice, leads, and polynya (including the well-known North Water) in northern Baffin Bay and the Foxe Basin. Detailed experimental ice analyses based on VHRR imagery were produced by a NOAA/NESS ice analyst and transmitted to the Alaska area for operational use by the National Weather Service. Research using digitized VHRR data has been initiated recently within NOAA/NESS to further development of ice enhancement methods and investigations of the information content of ice temperature variations. Available computer resources constrain the digitization of VHRR data to small areas. Rectification and mapping, which are requirements for production of Composite Minimum Brightness charts, are precluded for the near future.

The 100-meter resolution of the experimental ERTS data permits the most detailed and precise interpretation and mapping of ice features ever possible from an earth satellite, but the coverage provided by the 185 km wide swath and the 18-day return cycle precludes virtually all operational use, especially when cloudiness and data processing delays are taken into account. Fig. 5 is an ERTS image (0.8—1.1 μm) covering approximately the same area as the upper-left corner of the 14 January 1973 VHRR image in Fig. 4. The long, narrow tabular iceberg at the lower right is the same one shown in the VHRR image three days earlier. Smaller bergs are seen embedded in the fast-ice promontory that has built out from the ice shelf in the vicinity of Shepard Island. Overlapping ERTS coverage at high latitudes sometimes permits views of restricted areas for several successive days, and it has been possible at times to track specific ice features over short periods to gain information on wind, current, and ice interactions. ERTS data can be used to detect and delineate areas of pack ice where melting has occurred and thus the extent of puddling, i.e. melt water pools [20]. This capability, which was first demonstrated using the visible and near-infrared imagery from Nimbus 3 [21], derives from the circumstance that the reflectivity of ice (or snow) diminishes sharply in the wavelengths from 0.7—1.3 μm, but only slightly at visible wavelengths, when its surface becomes wet. Fig. 6 comprises two ERTS images of an area in the East Siberian Sea (73° N, 162° E). Image (A) is in the 0.6—0.7 μm band and shows a variety of floe and pack conditions, including some open water. Note particularly some low-reflectance ice, mostly in the left side of the picture. These areas are interpreted as melting or actually flooded ice because it is noted that they become as non-reflective as open water in image (B), which is in the 0.8—1.1 μm band. The failure of the second tape recorder on ERTS 1 in March 1973 precludes further coverage of Antarctica until ERTS 2 is placed in orbit, which could be as late as 1976. The data acquisition station in Alaska should permit continued coverage of some parts of the Arctic, however, by means of direct readout.

The success of the Electrically Scanning Microwave Radiometer (ESMR) on Nimbus 5 represents a significant advance in our capability to monitor sea ice throughout the year and in virtually all weather. The single greatest limitation on satellite sensing of sea ice at visible and infrared wavelengths is cloud cover, but natural emission from earth's surface at microwave wavelengths is attenuated only slightly by clouds typical of polar regions. Large brightness temperature contrasts, which are due to emissivity rather than physical temperature differences, are found at ice/water boundaries and these are readily observed through clouds. The ice surface itself has been observed to have emissivity variations,

Fig. 5. Reflective infrared band (0.8—1.1 μm) from MSS on ERTS 1 satellite, 11 January 1973, in Getz Ice Shelf area of Antarctica.

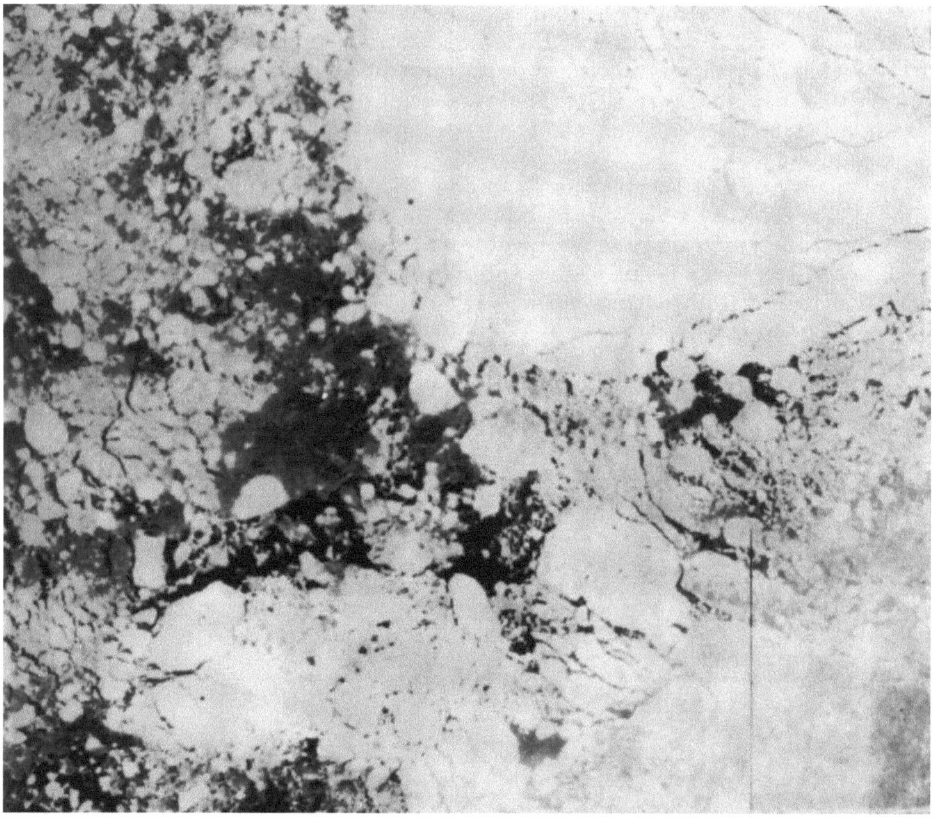

Fig. 6 (A)

Fig. 6. MSS images from ERTS 1 satellite, 19 September 1972, in East Siberian Sea area (A) visible band (0.6—0.7 μm); (B) reflective infrared band (0.8—1.1 μm).

first-year ice being somewhat more emissive than multi-year ice. The only drawback to the use of ESMR imagery for some types of ice studies is the relatively coarse spatial resolution of the data, viz. 25 km. Better resolution than this will require the orbiting of substantially larger antennas.

The IRLS on Nimbus 4 has been used to track six drifting buoys emplaced in the pack ice of the Beaufort Sea as part of the Arctic Ice Dynamics Joint Experiment (AIDJEX) in 1972 [22]. The data buoys were designed for an operational lifetime of up to one year, measuring atmospheric pressure and near-surface air temperature when interrogated from Nimbus 4. These data and the buoy's location information are telemetered to the data acquisition station near Fairbanks, Alaska. Five of the buoys in this array remained functional throughout the past winter.

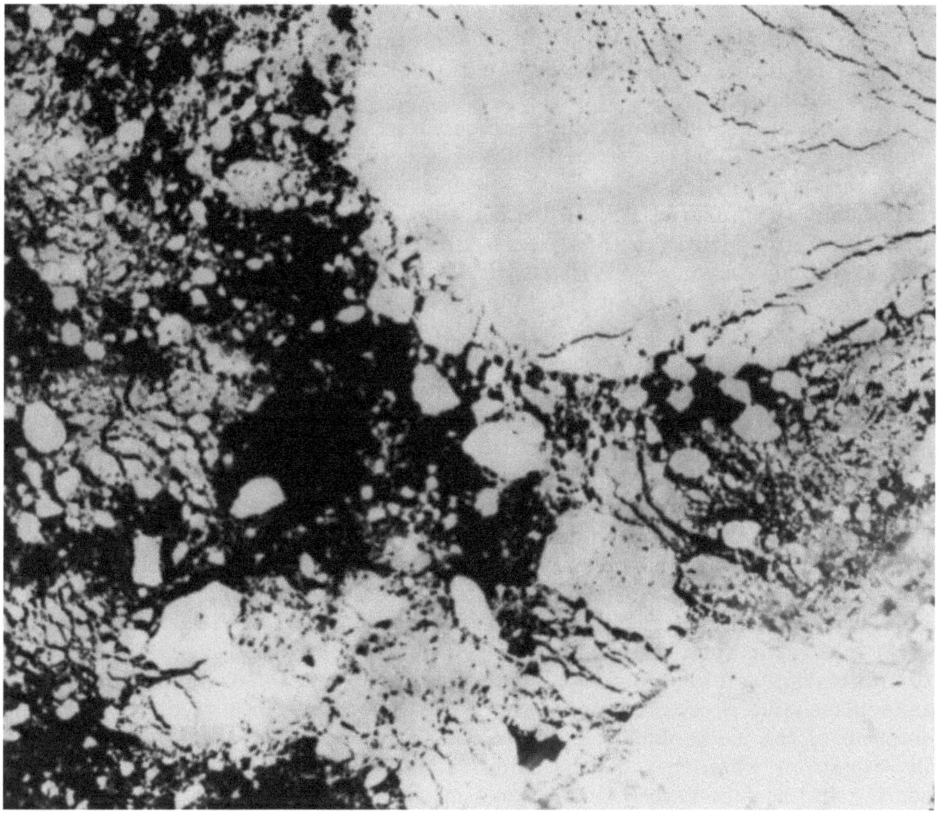

Fig. 6(B).

## 4. Future Potential Contributions of Satellite Technology to Sea Ice Studies

The sensor complement of the NOAA series of polar-orbiting operational satellites will remain unchanged through about 1977, except that the visible channel will be opened up to 0.5—1.0 μm on some of the Scanning Radiometers, The application of SR and VHRR measurements, especially the latter, to sea ice problems is likely to increase rather greatly as the number of ground stations equipped to receive these data continues to grow and as better equipment and techniques for processing, displaying, and applying these observations are developed. The trend toward more quantitative, objective, and automated application of satellite data, in contrast to interpretation of images, is already apparent. Planned-for improvements to ITOS after about 1977 include the replacement of the SR and the VHRR by the AVHRR, or Advanced Very High Resolution Radiometer, and a conversion to transmission of digital data directly from the satellite. This will be accomplished by an onboard data processor, which will also have the function of "degrading" the stored 1 km data to about 3 km resolution for global meteorological coverage purposes. Full-resolution measurements will be available for direct readout service to suitably equipped ground stations and for limited area coverage via the stored data mode. Conventional APT service will continue to be provided also. The AVHRR will have four channels: visible (0.5—0.9 μm), reflected IR (0.75—1.0 μm), water vapor absorption (6.5—7.0 μm), and thermal IR (10.5—12.5 μm).

The next two satellites in the Nimbus series, presently designated F and G, probably to be launched in mid 1974 and late 1976, respectively, will also likely carry sensor or data relay systems of interest to ice researchers. In addition to a THIR and an improved ESMR (0.8 cm wavelength), Nimbus F will be equipped with a Scanning Microwave Spectrometer (SCAMS: 1.35—0.465 cm wavelengths), and a system known as TWERLE/RAMS (Tropical Wind Energy Conversion Reference Level Experiment/Random Access Measurement System). The latter could be used to position and collect data from unmanned drifting stations in the ice pack. Final selection of the sensor suite for Nimbus G has not been made, but the secondary discipline objective of this satellite is stated to be oceanography.

ERTS 2, presently scheduled for 1976, will have the same RBV and MSS sensor systems as ERTS 1 except that a fifth channel will be added to the MSS. This extra detector will provide measurements in the thermal infrared with a nominal resolution at nadir of 250 m. ERTS 2, like ERTS 1, will have a data collection and relay capability but no ability to position moving instrument platforms.

This review has dealt exclusively with USA satellite systems, although it is known that the USSR's operational Meteor satellites and Cosmos research satellites have carried sensors useful for sea ice studies, and that there is naturally great interest in such studies in the USSR. It is hoped that increased exchange of information and ideas among all countries engaged in sea ice investigations will take place and accelerate effective application of satellite technology to the solution of sea ice problems. Satellite data are used most effectively in sea ice investigations when they can be calibrated and controlled with aircraft remote sensing and surface-based observational programs, such as in AIDJEX. Such an approach should be valuable, if not essential, to the upcoming POLEX (Polar Experiment) program. The potential results of such coordinated investigations will have application not only to the solution of sea ice problems, but to short-term weather and long-term climatological forecasting over areas far removed from the Arctic and Antarctic.

It would appear that there will be no dearth of satellite measurements becoming available during this decade, but that more of a problem will be the marshalling of sufficient data processing, human, and other resources to be able to cope effectively with this flood of new information. We could already be to the point where satellites are collecting data faster than we have the people and computers to ingest them, digest them and disseminate them in a usable form for the research or operations people in turn to apply them to their problems.

## References

[1] J. O. Fletcher, Rand Corp. Memo. RM-5793-NSF (1969).
[2] N. Untersteiner and G. A. Maykut, Rand Corp. Memo. RM-6093-PR (1969).
[3] F. I. Badgley, Rand Corp. Memo. RM-5233-NSF (1966).
[4] D. Q. Wark and R. W. Popham, in: Proc. 1st Int. Symp. on Rocket and Satellite Meteor., North Holland Publ. Co., Amsterdam 1962 (p. 415).
[5] A. Schwalb, NOAA Tech. Memo. NESS-35 (1972).

[6] National Aeronautics and Space Administration, Nimbus 5 User's Guide (1972).
[7] General Electric, ERTS Reference Manual, Philadelphia, Pa. (1972).
[8] R. W. Popham and R. E. Samuelson, Arctic **18**, 246 (1965).
[9] K. P. Vasil'ev, Hydromet. Res. Ctr., USSR, Trudy **36** (1968).
[10] G. T. Potocsky, US Naval Oceanog. Office IR No. 70-6 (1970).
[11] C. Swithinbank, The Polar Record **15**, 19 (1970).
[12] P. G. Aber and E. Vowinckel, Arctic **25**, 263 (1972).
[13] K. Strübing, Deut. Hydrog. Z. **23**, 193 (1970).
[14] A. L. Booth and V. R. Taylor, Bull. Am. Met. Soc. **50**, 43 (1969).
[15] E. P. McClain and D. R. Baker, ESSA Tech. Memo. NESCTM-12 (1969).
[16] E. P. McClain, Arctic **26**, 45 (1973).
[17] G. Wendler, J. Geophys. Res. **78**, 1427 (1973).
[18] J. C. Barnes, D. T. Chang and J. H. Willand, J. Geophys. Res. **77**, 453 (1972).
[19] J. C. Barnes, D. T. Chang and J. H. Willand, Allied Res. Assoc. Final Rep. No. 8G93-F (1972).
[20] E. P. McClain, in: Proc. Earth Resources Technology Satellite-1 Symposium, NASA GSFC, Greenbelt, Md, 1973 (p. 127).
[21] A. E. Strong, E. P. McClain and D. F. McGinnis, Mon. Wea. Rev. **99**, 828 (1971).
[22] P. Martin, AIDJEX Bull. No. 14, Univ. of Washington (1972).

Earth Survey Problems — Akademie-Verlag, Berlin 1974

# POLAR ALBEDO CHANGES AND THEIR CLIMATIC CONSEQUENCES

H. Flohn

University of Bonn, Bonn, FRG

Based on the Manabe-Wetherald model of radiational equilibrium, time-variations of the surface albedo produced by the varying extension of ice and snow in high latitudes are compared with the accompanying fluctuations of the equilibrium temperature. The results are quite realistic: they indicate a decisive climatogenetic role of the albedo temperature equilibrium. This supports the possible role of large-scale surges from the potentially unstable Antarctic ice sheet (after a modified version of A. T. Wilson's hypothesis) for the observed dramatic global coolings connected with the sequence of glaciations and deglaciations during the last $10^5$ years, and necessitates a continuous monitoring of the extensions of both polar ice caps from orbiting satellites.

In the radiation budget of the earth's surface the variation of surface albedo is the most effective: typical values of the sea surface are between 0.04 and about 0.10, of snow-free land surfaces between 0.05 and about 0.35, but the albedo of a freshly fallen snow cover rises to about 0.80, and the albedo of the polar sea ice varies only between 0.62 (during melting season) and about 0.80. The extent of snow and ice is therefore one of the most powerful climatogenetic factors. Its influence on the thermal equilibrium of the atmosphere can be estimated on the basis of the well-known radiation model given by Manabe and Wetherald [1], assuming convective adjustment, constant relative humidity and average cloudiness at 40° N. With representative values for the area and the average albedo of open and ice-covered oceans, and of snow-free, ice-covered and snow-covered (mostly forested) continents, we obtain (Table 1, top section) a realistic value for the average actual surface albedo of the earth (0.134). The hemispheric albedo difference indicates that the equilibrium temperature of the northern hemisphere is about 1 K higher than that of the southern hemisphere; the real temperature difference is 1.8 K.

The following short contribution deals with the global or hemispheric temperature variations and their correlation with the variations of the sea ice in the Arctic and Antarctic, including some more speculative considerations on the causes of ice ages and other palaeoclimatic features. It is well known [2] that the extent of the arctic sea ice has greatly varied during the last millenium: while between 860 and 1200 AD the average annual period of drift ice at the coasts of Iceland was zero or (rarely) 1—2 weeks, and 0—5 weeks in the following centuries, it increased during 1600 and 1900 to 12—25 weeks, decreasing again to nearly zero

Table 1

Surface Albedo ($a_s$) and Equilibrium Temperature ($T^*$) Deviations (areas in $10^6$ km²)

| Albedo | Oceans open | Oceans ice | Continents open | Continents ice | Continents snow | Average albedo $a_s$ | Deviation $T^*$ (°K) | Remarks |
|---|---|---|---|---|---|---|---|---|
| | 0.05 | 0.70 | 0.12 | 0.75 | 0.30 | | | |
| N. Hemisphere (NH) | 145 | 10 | 70 | 3 | 27 | 0.1294 | — | actual |
| S. Hemisphere (SH) | 190 | 16 | 33 | 13 | 3 | 0.1384 | — | (1901—50) |
| Earth (E) | 335 | 26 | 103 | 16 | 30 | 0.1339 | — | |
| Model NH 0 | 142 | 13 | 70 | 3 | 27 | 0.1373 | −0.95 | N. Hemis. 1890 |
| Model SH 0 | 188 | 18 | 33 | 13 | 3 | 0.1434 | −0.60 | S. Hemis. 1850 |
| Model E 4 | 317 | 30 | 108 | 49 | 6 | 0.1731 | −4.6 | Ice age (sea |
| Model E 5 | 307 | 40 | 108 | 49 | 6 | 0.1860 | −6.2 | level —100 m) |
| Model SH 3 | 165 | 41 | 33 | 13 | 3 | 0.2022 | −7.6 | Wilson surge |

around 1950. The areal loss during the warming period from 1900 to 1940 has been estimated to be nearly $3 \times 10^6$ km². Recent studies of the subantarctic oceans have revealed one or more large outbreaks of antarctic shelf ice (mainly from the Weddell Sea) of the order of (at least) $10^5$ km³ [3] between 1840 and 1900, accompanied by a drastic cooling of about 5 K at Deception Island [4]. Assuming an average thickness of 200 m shortly after the outbreak, and of about 40 m after some years, the average areal increase must have been near $2 \times 10^6$ km².

Assuming these areal variations of the arctic and subantarctic sea ice, we obtain from the Manabe-Wetherald model deviations of the equilibrium temperature of −0.95 and −0.6 K respectively (Table 1, center) for the maximum extent of ice during the last century. The value for the northern hemisphere is slightly higher than the observed value of −0.6 K [5]; at the southern hemisphere no representative averages are available before about 1900.

During the last glaciation, with the maximum extent of the ice near 18000 BP*, the area of the continental glaciers is well known [2]. The area with a seasonal snow cover could only be estimated (Table 1, bottom): here an annual average must be chosen. The extent of the sea ice cover is equally unknown; recent investigations on some Atlantic deep-sea cores between latitudes 42° N and 62° N [6] indicate that during a not too short interval the arctic sea ice reached the Bay of Biscay and 44° N in the open Atlantic; the low altitude of the snow-line in the mountains of northern Spain and southern Ireland, suggests a temperature drop of 12 K or even more. From this evidence the model E 5 (see Table 1) has been designed, while during the greater part of the last ice-age model E 4 (Table 1) seems to be more realistic. The deviation of $T^*$ of 5 or 6 K coincides with all realistic evidence during this period (25000—15000 BP) except the local surroundings of the large continental ice sheets and the Biscay coasts, but including the tropical Atlantic and the isolated peaks of equatorial mountains. The quite convincing coincidence between $T^*$ estimated from surface albedo $a_s$ and the actual hemispheric (or global) temperature $T$ seems to indicate that the actual average temperature variations correspond rather closely to the long-term variations of

* BP = Before Present (1900).

the global ice cover. (This certainly does not indicate that this feedback correlation between $a_s$ and $T^*$ can be interpreted as the "cause" of large-scale climatic fluctuations.)

Apart from all considerations of the genesis of climatic fluctuations [7, 8] we can safely state that the astonishingly large areal variations of the polar sea ice in both hemispheres are of highest importance for any realistic attempt to understand and to forecast climatic variations. According to a critical survey of recent quaternary research [9], the repeated occurrence of catastrophic hemispheric (or global) coolings of the order of 3–6 K during a time span of not more than a century or even less [10] is unequivocally derived from many independent sources of evidence. During the last $10^5$ years at least five occurrences of this kind have been observed, some of them lasting only a few centuries, others introducing a new ice age with a duration of more than $10^4$ years.

Events of this rapid development and intensity are certainly rare; a realistic physical interpretation leads to a modified version [11] of Wilson's Antarctic surge hypothesis [12] as trigger of rapid global cooling of the order of 4–6 K. Model SH 3 (Table 1) describes an antarctic surge according to Wilson's assumptions; its duration ought not to be longer than several decades to produce a strong impact on the earth's climate. Other simultaneous features, such as the temperature difference between Atlantic and Pacific coasts during glaciation and the widespread arid period reducing drastically the equatorial rain forests of Africa and South America, add to the high probability of this unorthodox interpretation. In fact, the antarctic ice sheet appears to be potentially unstable, evidenced also by bottom melting at 2164 m depth at Byrd station [13]. A model of the bottom temperature pattern has been recently outlined [14], suggesting that the greatest probability of a future surge is in the Weddell Sea area (Filchner ice shelf).

The close correlation between the extension of continental ice sheets (and also of sea ice) and global temperature variations gives a further strong argument in favour of a continuous and detailed monitoring of both polar ice caps from orbiting satellites, as in [15].

## References


[1] S. Manabe and R. T. Wetherald, J. Atmos. Sci. **24**, 241 (1967).

[2] H. H. Lamb, Climate: Present, Past and Future, Methuen, London 1972.

[3] H. H. Lamb, WMO Tech. Note **87**, 428 (1967).

[4] O. Orheim, Phil. Diss. Ohio State University (1972).

[5] J. M. Mitchell Jr, in: Global Effects of Environmental Pollution, Springer, New York 1970 (p. 139).

[6] A. McIntyre et al., Deep-Sea Research **19**, 61 (1972); Quaternary Research **2**, 350 (1972).

[7] Inadvertent Climate Modification, Report of the Study of Man's Impact on Climate. MIT Press, Cambridge, Mass. 1972.

[8] H. Flohn: Ann. Meteor. Neue Folge **6**, 59 (1973).

[9] J. Kukla et al., Quaternary Research **2**, 261 (1972); Science **178**, 197 (1972).

[10] W. Dansgaard et al., Quaternary Research **2**, 396 (1972); Nature **235**, 229 (1972).

[11] H. Flohn, Eiszeitalter und Gegenwart **20**, 204 (1969); unpublished lecture (Melbourne) 1972.

[12] A. T. Wilson, Nature **201**, 147 (1964); Canad. J. Earth Sci. **6**, 911 (1969).

[13] A. J. Gow et al., Science **161**, 1011 (1968).

[14] W. Budd et al., Polarforschung **6**, 293 (1970).

[15] N. A. Streten, Arch. Met. Geophys. Biokl. **A22**, 119 (1973).

# Water Resources Related Problems

Earth Survey Problems — Akademie-Verlag, Berlin 1974

# ASSESSMENT OF APPLICATIONS OF SPACE-BORNE REMOTE SENSING TO HYDROLOGY AND WATER RESOURCES: AN OVERVIEW

P. Bock

University of Connecticut, Storrs, Conn., USA

Space-derived imagery now exists in sufficient amounts to assess its present and potential usefulness to hydrology and water resources. This assessment is based on conceptual models of the hydrologic cycle and its processes. Overviews of hydrologic parameters vs. remote sensing capabilities are given. Future developments in sensor technology and data interpretation techniques are suggested. Examples of hydrologic applications, notably ERTS 1 experiments, are given.

In summary, recent developments suggest sufficient hydrologic benefits using remote sensing techniques to justify an enlarged research and development effort directed at yielding or improving the following capabilities: real-time; cloud-free and day-night; spatial resolution (25 m horizontal, less than 1 m vertical); camera pointability; and inexpensive APT-type readouts.

## 1. Introduction

The purpose of this review paper is to present to the space technology community an assessment of applications of space-borne remote sensing to hydrology and water resources problems. Emphasis is placed on the discipline of hydrology and on the applied problems of water resources rather than on problems of technology and remote sensing techniques. Remote sensing from aircraft and from space are mutually supportive, but the former is not discussed here.

The paper is divided into three main parts: (1) Hydrology and water resources, (2) Present status of remote sensing in hydrology and water resources, and (3) Potential future role of remote sensing in water resources. The first part outlines the main features of hydrology and major national and international water problems. The second part provides a summary of applications of space-borne remote sensing to hydrologic problems with emphasis on the recently launched ERTS 1. The third part attempts to relate the potential future role of space-borne remote sensing to the existing and emerging major water problems.

## 2. Hydrology and Water Resources

### 2.1. The Hydrologic Cycle

The basic concept of hydrology is the hydrologic cycle. At the planetary scale, the heat, energy, mass, and momentum transfers involve the atmosphere, the continents and the oceans. At the continental and drainage basin scales (usually of most practical engineering interest to hydrologists) emphasis is placed primarily on mass balances which express the occurrence and movement of water on and within the terrestrial sphere. Here runoff (stream flow) is the central theme. The other major hydrologic elements (evapotranspiration, precipitation, ground-water flow and changes in storages such as in lakes, snow and ice masses, soil moisture, ground water, biological waters) are usually studied with respect to runoff. Hydrologists also attempt to understand the fundamental geophysical phenomena such as precipitation processes, infiltration, flow through saturated and unsaturated porous media, sedimentation and erosion, etc. The present state of such knowledge requires integration. Conceptual models of the complete hydrological cycle stemming from basic physical principles have not yet been developed although Manabe [1] has proposed a simple model of the hydrologic cycle at the planetary scale and Crawford and Linsley [2] have developed watershed models incorporating physical-empirical relationships. M. I. Lvovitch [3] has compiled statistics about global water balances.

Although hydrology is mainly concerned with the earth's fresh waters, its systematic understanding relies on all the earth sciences, notably meteorology, oceanography, and geology. The recent emphasis on the "quality of the environment" has focused attention on the biochemical aspects and socio-economic considerations.

The schematic diagram of the hydrologic cycle at the macro scale shown in Fig. 1 indicates the relationships of the major hydrologic elements. The interrelated mass continuity relationships for the atmospheric-continental-oceanic branches are given below.

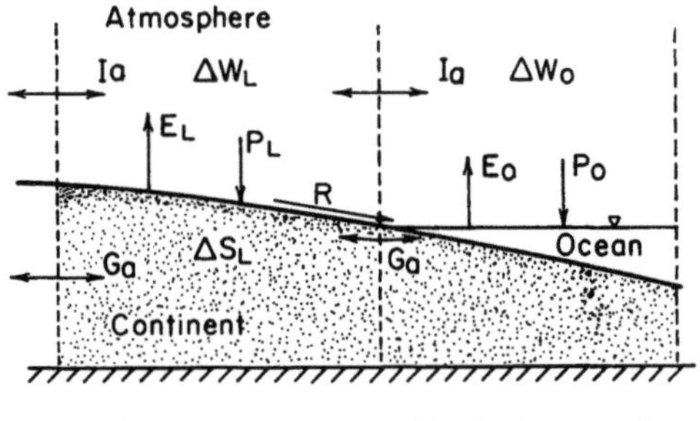


Fig. 1. Schematic of general hydrologic cycle.

Atmospheric/Continental

$$E_L - P_L \pm I_a \pm \Delta W_L = 0 \tag{1}$$

Atmospheric/Oceanic

$$P_O - E_O \pm I_a \pm \Delta W_O = 0 \tag{2}$$

Continental

$$P_L - E_L \pm G_a - R \pm \Delta S_L = 0 \tag{3}$$

Oceanic

$$P_O - E_O + R \pm G_a = 0 \tag{4}$$

where $P_L, P_O$ = precipitation over land and ocean respectively, $E_L, E_O$ = evaporation over land and ocean respectively, $R$ = runoff, $G_a$ = net ground water flow, $I_a$ = net advected atmospheric moisture, $\Delta W_L$ = change in precipitable water, $\Delta S_L$ = change in all forms of terrestrial water storage, including snow and ice, changes in lakes and reservoirs, soil moisture, ground water.

Further detail of the hydrologic cycle below the ground surface is shown in Fig. 2. The mass continuity equation for the saturated zone below the ground water table is:

$$G_r - E_g - B \pm G_a \pm \Delta G = 0 \tag{5}$$

where $G_r$ = ground water recharge, $E_g$ = evaporation component derived from saturated zone, $B$ = component of $R$ contributed from saturated zone, $\Delta G$ = change in ground water storage.

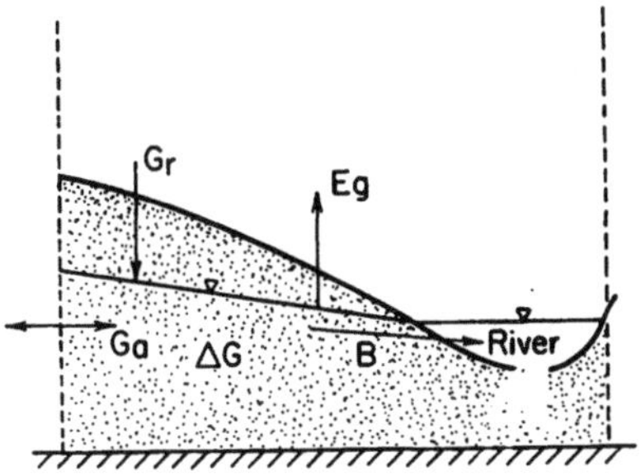


Fig. 2. Schematic of hydrologic cycle below ground.

Approaches to understanding the hydrologic cycles involve mass accounting procedures [4, 5], hydrodynamic methods, and statistical/stochastic procedures. The former two methods are concerned primarily with physically derived relationships, the later with statistical properties of the time series of the hydrologic parameters. All approaches depend on sufficient hydrologic data (time and space dependent) which generally are grossly inadequate at the planetary scales and inadequate at smaller scales [6]. Adequate theoretical constructs of the hydrologic cycles based on fundamental physical relationships (analogous to the General Atmospheric Circulation Models) are lacking. The development of more adequate hydrological data tends to provide the basis for better theoretical hydrologic understanding which in turn suggests an improved observational network leading to improved models for hydrologic forecasting and for water resources management purposes.

Space-borne systems can improve hydrologic understanding in three important ways by: (i) providing repetitive, synoptic, multi-spectral imagery on a world scale, (ii) providing standardized, centralized, hydrologic data in computer processable form, and (iii) providing the impetus and means for integrating fractionated hydrologic networks and for improving mathematical models of the hydrologic cycle. The role of space-borne remotely-sensed data in improving the hydrologic data base is outlined in § 3 and § 4.

## 2.2. Water Resources Problems

There are many different categorizations of water resources problems. For example, the US Water Resources Council [7] has summarized ten regional water management problems: adequacy of annual natural runoff 9*; ground-water storage depletion 6*; water quality: waste 10*, heat 5*, salinity 5*, sediment 9*; flood damage 10*, and wetlands 9*. These problems have been assessed for 20 regions comprising the 50 states. Each problem is assigned a rating of relative severity. The starred numbers (*) indicate the number of regions that are faced with existing and emerging regional water management problems that are classified as "major" or "severe" (for example, nine regions have major/severe problems associated with adequacy of annual natural runoff.) Only one region (Tennessee Basin) is without a major or severe problem. Nine of the twenty regions have five or more regional management problems that are classified as major or severe. Eighty-one of 200, or about 40% of all US regional water management problems, are classified as major/severe problems.

On a world scale recent UN Studies [8, 9] have indicated that irrigation may become the key issue of water resources development in the long-range perspective. Irrigation is inherently a consumptive use which denies direct re-use of the water. The mounting water demand is particularly acute for the developing countries which are initiating or expanding irrigation schemes to meet increasing population demands for agriculture production. The 1972 UN Conference on the Human Environment in Stockholm has called on the UN in cooperation with FAO and other interested agencies to develop projections on long-range irrigation demand and its impact on water-management schemes including assessing impacts of large-scale irrigation on the global environment. The recently acquired ERTS 1 data have demonstrated their potential usefulness for such regional water resources surveys and for country development schemes. Examples are cited in § 3.

The rural-urban water supply, rather than irrigation, is the most urgent near-term water resources problem in the developing countries. A 1972 WHO Annual Report indicates that 77% of the 1627 million population surveyed (including 88 developing countries) is not adequately served by community water supply services. In rural areas, only 12% of the population had adequate water supply in 1970. Even the attainment of the Second UN Development Decade Goals in 1980 would only provide no more than 25% of the rural population. Adequate hydrologic data must precede rational water development. Space-acquired data of water resources on the scales of watersheds, tributaries and reaches have proved their usefulness for supplementing conventionally derived data which are becoming progressively inadequate to meet the global needs.

Industrial water supply, on a world scale, is the third major water need. Water for manufacturing, cooling, waste disposal, or hydropower, is largely non-consumptive and used instream requiring regulation and storage of flow, water quality control, water pollution abatement, and the considerations of overall environment effects. Remotely sensed data to assess and manage industrial water supplies would likely require finer spatial resolution (for local projects), more advanced sensor technology (for assessing and monitoring water quality parameters), and the capability of near-real-time data management systems.

In summary, on a world scale, per capita withdrawal uses for all three major water uses are expected to increase. Hence total water withdrawal will increase at an even greater rate than the population growth rate. A more than several-fold increase in total water withdrawal is projected in three decades although there has been stationary or declining per capita uses in some industrially advanced nations such as USA and Israel. For the developing nations with high populations and projected high population growth rates, the degree of use of estimated available water resources is expected to increase dramatically during the next decades: for example, in India from 40 to 96% and in Mexico from 13 to 41%.

### 2.3. The Need for Comprehensive Resource Planning

As water resources development schemes become more complex and more costly, the need for comprehensive resource planning becomes apparent. Some potentially important programs leading to firmer bases of planning include the UNESCO International Hydrological Decade Program, the forthcoming International Hydrological Program, the WMO Global Atmospheric Research Program, The WMO World Weather Watch, The UNESCO Man and The Biosphere Program and the NASA Earth Resources Satellite Program.

The need for international cooperation in water resources is pervasive and increasing. About one quarter of the world's countries are located within international river basins (those embracing two or more countries). Of the nineteen largest river basins whose drainage areas are greater than one million square kilometers, fifteen are international basins involving two or more national jurisdictions. Recommendation 55 of the 1972 UN Stockholm Conference concerning a world registry of major or ortherwise important rivers, recommended in part that a revised and completed version of the register be prepared by the United Nations Secretariat. In a more general vein, Recommendation 51 of the Stockholm Conference concerning principles of cooperative actions for managing and developing water resources shared by two or more countries, recommended that the UN Secretariat, in cooperation with other interested international organizations, prepare a report describing and analyzing conceptual models and case studies, taking into account the great variety of physical and socio-economic conditions of international river basins as they affect the shares of the individual contributing countries. The inventorying and monitoring of the 200 international first-order river basins is within the demonstrated capabilities of a space-borne earth resource survey system such as ERTS 1.

## 3. Present Status of Remote Sensing in Hydrology and Water Resources

The first weather satellites launched in 1960, followed by TIROS and Nimbus satellites carrying TV cameras and radiometric instruments, were designed primarily for meteorological purposes. However, space-borne remote sensing for earth resources surveys is less than a decade old. The manned Gemini and Apollo satellites provided multi-spectral data of earth resources useful for experimental and planning purposes. Satellite applications for earth surveys and hydrology purposes have been reported previously [10—22].

The NASA Earth Resources Technology Satellite 1 (ERTS 1) launched on 23 July 1972 is the first satellite dedicated wholly to the task of obtaining imagery of the earth's surface for earth survey purposes. Some 34000 separate images of all major land masses of the world are now available for study. About 40 of 300 ERTS 1 investigations are primarily concerned with hydrology and water resources.

Two NASA-sponsored symposia, 29 September 1972 [23] and 5—9 March 1973 [18] comprising some 200 papers report on early and significant results obtained from ERTS 1. About 50 papers report on the latest studies using space-derived data for hydrology and water resources purposes and are summarized below.

### 3.1. ERTS Studies outside the US

Preliminary results of ERTS 1 experiments in South America, Spain, the Lower Mekong Basin, Mali, and Canada have been reported.

G. E. Stoertz and W. D. Carter [18] report that for desert and semi-desert areas in Bolivia, Argentina, Chile and Peru, hydrologic features using ERTS imagery were interpreted such as: principal types of salt crusts on major salars (salt-encrusted playas) some of which are subject to modification by seasonal flood waters; extent of flood waters and lakes on the floors of more than 120 closed basins (these provide a basis for more accurate mapping of hydrologic features in the central Andes); prominent former lake shores of Pleistocene lakes; mid-winter snow cover in the high Andes (these data are essential to the appraisal of water resources and runoff). The ERTS images provide for the first time a single source of data which permitted identification of these multiple features. Observations of synoptic snow cover was not possible prior to ERTS. The authors conclude that ERTS thus provides an "invaluable data base for earth resources survey and development in this large and relatively undeveloped area."

R. Espejo, S. Torrent, and C. Roquero [18], using ERTS data, have made estimates of future trends in the meander evolution of the River Ebro. This is the most important water course in Spain crossing the Iberian Penninsula in the southeastern direction. Major changes in river bed location cause considerable dislocation in the use of flood plain agricultural lands. ERTS data make possible better planning of agriculture lands subject to river meanderings and flooding.

W. J. Van Liere [18] using ERTS multi-spectral photography for the Lower Mekong Basin (Khmer Republic, Laos, Thailand and Viet-Nam) reports that the ERTS data have supplemented other data on forest cover of watersheds for inaccessible areas, refined general survey data concerning land forms and soils, and provided new information which would be almost or entirely impossible to detect with ground surveys or conventional photography. Such information concerns the mechanism of flooding and drainage of the delta, siltations and mappings of acidity of the Great Lakes and possibly also of salinity in the lower delta. The author also states that ERTS imagery fills gaps in the basic data which "are urgently needed in order to finalize a realistic post-war development program for the (Lower Mekong) Basin".

ERTS imagery is being used in resource management development in Sahelian Africa (Republic of Mali) in a study reported by N. H. MacLeod [18]. Hydrological analysis of ERTS imagery of the Niger and Beni River watersheds have shown the size, timing and areal extent of the annual flood. An inventory of the Malien resources is underway assisted by ERTS imagery showing the extent of the desert, mapping of soils, new and old drainage networks, feed and water for livestock and people, and land use.

Canadian studies using ERTS data are reported by L. W. Morley [18] and by L. Sagn-Wittgenstein [23]. About four orbits per day over Canada have provided some 3000 scenes as of 26 September 1972. Initial studies of ERTS frames (mostly MSS 6) have shown many hydrologic features and phenomena including the follow ing: silt-laden plume of the Fraser River discharging into the Strait of Georgia; many glaciers; the Birch River deltaic plain showing the winding channel and wet areas probably subject to periodic flooding, and bench areas; the Peace River flood plain whose river channel is well defined in the ERTS frames, islands, meanders, eroded channel of the Caribou River, a series of rapids and falls (Vermillion Chutes), man-made clearings on the south branch of the Peace River; the MacKenzie River Delta clearly showing river rapids, a 70 acre (28 hectare)* landslide, mudbanks, sand bars and eroded riverbanks. Arctic lakes greater than 400 feet (122 m) in diameter are visible, those less than 350 feet (107 m) are not visible on MSS 6.

## 3.2. ERTS Studies in the US

### 3.2.1. Flood Mapping, Survey of Surface Water

Several papers report on the problem of stream definition and flooding. E. F. Hollyday [18] in a study of two small basins in Maryland and Delaware, extracted the following hydrologic-related information from ERTS imagery; open water, infrared-reflecting vegetation, snow, and massed works of man. These ERTS-derived basin characteristics were then statistically related to stream flow characteristics using multiple-regression techniques. Preliminary results show improvement in correlations in 44 out of 101 test cases.

G. R. Hallbert, B. E. Hoyer and A. Rango [18] also used ERTS MSS imagery to map the flood of September 1972 on the East Nishnabotna River in Iowa. The flood boundaries mapped at a scale of 1:250000 agreed with boundaries mapped by low-altitude imagery and ground methods and allowed rapid appraisal of the areal extent of flood inundation (100 year flood) for the entire river system. The ERTS data were acquired seven days following flood recession showing that the time period critical for imagery acquisition to be at least one week.

Similar reporting of flood mapping by ERTS MSS imagery is reported by R. B. Morrison [18] for the Gila River, Arizona flood of 20, 21 October 1972, the second largest flood on record. Inundation limits mapped from ERTS imagery of 1, 2 November (one and one-half weeks after the flood) agreed well with those obtained by aerial photography and ground survey. Areas of severe sand and gravel deposition and local erosion were detected. A quick assessment of flood damage was made by comparing before and after ERTS images.

Surveying of open water, lakes and ponds using ERTS imagery have been reported. D. S. Gilmer, A. T. Klett and E. A. Work, Jr. [18], using the ERTS near-infrared MSS band have determined the number and distribution of prairie, ponds and lakes in North Dakota and compared the results with aircraft data. Ponds as small as one to two acres can be recognized sometimes, whereas ponds greater than two acres are consistently mapped. C. C. Reeves, Jr. in another study [18] of playa lakes in the Texas High Plains shows that satellite imagery can be used for a census of the thousands of lake basins which commonly exist in semi-arid areas, and which sporadically contain water at various times of the year. Over 6000 lakes were counted in a single ERTS frame; however, the number, size and distribution of these lakes vary depending on season of the year, storm tracks, etc. In another study in Alaska, B. J. VanTries [18] has interpreted ERTS

* 1 acre = 0.4 hectare.

data in the Yukon-Kuskokwin Delta which contains thousands of small ponds important to wild life and the production of water fowl. He reports minimum pond size measurable from ERTS data is approximately five acres.

### 3.2.2. Turbidity Observations

Hydrologic studies of Lake Ontario and its drainage basin are being undertaken as part of the International Field Year of the Great Lakes. A. Falconer [18] has reported that interpretations of ERTS data have yielded regional-scale maps of land use, soil and physiographic information which are required for modelling and behavior of river basins. F. C. Polcyn and T. W. Wagner [18] are evaluating the elements of the terrestrial water balance using image interpretation techniques to enhance and classify significant hydrological and limnological features. Data processing includes level slicing, ratioing of spectral bands and classification of features. E. S. Plukowski [18] has demonstrated the utility of ERTS imagery to monitor turbidity plumes generated by the Welland Canal and the Genessee and Oswego Rivers. The effluent from the municipal outfall sewer of the City of Rochester can be seen two miles off the Lake Ontario shore in ERTS imagery. Detection of turbidity dynamics in Tampa Bay, Florida using MSS imagery from ERTS is reported by A. E. Coker, A. Higer, and C. R. Goodwin [18]. The processed data integrated with other modelled parameters provide an overview of the dynamics of turbid material during dredging operations.

Water turbidity detection for two large reservoirs in Kansas using ERTS imagery reported by H. L. Yarger and J. R. McCauley [23] indicate good qualitative correlations between tone variations on MSS bands and Secchi disk and suspended load measurements. The authors suggest remote sensing methods for acquiring low-cost water turbidity data for large reservoirs. Sedimentation levels, mud flats of organic and mineral origin, are clearly defined for the Anacostia and Potomac Rivers waters in studies by J. S. Schubert and N. H. MacLeod [23].

### 3.2.3. Snow and Ice

The use of ERTS data for mapping snow cover is reported in studies by J. C. Barnes [18] and by M. F. Meier [18] for areas in the mountainous western US. Barnes reports that snow cover can be detected from the surrounding snow-free terrain by comparison of the high reflectance in the MSS 4 band. Snow line changes have been mapped on different dates using ERTS imagery. Differences in reflectance in the MSS 7 band of some glaciers is believed to be due to the existence of meltwater on the surface of the glacier at lower elevations in contrast to snow cover at higher elevations. M. F. Meier reports that percentages of snow-covered area on specific drainage basins measured from ERTS imagery are reported to within a few per cent of snow-covered areas. ERTS data combined with other hydrologic and meteorological observations in the North Cascade Range are used to calculate water stored in the mountain snowpack. This information can be used in subsequent years to forecast meltwater runoff. Frequent updating of expected inflow to reservoirs could be possible if similar satellite data were available in the future. Snowline altitude was mapped to an accuracy of 60 meters using enlarged ERTS imagery at 1:250000 scale.

Meier also reported that glacier accumulation area ratios were determined from ERTS imagery. Also subtle flow structures, undetected on aerial photographs, were visible. Surging glaciers were identified. Marked changes in tidal glacier termini were discovered between successive ERTS passes.

D. R. Weisnet [23] also studied ERTS imagery of snow conditions in the Cascade Mountains: A mean snowline elevation on Mt. Rainier for 28 July 1972 was estimated using ERTS data which compared reasonably with a mean firn line estimated by conventional methods. Melting snow conditions were indicated for several mountain areas by comparing the MSS 6 band with the MSS 7 band which showed greatly reduced snow reflectance. The author states that "the resolution of the ERTS MSS is certainly sufficient for determining snow-line elevation in mountainous terrain for all but the most stringent hydrological requirements."

Analyses of arctic ice features are reported by W. I. Campbell [23]. He states "the ERTS imagery is clearly the best sequential, synoptic data on ice ever taken by means of remote sensing for purposes of ice dynamic studies such as those being conducted as part of the AIDJEX Project".

### 3.2.4. Coastal Wetlands, Offshore Dynamics, River Plumes

Mapping of coastal wetlands using ERTS data has been reported by several investigations teams. R. R. Anderson, V. Carter and J. McGinness [23] have analyzed ERTS data in the Chesapeake Bay, Maryland and Savannah, Georgia to determine usability for wetland mapping and ecological studies using map scales of 1:1000000, 1:125000, and 1:24000. On the scale 1:1000000 (as delivered from NASA) the following were determined: marsh water interface and upper wetland boundary; large plant communities, 100 meter and above (including *spartina alterniflora*, *spartina patens*, and *spartina cynosuroides*, also *Juncus roemerianas* and *typha spp.*); tree islands as small as 160 m and small streams 16 m wide. On scale 1:125000 marsh water interface and upper wetlands boundary, and successional zones are clearly shown; smaller communities (less than 25 m in some cases) of the above species, open, non-vegetated ditches for drainage and agriculture. On scale 1:24000 all boundaries seen in above scales become blurred.

Studies of the Delaware Coast by V. Klemas and D. Bartlett [18] indicate that spatial discrimination capabilities of ERTS imagery compare favorably with those of aerial infrared photography and that spatial resolution is the dominant factor limiting the potential for detailed mapping of coastal vegetation species using ERTS imagery.

R. L. Mairs, F. S. Wobber and D. Garofalo [18] have reported that ERTS imagery from several successive passes over the New Jersey coast indicates that these data can be used in real-time decision making. These imageries have shown the extent, predominant drift, and dispersion characteristics of waste disposal in coastal New Jersey waters including the New York harbor tidal discharges. Also shown is the increased reflectance of the turbid waters due to large sewage effluent flow and the tidal dynamics which cause the turbid waters to emanate out to the populated bathing beaches in southern New Jersey shore area.

The monitoring of large dumping of municipal and industrial wastes in the New York Bight is reported by C. T. Wezernak and N. Roller [18]. ERTS imagery clearly shows the distribution of an acid-waste discharge, sewage sludge dump and major suspended sediment inputs into the dump area. Analyses indicate the ERTS data will provide information about environmental effects of marine dumping. Similar studies by R. Charnell and G. Maul [23] using ERTS imagery of the New York Bight show the Hudson River plume being easily distinguished.

Ground truth information shows a drop in surface water salinity of about 2% across the plume. Evidence of waste dumping is also seen. The authors suggest that satellites with sensors optimized to view the ocean in visible and infrared wavelengths, supplying synoptic wide-area data, will make management of the coastal zone on a broad scale much more realistic.

Several studies using ERTS imagery of estuarine and coastal waters are reported. R. T. Kerhin [18] studying the Chesapeake Bay has determined some general sedimentation patterns, growth of beach features and stability of the bar and trough topography. V. Kermas, R. Srna and W. Treasure [18] using imagery from three successive ERTS passes over the Delaware Bay and the Atlantic coastal region have identified strong suspended sediment patterns and several distinct types of aquatic interfaces; namely, strong change in color and turbidity, and discontinuities associated with salinity and temperature. These were correlated with a limited amount of data from boats and aircraft. Studies of correlations of ERTS MSS imagery with suspended matter and chlorophyll in the lower Chesapeake Bay are reported by D. E. Bowker, P. Fleischer, T. A. Gosink, et al. [18]. The results indicate the utility of ERTS MSS imagery in monitoring estuarine water for the assessment of siltation, productivity and water types. F. H. Ruggles, Jr. [18] has delineated Connecticut River plumes using ERTS frames. These provide the first input to a timelapse analysis of circulation patterns at the eastern end of Long Island Sound. ERTS imagery of nearshore northeastern Pacific ocean waters have been studied by P. R. Carlson, R. J. Janda and T. J. Conomos [18]. Variations of river effluent plumes of the Eel, Klamath and Smith rivers of northern California are related to differences in land forms and land use within the contributing river basins. The 18-day ERTS observational cycle allows definition of seasonal and storm-related variations in effluent dispersal patterns.

Sediment distribution and coastal processes in Cook Inlet, Alaska have been studied by D. M. Anderson, L. W. Gatto and H. L. McKim [18] using sequential ERTS MSS imagery. Visible in the MSS frames are current patterns of differential concentrations of suspended sediment and tongues of clear oceanic water. Fresh water plumes, heavily laden with sediment, are visible at the mouths of all major rivers. Tidal flats and a number of unmapped cultural features appear prominently in MSS bands 5 and 6. In other studies reported by R. E. Hunter [18], distribution and movement of suspended sediment in the Gulf of Mexico off the Texas coast using ERTS imagery proved very useful. The suspended sediment is used as tags on water masses to obtain information on water movements.

L. R. Lepley, G. Calderon and J. R. Hendrickson [18] using ERTS imagery of the Northern Gulf of California have mapped submerged shoals, current streamlines, and location of possible upwellings, downwellings and submarine springs. R. S. Williams, Jr. [18], studying ERTS MSS frames of the Plymonth and Duxbury area of Massachusetts have concluded that ERTS MSS imagery provide two-dimensional coastal subaerial and submarine information (bay mouth bars, shoals, dredged channels, etc.) superior to conventional map products at least to scale of 1:250000.

### 3.2.5. Water Depth Estimation

Estimation of water depth using ERTS imagery is reported by D. S. Ross [18] for the Great Bahama Bank. Where a reflective bottom and clear water is found, MSS 4 images can be used for estimating depths in about two, five and ten meter steps, while MSS 5 images would indicate very shallow water. Correlation of depth measurements at 1, 1.5 and 4 fathoms (1.8, 2.8 and 11.0 m) has been successfully demonstrated using ERTS MSS data by F. C. Polcyn [18] for the Little Bahamas Bank. The author states that satellite data will provide geographical evidence for verifying the existence or non-existence of doubtful shoal waters now appearing on world charts and considered to be hazardous to shipping.

#### 3.2.6. ERTS Data Collection System

Results of experiments using the ERTS Data Collection System (DCS) have been reported by a number of investigators. H. H. Schuman [18] using a single water level recorder connected to an ERTS data collection platform on the Verde River in Arizona successfully transmitted stream stage information 4.5 times per day on the average for a 43-day period. The amount of data received far exceeded the expected transmission rate (twice per day) from the DCS system. The DCS data permit daily stream flow to be calculated to within $\pm 5\%$ for slowly changing river flows and to within $\pm 10\%$ for rapid flow changes.

Preliminary analysis of ERTS-DCS data from 20 water stations (5 stream, 3 ground water wells, 12 water quality) in the Delaware River Basin indicate that the DCS system is performing well. The three to four transmissions received each morning period and each evening period provide daily operational data to headquarters of the Delaware River Basin Commission. These data are useful for the planning of a national DCS ERTS-type, near-real-time hydrologic relay system.

The most extensive ERTS DCS experiment is being conducted by the New England Division (NED) Corps of Engineers to determine the feasibility and usefulness of ERTS data-relayed information to the management and operation of the flood control and multi-purpose reservoir system operated by NED. This study serves as a prototype for assessing the DCS system at the national scale for day-to-day management of Corps of Engineers reservoirs projects. In New England, data collection platforms are operated on eleven stream gages, two precipitation gages and four water quality stations. In support of the relayed data, ERTS imagery analysis is also being conducted to interpret areal and spatial variability of dynamic hydrologic basin parameters as these relate to operational missions.

### 3.3. Matrices of Remote Sensing Capabilities for Hydrologic Elements

R. H. Burgy et al. [24] have analyzed hydrologic systems from the point of view of water data requirements and the relative merits of data acquisition by remote sensing techniques. The hydrologic system is modelled as seven interrelated subsystems: (i) evaporation (evapotranspiration), (ii) vegetation (includes soil-vegetation interrelation), (iii) precipitation, (iv) streams and estuaries, (v) reservoirs and lakes, (vi) unsaturated soil zone, (vii) saturated zone. For each subsystem, the important hydrologic parameters and applications are identified. Measurement criteria based on hydrologic requirements are specified for each parameter: frequency, ground resolution, estimated range, desired accuracy, and point estimate or averages. Standard user techniques and standard acquisition methods are noted. Information on instrumentation and range in the electromagnetic spectrum are listed for each parameter. Matrices for the various hydrologic subsystems are given in [24]. These contain valuable information gathered from multiple sources and assist the systematic assessment of the role of remote sensing to hydrology.

Robinove [16] has evaluated remote sensors for hydrological purposes. He includes for photography: panchromatic, multi-spectral, infrared, color, color infrared and radar imagery; also microwave radiometry and imagery and radar imagery. These are evaluated for: measurement of physical characteristics of water surfaces, snow surveying, and mapping, mapping and description of groundwater features, glaciology, geomorphology and measurement of the liquid-vapor transfer in the hydrologic cycles.

### 3.4. Summary of Assessments of Space-Derived Imagery for Hydrology and Water Resources

Based on the experience to date using space-derived imagery, the following assessments can be made:

1. Space-derived imagery now exists in sufficient amounts to assess its present and potential usefulness to hydrology and water resources.

2. Most applications of remote sensing are at the descriptive stage. However, developments leading to quantification of many hydrological aspects can be expected.

3. The major characteristics of present space-borne remote sensing systems which have demonstrated usefulness to hydrology are: (a) repetitive coverage in the visible and near-IR range of large areas at resolutions sufficiently fine (spatial, spectral, temporal) to provide significant benefits for many applications; (b) reliable, near-real-time data-relay capabilities from hydrologic gages on the ground; (c) computer-ready, standardized, and merged digitized input data from multiple sources, collected at a centralized data processing center.

4. Space-borne imagery has been demonstrated to be useful for interpreting general hydrologic information, especially for areas where mapping information is inadequate (in remote arctic regions and in many developing countries).

5. Space-borne remote sensing techniques have demonstrated the potential capability for yielding better scientific understanding of the hydrologic cycle (particularly at regional, continental and planetary scales) in terms of mass budgets, better definitions of surface water bodies (including dynamic changes), for improving input data to mathematical models of hydrologic systems, for assessing atmospheric water quantities, for assessing certain terms in evaporation (evapotranspiration) equations, and for gaining better understanding of snow and ice-covered areas.

6. Space-borne remote sensing techniques have demonstrated the potential capability for improving the design, planning and management of various water-resources projects. Important hydrologic objectives which are within the present technology capabilities are: (a) general reconnaissance, monitoring and surveying of dynamic hydrologic features and phenomena over sizeable areas on a routine basis; (b) mapping of open surface water bodies as small as several acres in size; (c) surveying of estuarine and coastal hydrologic features (including submarine features), and monitoring of dynamic water circulation patterns (including identification of sediment-laden waters, off-shore waste dumping sites and plumes of outfall sewers); (d) mapping of areal extent of snow-covered and ice-covered

areas (including some discrimination of melt water vs snow) and mapping of snow line elevations in mountainous terrain for runoff estimation purposes; (e) mapping of flooded areas on a regional scale; (f) mapping of coastal wetlands at scales as fine as 1:125000 for planning and administration purposes; (g) detection of water turbidity characteristics in large reservoirs, lakes and at mouths of rivers.

7. Advances in the general field of earth observations from space are expected to provide significant benefits to hydrology. These include better weather forecasts, improved estimates of atmospheric water vapor; improved knowledge of radiation fluxes at the surface (applicable to evaporation studies); improved thematic mapping of land use, vegetative cover, topography, geology, soils and soil moisture; and improved precipitation observations.

## 4. Potential Future Role of Remote Sensing in Water Resources

### 4.1. Deficiencies in Networks and Data

Space technology comes at a time when scientific knowledge of the hydrologic cycle and management knowledge of water projects are becoming vital to rational planning. Present hydrologic networks are inadequate to provide data for quantitative understanding of the global hydrologic cycle, even though there are an estimated 250000 hydrologic gages of all types scattered throughout the world. Two-thirds of the world's land area possess marginal or less than minimal hydrologic observational networks according to WMO standards [6].

Elementary facts of the earth's waters are inadequately known. We have been unable to measure routinely the discharge to the sea of the major rivers of the world. We have been unable to inventory the major lakes of the world. We are not able to guess at the number of lakes greater than 100 km$^2$ in size. We have little knowledge of precipitation over major drainage basins of the world. We have little knowledge of the amounts of waters below the ground water table, in the soil zone, and in galciers, snow and permafrost. We are unable to measure evaporation adequately and have hardly begun to estimate transpiration rates. We are unable to measure the water-borne sediments transported to the sea.

### 4.2. Concepts of Water Information Systems

Hydrology needs new water information systems with capabilities of yielding types of information that cannot be obtained from conventional networks. Large-scale operational systems will be needed that include ground and remote sensing subsystems, communications networks, and real-time data processing facilities.

Concepts of water information systems using space technology and remote sensing have been proposed previously by Bock [25—27], Kennedy [28], Tepper [19] and in reports of the interagency Earth Observatory Satellite Mission Review Group [29]. A report of the National Academy of Sciences [21] has proposed a package of hydrologic satellites. These consist of systems for hydrologic data relay, called hydrologic communications satellites (HCS), and hydrologic remote sensing HSS 1 incorporating presently feasible systems and HSS 2 requiring further research and development. HCS would relay data from ground hydrologic stations to satellite-data receiving stations (SDRS). HSS 1 and later HSS 2 would provide remotely sensed data to SDRS. Data processing and interpretation would be performed at Regional Hydrologic Centers (RHC) where short period applications (flood warning, hydropower operations, etc.) and long period applications (water resources planning, basic data inventory) would be undertaken. These hydrologic data would then be sent to user agencies who would provide further analyses and interpretation. The overall scheme is conceived to be part of an operational earth survey system related to meteorology, oceanography, geology and other environmental monitoring systems.

M. W. Molloy and V. V. Salomonson [30] have reported on the overall development of programs in remote sensing related to water resources. The various weather satellites are described: TIROS, Nimbus, ATS, ITOS, SMS/GOES and the multinational GARP and GATE programs. Hydrologic phenomena are identified according to their space-time scales. These are superimposed on the time-space capabilities for present and proposed remote sensing systems ERTS, Synchronous Earth Observatory Satellite (SEOS), Earth Observatory Satellite (EOS), and ground sensors-satellite relay systems. Fig. 3 [30] shows that: (i) hydrologic phenomena involve time scales measured from seconds to centuries (9 orders of magnitude) and have spatial dimensions ranging from continental to less-than-one meter (7 orders of magnitude); (ii) present ERTS capabilities cover only a small space-time area of hydrologic phenomena ($3 \times 2$ orders of magnitude); (iii) ERTS B, SEOS and EOS enlarge the space-time hydrologic domain to $5 \times 7$ orders of magnitude but important gaps remain. It is evident that space-borne systems provide capabilities in the hydrologic space-time domain which are necessary but unattainable by direct sensing methods.

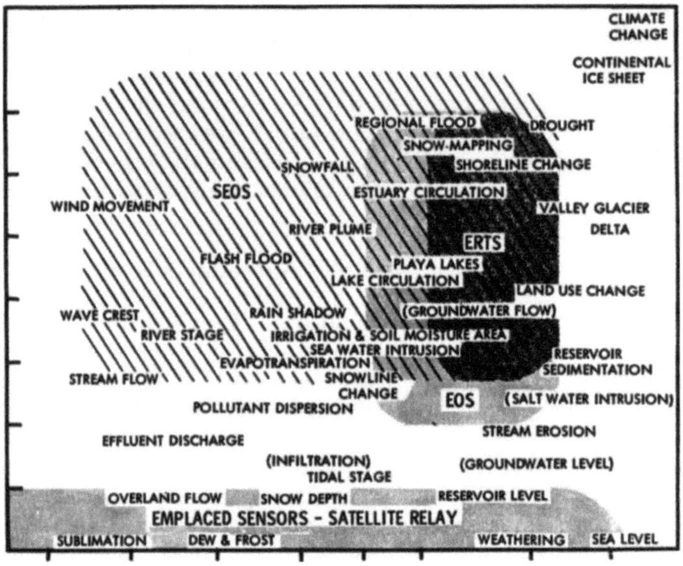


Fig. 3. Effective uses of water resources satellites.

### 4.3. Future Developments in Remote Sensing for Hydrological Purposes

The future pattern of development of space technology for hydrological purposes may be guided by progressions of the meteorological satellite programs: "from research to operational systems; orbital to synchronous satellites; visible to infrared, microwave and possibly radar sensors; imagery to temperature, altitude and composition measurements; and, from cloud photo interpretation to machine processing and numerical modeling. The weather program now includes data relay from instruments on the ground, on buoys and balloons, and plans for international systems" [30].

The development of satellite programs for hydrological purposes is likely to follow a somewhat different pattern. Multiple hydrologic objectives tend to diffuse research into numerous local efforts in numerous governmental operations. General hydrologic models are only in the formulative stages. Except for the IHD started in 1965, there has been little international experience in undertaking large-scale hydrologic research projects and operational systems. Priorities for country development, for water resources development, for environmental protection seem to be less understood than the more direct linkages of aviation-military-meteorological space efforts. Hence adequate funding for hydrology (and the other non-meteorological earth sciences) has lagged.

#### 4.3.1. Future Developments

Future developments important to hydrology and water resources are enumerated below.

*General*

1. Strengthening of the academic and scientific basis of physical hydrology is required to guide space programs. Remote sensing knowledge in the general hydrologic community should be upgraded. In particular, physically based mathematical models of the hydrologic cycle must be developed to guide experimental and operational hydrologic space programs. This is a long-term undertaking.

2. Fractionated hydrologic programs must be consolidated to provide a focus of efforts and economies of scale. Space systems are inherently expensive and global in nature. International cooperation should be sought.

3. Commonalities must be sought with meteorological space programs and other earth survey programs.

4. In the early stages, qualitative, descriptive approaches for use of space-derived imagery should be sought for early, serendipidous, and high-yield pay offs. For example the qualitative description of the areal extent of snowpack provides useful, immediate improvements of estimates of snowmelt runoff, although ultimately the quantitative measurement of water content of snowpack on an areal basis would be more useful. Fast regional mapping of on-going floods is extremely useful; mathematical models coupled to data from improved remote sensors would be more valuable for predicting the dynamics of floods and for management decisions and for enumerating flood damages.

5. Space technology programs should reflect the multiple objectives and fractionated nature of hydrology and water resources. In contrast to meteorology, hydrology programs require greater breadth, more flexibility, more opportunism,

more alternatives to be evaluated, and a highly iterated development program based on on-going developments.

6. Institutional arrangements should be simplified. These are complex at national and international levels because water management practices are constrained by political, historical, legal and social considerations.

7. Hydrology should seek a clear focus (such as the meteorologists' "to provide better forecasts") to marshall public support for sizeable hydrologic efforts in space programs. Molloy and Salomonson [30] report annual benefits of $ 700 millions attributed to better management of water resources (mostly minimizing flood damage and improving forecasts of irrigation water) derived from a proposed earth resources survey satellite system. This represents 50% of the estimated total $ 1.4 billion annual benefits to the US economy derived from the proposed satellite system.

*Multi-disciplinary*

8. Attainment of all-weather, day-night, pointing and location systems.
9. Quantification of temperature structures.
10. Quantification of water quality parameters (or indicators).
11. Attainment of ground resolutions as fine as 25 m horizontal, less than 1 m vertical.
12. Deployment of inexpensive APT-type readouts.

Finally of extreme importance:

13. Development of inexpensive, reliable, real-time data collection, data relay systems.

*Surface Waters*

1. Stream: flow depth, velocity, flow rate, sediment.
2. Coastal waters: vertical distributions, quality indicators.
3. Snow: water content, depth.
4. Ice: thickness.
5. Lakes: depth, internal circulations, quality.

*Subsurface Waters*

Soil moisture: extent and amount.

Ground water: indicators of fluctuation of ground water table and seepage.

*Atmospheric Waters*

Precipitation: rate, extent.

Vapor flux: precipitable water, flux of water vapor.

The space age has changed man's thinking about the earth and himself. Hydrologists are now beginning to turn their attention from man's ancient water wheels to nature's hydrologic cycle. Space technology and remote sensing are providing this new vision.

## References

[1] S. Manabe and J. Smagorinsky, Mon. Wea. Rev. **95**, 155 (1967).

[2] N. H. Crawford and R. K. Linsley, Tech. Rep. 39, Dept. of Civil Engineers, Stanford University 1966.

[3] M. I. Lvovitch, EOS Transactions, Amer. Geophys. Un. **54**, 28 (1973).

[4] Publ. No. 93, Int. Assoc. Scient. Hydrology, UNESCO 1971.
[5] G. P. MALHOLTRA and P. BOCK, Symp. on World Water Balance, Publ. No. 93, Int. Assoc. Scient. Hydrology, UNESCO 1971 (p. 501).
[6] World Meteorological Organization, WMO, No. 168, T. P. 82, WMO, Geneva, Switzerland.
[7] 1st National Assessment of the Water Resources Council, Water Resources Council, Washington, D.C. 1968.
[8] Note of the Secretary General, UNEC 7/35, 25 September 1972.
[9] Triennial Rep. on Water Resources Development 1968—1970, Rep. ST/ECA/143, Dept of Economics and Social Affairs, UN, N.Y. 1971.
[10] Abstracts for 4th Ann. Earth Resources Prog. Rev., Jan. 1972, NASA Manned Spacecraft Center, Houston, Texas.
[11] J. C. ALISHOUSE, D. R. BAKER, E. P. MCCLAIN and H. W. YATES, NOAA Tech. Memo. NESS 26, Washington, D.C., March 1971.
[12] Bibliography of Remote Sensing of Earth Resources for Hydrological Applications 1960—67, US GSWRD, Interagency Report, NASA-134 (1968).
[13] E. P. MCCLAIN, 3rd Ann. Earth Resources Prog. Rev. **3**, Hydrology and Oceanography, NASA, Houston, Texas (1970).
[14] Proc. 7th Int. Comp. on Remote Sensing of Environment, Vols 1—3, Center for Remote Sensing Information and Analysis, Willow Run Laboratories, Univ. Michigan, Ann Arbor, Michigan, May 1971.
[15] Proc. before Comm. on Science and Astronautics, House of Representatives, 92nd Congr., Jan. 1972 and Panel on Science and Technology, 13th Meeting, US Government Printing Office, Washington, D.C. 1970.
[16] C. J. ROBINOVE, paper presented at UN Conf. on Exploration and Peaceful Uses of Outer Space, Vienna, August 1968.
[17] V. V. SALOMONSON, paper presented to US/IHD Comm. on Remote Sensing in Hydrology, Goddard Space Flight Center, NASA 1972.
[18] Abstracts of Symp. on Significant Results obtained from ERTS 1, March 1973, NASA/ Goddard Space Flight Center, Greenbelt, Md, March 1973.
[19] M. TEPPER, unpublished.
[20] TRW Systems Group, Coastal zone oceanographic requirements for earth observatory satellites A and B, Feb. 1971, prepared for NASA Langley Research Center, Hampton, Va.
[21] Useful applications of earth-oriented satellites-hydrology, prepared by Panel 3 of Summer Study on Space Applications for NASA, Nat. Acad. Sci., Washington, D.C. 1969.
[22] T. WAGNER, unpublished rep., Willow Run Lab., Univ. of Michigan, undated.
[23] Proc. Symp. on Earth Resources Technology Satellite-1, September 1972, NASA, X-650-73-10, Jan. 1973.
[24] R. H. BURGY, D. R. STORM, M. L. HORTON and J. P. MALINGREAU, in: Ann. Prog. Rep., An Integrated Study of Earth Resources in the State of California Using Remote Sensing Techniques, May 1972, Space Sciences Lab., Univ. California, Berkeley, Chap. 3.
[25] P. BOCK, report for review by US Nat. Comm. for the IHD, and WMO Comm. for Hydrometeorology, Sept. 1965.
[26] P. BOCK, paper presented at 13th Ann. Amer. Astronaut. Soc., Dallas, Texas, AAS67-109, May 1967.
[27] P. BOCK, Proc. 2nd Ann. Amer. Water Resources Conf., Nov. 1966 (p. 405).
[28] J. M. KENNEDY, paper presented to Committee on Remote Sensing Programs for Earth Resource Surveys, Nat. Res. Counc., Washington, D.C., March 1972.
[29] Earth Observatory Satellite Mission Review Group (EOSMRG), Final Rep., November 1971, NASA, Washington, D.C.
[30] M. W. MOLLOY and V. V. SALOMONSON, paper presented at Symp. Remote Sensing of Water Resources, Burlington, Ont., Canada, June 1973.

Earth Survey Problems — Akademie-Verlag, Berlin 1974

# APPLICATIONS OF THE ERTS 1 SATELLITE IN REMOTE SENSING OF WATER RESOURCE DATA IN CANADA

J. D. Mollard[a] and P. A. Carr[b]

[a]J. D. Mollard and Associates Limited, Regina, Saskatchewan, Canada
[b]Inland Waters Directorate, Department of Environment, Ottawa, Ontario, Canada

In Canada the ERTS 1 satellite is being used to help solve hydrologic problems in two principal ways. First, the space imagery is being interpreted to facilitate surveys that explore, measure, develop, and conserve Canada's surface and underground waters. Secondly, the ERTS 1 satellite is being used to re-transmit data on water levels, ice break-up, and chemical quality obtained from platforms installed in remote Canadian lakes and rivers. Applications discussed include rapid and inexpensive preparation of schematic terrain and hydrogeological maps of northern Canada, mapping surface-water bodies and certain data on the quality of water in them, detecting underground water supplies and identifying ground-water recharge and discharge areas, detecting faintly expressed manifestations of geological structures and discontinuities that control ground-water flow and localize mineral and petroleum deposits, and mapping sedimentation in rivers and lakes that present construction, fish, and wildlife environmental problems. Other ERTS 1 applications discussed concern mapping muskeg and permafrost hydrology problems, studying the effect of snowfall and its melting on spring runoff, flooding hazards and soil moisture conditions for crop germination, and finally the tracking of sea ice break-up and movement affecting Arctic shipping and transportation, development of new port facilities, and the search for favorable channel crossings for large-diameter Arctic gas pipelines.

## 1. Introduction

Because so little interpreted ERTS imagery is available for study by those wanting to improve their skill, the authors have devoted a major part of the paper to the analysis of space imagery examples. Past water-resource studies and present and future uses, significance, and research are mentioned briefly in some of the captions. Also, because the ERTS 1 satellite offers such a great number and variety of applications in water-resources surveys, it is necessary in a paper of this length to select certain applications to emphasize, listing others that have been studied in ERTS images but cannot be discussed in as much detail for space limitations.

## 2. Six Main Hydrological Applications emphasized in the Paper

Six applications emphasized in this paper relate to studies in which the ERTS 1 satellite can be used to facilitate:

1. Rapid and inexpensive preparation of small-scale schematic surficial geo-

logical, hydrogeological and terrain maps covering sparsely inhabited arctic and subarctic regions (Figs. 1 and 2). These small-scale maps provide a good appreciation of northern terrain and hydrological conditions and problems, particularly those associated with permafrost and peatland conditions (Plates 1 and 2).

2. Location, exploration, and development of major ground-water sources for municipal and industrial use (Figs. 3 and 4 and Plates 3, 4 and 5).

3. Prediction of local, regional, and basin groundwater flow systems (Plates 4, 5 and 6, and Fig. 5).

4. The mapping of peatland types and peat depths, which are closely related to surface water and groundwater conditions (Plates 7 and 10 and Fig. 6).

5. Detection and plotting of weakly expressed structurally controlled lineaments in areas where the bedrock surface is widely masked from direct observation by glacial deposits. Their identification and interpretation aids in prospecting for higher yielding water wells and in the exploration for mineral and petroleum deposits (Plates 8 and 9, and Fig. 7).

6. Monitoring water levels, ice break-up, and water quality in remote Canadian lakes and rivers by ERTS 1 retransmission (Figs. 8, 9).

The figures have been selected to illustrate these six uses of the ERTS 1 satellite while also including terrain data that are relevant to the additional applications isted in the following section.

## 3. Other Water Resource Surveys aided by ERTS 1 Imagery Interpretation

Other Canadian water resource surveys evaluated in ERTS 1 multi-spectral imagery [1], many of them referred to in captions to the illustrations, include:

1. Mapping surface-drainage courses and river-basin networks, eroded badland areas, and classifying and mapping river-system and channel morphology on a regional basis (Plate 1).

2. Studies of mixing of different qualities of water at the confluence of large rivers and where major rivers enter large lakes or the ocean.

3. Charting sequential changes in depths of water in large shallow lakes based on discernible changes in their shape and size.

4. Detecting changes in the concentration of mineral salts in saline lakes from changes in lake levels and extent of salt deposits around their shore zones (Plate 4).

5. Mapping sedimentation patterns in large lakes and in the ocean and relating them to natural phenomena and to changes in land use in the watersheds contributing runoff water and sediment discharge (Plate 2).

6. Detecting pollution in large lakes (e.g. algae blooms).

7. Lake and shoreline classification for recreation development (Plate 3).

8. Charting the accumulation and melting of snow in sequential ERTS 1 imagery in relation to storage of ponded surface waters and wetland habitats, flooding hazards during spring runoff, and the effects of differences in spring runoff on soil moisture available for crop germination and recharge to the ground-water system (Plates 4 and 5).

9. Mapping areas of higher moisture contents of soils owing to differing moisture-retention qualities of these soils (e.g., fine- vs coarse-textured soils), and areas of relatively higher or lower ground-water tables.

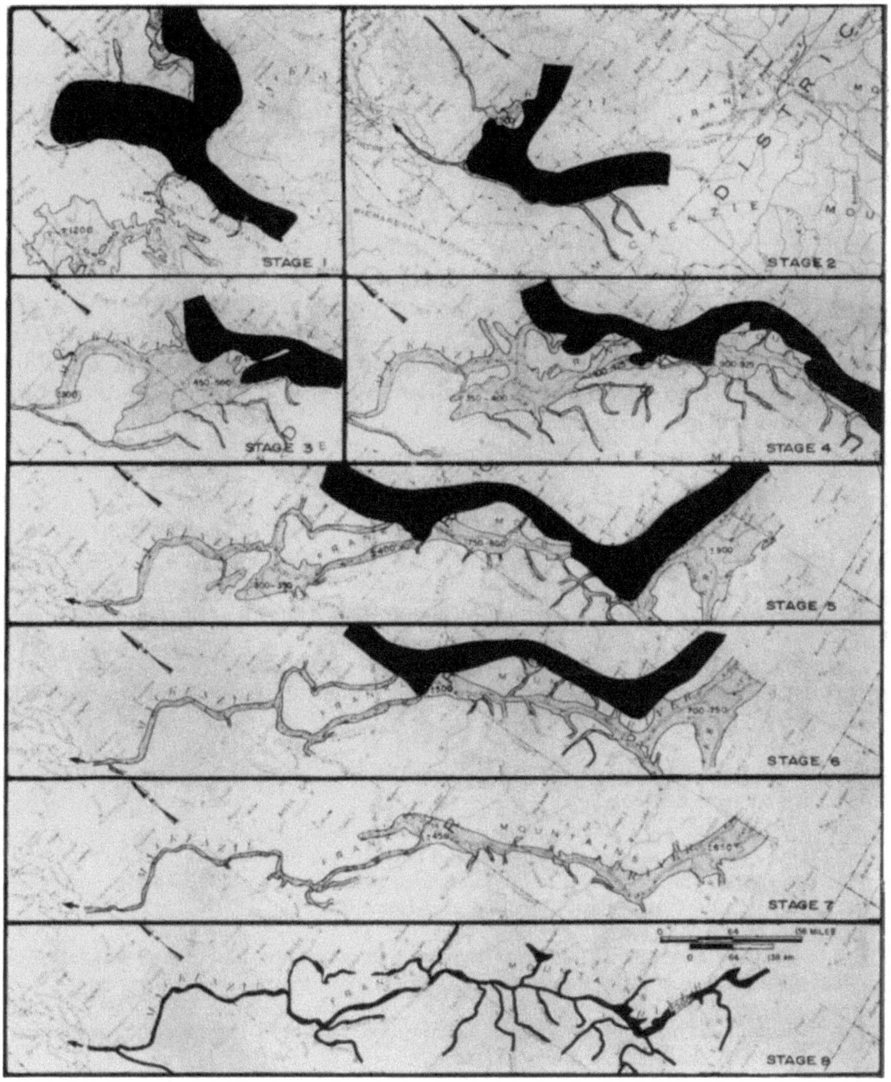


Fig. 1. Schematic map showing inferred successive Wisconsin glacier ice-front positions and associated proglacial lake outlines and elevations along their shorelines. Mackenzie Valley, Northwest Territories and northern Yukon, Canada.

10. Mapping large springs entering snow-covered lakes and detecting large structurally controlled springs originating from bedrock faults or other geological discontinuities.

11. Preliminary surveys associated with the location of major dam sites along with, in some cases, the study of associated river diversions (Plate 10).

12. Charting sea ice break-up and movement, and assessing its significance in relation to coastline geomorphology—mainly as applied to arctic and subarctic shipping, navigation, port development, and the search for gas pipeline routes (Plate 11).

## 4. Presentation of Data in Illustrations

Certain bands of ERTS 1 multi-spectral imagery have been selected to illustrate the use of space vehicles in making water-resource surveys in Canada; these are presented as "Plates". Usually only a portion of the total area shown on a given ERTS frame is reproduced. Each is shown with a surrounding grid of letters and numbers to identify specific features on the scene imaged, and is accompanied by a detailed caption. Line drawings are included where they are considered to provide useful ground-truth information.

Fig. 2. →

LEGEND OF PRINCIPAL LANDFORMS AND SURFICIAL MATERIALS

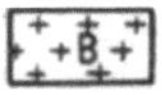

Large hills, ridges, and mountains. Mainly folded and eroded sedimentary rocks.

Glaciated bedrock hills and ridges. Mainly thin till deposits over smooth glacier-eroded bedrock.

Smooth pediment and piedmont slopes. Mainly thin silt, fine sand, and scattered rubble over bedrock.

Undulating to rolling plains with and without ice-flow features. Mainly lodgement till. Mo in the northern Yukon indicates pre-Wisconsin till and Mh northeast of Inuvik indicates a heavy-textured clay-rich till.

Gently sloping plains. Mainly thin and discontinuous lacustrine silt clay over sheet-eroded ground moraine.

Glacial-lake basin plains. Mainly widespread peat and thermokarst developed on thick lacustrine silt and clay deposits.

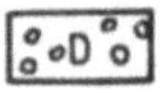

Large deltaic plains. Mainly fine sand and silt with minor clay overlain by peat and dotted with thermokarst features. Marine deltas are shown as Dm; estuarine deltas, De; and glaciallake deltas as D, often with scattered sand dunes.

Low lakebed plains. Mainly interbedded deltaic fines and and silt with frequent lacustrine clay strata.

Low lakebed plains. Mainly thin and discontinuous interbedded deltaic and lacustrine fine sand, silt, and clay with local till "islands".

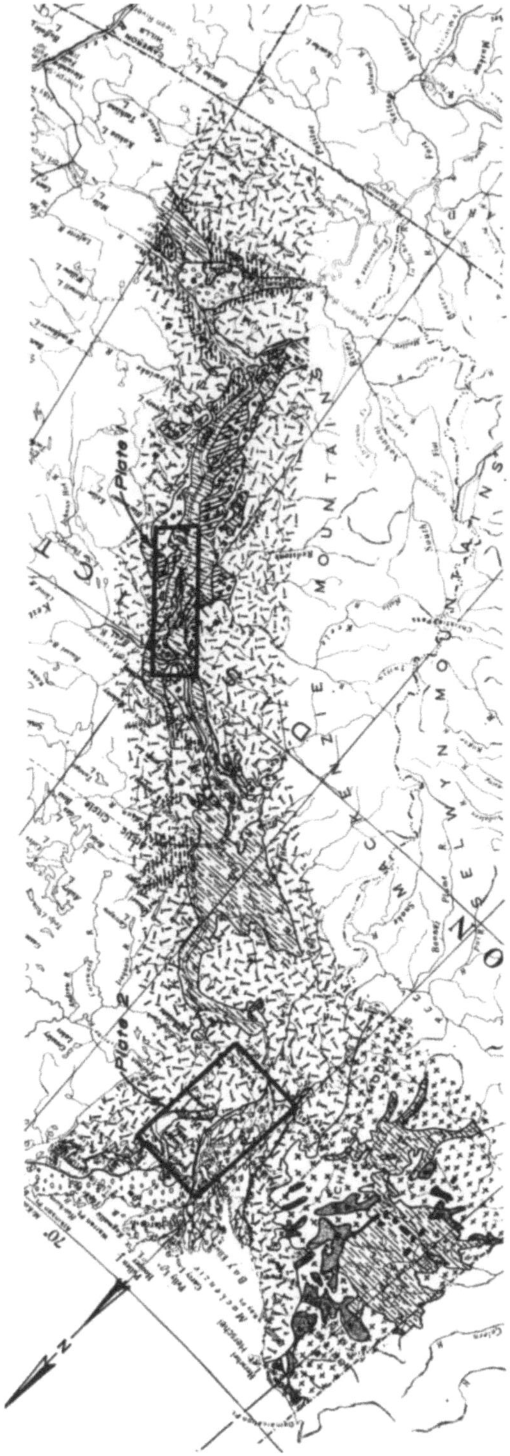


Fig. 2. Schematic map showing dominant landforms and surficial materials in the Mackenzie Valley, Northwest Territories, and in the northern Yukon, Canada. (For details see overleaf.)

## 5. Schematic Terrain and Hydrological Maps

Figs. 1 and 2 show generalized geomorphological and hydrogeological data prepared originally from federal government 1 inch = 12 mile mosaics, 1 inch = $1^{1}/_{2}$ mile contact aerial photographs, and 1:250000 and, where available, 1:50000 topographic maps [2]. These maps were prepared before ERTS 1 satellite multi-spectral imagery became available. They were developed mainly to give a first appreciation of (i) the effects of the last Wisconsin glacier retreat on the location and extent of proglacial lakes and their approximate maximum levels (Fig. 1), and (ii) the distribution of dominant landforms and surficial materials along with their hydrogeological implications (Fig. 2). In particular, they were intended to serve as a logical basis for more detailed follow-up mapping of hydrogeological information.

The original maps were prepared at a scale of 1:640000 and took several months to complete. Figs. 1 and 2—at scales of 1:6700000 and 1:5700000 respectively—were synthesized from the 1:640000-scale maps. The important point here is that the time and expense required to make these maps could have been greatly reduced through the use of good quality ERTS 1 imagery, with the satellite passing over the mapped area once every 18 days. It is also worth noting for comparison that a single ERTS image covers the same area covered by 16000 contact aerial photographs flown at the common scale of 1:12000 using a 6-inch lens camera and 60% end lap and 30% side lap. Plates 1 and 2 show relatively small areas of Figs. 1 and 2 (see inset areas marked Plate 1 and Plate 2 on Fig. 2).

The regulating effects of runoff waters stored in large lakes on river channel type, channel stability, and on fluvial processes and sedimentation [3], along with their engineering and environmental ramifications, may be inferred from fluvial

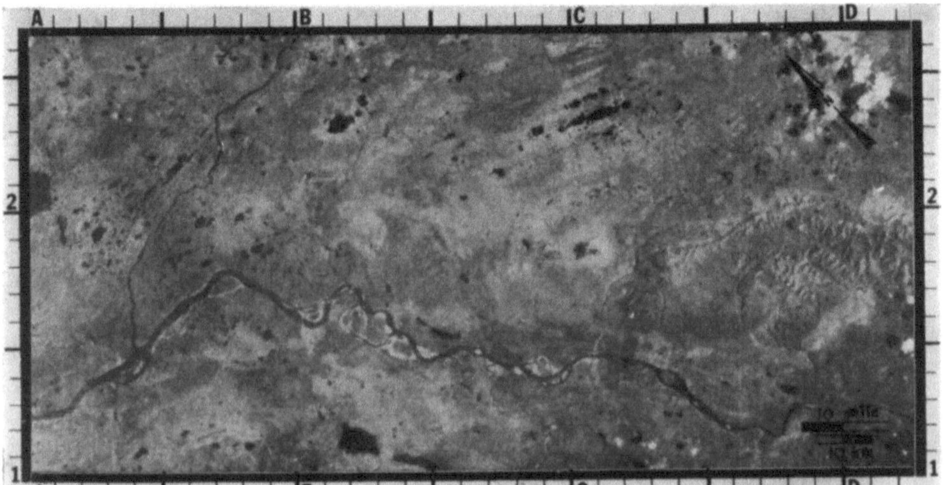


Plate 1. ERTS 1 band showing inset of Plate 1 area on Fig. 2. Confluence of the Great Bear River and Mackenzie River, NWT, Canada.

morphological features ("river-scape morphology") recorded in ERTS 1 imagery. A noteworthy example of this can be seen in the portion of original ERTS 1 scene shown in Plate 1.

In parts of the arctic and subarctic studied the satellite imagery shows the approximate (i) distribution of wet and dry tundra areas, and (ii) extent of thermokarst development. Both are expressions of surface and subsurface hydrological conditions and can often be correlated with the texture of surficial geological materials as well as with regional and local topography and associated surface and subsurface drainage conditions in permafrost regions (Plates 1 and 2).

The accelerated development of a particular variety of landslide variously called tundra mudflows, active-layer glides, or detachment slides can often be

←

Plate 1.

Primary application: Permafrost hydrogeological mapping.
Secondary applications: River channel classification studies; prediction of significantly different streamflow patterns in rivers located on the east and west side of Mackenzie River.
Latitude and longitude of prominent physical feature and its coordinates on surround: Confluence of the Keele and Mackenzie Rivers B.7:1.4$^1/_2$; lat. 64°25′; long. 124°50′.
ERTS imagery designation: E-1033-19175.
Date of imagery: 25 August 1972.
Band shown: 6
Past water-resource studies in area shown by ERTS imagery: Extensive terrain mapping; regional groundwater geology and hydrology studies.

*Interpretation of imagery*

The morphology of the rivers, as seen in the original ERTS imagery from which this small portion was taken, suggests that streamflows in larger tributary rivers originating on the east side of the Mackenzie River are considerably regulated by storage in large lakes, namely, Great Bear and Blackwater Lakes. However, no similar runoff-regulating lakes exist on the west side of the Mackenzie River and this is evident from the channel patterns. Streamflow in rivers on the west side of the Mackenzie is therefore more "flashy" and channel beds there are more conspicuously braided as a result. Whitish areas at B.9 to C.0:1.0 to 1.2 on the Keele River are exposed coarse gravels and cobbles in the bed of an actively shifting braided river. Because of the large storage capacity of Blackwater Lake, and its effect on river regulation and sediment load, Blackwater River runs clear; and in conventional panchromatic photography the Blackwater shows black at its confluence with the silt-laden Mackenzie River (C.9:1.3). Note, also, that the Mackenzie River between A.9 and C.4 exhibits an anastomosing channel pattern (islands surrounded by branching channels).

Poor surface drainage conditions associated with perennially frozen ground (permafrost) are indicated by the presence of thermokarst lakes and large ponds. Thermokarst drainage features commonly occur on bedded fine-grained (silt, clayey silt, silty clay, and silty fine sand) alluvial, lacustrine, and deltaic deposits, along with scattered wind-deposited sand dunes, as at A.1 to A.7:1.7 to 2.2; A.7 to B.3:1.7 to 1.9; and A.4 to A.8:1.4 to 1.7. However, it should be noted that better-drained sloping ground underlain by these same surficial sediments shows relatively little thermokarst development. The light gray thermokarst areas support Sphagnum and very short (dwarfed) black spruce scrub vegetation.

Bedrock rapids and springs in the Great Bear River at A.7:2.2 to 2.4 are not apparent in band 6, the only band of ERTS imagery available for examination covering this portion of the Mackenzie River Valley.

correlated with recent fire burns in northern permafrost regions. Continuous ERTS 1 coverage would give an insight into where and when this type of slope instability may be expected to take place—something that would assist in improved ecological and land-use planning, including better prediction of the rates of geological processes such as river and lake sedimentation, which have an effect on construction activities and fish and wildlife ecology.

Although space imagery is useful in schematic mapping of larger Pleistocene and Holocene landforms and inferring the kinds of surficial materials in them, the amount of information that can be extracted is significantly enhanced when it is combined with stereoscopic study of different types of aerial photography flown at scales of 1:50000 to 100000. Study of terrain features in colour, false-colour (colour IR), black-and-white and thermal infrared, and panchromatic airphotos at scales of 1:10000 to 1:30000 following ERTS imagery interpretation increases the capabilities of the space imagery even more. Specific localized areas requiring detailed attention of the observer may in many cases be pinpointed originally from study of the ERTS multi-spectral imagery.

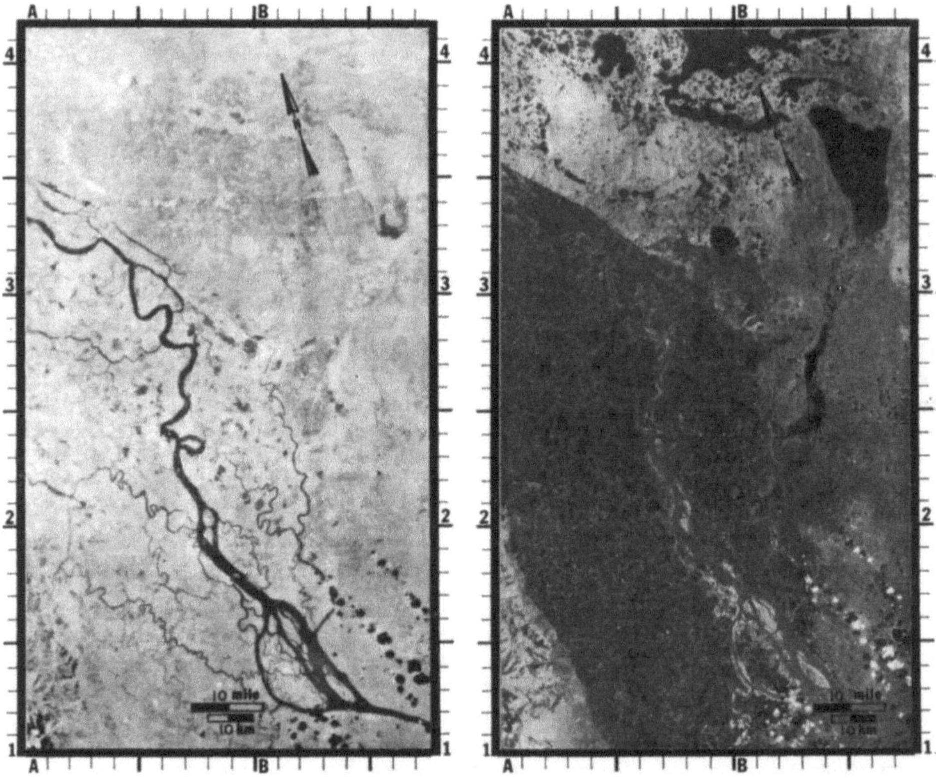


Plate 2. ERTS bands 5 and 6 showing inset of Plate 2 on Fig. 2. Mackenzie Delta area, Northwest Territories.

## 6. Location of Municipal and Industrial Ground-water Supplies

Fig. 3 shows the locations of 44 aquifers located and tested in southeastern Saskatchewan before the first ERTS imagery became available in late July 1972. Most of these aquifer investigations were carried out in the decade 1962 to 1972 and dealt with the location and development of ground-water supplies for towns, cities, and industries. Safe yield requirements typically range between 20 and 500 Imperial gallons a minute (90—2270 litres per minute). In carrying out each of these 44 investigations, remote-sensing studies were made from the interpretation of 1:16000 to 1:60000 panchromatic aerial photographs. In a significant proportion of cases they resulted in the location of favorable-looking exploratory test-drilling sites and in useful preliminary appraisals of aquifer size and recharge possibilities [4].

The locations of these 44 aquifers were plotted on ERTS 1 black-and-white positive paper prints, commonly on all four wavelength bands imaged during each pass, where all bands were available. Eight of the 44 aquifers located and tested prior to receipt of the ERTS 1 imagery could be discerned on the limited number of space-imagery scenes available for examination at the end of March 1973.

←

Plate 2.

Primary application: Permafrost terrain and tundra hydrogeological mapping.
Secondary application: Delineation of drainage features on a large estuarine delta characterized by widespread thermokarst development.
Latitude and longitude or prominent physical feature and its coordinates on surround: Mackenzie Delta A.5:2.0; lat. 68° 30′; long. 134° 30′.
ERTS imagery designation: E-1007-20140.
Date of imagery: 30 July 1972.
Bands shown (left): 4 or 5; (right): 6.
Past water-resource studies in area shown by ERTS imagery: Extensive terrain mapping; study of the relation of ground-ice occurrence to surface geological materials and to hydrological conditions in a continuous permafrost region.

*Interpretation of imagery*

In band 6 it is difficult to confidently identify the numerous sinuous streams traversing the Mackenzie Delta (A.0 to B.0:1.0 to 3.5). Even the smallest serpentine streams are, however, easily delineated in the left image (probably band 4 or 5). Colour composites may yield useful data on changes in thickness of the active layer and depth to the permafrost table associated with standing water (where the permafrost has thawed up to several hundred feet locally, commonly places where the water depth is greater than the thickness of winter ice in lakes and rivers, a figure of some 6 to 9 feet). Colour composites should also show silt plumes in the west ends of both Campbell Lake and Sitidgi Lake (B.$3^1/_2$:2.5 and B.6:3.$2^1/_2$ to 3.$3^1/_2$, respectively).

Notice, too, that lighter-toned areas in the tract A.0 to B.3:3.2 to 4.1 correspond to higher, better-drained ground (above about elevation 400 feet) and mainly to dry tundra areas versus wet tundra areas lying below 400 feet. The tundra below 400 feet elevation exhibits much greater development of thermokarst features. Depth to the permafrost table commonly ranges from a few inches to a few feet in virtually all sites that are not overlain by deep standing water. A recent fire burn shows up clearly in vicinity of Inuvik, and south of Noel Lake (see the outline at A.9:3.1 on band 6).

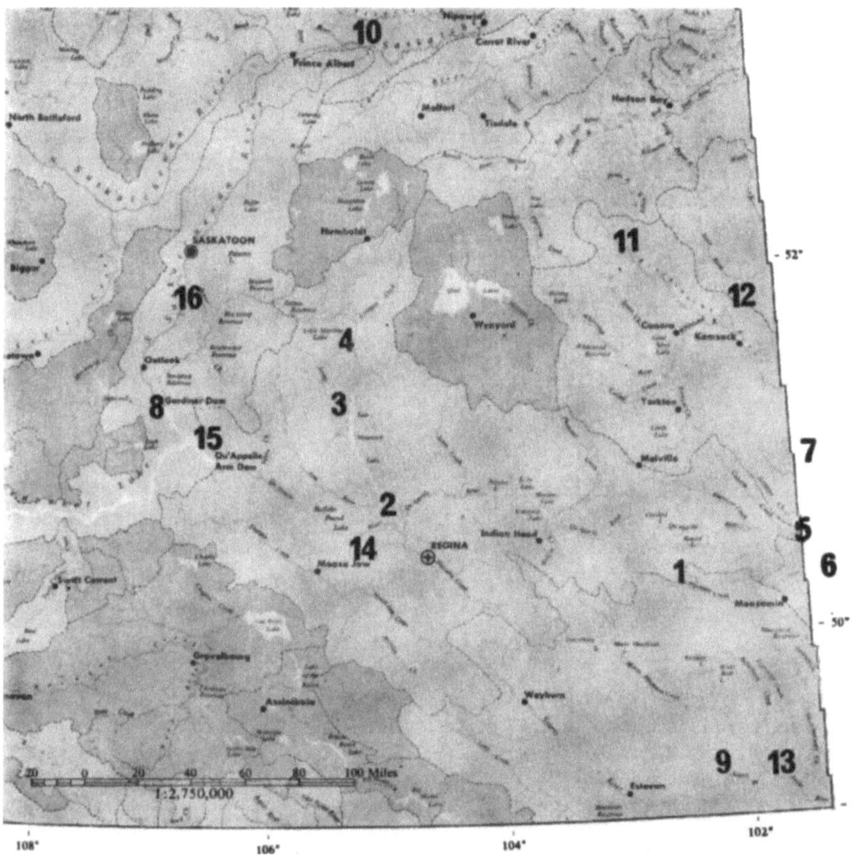


Fig. 4. Key map showing geomorphic settings and land-use patterns that in ERTS imagery correlate with upland sandy and gravelly outwash and glacial-lake deltaic plains, major water-table aquifer systems, and local spring occurrences. Southeastern Saskatchewan and adjoining western Manitoba, Canada.

Geomorphic settings and land-use patterns that in ERTS imagery correlate with upland sandy or gravelly outwash and deltaic plains, water-table aquifers, and spring occurrences, Location: southeastern Saskatchewan and adjoining western Manitoba. Surficial water-table aquifers and associated hydrogeological environments that were correlated with ERTS imagery occur at the following river and/or creek valley confluences, which valleys also functioned as major pro-glacial meltwater channels: 1) Pipestone River and Ekapo Lake channel, south of Broadview; 2) Arm River channel and Last Mountain Lake, south of Regina Beach; 3) Lewis Creek channel and Last Mountain Lake, south of Imperial; 4) Little Manitou Lake channel and Lanigan Creek, northwest of Venn; 5) Qu'Appelle Valley and Assiniboine Valley, in vicinity of Welby and Fort Ellice; 6) Birdtail Creek and Assiniboine River, east of Treat, Manitoba; 7) Shell River and Assiniboine River, east of Shellmouth, Manitoba; 8) Anerley Lakes channel and the South Saskatchewan River, north of Dunblane; 9) Moose Mountain River and Souris River, northeast of Alameda; 10) the North Saskatchewan and South Saskatchewan Rivers, east of their confluence; 11) Lillian River and Assiniboine River, east of Preeceville; 12) Swan River and Spruce Creek, east of Norquay; 13) Antler River and Lightning Creek, east of Carnduff; 14) Moose Jaw Creek and Qu'Appelle River, northwest of Stony Beach; 15) the Qu'Appelle Valley and the South Saskatchewan River, east of Elbow; 16) Beaver Creek and the South Saskatchewan River, west of Dundurn. The localities were identified originally in airphoto-interpretation studies carried out in search of a) groundwater supplies for towns and industries, and b) sand and gravel supplies for construction purposes.

One geomorphic setting that can be identified in the ERTS imagery and can be correlated with major aquifers is the confluence of major glacial meltwater channels (Fig. 4 and Plate 3). These settings are commonly associated with sand and gravel in large outwash and glacial-lake deltaic plains. The aquifer locations shown in Fig. 4 were identified originally in airphoto studies carried out in search of (a) ground-water sources, and (b) sand and gravel supplies for construction purposes. Many of the larger aquifers consist of uniform fine sand that has been eroded and re-deposited by the wind. The aquifer sites are usually characterized by large springs where the present valley bottom is situated below the base of the aquifer, usually the contact of surficial sand and gravel overlying till in upper valley sides. Water wells in these aquifers often yield 500—1000 Imperial gallons per minute (2270—4540 litres per minute) [5].

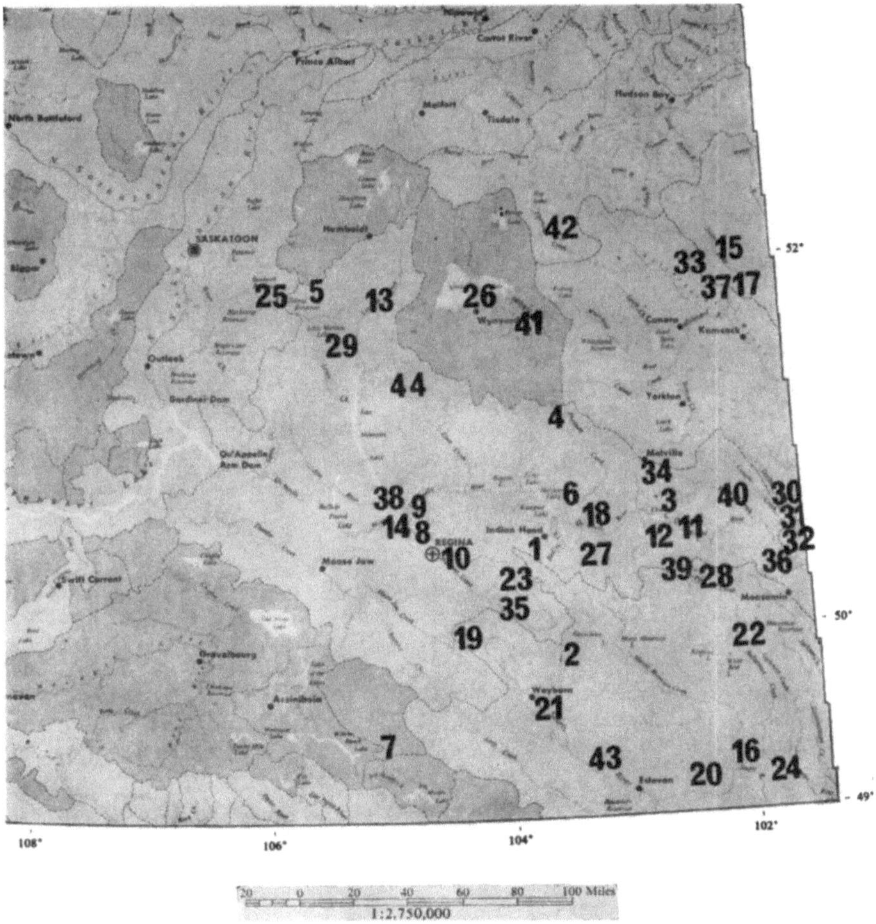


Fig. 3. Key map showing locations of tested aquifers in southeastern Saskatchewan, southeast of Saskatoon, Canada. Used to correlate with and evaluate ERTS 1 multi-spectral imagery.

Two other aquifer sites commonly detectable on the space imagery include (i) chains of elongate lakes located at intervals along abandoned meltwater channels, and (ii) large alluvial fans where major tributaries discharge into lakes or into major meltwater channels (Plates 3 and 4). The meltwater channels may have formed superglacially, englacially, or proglacially, each having a slightly different expression in the landscape.

In summary, good quality ERTS imagery can disclose potentially favorable major water-table aquifers for further inspection in conventional aerial photography and investigation in the field. Increased or decreased recharge to these aquifers may also be indicated from sequential winter and early spring satellite

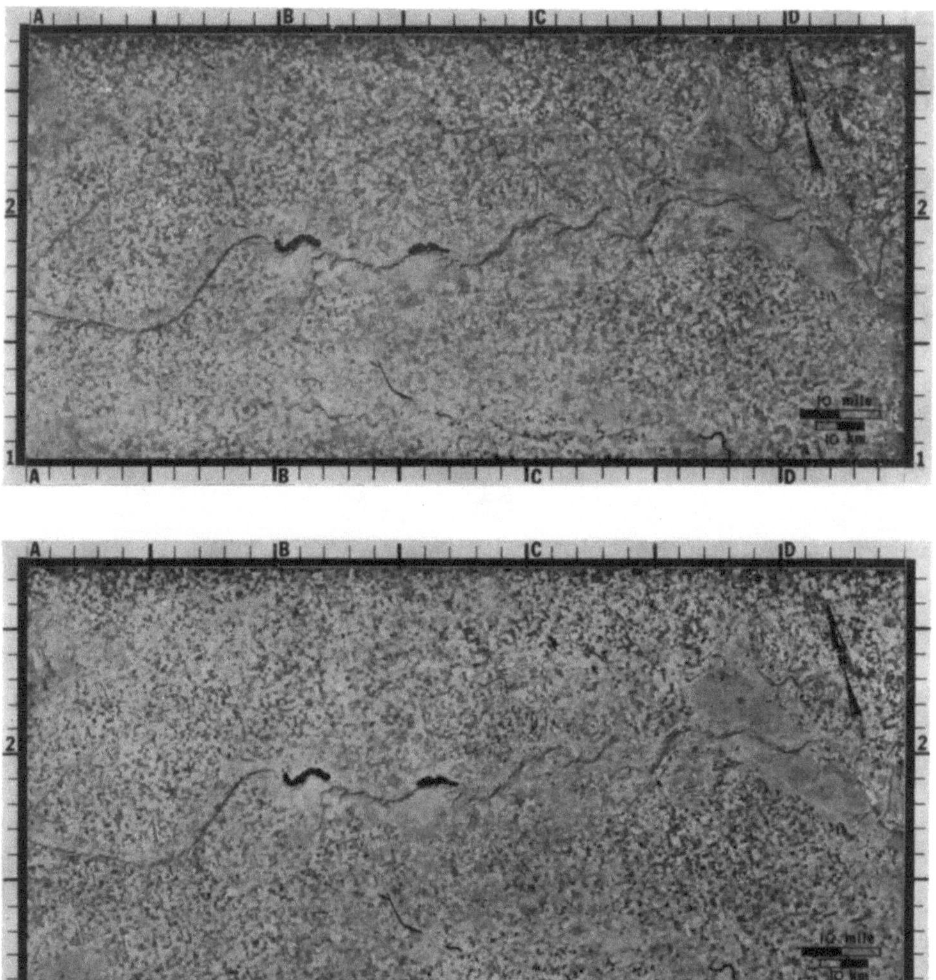


Plate 3. ERTS imagery scene showing one particular geomorphic and land-use setting that correlates well with major water-table aquifers.

imagery showing snow cover and its pattern of melting, runoff, and infiltration. Late spring imagery showing hummocky moraine landscapes commonly discloses ponded snowmelt waters in ponds and small lakes.

In making satellite imagery studies as useful and significant as possible, it is emphasized that the interpretation of conventional panchromatic and colour infrared photography by experienced specialists is almost always desirable [6]. The interpretation of suitably scaled panchromatic photography should be followed in turn by routine field reconnaissance and by a proper exploratory test-drilling program.

## 7. Evidences of Ground-water Recharge and Discharge

ERTS imagery scenes from several areas were examined to see if space-derived data could be used to help distinguish local, regional, and basin patterns of ground-water flow. It was found that local ground-water discharge areas are revealed by irregular whitish shore-zones around saline lakes and by small and large salt flats

←

Plate 3.

Primary application: Groundwater development, chiefly major water-table aquifers.
Secondary applications: Lake and shoreline classification for recreation development.
Latitude and longitude of prominent physical feature and its coordinates on surround: Round Lake B.6:1.9; lat. 50° 32′; long. 102° 20′.
ERTS imagery designation: NASA E-1083-17163.
Date of imagery: 14 October 1972.
Bands shown (top): 5; (bottom): 7.
Past water-resource studies in area shown by ERTS imagery: Twelve industrial and municipal groundwater location and development investigations; shoreline classification and mapping studies on Crooked and Round Lakes.

*Interpretation of imagery*

Ground-water supplies occur along linear sections of streams that appear to be bedrock-controlled (i.e. structurally controlled). Three of these linear stream valley segments are given by the following coordinates on band 5 (top): A.0:1.8 to A.3$^1/_2$:2.1; B.9:2.2$^1/_2$ to C.4$^1/_2$:1.9; and B.8:2.7 to C.6:2.1. Aquifers discovered in these valleys are usually artesian and typically occur beneath a confining layer of thick, relatively impervious, fine-grained alluvium.

Large water-table aquifers, commonly with associated springs locally, occur in thick sandy outwash and deltaic sediments on the upland at the confluences of major glacial meltwater rivers (see area 5, Fig. 4). These former drainage channels may or may not be occupied by misfit creeks and rivers. Two examples identifiable in Plate 3 are located at B.5 to B.6:1.2 to 1.3, south of the Town of Broadview; and the rudely parallelogram-shaped area bordered on the west by C.6:2.1 to 2.3 and running southeast to D.3:1.6 to 1.9, at the confluence of the Qu'Appelle and Assiniboine Rivers. These particular geomorphic settings, which show up quite well on bands 5 and 7, are attractive environments to prospect for major industrial or municipal groundwater supplies. Only slightly less attractive sites are the large sandy and gravelly alluvial fans occurring at the mouths of major tributary creeks draining into the Qu'Appelle Valley (see four such sites in the area A.7 to B.8:1.8 to 2.0).

Two brine-storage reservoirs (potash plants) are easily distinguished in band 7 (C.1:2.4$^1/_2$ and C.2$^1/_2$:2.3$^1/_2$). Even quite large springs at C.8$^1/_2$:2.1 do not show up on any of the four spectral bands imaged.

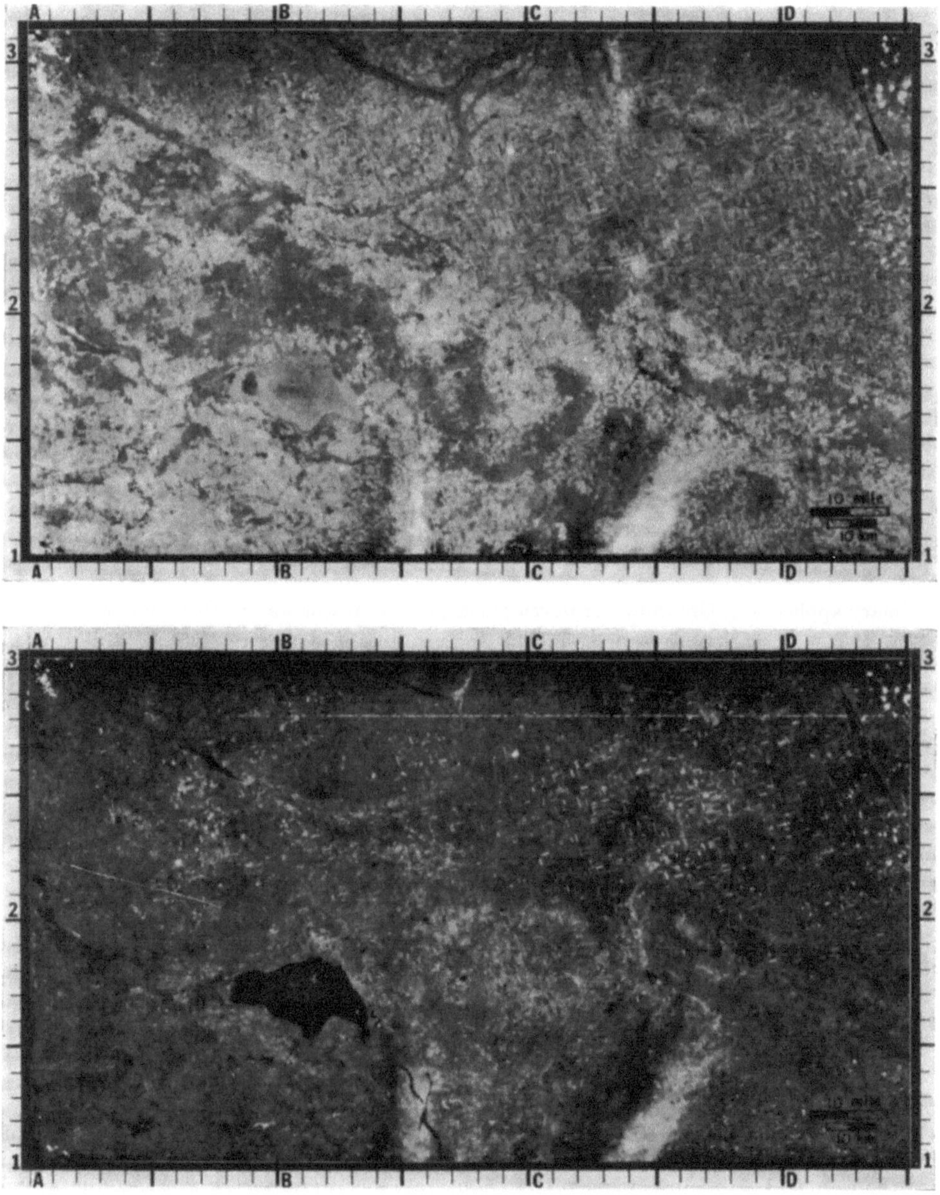


Plate 4. Two ERTS imagery bands of an area in southeastern Saskatchewan showing dominant surficial geological deposits, major hydrographic features, and local hydrogeological features (e.g. local areas of high salinity, and surficial lineaments emphasized by drainage elements that appear to be structurally controlled).

(Plate 4). In one area examined south of Winnipeg, Canada, a lighter pinkish signature on an MSS false-colour composite coincides approximately with a region of flowing artesian wells [7]. They also correspond with widespread salinized and alkalinized soil areas shown on provincial soil-survey maps [8].

In the Winnipeg area, the pinkish false-colour composite shows minor water courses that are barely discernible on the ground, whereas the yellow MSS false-colour composite brings out shallow-water features in lakes—as at the southern end of Lake Winnipeg. Moreover, if for instance pink and blue false-colour over-

←

Plate 4.

Primary application: Ground-water investigations.
Secondary applications: Surface-water development and mapping potential canal seepage problem-areas.
Latitude and longitude of prominent physical feature and its coordinates on surround: Old Wives Lake B.1:1.7; lat. 50° 05′; long. 105° 55′.
ERTS imagery designation: E-1031-17274.
Date of imagery: 23 August 1972.
Bands shown (top): 5; (bottom): 7.
Past water-resource studies in area shown by ERTS imagery: Mapping and classification of Buffalo Pound Lake shoreline (B.5:2.9) from the standpoints of potential erosion, slumping, and flooding hazards; municipal groundwater studies; surface-water studies (canals, reservoirs) on bed of former Glacial Lake Regina, a heavy clay lakebed.

*Interpretation of imagery*

Being highly cohesive, the heavy clay soils on the level to gently undulating bed of former Glacial Lake Regina have high moisture-retention capacity; and in a semiarid climate these soils are almost completely cultivated. The dominant crop is wheat. Southern boundary of the heavy clay lakebed can be detected on bands 5 and 7. In detail the southern margin of the former glacial-lake basin runs from A.0:3.0 to B.7:2.1 to D.5:1.6, approximately. The blanket of uniform lake-bottom sediments, commonly ranging up to 30 feet in thickness, tends to mask surficial indicators of municipal ground-water prospects except along the larger drainage channels (A.9:2.5) and where sandy deltaic deposits mark the positions of former shorelines and lake margins (A.8:2.4 on band 5).

Two prominent lineaments in the Avonlea area (C.4:1.8 to C.9:1.5 and C.9:1.6 to D.4:1.3 on band 5) are believed to represent deep-seated (possibly basement) faulting that carries up into the Paleozoic and Mesozoic strata. Certain of the deep-lying geological structures are believed to have faint rectilinear surface expressions (Dr. W. O. Kupsch, professor of geology, University of Saskatchewan, and Mr. R. Rudichuk, consulting geologist, Regina, private communications). The darker-toned surficial lineaments are accentuated by groundwater and surface moisture and are not noticeable on band 7. These and other less prominent lineaments are better expressed in Plate 4 (band 5).

Note that spring-fed Shoe Lake, an "alkali" lake, is obvious in band 5 (C.1 to C.2:1.1 to 1.2) but is not evident in band 7. The large dark-toned area north and east of Shoe Lake is a thick sand and gravel outwash deposit containing a water-table aquifer. Most darker-toned areas seen in band 5 correspond to areas of hummocky disintegration moraine having water in numerous kettleholes (black specks at B.0:2.1), which are ground-water-recharge areas. Smaller dark-toned areas are commonly surficial outwash sand or gravelly sand deposits (potential water-table aquifer). Note also, that Old Wives Lake (B.1:1.7) is very shallow (compare bands 5 and 7); its salt content varies widely depending on the volume of spring runoff that enters the lake.

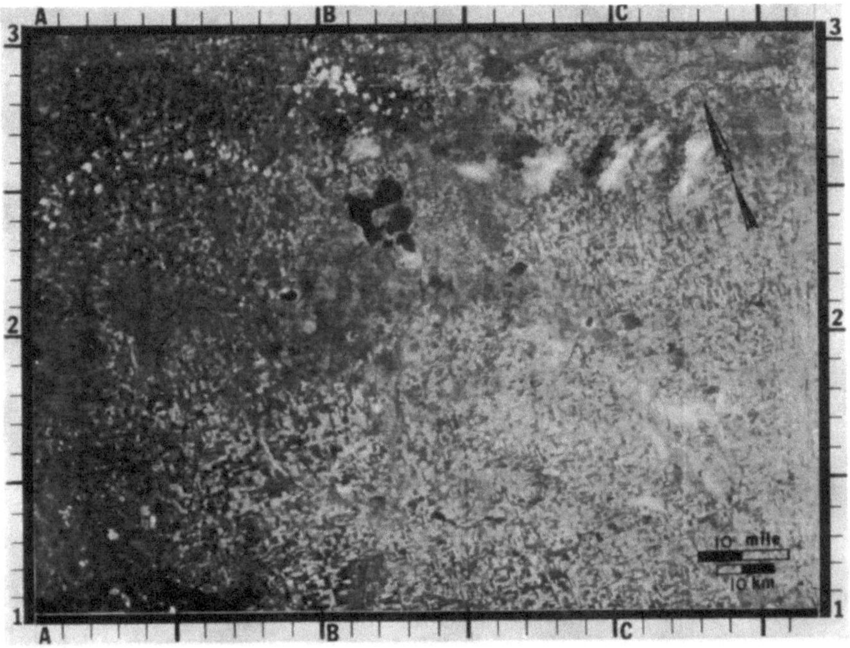


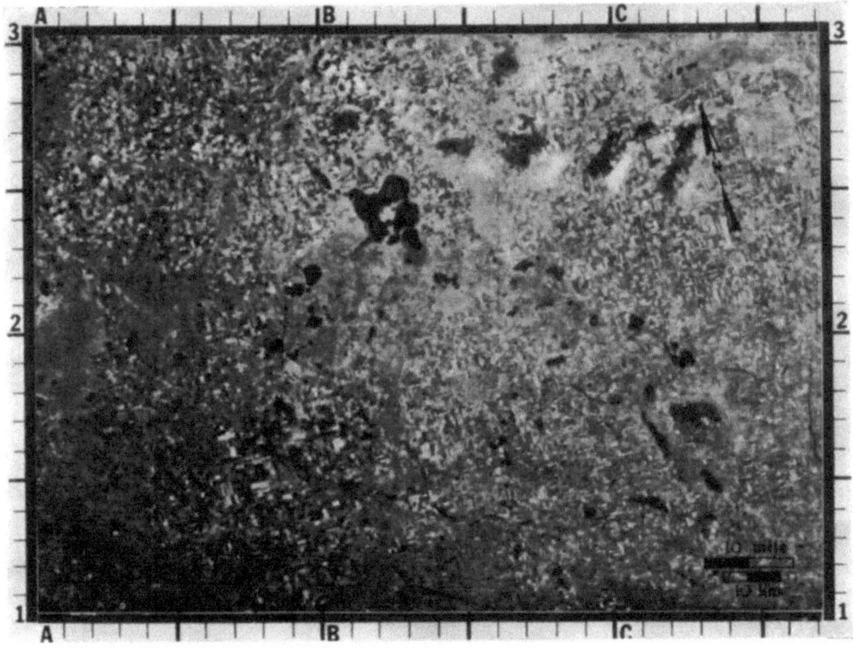


Plate 5. ERTS imagery bands 5 and 7 (below) showing terrain indicators of potential municipal groundwater supplies. West-central Saskatchewan, Canada.

lays are put together so that they do not quite coincide, a pseudo three-dimensional effect is apparent when the image is viewed through a hand magnifying glass.

Black-and-white positive prints of MSS band 6 and false-colour composite prints from bands 5, 6 and 7 taken on 30 July 1972, show the Lake Agassiz glaciolacustrine clay plain, ice-contact stratified drift hills composed of sand and gravel, ground moraine (till plains) sand dunes, and extensive marshes and swamps (Plate 6) [9].

In future projects it is possible that certain MSS false-colour composite imagery may also be used to monitor changes in the retreating (or advancing) positions of flowing artesian zones in unconfined aquifers—as, for example, where increased groundwater withdrawals are made by municipalities and industries (Fig. 5). Reduced toxicity from salts and improved fertility of agricultural soils might also be expected to take place over time with man-induced reduction in artesian pressures, hopefully monitored by ERTS imagery at some point in the future.

←

Plate 5.

Primary application: Groundwater investigations.
Secondary applications: Channel amelioration and bridge-site selection studies.
Latitude and longitude of prominent physical feature and its coordinates on surround: Manito Lake B.2:2.4; lat. 52°45′; long. 109°42′.
ERTS imagery designation: E-1017-17494.
Date of imagery: 9 August 1972.
Bands shown (top): 5; (bottom): 7.
Past water-resource studies in area shown by ERTS imagery: Four ground-water studies and one study of fluvial morphology and meander-cutoff processes.

*Interpretation of imagery*

As noted previously, favorable municipal groundwater development sites often occur along abandoned glacial meltwater channels occupied by a chain of elongate lakes. In one such depression, two municipal groundwater supplies were developed in a water-table aquifer at B.1:1.5. Note the chain of elongate lakes following the trend A.8$^1/_2$:1.6; B.0$^1/_2$:1.5; and B.4 to B.7:7.3$^1/_2$ to 1.4. In the field, only one of these lakes was found to be deep and free of detrimental salts and algae. This particular lake, called St Lawrence Lake (B.0$^1/_2$:1.5), is bordered on the south by deep outwash sand and gravel (up to 115 feet, of which 50 feet is water-bearing). Good quality water in the lake serves to recharge two nearby municipal wells that supply water to the Towns of Provost, Alberta, and Macklin, Saskatchewan.

Of 15 rather easily distinguished small lakes in the areas B.9 to C.5:1.3 to 2.1, seen on near-infrared band 7, the lake that shows up most conspicuously on band 5 is Akerlund Lake, at B.9$^1/_2$:2.0$^1/_2$ west of Unity. In the Akerlund Lake depression, springs are currently being tapped to supply the municipal water needs of the town. A crescent-shaped white halo (high reflectance) around the southern shore of Akerlund Lake is associated with long-continued evaporation of ground-water discharge and consequent precipitation of alkali salts at ground surface. This lake, which is better seen stereoscopically in high-altitude air photographs is located in what is considered to be a major salt-solution collapse feature.

The dark-toned lake at A.9:2.1$^1/_2$ on band 5 (red) is deeper and more saline than nearby lakes that are barely discernible (see band 7).

Plate 6. Portion of ERTS imagery of Winnipeg area showing diverse surficial deposits and an area south of Winnipeg that appears lighter-toned in pinkish MSS false-colour composite positive prints. This lighter area corresponds roughly to an area of flowing artesian wells and with salinized and alkalinized soil areas outlined on provincial soil-survey maps [8].

## 8. Peatland Hydrology

It is said that there is over 500000 square miles of peatland (muskeg) in Canada. Owing to accelerating resource development in northern Canada and particularly along a mid-Canada corridor, where the most extensive peatland is distributed, any economic technique that will facilitate mapping of peatland types and give any indication of their extent, relative depth, and hydrology is highly significant. To date most of the remote-sensing techniques employed for this purpose have concentrated on the interpretation of panchromatic aerial photography [10].

Plate 7 and Fig. 6 show striking correspondence between peatland patterns imaged by the ERTS satellite (band 5) and varying depth ranges of peat [11]. Peat type and depth are closely related to surface water and ground-water flow and chemistry and produce different spectral signatures on ERTS images (Plate 10).

Preliminary studies of ERTS 1 imagery suggest that diagnostic features of different peatland terrain types may be emphasized in snow-covered scenes taken when the sun elevation is relatively low.

## 9. Structurally Controlled Surficial Lineaments

Lineaments delineated on small-scale mosaics and thought to be associated with submasked geological structures have been studied for some time [12, 13]. However, these linear elements of the landscape seem to show up much better

←

Plate 6.

Primary application: Groundwater recharge-discharge interpretation.
Secondary applications: Mapping surficial deposits; wetland and waterfowl-habitat surveys and wildlife management.
Latitude and longitude of prominent physical feature and its coordinate on surround: Red River Delta B.3:3.4; lat. 50° 15′; long. 96° 30′.
ERTS imagery designation: E-1007-16531.
Date of imagery: 30 July 1972.
Band shown: 6.
Past water-resource studies in area shown by ERTS imagery: Flood mapping, floodway planning and location studies, groundwater studies, lake-shore classification for recreation, peatland hydrology, river-type classification, mapping of surficial deposits.

*Interpretation of imagery*

The area situated to the south, southwest, and southeast of B.5:2.0 on this first near-infrared band corresponds to lighter tones on a pinkish ERTS 1 MSS false-colour composite positive print and to an area of flowing artesian water wells (groundwater discharge area). The light tones may reflect residual salt concentrations in surface and subsoil horizons resulting from upward movement and evaporation of strongly artesian groundwaters.

The sinuous channel habit at B.3:2.1 to 2.3 suggests a low-gradient river flowing in cohesive deposits and carrying very little sediment larger than fine-sand size. Note the generally darker spectral signature over the Lake Agassiz clay plain surrounding Winnipeg (at B.3:2.3), the large kame-delta deposit at B.5:2.7, wave-bevelled till plains at C.1:1.4 and A.4: 3.5, marshland at the Red River Delta at B.4:3.4, silt and fine sand in floodplain and deltaic deposits at A.2:1.3, and strandlines and beach ridges at C.7:1.5 [9].

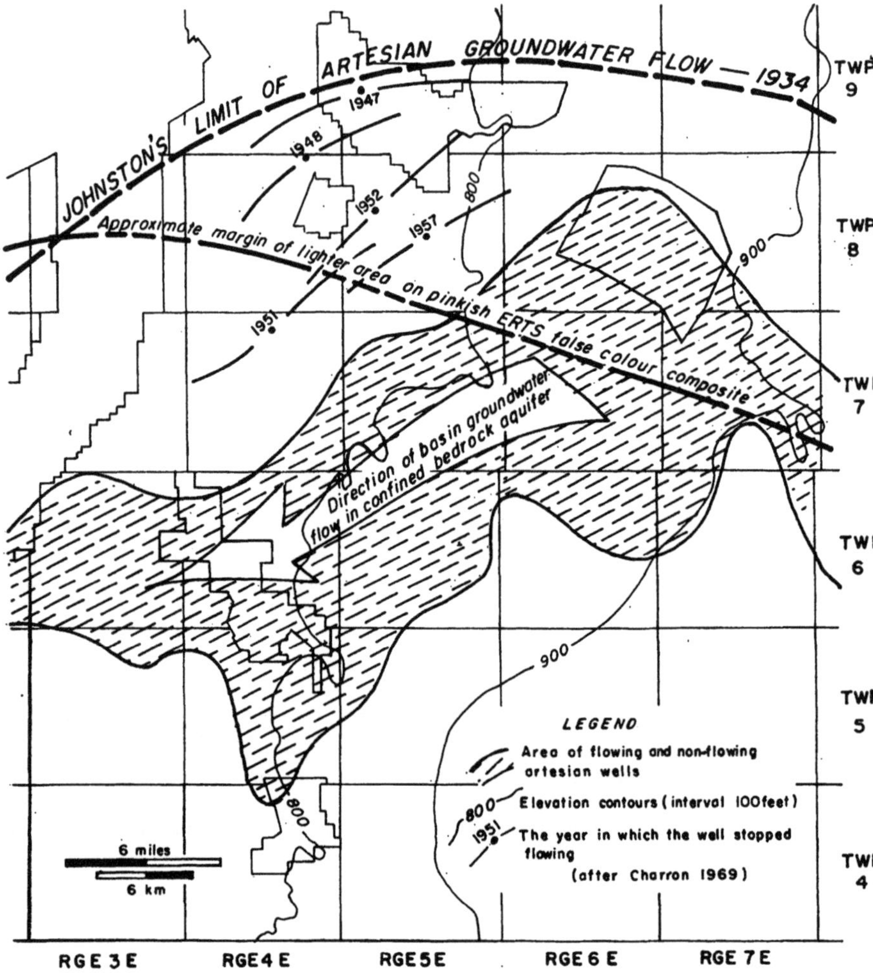


Fig. 5. Key map showing the retreating margin of flowing artesian wells as a result of large scale industrial groundwater withdrawals over the period 1947 to 1965 (after J. E. Charron [7]). Represents a small portion of the ERTS scene shown in Plate 6. South of Winnipeg, Manitoba.

in space imagery than they do on conventional mosaics produced from contact aerial photographs because of the continuity afforded by the small scale and the absence of contact photograph margins and corners inherent in uncontrolled mosaics [1, 14, 15]. Two areas, portions of ERTS frames only, showing alleged (morphotectonic) lineaments are illustrated in Plates 8 and 9.

Mapping of these lineaments and their intersections is not only significant in the search for localities of higher secondary (fracture) porosity and permeability for siting higher yielding water wells, they are localities having a higher potential for groundwater contamination from pollutants. In addition, their detection and mapping may assist in planning geological and geophysical exploration programs for oil and gas and mineral deposits [16—18]. Compare, for instance, the location and orientation of lineaments on Plate 8 with those of oil and gas fields north of Calgary, seen in Fig. 7.

VP - Very deep peat (> 9 feet deep) (> 2.7 m)

DP - Deep peat (6 - 9 feet deep) (1.8 to 2.7 m)

MP - Medium peat (3 - 6 feet deep) (1 to 1.8 m)

SP - Shallow peat (0 - 3 feet deep) (0 to 1 m)

WS - Wooded mineral soils

Fig. 6. Key map showing differences in peat-depth ranges and large whitish groundwater-discharge areas ("Saline", surface salt accumulations), both terrain conditions shown in ERTS imagery (band 6) in Plate 7. Saskatchewan River delta region, Canada.

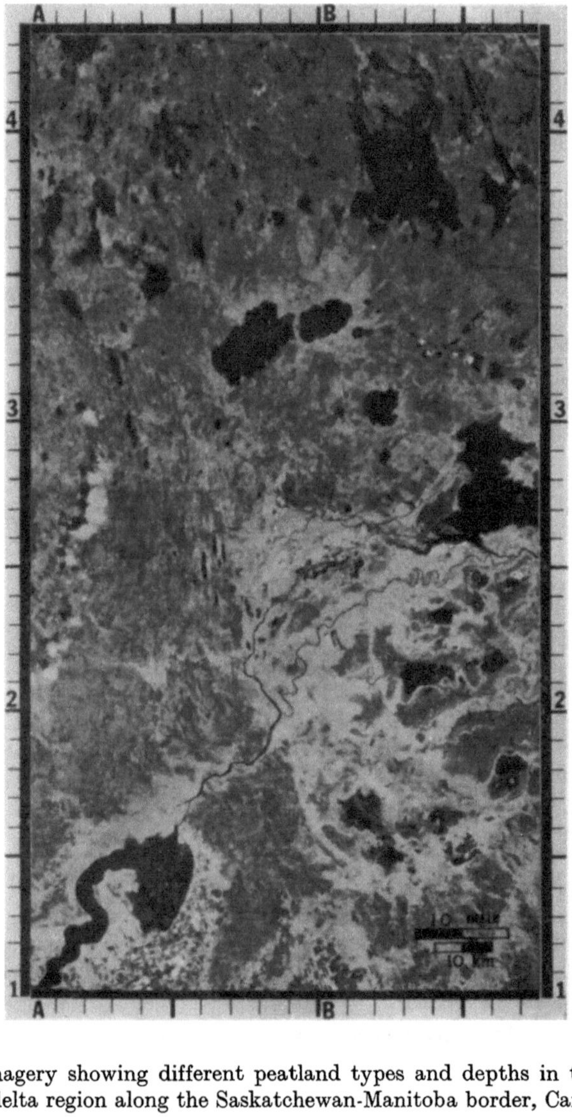


Plate 7. ERTS imagery showing different peatland types and depths in the Saskatchewan River delta region along the Saskatchewan-Manitoba border, Canada.

←

Plate 7.

Primary application: Peatland hydrology and associated peat type and depth.

Secondary application: Reservoir-surface studies for shoreline recreation development.

Latitude and longitude of prominent physical feature and its coordinates on surround: Tobin Lake A.5:1.4; lat. 53°35′; long. 103°30′.

ERTS imagery designation: E-1031-17263.

Date of imagery: 23 August 1972.

Bands shown (left): 6.

Past water-resource studies in area shown by ERTS imagery: Dam-site location studies; studies of drainage of the past recurrently flooded Carrot River-Cumberland marshland; shoreline studies of Tobin Lake; studies of delta growth into Cumberland Lake; reconnaissance river-diversion studies for power generation (Churchill River diversion via Sturgeon-Weir River to the Saskatchewan River).

*Interpretation of imagery*

The following significant information may be interpreted from band 6:

(1) Differences in peatland types in the area A.7 to B.3:1.3 to 1.8, primarily differences in hydrology and related vegetal cover (see Fig. 6 for correlation with peat-depth ranges and saline areas).

(2) Differences in peat depths in the area B.0 to B.7:2.0 to 2.4. The deepest peat (VP) images darkest in gray tone. The shallowest peat (SP) correlates with light-gray tonal areas (compare with Fig. 6). Peat depth ranges between these extremes image in intermediate gray tones. "Saline" is lightest gray of all except for narrow ribbons of ground that follow small winding creeks bordered by wooded natural levees composed of silty and fine sandy mineral soils (WS). The significance of being able to outline, even approximately, and on a regional basis, different peat depths is great when planning large-scale drainage schemes, locating and constructing dike lines, and locating new transportation routes.

(3) Differences in the pattern of lakes in the Precambrian Shield (north of Y-coordinate 3.5) and in the drift-mantled gently dipping Paleozoic strata to the south.

(4) Salt springs outlining a major groundwater-discharge area at B.4 to B.7:1.3 to 1.7.

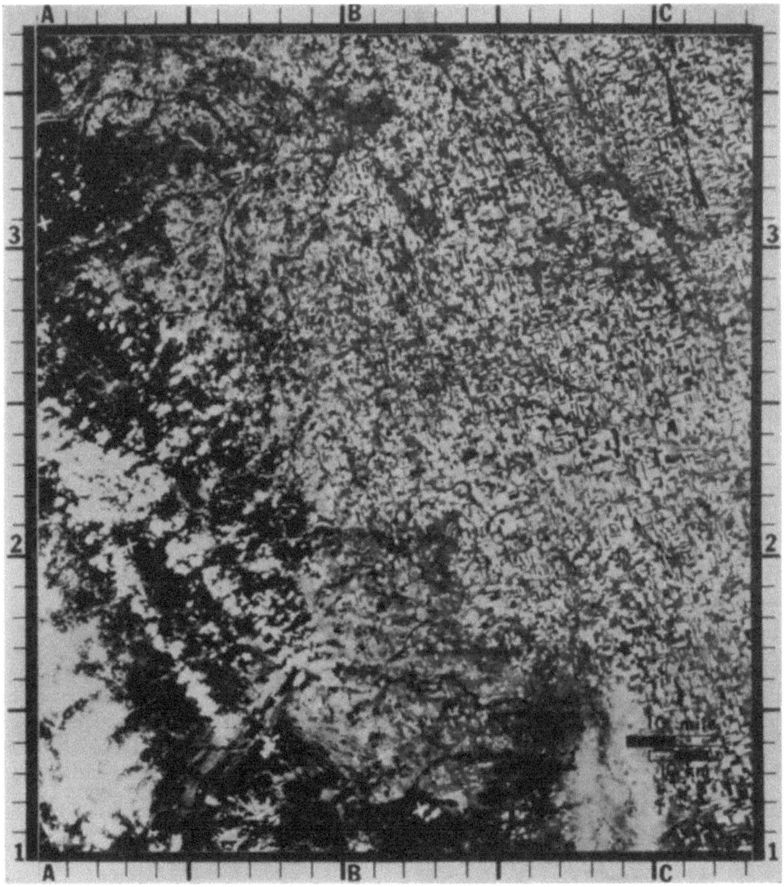


Plate 8. Map showing surficial lineaments believed to be an expression of submasked geological structure, and probable corresponding zones of higher secondary (fracture) porosity and permeability, higher well yields locally, and narrow linear zones cosidenred to represent greater possibilities for groundwater contamination. The trends also appear to correlate with elongate oil and gas fields (see Fig. 7). North of Calgary, Alberta.

Surficial lineaments in the Avonlea area of Saskatchewan (Plate 9) coincide with faults in Mississippian strata lying some 3000 feet below drift-covered Mesozoic rocks [18]. These faults are believed to extend down to Ordovician beds at a depth of approximately 8000 feet below ground surface and may even be basement-controlled (R. Rudichuk, geologist, personal communication).

Such localized linear trends may also yield evidence of deep regional groundwater flow systems across the Prairies and thus mark sites where surface and near-surface water samples may be collected for detailed chemical analyses and hydrochemical interpretation of groundwater movement [7].

The use and significance of all of the foregoing applications of ERTS imagery is greatly enhanced by studying the scenes obtained from spacecraft along with the interpretation of conventional panchromatic, colour, and black-and-white and colour-infrared photography taken from aircraft, particularly when several different scales of black-and-white photography are used [6]. At this point in time one of the most helpful remote-sensing tools that can be used to increase the use and significance of the space imagery in addition to existing map data is high-level stereoscopic aerial photography taken at scales varying between 1:50000

←

Plate 8.

Primary application: Surficial lineament study related to bedrock-fracture porosity and permeability and therefore groundwater movement and exploration.
Secondary application: Oil and gas field exploration and development.
Latitude and longitude of prominent physical feature and its coordinates on surround: Calgary, Alberta B.8:1.4$^1/_2$; lat. 51° 02′; long. 114° 04′.
ERTS imagery designation: E-1092-18075.
Date of imagery: 21 October 1972.
Band shown: 5
Past water-resource studies in area shown by ERTS imagery: Ground-water studies at Canmore, Cochrane, Calgary (two major industrial ground-water studies), and Sundre.

*Interpretation of imagery*

This area of Alberta, located mostly north of Calgary, shows several prominent surficial lineaments (dark tonal alignments) that are believed to be structurally controlled (possibly even basement controlled). The lineaments are seen best by holding the original ERTS image about a foot from the eyes, at an oblique viewing angle of 30° or less. The image should then be rotated slowly until the threadlike tonal alignments come into foreshortened perspective. The space-image interpreter must be careful to distinguish these fine natural features from man-made cultivation features and from transportation routes or other sublinear man-made features. Length, density (incidence), orientation (rose diagram), and intersection as well as change in direction of the lineaments may be significant in different localities and in geological situations. Compare these lineaments with those apparent in the Avonlea region of south-eastern Saskatchewan (Plate 9). The most prominent tonal lineaments and elongate oil and gas fields (Fig. 7) trend between N 10° W and N 45° W with trends of N 15° to 30° W being the most common [16, 17]. In the present study, our primary interest was not petroleum exploration but, rather, the identification of zones of secondary (fracture) porosity and permeability and their influence on the localization of ground-water movement in near-surface sandstone strata. However, insufficient subsurface ground-water hydrological data are available for making detailed correlations between local changes in bedrock permeabilities and changes in the yield of water wells situated over and away from the prominent surficial lineament trends indicated on the ERTS imagery. Detailed correlation study is considered to be a desirable project for future research study.

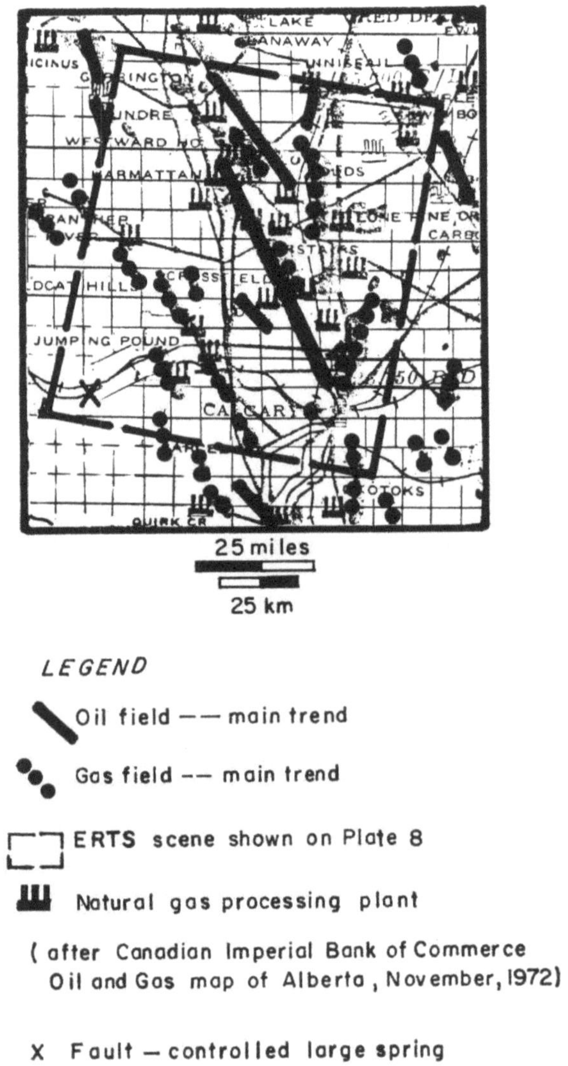


Fig. 7. Key map showing the dominant linear trends of oil and gas fields in the area shown in Plate 8.

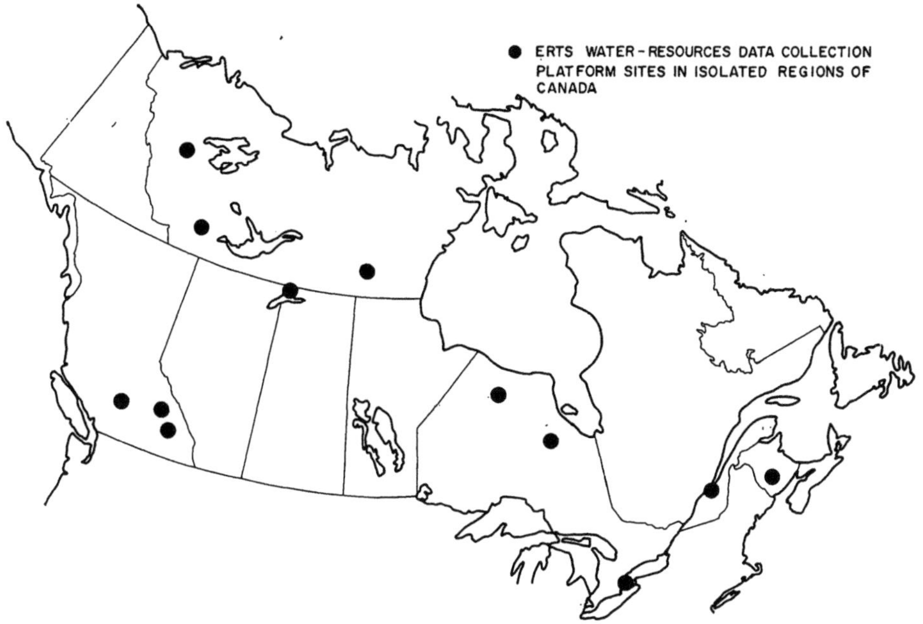


Fig. 8. Map of Canada showing the general location of Water Survey of Canada's platform-installation sites for water data collection and transmission to ERTS 1 satellite for re-transmission.

and 1:100000. Pollution in lakes and rivers, and detection of cold groundwater discharging into warmer lakes or the ocean are often best sensed from thermal infrared imagery [19].

The space imagery analyst is encouraged by the fact that the longer he studies this imagery the more confident he becomes of his analyses and the greater the number and variety of applications he sees.

## 10. Monitoring Hydrological Data in Remote Lakes and Rivers

Recent studies indicate that sensing devices placed in platforms in isolated Canadian lakes and rivers can transmit useful hydrological data to the ERTS 1 satellite, which is then re-transmitted to a receiving satellite station on the earth's surface [20]. Locations of experimental platforms being operated by the Water Survey of Canada are shown in Fig. 8. Devices at most of these platforms sense water levels; but others can detect ice thickness and ice break-up in lakes and rivers, and some monitor chemical data on water quality. These data are re-transmitted via the ERTS 1 satellite and are received at the Canada Centre for Remote Sensing in Ottawa within 24 hours.

As one can see from inspection of Fig. 8, almost all of these platforms are situated in remote places in Canada. As a result, they are accessible only by chartered plane, and are visited only once or twice a year.

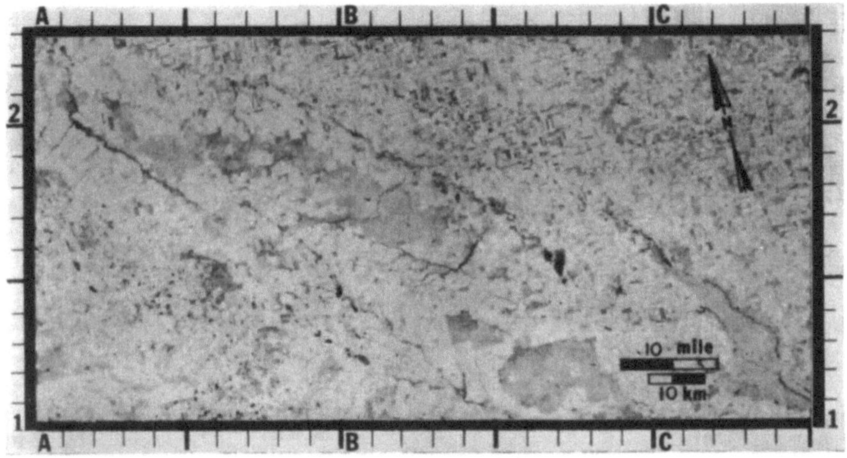


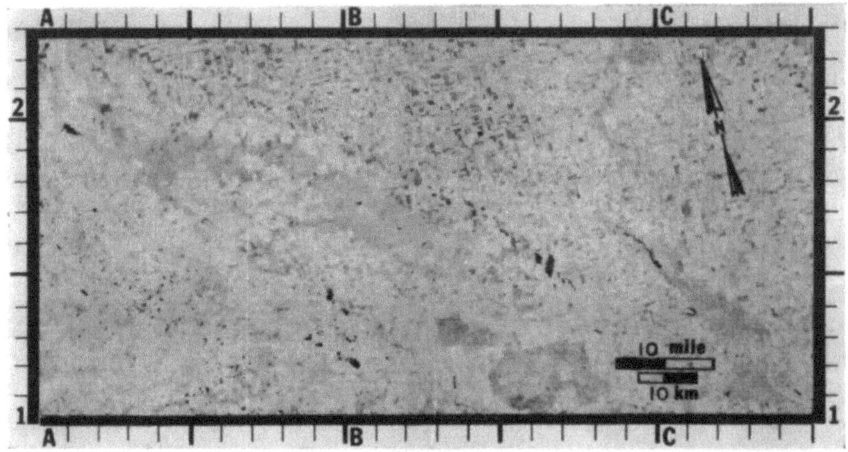


Plate 9. Surficial lineaments in vicinity of Avonlea, Saskatchewan, Canada. The lineaments are similar to those seen in Plate 8.

Plate 9.

Primary applications: Drainage delineation; ground-water studies.
Secondary application: Surficial lineament study for oil and gas exploration.
Latitude and longitude of prominent physical feature and its coordinates on surround: Souris River C.0:1.6 to C.4:1.2; lat. 49°30′; long. 103°40′.
ERTS imagery designation: NASA E-1048-17220.
Date of imagery: 9 September 1972.
Bands shown (top): 5; (bottom): 6.
Past water-resource studies in area shown by ERTS imagery: Tonal lineament studies for oil and gas exploration using semi-controlled mosaics; seven groundwater evaluation studies; drainage mapping for litigation cases; flooded area delineation.

*Interpretation of imagery*

Band 5 shows up drainage features and associated tonal lineaments much better than does band 6. The darker-toned lineaments commonly indicate deeper or darker organic material, finer-grained soils, higher moisture conditions at ground surface, or slightly thicker or taller

Fig. 9. Typical water survey installation hut with antennae from which data are transmitted to ERTS 1 satellite for re-transmission to a satellite receiving station on the earth's surface.

←

vegetal cover. Some of the striking lineaments imaged here correspond to deep-seated faulting that, despite a thin drift cover, are believed to extend downward in the subsurface to Mississippian strata and even deeper. The multiple lineament pattern seen in band 5 is thought to reflect the influence of a series of fault structures. Two particularly strong lineaments strike from A.1:2.0 to A.6:1.7 and from A.8:$1.7^1/_2$ to B.3:1.5; but many other lesser linear features can be seen in general area imaged [18].

None of the seven groundwater-development sites investigated previously appear to have any surface expression in bands 5 or 6. Most of the ground-water supplies discovered occur as small artesian aquifers in intertill glacial sand.

The quality of this ERTS imagery is particularly good. For example, changes in topography between the Missouri Coteau, a hummocky moraine with partly water-filled kettleholes (sloughs), and the Regina clay lakebed lying west of the north point are easily detected. The different patterns of present land use reflect fundamental differences in surface-water and groundwater hydrology, particularly in regard to the proportion of snowmelt and rainfall that runs off the land to enter surface drainage systems and the proportion that remains trapped in closed depressions to evaporate or replenish groundwater storage.

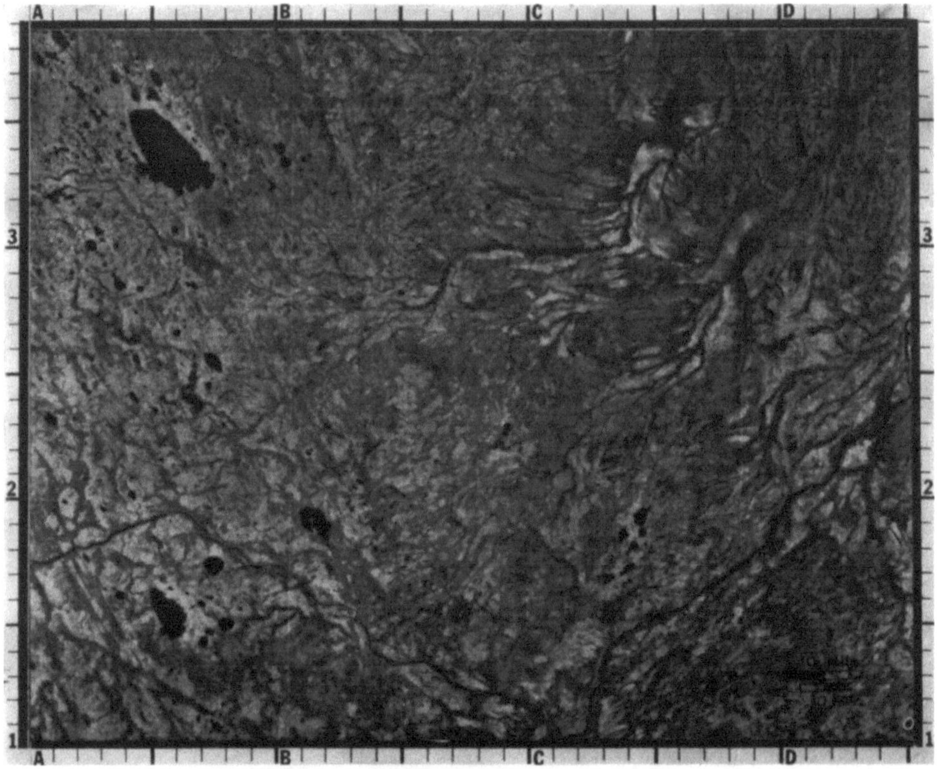

Plate 10. ERTS image (band 6) showing surficial geological features, three dam sites studied in the field, and regional peatland types. Confluence of the Kenogami and Albany Rivers in northern Ontario, Canada.

Fig. 9 shows a typical installation hut with antennae. Climatic conditions in northern Canada are extreme and, in the case of this installation, the antenna continued to operate successfully despite snow cover. The only place where any difficulty was experienced in transmission was from a platform situated in the mountainous regions of British Columbia.

The remote-sensing system indicates almost immediately when there is instrument malfunction. Thus a visit to the field can be planned, the problem can be corrected, and a long period for which normally no data would be available can be avoided. Accordingly, unnecessary visits to these sites can be avoided.

The data re-transmitted by the ERTS 1 satellite are used to predict river and lake ice break-up in the spring, control water levels in the International Treaty dams on the Columbia River in British Columbia, and predict water levels on the Mackenzie River in northern Canada. The Mackenzie River is a large waterway and is a vital link in northern transportation and shipping. In fact, the principal method of transporting bulk cargo into the northwestern Arctic is by means of barges on the Mackenzie River. Because the shipping season is short, it is essential that barges carry a maximum load and yet not be subject to the risk of running

←

Plate 10.

Primary application: Dam-site and reservoir studies.
Secondary applications: Studies of ground-water discharge into the Albany River; mapping of surficial lineaments.
Latitude and longitude of prominent physical feature and its coordinates on surround: Missisa Lake A.6:3.3; lat. 52° 20′; long. 85° 15′.
ERTS imagery designation: E-1035-16063.
Date of imagery: 27 August 1972.
Band shown: 6.
Past water-resource studies in area shown by ERTS imagery: Dam-site selection; reservoir leakage assessment; prediction of sites of groundwater discharge into the Albany River; location of bedrock outcrops in the river bed.

*Interpretation of imagery*

Three dam sites studied during the course of office airphoto interpretation and follow-up field reconnaissance investigations are located as follows:
Site 1—A.5:1.9
Site 2—B.5$^1/_2$:1.3$^1/_2$
Site 3—C.8$^1/_2$:1.7$^1/_2$
Karstic carbonate bedrock occurs in the river bed and along the shoreline at Site 1 and Site 3; siltstone and mudstone strata overlie dolomite at Site 2. It appears that the occurrence of bedrock locally in the riverbed and along shorelines might be associated with relatively straighter reaches of the Albany River.

On the upland, several large areas of "string bogs" (botanically, patterned fens) are located at B.0:2.2$^1/_2$; A.6:1.7$^1/_2$; A.9:2.6; A.2$^1/_2$:3.1$^1/_2$.

Most of the vegetal cover on the upland consists of stunted black spruce and Sphagnum, where the water table is situated at or above ground surface. These extensive Sphagnum-covered bogs appear uniform light gray in tone and show little evidence of flow lines.

Upland mineral soils beneath peat consist mostly of thin, discontinuous marine silt overlying dense calcareous silty till that has been derived mainly from underlying or nearby carbonate bedrock strata. In places, a layer of glaciolacustrine clay occurs between the marine silt and underlying till.

The north-northwest/south-southeast direction of glacial fluting can be seen west of the X-coordinate B.0. Glacial fluting can be seen again, but trending northeast/southwest, south of the Albany River at C.5 to D.0:1.0 to 1.4. Abandoned strandlines are evident at B.6:2.6.

No reliable explanation can be given for the anomalous-appearing gently curvilinear lineament that passes through coordinates D.3:3.9; D.1:2.7; C.9:2.0; and C.5:1.3. This curvilinear trace may be a relict ice-scour marking on the bed of Tyrrell Sea, scribbed on the ocean floor in early post-glacial times by an ice keel on floating sea ice or on an iceberg. Assuming the vestigial trace was scribbed by an iceberg calved off a glacier to the north that terminated in Tyrrell Sea, in its journey south the iceberg would tend to deflect to the right under the influence of the Coriolis force and thus show a trace curving to the right travelling in a southerly direction.

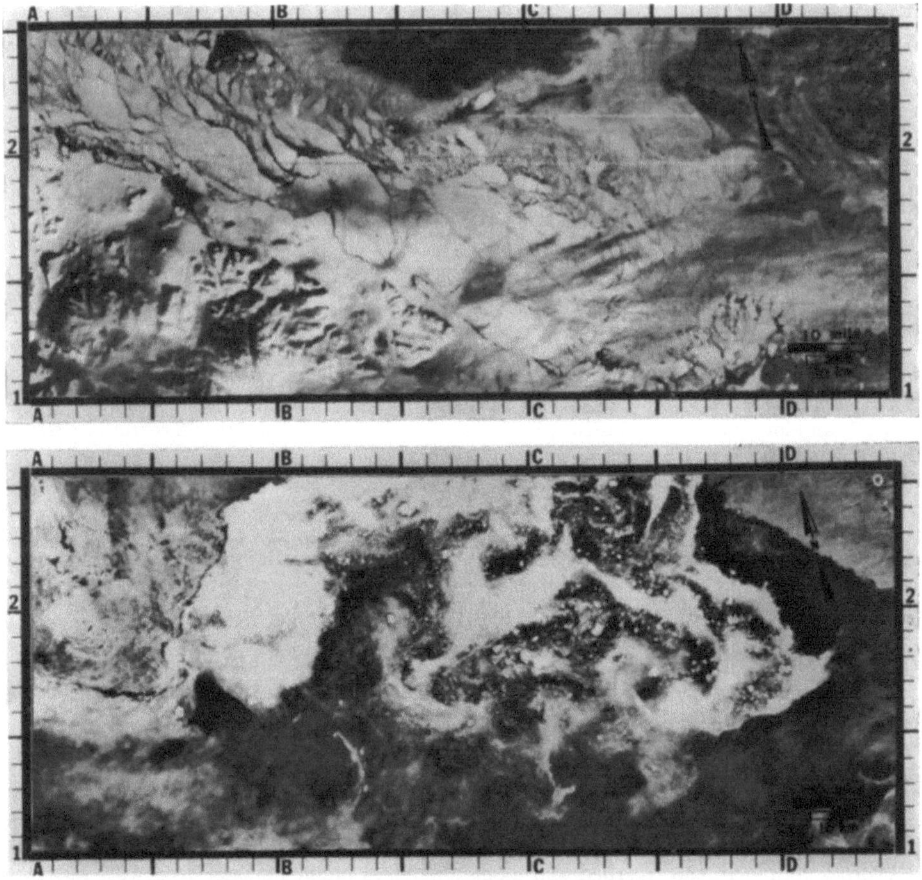

Plate 11. Two ERTS images (band 6) showing Arctic sea ice break-up and movement patterns. Baffin Island and Southampton Island, Canada.

Plate 11.

Primary application: Arctic sea ice break-up and movement.
Secondary application: Differentiating different types of sea ice (e.g., first year ice, multiyear ice, etc.).
Latitude and longitude of prominent physical feature and its coordinates on surround: (Top): Cumberland Peninsula (Baffin Island) lat. 64° 45′; long. 64° 30′; (Bottom): Bell Peninsula (Southampton Island) lat. 63° 30′; long. 81° 00′.
ERTS imagery designation: (top): E-1083-15275; (bottom): E-1037-16151.
Date of imagery: (top): 14 October 1972; (bottom): 28 August 1972.
Bands shown (top): 6; (bottom): 6.
Past water-resource studies in area shown by ERTS imagery: Pipeline route selection studies.

*Interpretation of imagery*

This illustration shows two different sea ice break-up patterns. Of interest are the varying break-up and movement patterns in different geographic localities in the Arctic and, particularly, the effect of different shoreline geomorphology and offshore hydrography on the time and rate of ice melting, break-up, and movement.

aground. For such an important waterway the Water Survey of Canada uses the ERTS 1 satellite to make predictions of water levels for navigational purposes three times weekly during the shipping season.

Water-level rises and ice break-up on remote northern lakes and rivers can be obtained regularly so that pilots of light aircraft flying in the north will possess up-to-date information, know where they will be able to land and whether, during ice break-up in the spring, their aircraft should be equipped with wheels, skis, or floats.

Remote sensing of chemical constituents of water, including pollutants, in isolated lakes and rivers also holds promise and experimental research in this connection is continuing.

## References

[1] J. D. Mollard et al., 1973. Preliminary Evaluation of ERTS 1 MSS Imagery for Water Resources Studies, with Special Reference to Groundwater Investigations in Southeastern Saskatchewan, Canada: Report prepared for the Inland Waters Directorate, Department of Environment, Ottawa, Canada.

[2] J. D. Mollard et al., 1972. Reconnaissance Study of Hydrogeology of the Mackenzie Valley: Report prepared for Inland Waters Branch, Hydrologic Sciences Division, Environment Canada, Ottawa.

[3] J. D. Mollard 1973. Airphoto Interpretation of Fluvial Features: Paper presented at the Ninth NRC Canadian Hydrology Symposium on Fluvial Processes and Sedimentation, Edmonton, May 8 and 9 (in press).

[4] J. D. Mollard 1969. Photo-Interpretation Studies in the Location of Prairie Groundwater Supplies: Paper presented to 22nd Annual Canadian Soil Mechanics Conference, Kingston, Ontario, December 8 and 9; reprinted in Canadian Geotechnical Journal, 7, No. 2 (1970).

[5] V. G. Beckie and J. E. Balzer, 1970. Industrial Groundwater Investigation at Welby, Saskatchewan: Paper presented at the Joint Annual Meeting of The Geological Association of Canada and The Mineralogical Association of Canada, Winnipeg, Manitoba, August 30 to September 2, 13 pp., 5 figures, 1 table (unpublished).

[6] J. D. Mollard, 1972. Airphoto Interpretation Manual—Landforms and Landscapes of Canada: 580 interpreted stereograms with captions and exercises, and a glossary of 2400-interdisciplinary terms used in remote sensing of terrain (available from the author), Regina, Saskatchewan.

[7] J. E. Charron, 1969. Hydrochemical Interpretation of Groundwater Movement in the Red River Valley, Manitoba: Inland Waters Branch, Department of Energy, Mines and Resources. Scientific Series 2, Ottawa.

[8] W. A. Ehrlich et al., 1953. Report of Reconnaissance Soil Survey of Winnipeg and Morris Map Sheet areas: Soil Report No. 5, Canada Department of Agriculture, Winnipeg.

[9] J. Thie and V. Woo, 1973. Preliminary Interpretation of an ERTS Satellite Image of the Winnipeg Area: Report of a Workshop Symposium on Remote Sensing Applications in Manitoba, Winnipeg, February 5 and 6.

[10] J. D. Mollard, 1960. Guides for the Interpretation of Muskeg and Permafrost Conditions from Aerial Photographs: *Oilweek*, Calgary, Alberta, July issue, 8 pp. (reprinted).

[11] K. W. Ayres et al., 1967. Preliminary Soil Map of the Saskatchewan River Delta Project: Saskatchewan Institute of Pedology, Publication SP 1, Regina.

[12] J. D. Mollard, 1957. Aerial Mosaics Reveal Fracture Patterns on Surface Deposits in Southern Saskatchewan and Manitoba: *Oil in Canada*, August, 11 pp. (reprinted).

[13] J. D. MOLLARD, 1959. Photogeophysics—Its Application in Petroleum Exploration Over the Glaciated Plains of Western Canada: *The 2nd Williston Basin Conference Proceedings*, Conrad Publishing Company, Bismarck. North Dakota (reprinted).

[14] G. F. OZORAY, 1967. The First Hydrogeological Application of a Space Photograph, Tamil Nadu, India: Research Council of Alberta. Contribution No. 611.

[15] B. B. BANNATYNE, 1973. A Preliminary Analysis of Lineaments in the Paleozoic Section of the Red River Valley: Report of a Workshop Symposium on Remote Sensing Applications in Manitoba, Winnipeg, February 5 and 6.

[16] P. J. HAMAN, 1972. A Regmatic Lineament Pattern, Innisfail Area, Alberta: Western Canadian Research Publications of Geology and Related Sciences, Series 2, No. 3, Calgary, Alberta.

[17] Canadian Imperial Bank of Commerce, 1973. Map Showing Oil and Gas Fields in Alberta: Maps prepared by the Petroleum and Natural Gas Department of CIBC, Calgary.

[18] W. O. KUPSCH and J. WILD, 1955. Lineaments in the Avonlea Area, Saskatchewan: Manuscript prepared at the Department of Geology, University of Saskatchewan, 12 pp. (unpublished).

[19] P. A. CARR and H. GROSS, 1972. Investigation of Infrared Anomalies in the Lac des Deux Montangues Area, Quebec: Scientific Series No. 19, Inland Water Directorate, Water Resources Branch, Environment, Ottawa.

[20] R. A. HALLIDAY, 1972. Applications of ERTS-A for Retransmission of Water Resources Data: Report for July to December 1972, Applied Hydrology Division, Water Survey of Canada, Department of Environment, Ottawa.

Earth Survey Problems — Akademie-Verlag, Berlin 1974

# TÉLÉDÉTECTION DE PHÉNOMÈNES HYDROLOGIQUES

Y. VUILLAUME

Service géologique national, Département Géologie de l'Aménagement, Orleans, France, and Association Internationale des Hydrogéologues (AIH)

Dans la télédétection de phénomènes hydrologiques, les principaux résultats acquis montrent pour l'instant l'intérêt des détecteurs passifs dans le visible (réflectance) et l'infrarouge (réflectance, émissivité) comme moyen de *detection* de phénomènes hydrologiques. Les thèmes abordés sont: *en hydrologie souterraine*, la dynamique de l'alimentation des nappes par les pluies, la détection des résurgences sous-lacustres ou sous-marines, *en hydrologie de surface*, la dynamique des mélanges de masses d'eaux, de transport de sédiments et de polluants. Les résultats prévisibles des recherches en cours ou à engager devraient permettre la *mesure* répétitive adaptée et efficace de phénomènes hydrologiques.

Cet objectif peut être atteint:

(i) Par une analyse multispectrale des phénomènes dans des fréquences «spécifiques» préalablement sélectionnées au cours d'expériences systématiques réalisées au sol, puis en l'air; ces recherches devraient être sensiblement valorisées par le moyen d'un avion «multispectral» du type ERAF. Ceci aurait pour conséquence de limiter le volume des données acquises, d'en faciliter le traitement et d'en permettre la transmission accélérée comme l'exigerait la surveillance de certains phénomènes hydrologiques.

(ii) Par un déplacement des fréquences captées vers les hyperfréquences, ce qui permettrait d'accéder à une donnée influencée par un facteur hydrologique: la constante diélectrique du milieu et de pénétrer plus profondément dans celui-ci.

(iii) De définier enfin les paramètres de mission optimaux pour les thèmes d'application considérés et de dimensionner ainsi les systèmes spatiaux et d'en évaluer la complémentarité et/ou la compétitivité avec les méthodes conventionnelles.

## 1. Introduction

L'éloignement relatif des deux secteurs d'activités en présence: l'espace et l'hydrologie, a pour principale conséquence que les hydrologues ont relativement peu contribué au développement de cette nouvelle approche de leur domaine; la comparaison du nombre des publications traitant de ce sujet publiées dans les revues spécialisées des deux origines, est significative à cet égard. Un rapprochement de ces points de vues complémentaires devient nécessaire pour mieux formuler les thèmes d'application, et focaliser les actions dans les domaines où elles sont les plus nécessaires et les plus performantes. Sur le plan géographique enfin, il apparaît une hétérogénéité significative dans l'état des connaissances qui résulte de l'avance technologique prise par certains pays.

## 2. Les possibilités de la télédétection en hydrologie

Une difficulté existe actuellement dans la formulation des problèmes hydrologiques dans l'optique de la télédétection. Nous tenterons l'approche de l'hydrologie sous les trois aspects fondamentaux de la télédétection: énergétique dans le spectre électromagnétique, spatial, temporel.

### 2.1. Particularités de l'hydrologie

L'hydrologie s'intéresse au domaine de l'eau dans son ensemble représenté par le cycle de l'eau (figure 1). Cependant, les aspects de nivologie et glaciologie seront cités ici accessoirement.

#### 2.1.1. Phénomènes hydrologiques et variables mesurées

*Aspects quantitatifs: volumes et flux.* Un bilan mondial des masses d'eaux de l'hydrosphère est fourni dans et tableau suivant:

Bilan hydrique mondial d'après [2]

| | | Volume stationnaire ($10^3$ $km^3$) | Renouvellement* (an) |
|---|---|---|---|
| | Océans | 1370000 | 3000 |
| eau douce | Eaux souterraines | 60000 | 5000 |
| eau douce | Glaciers | 24000 | 8600 |
| eau douce | Interface | 4000 | 330 |
| eau douce | Lacs | 230 | 10 |
| eau douce | Humidité du sol | 82 | 1 |
| eau douce | Vapeur atmosphérique | 14 | 0,027 |
| eau douce | Rivières (volume instantané) | 1,2 | 0,032 |
| | Hydrosphère totale | 1458327 | 2800 |

$$* \text{ Renouvellement} = \frac{\text{volume en place}}{\text{débit}}$$

Ce tableau illustre l'importance des eaux souterraines et des glaciers et de l'interface des nappes en tant que réserve planétaire d'eau douce. Le volume d'eau en place dans les lacs, rivières, l'atmosphère et le sol est infiniment plus réduit. Le renouvellement naturel de ces réserves fournit une appréciation de la dynamique des processus: elle est la plus importante dans les rivières et l'atmosphère, beaucoup plus réduit dans le sol et les lacs. Les transferts entre masses d'eau en place se font par les processus thermodynamiques (vaporisation, condensation, infiltration) ou gravitaires (précipitation, ruissellement, infiltration).

Les méthodes de mesures mises en oeuvre sont:

(i) thermodynamiques pour l'évaporation: avec une précision qui varie sensiblement, mais atteint couramment, en fonction des méthodes de calcul qu'il est possible d'utiliser, de l'ordre de 50 à 10%: les moins précises (Turc et Thorn-

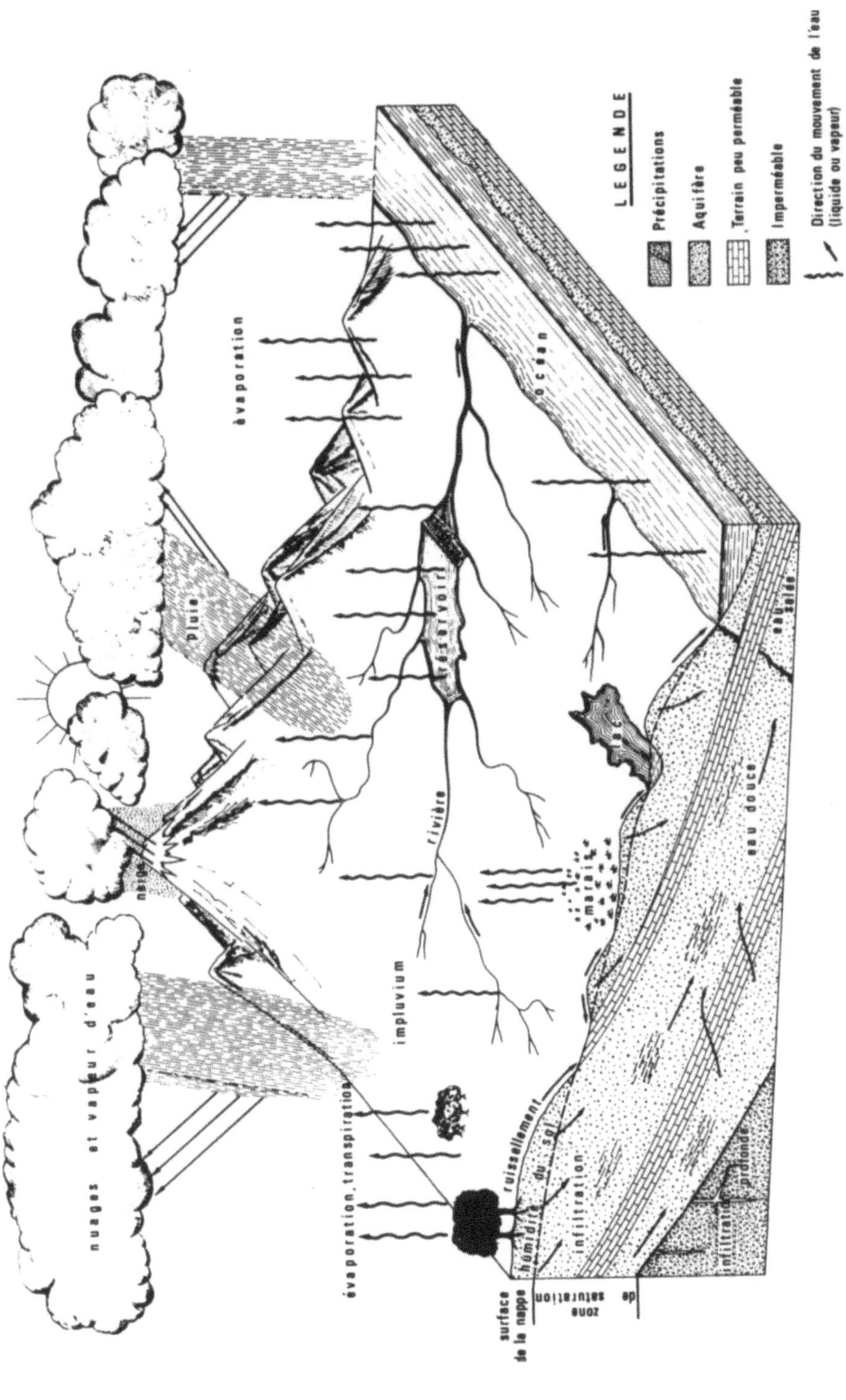


Fig. 1. Le cycle de l'eau.

thwaite) utilisent essentiellement des données de température et de précipitations; les plus précises (Penmann) sont des méthodes de calcul du bilan énergétique des échanges atmosphère/surface qui utilisent en plus des variables directement accessibles par télédétection (albédo du milieu, rayonnement net global du soleil, température).

(ii) gravitaires pour les précipitations et le ruissellement: précipitation: mesure de volume tombé, imprécision sur la mesure, quelques pourcents; ruissellement: mesure de niveau d'eau dans des sections étalonnées et conversion en débit à partir d'une courbe de tarage, imprécision sur la mesure de 5 à 10%; infiltration: l'alimentation des nappes n'est pas couramment mesurée. Elle est déduite le plus souvent par différence dans l'équation du bilan hydrologique. Estimée à 50 ou 100%.

Une méthode de base pour l'évaluation des ressources en eau d'une région pendant une certaine période repose sur l'égalité:

$$P = E + R + I$$

$P$ = précipitations; $E$ = évaporation; $R$ = ruissellement; $I$ = infiltration.

Il semble surtout possible d'accéder par télédétection aux phénomènes d'ordre thermodynamique (émissivité I. R. → température, émissivité en hyperfréquence → constante diélectrique du milieu).

*Aspects qualitatifs.* Les caractéristiques de la qualité de l'eau sont très nombreuses comme en rend compte la figure 3. Les phénomènes mis en jeu très complexes au niveau moléculaire et atomique, sont d'ordre chimique ou biologique. La perturbation de l'état naturel, qui survient en cas de «pollution», peut conduire à des états irréversibles dont certains sont très dommageables pour notre environnement: entrophisation des lacs par exemple. Au niveau hydrologique, les phénomènes sont essentiellement conditionnés par la dynamique propre du support que constitue l'eau. Les méthodes de mesures actuellement utilisées sont très variées; certaines des plus modernes font appel à la spectrométrie d'émission ou d'absorption. Les gammes de concentration à doser sont en général très faibles de l'ordre du ppm ou du ppb, si bien de la plupart des mesures sont faites au laboratoire. Certains éléments chimiques pourraient être identifiés et dosés par télédétection dans la mesure où ils possèdent une raie d'émission ou d'absorption dans la partie du spectre couverte. En ce qui concerne les transferts de polluants, des variables nouvelles, facilement accessibles par télédétection, pourraient être utilisées.

### 2.1.2. Régionalisation des phénomènes hydrologiques

*Manifestation géographique et échelle des phénomènes.* La variable spatiale est fondamentale en hydrologie, car elle implique la notion d'unité naturelle (souvent différente de l'unité politique) qui est la seule valable à considérer pour l'évaluation des ressources en eaux. Les unités naturelles ont des tailles sensiblement différentes ce qui implique la notion d'échelle du phénomène, c'est-à-dire: d'une part la taille de l'information minimale à recueillir (pas d'espace) et d'autre part la vision globale de cet ensemble.

*Existence d'un réseau d'observation.* Les disparités de répartition spatiale des composants du cycle de l'eau sont importantes et rendent nécessaire l'implantation d'un réseau d'observation adapté si possible à ces variations spatiales ou temporelles au long des cycles hydrologiques. Les fréquences d'observation vont en général dans l'ordre décroissant: précipitation, évaporation, ruissellement — fluctuation des eaux souterraines, et varient selon l'importance de la station d'observation. Les données ponctuelles fournies par chaque station sont ensuite regroupées et analysées afin de régionaliser la variable examinée.

L'interprétation des données passe par les étapes préalables contraignantes de: collecte, critique de cohérence, mise au format, régionalisation (courbes isovaleur). A titre d'exemple, voici d'après [3] des éléments du réseau de Grande-Bretagne, pays dans lequel l'intégration des différents réseaux hydrologiques est très avancée: Pour une surface de 230000 km$^2$:

Précipitations: 7000 points de mesure dont 250 avec enregistrement
Evaporation: 80 stations permettant l'évaluation de l'ETP mensuelle
Ruissellement: 530 stations dont 300 avec enregisteur (fréquence: 15')
Eaux souterraines: 1500 points dont 100 avec enregistreur
Qualité des eaux de surface: 30 stations à 10 variables en 1970 frequence: 1 h
100 prévues en 1975
Qualité des eaux souterraines: 500 points (2 à 4 mesures/an).

Les principales limitations de ces réseaux sont les suivantes:

densité spatiale: pour des raisons de coût de gestion et d'accès au lieu potential de mesure, ou de variabilité imprévisible des phénomènes observés
fréquence: certains phénomènes aléatoires sont trop rapides
méthode de mesure: nature de phénomènes appréhendés.

La différence de densité de ces réseaux montre que certains éléments importants du cycle de l'eau sont très mal connus: évaporation, infiltration, qualité des eaux notamment.

### 2.1.3. Evaluation des ressources en eau — bilans hydrologiques — intèrêt et limitations actuelles

*Méthode d'évaluation des quantités.* Elle repose sur l'équation du bilan. Pour un bassin européen convenablement équipé, la répartition relative des masses est celle de la figure 2 d'après [4]. On notera que le degré de précision dans cette évaluation est fonction: du principe des méthodes de mesure (cf. 2.1.1.); de la densité spatiale du réseau d'observation (cf. 2.1.2.); de la fiabilité du réseau et de sa permanence (cf. 2.1.1.) et que les mesures sont moyennées pour lisser l'influence des fluctuations interannuelles (les moyennes peuvent porter sur 30 années d'observation).

Cette figure montre surtout que: la proportion d'eau de pluie tombée sur un bassin et techniquement utilisable est de 25% dont le plus grand part suit le trajet souterrain et que plus de 60% de ce volume d'entrée sont perdus par évaporation à la surface d'eau libre ou par évapotranspiration des plantes. Au total les 3/4 de ce volume d'entrée sont connus avec une imprécision telle qu'il n'est pas possible d'en tirer partie autrement que pour les exploitations de longue durée.

*Méthodes globales d'évaluation de la qualité.* Elles sont encore en cours d'éla-

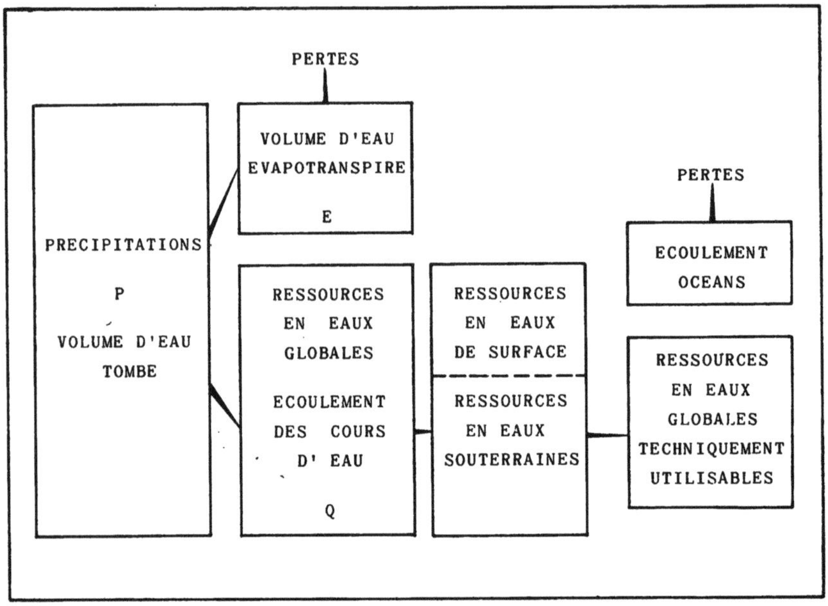


Fig. 2. Répartition des éléments du bilan [4].

boration et concernent surtout les eaux de surface: rivières, lacs, côtes. La principale difficulté rencontrée à ce jour tient à l'inexistence du réseau d'observation qui représente une charge financière considérable: du fait du nombre des variables à considérer: celles globales sont de l'ordre de 5 à 10, celles particulières sont de l'ordre de 3 à $n$: du fait des variations locales très rapides des pollutions et de l'incertitude sur les pas de temps à observer. La figure 3 d'après [7] met en évidence le nombre des variables à considérer et l'importance relative de leur perturbation dans le cas du rejet d'un effluent dans une rivière depuis la bouche de l'émissaire jusqu'à la zone aval de retour à la normale.

*Différentes unités naturelles*

*A l'échelle mondiale:* Le bilan des masses (tableau § 2.1.1) a été tenté par plusieurs auteurs [2]. Le bilan des flux essentiellement d'ordre atmosphérique entre hémisphères, qui survient du fait du cycle été-hiver a été également tenté [5], même avec des données insuffisantes, faute de réseau d'observation convenable. Le transfert de masse entraînerait une oscillation de l'axe de la terre dont l'amplitude serait de l'ordre de 6 à 7 m.

*A l'échelle des bassins* hydrologiques français, une évaluation a été faite en [6] et ramenée à l'ensemble du territoire national. Le bilan hydrologique moyen interannuel de 1958 à 1967 défini à parti des observations réalisées sur 65% du territoire et extrapolé à toute la surface (550000 km²) est le suivant:

| | mm |
|---|---|
| Precipitations | 950 |
| Ecoulement (surface + souterrain) | 410 |
| Déficit d'écoulement (Evaporation + ET) | 540 |

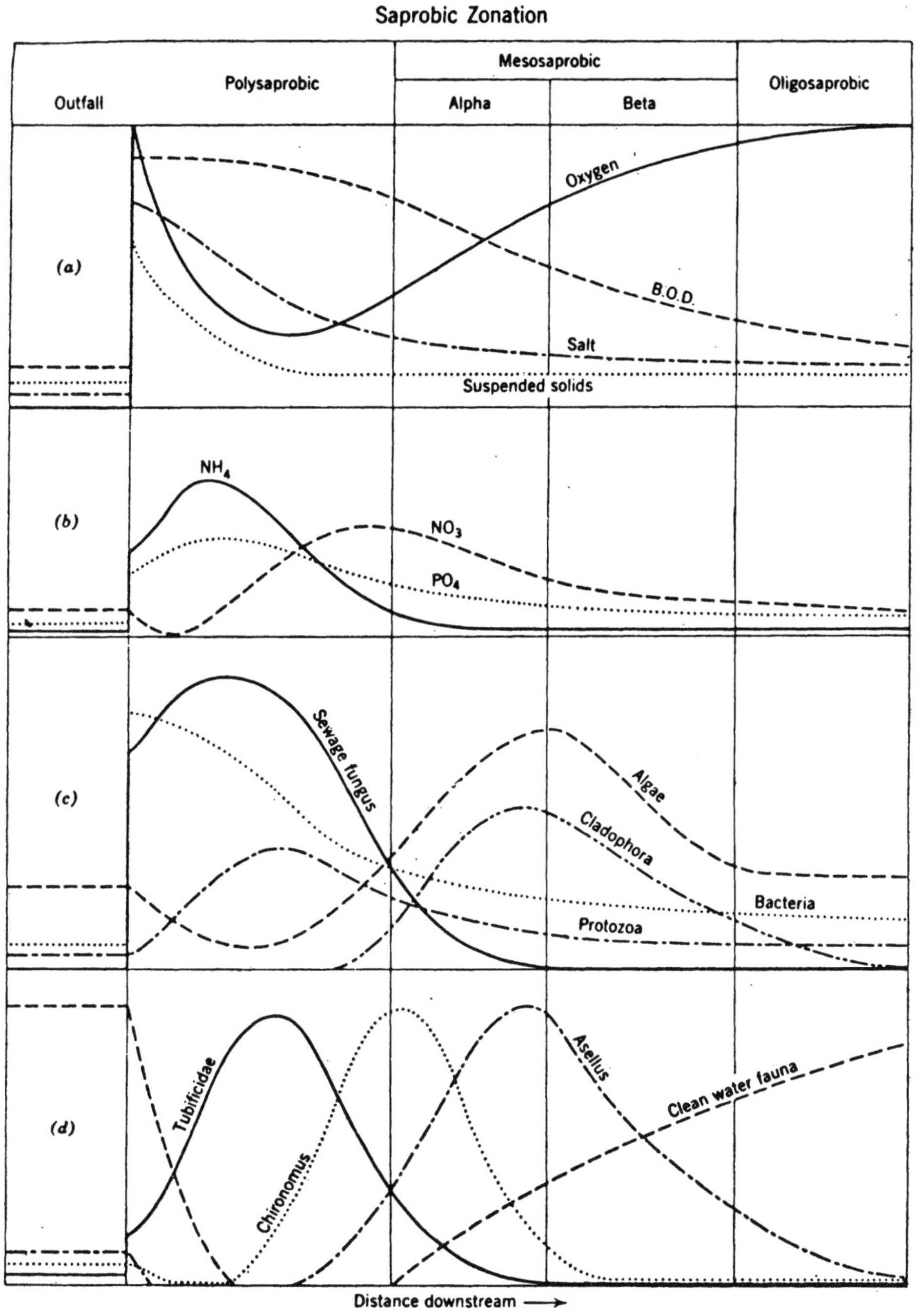


Fig. 3. Variables chimiques et biologiques de la qualité de l'eau [7].

### 2.1.4. Limitation actuelle des méthodes d'études hydrologiques et possibilités potentielles d'applications de la télédétection

Compte tenu de l'analyse des phénomènes hydrologiques réalisée en § 2.1, il est possible de définir les principaux axes d'application de la télédétection dans cette discipline et d'en déduire les principaux thèmes susceptibles de développements: aux niveaux énergétiques, temporels, spatiaux dans les domaines quantitatif et qualitatif.

*Evaluation de la quantité.* La liste du défaut de connaissance actuel en hydrologie a été fait par [8] et concorde avec nos observations précédentes. Les méthodes de mesure à améliorer sont avant tout celles relatives à l'évaporation et l'évapotranspiration et l'infiltration qui constituent des termes essentiels de l'équation du bilan et sont les plus mal connues. Ces deux domaines constituent des secteurs prioritaires aux différentes échelles planétaire à local.

La recherche des zones géographiques mal connues actuellement a été effectuée [8] et [6].

(i) Défaut de connaissance des ressources en eau réelles des régions de socle arénisé et des régions karstiques surtout par limitations techniques des méthodes conventionelles ou densité de mesures représentatives;

(ii) défaut de connaissances comparables dans les zones montagneuses et les régions méditerranéennes par défaut des réseaux actuels *résultant des difficultés d'accès* aux points de mesure et de la grande variabilité des évènements hydrologiques, situation comparable dans les régions côtières par défaut d'équipement;

(iii) nécessité d'améliorer globalement le temps de réponse du système entre acquisition, normalisation, intégration et synthèse des résultats pour répondre aux objectifs nouveaux: de l'aménagement des ressources en eau qui tendent à une modulation des réserves en place selon un pas de temps de plus en plus fin; ceci est particulièrement nécessaire dans les zones *irriguées* où un défaut actuel de contrôle efficace du bilan d'eau et de sol limite rapidement les possibilités de développement de ces pratiques agricoles;

(iv) défaut de connaissance dans les zones *marécageuses* et des deltas dont la réserve en eau est difficile à évaluer, surtout en période de transition (fonte des neiges ou fortes crues d'alimentation) et importante à connaître pour la prévision des débits dans les régions aval;

(v) méconnaissance de l'incidence des activités humaines sur le régime et le débit des cours d'eau résultant d'aménagements dont la correction prend rapidement une grande importance économique, tel que: drainage des marais, mise en exploitation des forêts, pratiques agrotechniques/surexploitation des ressources en eau, urbanisation;

(vi) modifications *locales* des conditions hydriques (pour des raisons naturelles ou artificielles) ayant une incidence sur l'occupation du sol: glissements de terrain, éboulements de falaises en montagne, subsidence du sol par compaction des terrains deltaïques ou par surexploitation des aquifères captifs et réservoirs de pétrôle sous-jacents.

*Evaluation de la qualité.* Ici les besoins sont considérables par défaut de: réseau de mesure, les problèmes étant relativement nouveaux et le coût d'installation et de gestion de tels réseaux étant bien supérieur à celui destiné à évaluer les quantités d'eau; inventaire des zones «d'intérêt». Certains états irréversibles de pollution des eaux sont déclenchés à partir de faibles élévations de concentration des polluants (l'entrophisation de certains lacs). Il serait extrêmement précieux de détecter ces phénomènes au début de leur installation alors qu'il est possible de les combattre. Une recherche des zones «potentielles» de pollution est à faire. On notera à ce sujet que la frange côtière des continents est le principal réceptacle des polluants. Une surveillance particulière de cette zone s'impose. A l'échelle de la France une première étude synthétique avec des méthodes conventionnelles fut réalisée [9] sur ce thème qui confirme bien cette tendance. Dans ce domaine nouveau appelé à prendre une grande importance économique, des structures de programmation et de gestion commencent seulement à se mettre en place au niveau des états et au niveau mondial [10].

### 2.2. Possibilités de la télédétection de phénomènes hydrologiques

L'examen des particularités de l'hydrologie réalisée en § 2.1 montre effectivement les besoins actuels de l'hydrologie en fait de méthodes et de mesures dont certains peuvent être satisfaits.

#### 2.2.1. Avantages de la télédétection

Ils résultent des trois variables fondamentales de la télédétection: sur le plan de l'énergétique dans le spectre électromagnétique: accès aux phénomènes énergétiques thermodynamiques de l'hydrologie (évaporation et infiltration) dans une large bande du spectre, avec une précision de mesure radiométrique très convenable; accès à l'identification de certains corps par la spectroscopie de raie dans le domaine de la qualité des eaux.

*Spatial:* L'obtention de données spatiales synoptiques dans lesquelles l'information unitaire permet des études régionales et parfois locales est dans la ligne directe de la réalisation de cartes thématiques très utiles. Cette approche devrait pallier les insuffisances des réseaux de mesure actuels et permettre de réduire les investissements en nouveaux réseaux de surveillance de la qualité des eaux. Les différentes tailles de phénomènes hydrologiques allant du niveau planétaire au bassin de quelques dizaines de km² pourraient être traitées à l'aide de toute la panoplie des vecteurs spatiaux.

*Temporel:* La répététivité constitue en hydrologie une variable de grande importance: cycles annuels, le plus souvent avec fluctuations interannuelles, cycles saisonniers d'alimentation — vidange, cycles journaliers, aléas de distribution à tous les niveaux et certains aspects d'urgence ou de prévision: crues subites, sécheresses catastrophiques. D'autres avantages résultent de variables accessoires: facilité d'accès aérien dans certaines régions peu accessibles au sol; vitesse de réponse du système.

#### 2.2.2. Limitations d'emploi de la télédétection

(i) Filtration atmosphériques: Les nuages ne sont transparents que pour les grandes longueurs d'ondes, ce qui limite le nombre des zones effectivement «couvertes» par certains senseurs. Des systèmes de correction de l'alternation des signaux émis ou réfléchis par les cibles devront être réalisés en fonction de l'état instantané de l'atmosphère. Les zones à couverture nuageuse permanent devront être répertoriées, car elles excluent l'emploie de certains senseurs.

(ii) Impossibilité d'un accès direct au milieu intéressant par l'existence de couverture végétale dont la répartition devra être connue.

(iii) Limitation de la profondeur de pénétration des ondes E. M. dans le milieu. Le pouvoir d'investigation réel, fonction de la dynamique propre du phénomène appréhendé devra être soigneusement évalué et constitue une inconnue importante pour l'étude des eaux souterraines, et à un moindre degré pour celle de l'humidité du sol.

(iv) Manque de souplesse de systèmes dont la mise en oeuvre est programmé à moyen terme, et doit donc de ce fait être soigneusement préparée sur le plan de l'objectif.

(v) Méthodes d'interprétation des données au niveau expérimental le plus souvent; elles demanderont de nombreuses années d'évaluation avant de passer au stade opérationnel.

(vi) Dialogue encore limité entre «utilisateurs» et «techniciens» et défaut d'information sérieuse et de formation des «utilisateurs» à la télédétection.

(vii) La rentabilité économique de tels systèmes ne pourra être évaluée qu'après des investissements importants.

## 3. Etat actuel des applications

Les applications en sont au stade de faisabilité ou d'expérimentation. Elles sont le fait de chercheurs éloignés du domaine de l'hydrologie et souvent réalisées à l'instigation des agences spatiales. Notre expérience pratique en ce domaine est actuellement limitée.

### 3.1. Thèmes d'application

Ils sont définis comme les sujets principaux d'activité hydrologique. Ils sont donc «hétérogènes» du point de vue de la télédétection, soit multiparamétriques. Une recherche exhaustive au niveau européen a été entreprise à ce sujet [1].

Evaluation des flux: évaporation; humidité des sols.

Qualité des eaux de surface: état naturel; pollutions primaires; autoépuration-entrophisation.

Hydrologie régionale: marges continentales; deltas; karst; régions de socle arénisé; montagnes.

### 3.2. Paramètres physiques du milieu

#### 3.2.1. Paramètres de flux

*Température de surface.* Une précision de $+0{,}2\,°C$ est atteinte à l'altitude de 300 m si l'on effectue les corrections d'absorption et d'émission de l'atmosphère, ainsi que celles relatives à la brume en surface de l'eau. La comparaison des températures de surface de l'eau (le résultat est la moyenne des températures sur une tranche de 0,5 à 2 cm) et par un radiomètre aéroporté (pouvoir de pénétration 20 $\mu$m sous la surface) a été réalisée [11] afin de mesurer le refroidissement exercé par l'évaporation à la surface de l'eau (de 0,4 à 0,9 °C).

*Teneur en eau des sols.* Les expériences [12] sur terrain expérimental ont permis de mesurer l'épaisseur de la zone saturée (au sommet de la nappe) jusqu'à 4 feet en utilisant un radiomètre PRT 5 (figure 4) avec scanner IR. En conditions naturelles, il est possible de cartographier l'épaisseur de la nappe d'eau souterraine. La mission fut réalisée après plusieurs jours de chaleur constante [13]. Un radar latéral (K-band) à double polarisation fait apparaître la sensibilité de réponse des ondes radar à l'humidité du sol [14].

*Turbidité et couleur.* Elle concerne le transport de matières solides, minérales ou organiques, en suspension dans l'eau. Les observations sont faites dans le

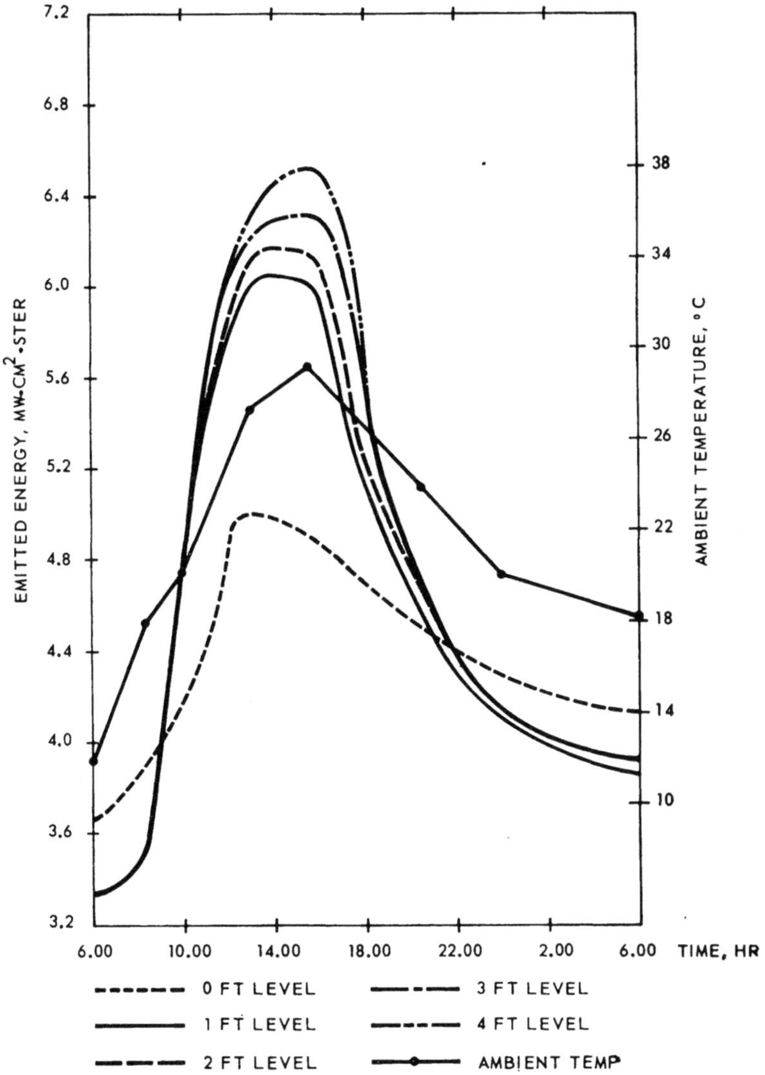


Fig. 4. Energie émise par le sable pour des niveaux d'eau différents [12].

visible: le pouvoir de pénétration est de l'ordre de 10 m). La turbidité est difficilement séparable de la couleur de l'eau. Ces deux paramètres sont souvent traités ensemble. Les senseurs utilisés sont nombreux, émulsion à spectre limité, spectromètre, laser.

«*Rugosité*». Ce paramètre concerne la morphologie de la surface de l'eau en relation avec les vents et les courants. Il est surtout utile en océanologie mais aussi en hydrologie lacustre. L'emploi d'un laser «laser wave profilometer» à hélium néon (6328 Å) a permis d'étudier des creux supérieurs à 5 mètres en relation avec des vents de l'ordre de 25 m $s^{-1}$ [15]. Il existe une relation entre la polarisation horizontale de la brillance en ondes millimétriques avec la vitesse du vent (á 19,34 GHz, et 55° d'incidence, la brillance croît de 1,1 °K (m $s^{-1}$) de vent, en polarisation horizontale). Un radiomètre multifréquence à 1,41, 8,36 et 19,34 GHz a été utilisé à cette occasion [16].

### 3.2.2. Paramètres de qualité

*Salinité*. La teneur en ClNa varie entre eau continentale «douce» (moins de 1 g/l) et eau de mer (environ 30 g/l), ainsi que dans les effluents de certaines industries, où les concentrations sont alors proches de la saturation ($\simeq$ 300 g/l). La meilleure réponse enregistrée pour NaCl est celle de 1,4 GHz, par rapport à 10,6, 22,2 et 31,4 GHz [17]. Dans la mesure où la température de surface du milieu ne varie pas sensiblement (ce qui exige l'emploi simultané d'un radiomètre I. R.), on peut transformer les températures de brillance observées en salinité, au moyen d'une courbe d'étalonnage. Dans le cas précis, à une amplitude de 13 °K $\pm$ 2 °K, correspondait une variation de salinité de 17 $\pm$ 5‰.

La propagation des ondes électromagnétiques de $10^6$ à $1{,}3 \cdot 10^{10}$ Hz dans l'eau dont la teneur en ClNa varie de 0‰ à 35‰ (sur la figure 5) a été vérifiée [18]. Il apparaît que la conductivité ionique du milieu, directement influencée par la concentration en NaCl, varie sensiblement dans le domaine de $10^6$ à $10^9$ Hz. La température influence peu la réponse du milieu en ondes électromagnétiques. La précision des mesures de NaCl pourrait être améliorée en travaillant vers les fréquences de $10^6$ à $10^8$ Hz.

*Hydrocarbures*. Ils interviennent, avec les détergents et les produits organiques très nombreux, dans le thème «pollutions primaires». La plupart des senseurs peuvent être utilisés: dans la bande U. V. les émulsions; le scanner infrarouge a été utilisé sur la côte californienne pour détecter la présence de nappe: le phénomène analysé est ici la différence de température de surface que crée la couche d'hydrocarbure en empêchant l'évaporation de l'eau. On peut aussi détecter cette présence par la différence de «rugosité» qu'elle crée à la surface de la mer. Une telle expérience a été tentée en Californie avec un side-looking radar. L'épaisseur d'hydrocarbures a pu être mesurée avec un radiomètre à 38 GHz, la mesure étant meilleure en polarisation horizontale (figure 6) [19].

*Traçage d'effluents*. On peut mesurer la concentration d'un traceur (rhodamine, fluoresceine) à l'aide de détecteurs dont les filtres sont centrés sur 5900 Å, fluorescence maximale du traceur, et 7000 Å, domaine de référence. Le système complet a une sensibilité qui permet de mesurer des concentrations inférieures au ppm. Une carte d'isoconcentration peut être directement réalisée en sortie de calculateur avec les informations enregistrées sur bande magnétique [20].

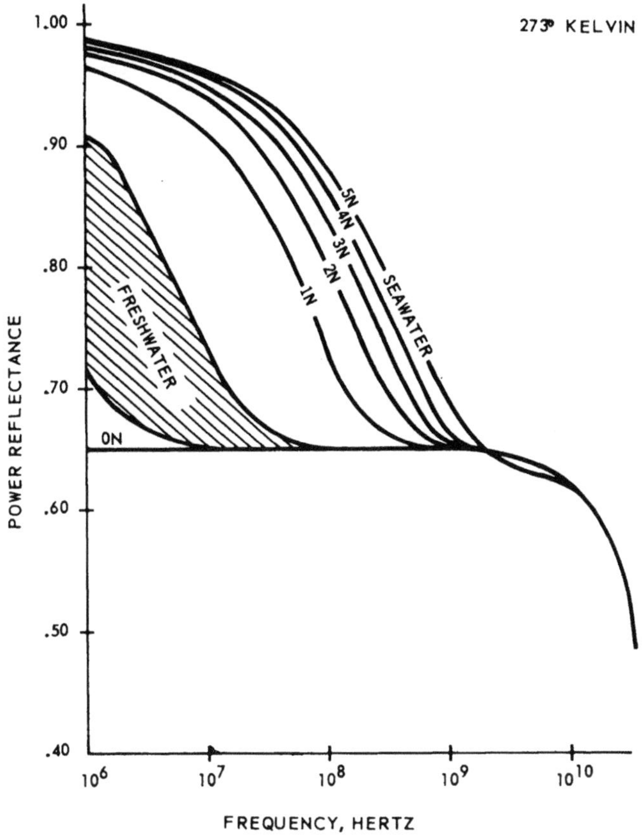


Fig. 5. Réflectance de l'eau pour différentes fréquences et teneurs en NaCl [18].

*Paramètres biologiques.* Il s'agit là d'un paramètre complexe, comparable à la turbidité-couleur dans lesquels il intervient d'ailleurs. Le visible est le domaine préférentiel de mesure (de 0,3 à 0,7 μm essentiellement). Les senseurs utilisés sont: les émulsions à bande limitée, le scanner multispectral à faible largeur de bande, le spectromètre. La radiance de l'eau diminue lorsque sa teneur en chlorophylle augmente, en particulier de 0,400 à 0,500 μm [21]. A partir de spectres de réflectance étalons établis de 0,4 à 0,75 μm, on recherche une corrélation entre une combinaison des réponses pour plusieurs longueurs d'ondes choisies et différencie ainsi algues et effluents [21].

### 3.3. Conditions actuelles d'opération

#### 3.3.1. Thème d'hydrologie

Le delta du Rhône est la zone-test choisie; elle permet l'étude de la salinité et du mélange des eaux de surface ainsi que celle des nappes libres peu profondes. Le réseau de contrôle est important; mais la synchronisation des contrôles volants avec les expériences aéroportées est toujours délicate.

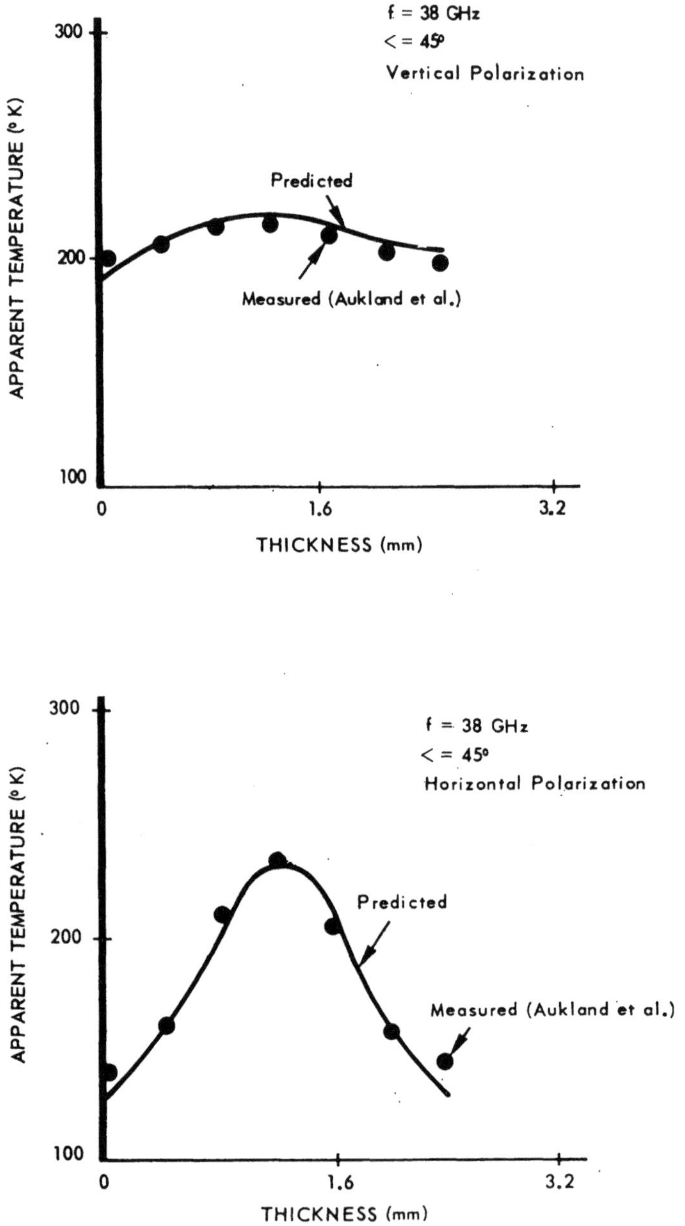


Fig. 6. Température apparente de surface de l'eau en fonction de l'épaisseur du film d'hydrocarbures [19].

#### 3.3.2. Acquisition des données

Les trois scanners aéroportés mis à notre dispositions par le CNES couvrent le domaine 0,4—15 µm. La conception de ces appareils étant différente, il en résulte une hétérogénéité des données qui accroît la difficulté de leur interprétation.

#### 3.3.3. Traitement et interprétation

L'interprétation s'est fait jusqu'alors à partir d'image et sous forme visuelle pour le repérage d'anomalies thermiques correspondant à des résurgences sous marines de la côte de la Méditerranée [23]. Les premiers traitements par calculateur sont en cours sur les données multispectrales fournies par le CNES.

## 4. Developpements possibles à moyen terme

### 4.1. En ce qui concerne l'objectif

#### 4.1.1. Phénomènes physiques

La recherche doit être orientée vers l'étude des phénomènes et variables hydrologiques accessibles dans le spectre électromagnétique. Ces recherches sont et seront conduites au laboratoire mais aussi sur des terrains expérimentaux dont toutes les conditions hydrologiques seraient contrôlées. Les deux orientations à retenir sont: étude de flux dans l'ensemble du spectre et identification par localisation de raies spécifiques d'émission ou d'absorption. Le pouvoir d'investigation dans le milieu devrait être étudié systématiquement, par le moyen de l'approche multispectrale.

#### 4.1.2. Phénomènes spatiaux

La recherche des régions susceptibles d'applications préférentielles devra être menée en tenant compte des avantages et limitations intrinsèques de la télédétection ainsi que des contraintes particulières aux phénomènes hydrologiques de ces régions, selon l'approche amorcée en § 2.2.
Des zones prioritaires d'expérimentations avancées seraient ensuite retenues en accord étroit avec les hydrologues dont la collaboration est indispensable.

#### 4.1.3. Phénomènes temporels

Une recherche des fréquences optimales de mesure doit être réalisée selon une procédure analogue en considérant les phénomènes à étudier dont la dynamique diffère sensiblement (précipitation, évaporation, infiltration, ruissellement, transfert de polluants) et les particularités de leur régionalisation. A cet égard la conduite d'observations systématiques au sol des principales variations temporelles de ces phénomènes dans le spectre électromagnétique paraît indispensable.

### 4.2. En ce qui concerne les moyens à mettre en oeuvre

Un dialogue permanent entre techniciens et utilisateurs permettra de dimensionner les systèmes de façon optimale et de les gérer efficacement. Dès à présent, une série de senseurs à hyperfréquences (passifs et actifs) paraît nécessaire. Des stations permanentes de ce type de senseurs se développeront rapidement. L'homogénéisation des stations modernes de mesure et l'enregistrement de paramètres d'environnement avec possibilité d'interrogation à distance devra être entrepris. L'acquisition de ces donnés *in situ* facilitera le traitement et l'interprétation de celles de télédétection.

### 4.3. Modalités de mise en oeuvre

Un programme cohérent de télédétection de phénomènes hydrologiques devra être élaboré couvrant la plus grande zone géographique possible. Des expérimentations systématiques des paramètres fondamentaux de la télédétection hydrologique seront prévues. L'objectif visé serait l'établissement de systèmes d'observations globaux. La correspondance avec les grands programmes mondiaux devrait être assurée: UNESCO/DHI et WMO, O. N. U. (Environment). La formation des utilisateurs potentials devra être assurée ainsi que leur information sur les possibilités actuelles des moyens: vecteurs, senseurs, missions, data processing, coûts.

## 5. Conclusion

La télédétection des phénomènes hydrologiques en est au niveau de la faisabilité et des premières expérimentations. Les développements attendus risquent d'être limités par défaut de dialogue et de coopération étroite entre hydrologues, utilisateurs et techniciens.

La base d'une activité commune fructueuse serait constituée par un programme d'application cohérent à l'échelle mondiale. Des priorités y seraient définies en ce qui concerne les phénomènes et les zones géographiques à étudier. La principale difficulté actuelle provient de l'absence de structuration des utilisateurs.

## Remerciements

Je remercie MM. J. Noel, J. Plevin, R. Tessier, pour leur aide pour la préparation de ce travail.

## Bibliographie

[1] D. Cerceau, J. Noel, P. Ogier, Y. Vuillaume and G. Weecksteen, in: ESRO (ERAF) Contractant Rep. No. 97 (1972).
[2] M. I. Lvovitch, in: Symp. on world water balance, AIHS-UNESCO 1970 (p. 401).
[3] R. G. Sharp and J. Harding, WMO Tech. Rep. No. 304 (1971).
[4] G. Castany, J. Margat, M. Albinet and O. Bouillin, in: Convegno internazionale sulle acque sotterrane, Palermo 1970.

[5] T. E. A. Van Hycklama, in: Symp. on world water balance, AIHS-UNESCO, 1970 (p. 434).
[6] O. Bouillin, Service géologique national Tech. Rep. No. 299 (1970).
[7] H. Hynes, in: Water quality analysis, Photog. Engng, March (1966).
[8] A. A. Sokolov, in: Symp. on world water balance AIHS-UNESCO 1970 (p. 10).
[9] M. Aubert, J. Aubert and S. Daniel, Inventaire national de la pollution des eaux de surface, CERBOM (1968).
[10] Conférence on the human environment — U. N. 48/7 (1972).
[11] R. L. Grossman, B. R. Bean and W. E. Marlatt, J. Geophys. Res. **74**, 2471 (1969).
[12] M. A. Hady and H. H. Kards, J. ASCE **9** (1971).
[13] V. I. Myers, NASA, Tech. Rep. No. MSC. 03742 (1970).
[14] H. C. Donald and W. P. Waite, Water Resource Res. **7**, 2 (1971).
[15] D. B. Ross and V. Cardone, NASA Tech. Rep. No. MSC. 03742 (1970).
[16] J. P. Hollinger, in: VII Symp. University Michigan 1971 (p. 1807).
[17] J. D. Droppleman and R. A. Meneua, J. Geophys. Res. **75**, 30 (1970).
[18] L. K. Lepley and W. M. Adams, Water Resources Res. **7**, 6 (1971).
[19] J. A. Richerson and J. W. Rouse, in: VII Symp. University Michigan 1971 (p. 1847).
[20] J. R. Eliason, H. P. Foote and M. J. Doyle, in: VII Symp. University Michigan 1971 (p. 731).
[21] G. L. Clarke and G. C. Ewing, in: NASA Tech. Rep. No. MSC. 03742 (1970).
[22] P. G. White, in: NASA Tech. Rep. No. MSC. 03742 (1970).
[23] J. Y. Scanvic and G. Weecksteen, in: Symp. télédétection, CNES, Paris 1971 (p. 189).

Earth Survey Problems — Akademie-Verlag, Berlin 1974

# NEW WAYS TO MONITOR THE MASS AND AREAL EXTENT OF SNOW COVER

M. F. Meier

US Geological Survey, Tacoma, Washington, USA

About 80% of all the fresh water in the world is stored as snow and ice on the land, a mantle of snow covers a fourth of the world's land more than four months out of a year, and in some regions virtually all of the streamflow is derived from snowmelt. Synoptic data on the snowpack are of great importance to hydrology. Continuous monitoring of the highly variable snowcover is possible only from satellites.

Snow can be readily distinguished from bare land on meteorological satellite images, but can be distinguished from clouds only with difficulty. Minimum brightness composites partly overcome this limitation. Higher resolution images such as those from ERTS 1 allow improved recognition of snow. For cloud-free, non-forested terrain, use of ERTS imagery with an interactive electronic console allows rapid measurement of snow area in specific drainage basins and these data are of immediate value in streamflow simulation models and forecasts of reservoir inflow.

Data on snowcovered area and snow water equivalent are both needed by hydrologists, and these data are especially important when the sky is cloudy. Thus microwave or radio wavelength sensors are required. Active systems are now used from aircraft to measure the thickness and internal structure of polar ice sheets. The feasibility of measuring seasonal snowcover in this way from satellites is uncertain. Passive microwave emission techniques appear to offer the potential of continuous monitoring of snowcovered area, snow wetness, and possibly snow water equivalent.

## 1. The Problem

One of the most dynamic geophysical features of Planet Earth is the ever-changing snowcover. No other widespread solid material on the earth's surface is deposited, and is removed, so quickly. No other widespread material on the earth's surface has such a high reflectivity to solar radiation yet is a near-perfect emitter of long-wave radiation, and thus has such a vital role in the earth's radiation balance and in driving atmospheric circulation. Snow annually covers about 22% of the earth ($1.2 \times 10^8$ km$^2$); about two-thirds of this is on land and the remainder is on sea ice [1]. The snow mantle covers one-fourth of the land surface more than four months out of a year [2]. Thus snow is quantitatively a most important feature of the world's surface.

Snow and ice also have an important role in the world's water balance. An enormous reserve of water is in temporary storage on the land surface in the form of snow and ice, amounting to 75—80% of all of the fresh water on, in, and above the earth [3]. Much of this water reserve is firn and ice of glaciers and ice sheets, only a small part of which is formed and released seasonally. The seasonal snowcover, however, is of critical importance to streamflow in many areas, especially in widespread regions such as Western United States and Central Asia where the bulk of the precipitation falls as high mountain snow, the melting of which produces the streamflow utilized for irrigation and hydroelectric power. In some areas virtually all of the streamflow originates from snowmelt. Even where perennial ice melt is important for the needs of man, it is the reflective property of the seasonal snowcover which causes the remarkable natural regulation of this glacier runoff [4].

Monitoring of the snowpack is essential to practical hydrology in many regions. This is true partly because of the quantitative importance of snowmelt in producing streamflow, soil moisture, and ground water recharge; partly because rapid melt of snow can cause disastrous floods; but most especially because measurement of snow in temporary storage on the ground allows prediction of subsequent runoff. According to a Panel of the United States National Academy of Sciences "The changing extent, surface temperature, thickness, water equivalent, and liquid-water content of seasonal snowpack are necessary for engineering design and operation and planning of water projects large and small. In the mountainous areas of the United States, millions of dollars are spent each year at fixed locations to measure snowpack for forecast purposes. Improved forecasts are estimated to be worth $10^7$ to $10^8$ dollars per year to water users in the western United States alone" [5].

Routine measurement of snow at individual points or stations is the basis for the well-known snow surveys used in many parts of the world to forecast streamflow. Advanced, computerized models are now routinely used to predict inflow to reservoirs; these models require data on snow mass and snowcovered area as input [6—8]. However, the snow mantle can form rapidly with a complex spatial pattern, and be removed even more rapidly, also in a complicated way. It is usually not possible to monitor these ever-changing complex patterns with a limited number of observation stations; so much of the input data to the numerical models is extrapolated and uncertain. As population expands and the per capita usage of water and hydroelectric power increases, better knowledge of our limited water resource will be vitally needed. Thus, the hydrologic community urgently needs a continuous, synoptic snow monitoring program. Because coverage over large areas is required this monitoring can only be done from space.

The monitoring of snow by remote sensing is important for other reasons. Many areas of major snow accumulation are inaccessible to instrumentation. These may be roadless areas at high latitudes or in high rugged mountains. Also of importance today is the fact that much of our snowy alpine country is being set aside for future generations as wilderness. For aesthetic and legal reasons it may not be possible to instal permanent installations such as telemetering snow gages in these areas, so an additional need for remote sensing exists.

## 2. Remote Sensing Techniques of the Present

One existing technique for the remote measurement of snow is unique in that it measures the total mass of snow lying on the ground, and this is the single quantity most needed by hydrologists. This remote sensing technique involves measurement of the absorption by the snow of the natural gamma radiation of the underlying soil or rock [9, 10]. This technique, however, requires that the sensor be flown within a few hundred meters of the ground, restricting its use to aircraft flights over relatively flat terrain. As it appears to be impossible to use such a scheme from satellite altitudes, no truly synoptic data are likely to be forthcoming.

Aerial photography is now used routinely to measure the extent of the snow-cover at specific times [11, 12]. This method, of course, provides data only on the area of snow, not its mass or hydrologic properties. In mountainous regions the relief causes image distortion and this makes automated data processing very difficult, if not impossible. Also, daytime and good weather conditions are required. Perhaps most important, the cost of repeated overflights strictly limits the potential usefulness of this technique.

The first meteorological satellites showed that snow on earth was clearly visible from spacecraft altitudes, unless a cloud cover interfered [13]. More than a decade of study of the snowcover on meteorological satellite imagery has clearly shown the potential, and the limitations, of this snow survey technique [14]. In flat non-forested terrain the approximate thickness of the snowcover can be inferred from the tone on images if the snow is not over about 10 cm thick [15]. In mountainous areas snow can be identified more reliably because the drainage and terrain patterns are clear and stable, and thus the area of snowcover can be accurately located and measured [16]; but very little data can be obtained on snow depth.

More recent meteorological satellites with higher resolution and infrared channels provide additional information to aid snow surveys. Increased resolution permits more accurate location of mountain snowlines. Because the near-infrared reflectivity of water is very low, wet snow can be identified by comparing visible and near-infrared images [17]. Thermal infrared imagery provides snow surface temperature and allows information to be obtained at night. Thermal infrared data appear to be especially useful in delineating snowcover on relatively flat terrain; in mountain areas the normal drop in temperature with increasing altitude may disguise the effect of snow, so that the brightest tones may cover a different area from that known to be snowcovered [18]. Fig. 1 shows simultaneous images of the Alps taken in the visible and thermal infrared wavelengths; note that the apparent "snowcover" appears to be more extensive on the infrared image.

Meteorological satellites may produce one or more images per day, and this repetitive coverage can be used to differentiate clouds from snow because clouds move, but snow patterns are relatively fixed in position. In order to suppress clouds, digitized images from several days may be composited retaining only the minimum brightness value at each map position. These composite minimum brightness (CMB) charts thus suppress or filter out relatively transitory cloud masses while retaining the imagery of the snow [19]. Persisting clouds will not be filtered out, but these generally appear less bright than snow.

The ERTS 1 satellite represents a major advance in ability to measure snow on earth: resolution is higher than that of existing meteorological satellites, and four-channel multi-spectral imagery in the visible and near-infrared wavelengths is provided. Many workers are now exploring the usefulness of this improved imagery to measure snowcovered area and snowline altitude [20—22]. The increased resolution allows excellent definition of the snowline in flat as well as mountainous areas (Fig. 2a). On smooth non-vegetated ground the planimetric accuracy of snowline mapping may approach 100 m, and the accuracy of measurement of snowline altitude (if good topographic maps are available) may be of the

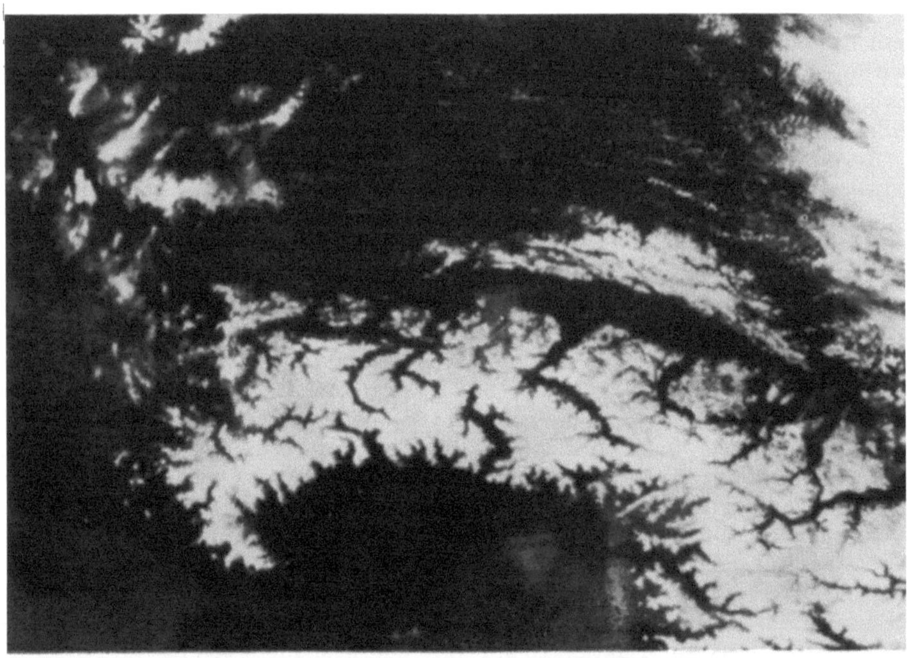

a

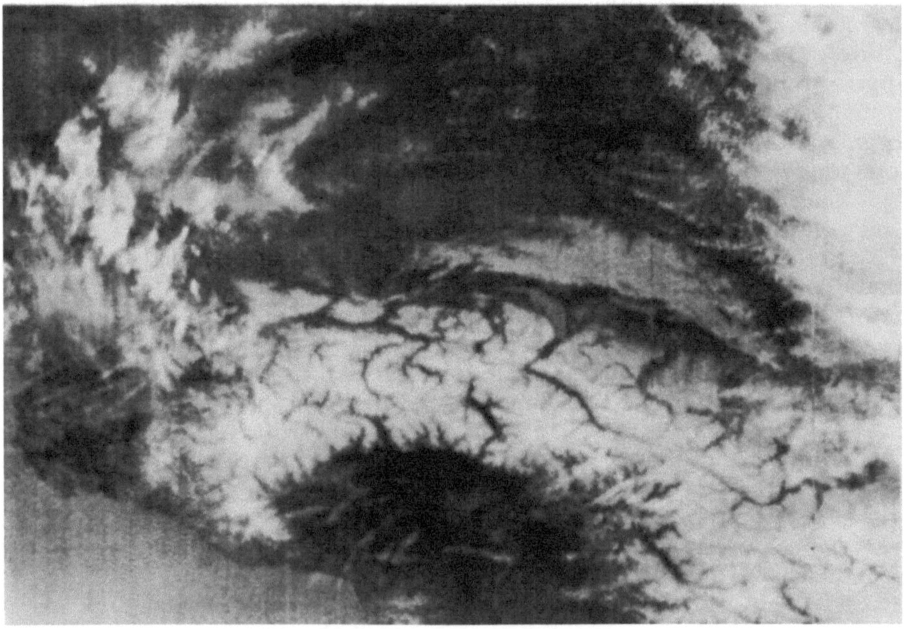

b

Fig. 1. Images of the Alps taken 16 April 1973, with visible (above) and infrared (below) radiation by the Very High Resolution Radiometer (VHRR) above the satellite NOAA 2. Photographs courtesy NOAA-NESS.

same order of magnitude or better. However, if the snow margin is highly irregular or if it is partly concealed by vegetation, the accuracy of mapping drops markedly (Fig. 2b). Differentiation of dry snow, wet snow, and firn or ice by study of their multi-spectral signatures is a subject which is still virtually untouched.

Satellite data can be used in another indirect way to measure precipitation and snowfall. Sequences of images can be used to categorize cloud types, e.g. bright shield, bright band, convective, etc. These cloud types can then be related to

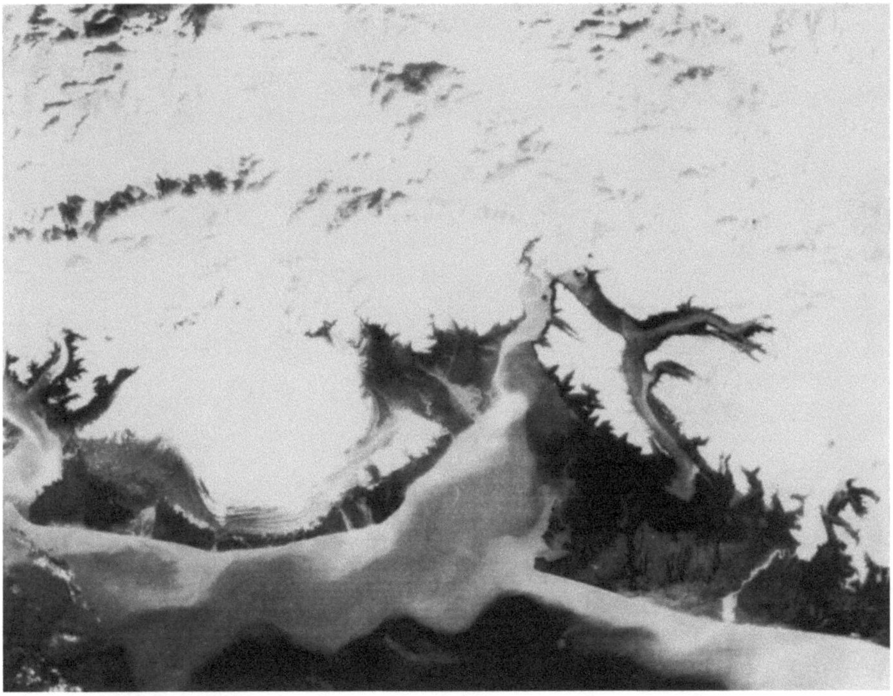

Fig. 2a. ERTS 1 image of snowcover in the vicinity of Yakutat, St. Elias Mountains, Alaska. Note the sharp delineation of the snowline. NASA ID No. 1077-20053-4.

observed precipitation at stations or as seen on meteorological radar. Then the relations of cloud types to precipitation amounts can be used to predict precipitation (and snowfall if temperature information is available) in ungaged areas [23, 24].

In summary, visible, near-infrared and thermal-infrared data from existing meteorological and earth resources satellites can be used, under certain circumstances, to map snowcovered area. The two main problems are vegetation and cloud cover, either of which may totally obscure the snow. Although CMB charts reduce the effect of clouds other desirable qualities such as resolution and sharp time definition are sacrificed. High resolution helps in differentiation of snow from clouds and vegetation, but a complete solution to this difficulty seems to be impossible.

Fig. 2b. ERTS 1 image of snowcover in Cascade Range, Washington. Virtually all land in the image area is snowcovered, but snow can only be seen above timberline on two volcanoes and in rectangular blocks where trees have been removed. Note also degradation and confusion of information content due to occasional clouds and fog. NASA ID No. 1203-18271-5.

## 3. Future Systems

Present observational systems do not, and cannot, supply the hydrologist with the snowcover data he needs for operational, management, or scientific purposes. Needed are all-weather synoptic data on an essentially real-time basis of (1) snowcovered area, (2) snow mass (water equivalent), and (3) snow wetness (free moisture content). Some techniques now under development may be able to meet these needs.

In order to obtain an all-weather data flow, lower frequency (cloud penetrating) electromagnetic sensors are of interest. Many types of active radio and microwave "radar" techniques have been applied to the study of snow. Side-looking radar (SLAR) images have been studied with conflicting results: new snow is not seen, and recrystallized old snow may be unobservable or may produce an exceptionally bright reflection [25, 26] apparently depending on surface roughness and grain size parameters. Few scatterometer data have been obtained from snow; whether this will be a useful technique for snow research is not known. Short-pulse or FM radar appear to have considerable promise for measuring the density or thickness of the snow.

A swept-frequency radar system ranging, say, from $10^7$ to $10^8$ Hz, and directed vertically downward might allow measurements of snow thickness *and* density (thus the mass, or water equivalent), and also provide information on the material below the snow [27]. This would work only if the snow exists as a horizontal planar layer. The interference patterns required for the success of this technique have been observed with passive radiometers viewing prepared samples of natural materials [28] but have not been observed viewing natural snow *in situ* [29]. Thus the feasibility of this technique and its application to spacecraft has not yet been proven, but the possibilities are exciting.

The potential of radar systems for observing snow and ice is perhaps best demonstrated by results of recent radio echo sounding experiments on the Antarctic and Greenland Ice Sheets and on arctic glaciers [30—32]. These relatively simple radar systems, operating at 35 to 400 MHz, have been used from aircraft or over-snow vehicles to measure rapidly the depth of ice many kilometers in thickness with a vertical resolution of about 10 m. In addition, internal layering of unknown origin has been discovered as has the presence of pools of water at the bed. The character of the echo "tail" has been used to measure the horizontal location relative to elements of bed roughness so that a similar measurement at a later date can be used to determine the horizontal ice velocity [33]. Proposals have also been made to measure the vertical velocity of the surface [34], and in combination with passive sensors to measure the mean annual temperature of the snow (the "10-meter temperature"). The obvious possibility of application of these techniques to certain kinds of snowpacks in temperate latitudes deserves investigation; however, the attenuation due to water in the snow is a problem which requires much further study.

Several general problems currently exist which may limit the usefulness of orbiting radar systems. For reliable penetration of thick wet snow or ice a relatively low frequency is required; thus high spatial resolution is impossible unless antennas of enormous size can be flown. High resolution is a requirement as radar returns from heterogeneous terrain usually cannot be interpreted unambiguously. All active systems also have relatively high power consumption, which may make their use in operational satellites difficult.

Passive microwave techniques may have the greatest potential for use in an operational snow survey satellite. Controlled experiments on the ground and from aircraft indicate that snow can be easily distinguished by its characteristic microwave radiation from most other land surface materials (Fig. 3). The microwave emission from snow depends primarily on snow wetness, depth and density [35]. Temperature, surface roughness, slope or terrain roughness, and the character of the underlying material appear generally to be of secondary importance; thus the potential exists to obtain much hydrologically useful information. Some studies indicate that snow can be seen through vegetation [36] but this needs further study.

The Electrically Scanned Microwave Radiometer (ESMR) operating at 19.35 GHz (1.55 cm) [37] on the satellite Nimbus 5 is providing the first micro-

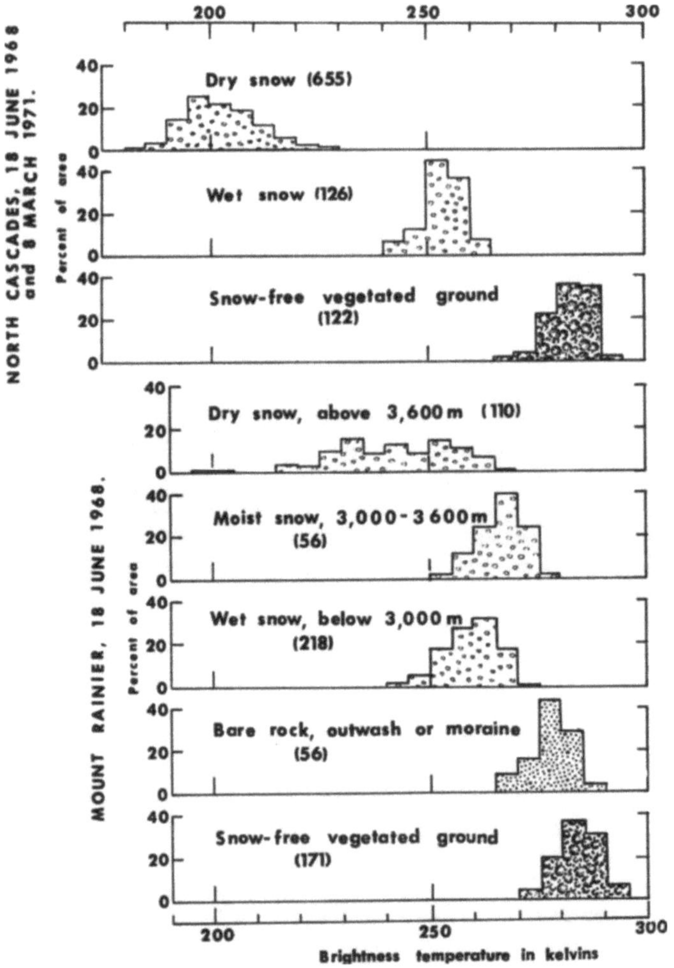


Fig. 3. Distribution of microwave brightness temperature, at 19.35 GHz, Cascades, Washington. Figures in parentheses indicate number of observations (resolution cells). From [36].

wave images from space. These show remarkable patterns of differing snow and types in Greenland and Antarctica. The differences in emission from place to place on these large snowcovered areas are large, more than one order of magnitude greater than the absolute or relative errors. The emission pattern can be qualitatively identified with terrain (e.g. snow on low ice shelves appears very different from snow on the high polar plateau of Antarctica), but it has not yet been possible to make a careful comparison of the microwave emission and the snow type as measured on the ground. No degradation of these images due to clouds can be seen.

Difficulties exist, however, in the interpretation of microwave emission data. These are due to two main problems: (i) the physics of microwave emission from snow (especially wet snow) is not yet well understood especially in regard to volume scattering effects, and (ii) several parameters affect the observed emission, so a multifrequency system will be required to provide unambiguous results. Also, the resolution of microwave radiometers flown in space will necessarily be crude by visual or infrared standards, but it has been shown that important results, such as amount of snowcovered area, may be calculated from the averaged brightness temperature for any field of view [36]. Thus the low resolution limitation of microwave radiometers or imagers may not outweigh the additional information value they may provide.

## 4. Conclusion

The need for synoptic data on snow is high, both from the point of view of science and in terms of practical application. Yet no operational method exists to measure the extent of snowcover, let alone the mass or wetness of the snowpack. ERTS-type satellites represent a great step forward, but systems employing only visible or infrared sensors are seriously limited by cloud cover and vegetation and provide no data from below the surface of the snow.

Eventually a snow survey satellite (or package aboard a hydrological or earth resources satellite) will be feasible, financially viable, and put into operation. It will probably rely heavily on microwave sensors, perhaps a combination of passive and active systems and probably working at two frequencies. But before this can be planned much more needs to be learned about the physics of emission, absorption and reflection of radiation from snow.

## References

[1] P. A. Shumzkiy et al., in: Research in Geophysics 2, MIT Press, Cambridge, Mass. 1964 (p. 425).
[2] J. Corbel, Neiges et glaciers, Librarie Armand Colin, Paris 1962.
[3] H. Hoinkes, Umschau in Wissenschaft und Technik **10**, 301 (1968).
[4] M. F. Meier, J. Am. Water Works Assoc. **61**, 8 (1969).
[5] Nat. Acad. Sci. Panel 3, Summer Study, Useful Applications of Earth-Oriented Satellites: Hydrology 1969 (p. 28).
[6] D. M. Rockwood, U. S. Corps of Engineers, North Pacific Division Tech. Bull. 22 (1964).
[7] V. P. Schermerhorn and D. W. Kuehl, IASH Symp. on the Use of Analog and Digital Computers in Hydrology I, 317 (1968).
[8] T. S. Abal'yan et al., Soviet Hydrology, 151 (1971).

[9] R. M. Kogan et al., Soviet Hydrology, 183 (1965).
[10] E. L. Peck et al., Water Resources Res. **7**, 1151 (1971).
[11] C. F. Leaf, Western Snow Conf. Proc. **37**, 19 (1969).
[12] J. Martinec, Int. Symposia on the Role of Snow and Ice in Hydrology, Banff (1972).
[13] S. Fritz, Archiv für Meteorol., Geophys. u. Bioklimat. Ser. A, 1962 (p. 186).
[14] R. W. Popham, World Meteorol. Organization, WMO/IHD Project Rep. 7 (1968).
[15] J. C. Barnes and C. J. Bowley, Allied Res. Assoc., Inc. Rep. 8G48-F (1968).
[16] J. C. Barnes and C. J. Bowley, Allied Res. Assoc., Inc. Final Rep. Contr. No. E-196-68 (1969).
[17] A. E. Strong et al., Mon. Wea. Rev. **99**, 828 (1971).
[18] J. C. Barnes and C. J. Bowley, Allied Res. Assoc., Inc. Final Rep. 8G92-D to NOAA and NESS (1972).
[19] E. P. McClain and D. R. Baker, ESSA Tech. Memo NESCTM 12 (1969).
[20] J. C. Barnes, ERTS 1 Symp., Greenbelt, Md., March 1973.
[21] H. Haefner et al., this volume, p. 351.
[22] M. F. Meier, ERTS 1 Symposium, Greenbelt, Md., March 1973.
[23] P. A. Davis, E. J. Wiegman and S. M. Serebreny, Stanford Res. Inst. Final Rep. Contr. 14-06-D-7047 (1971).
[24] P. A. Davis and E. J. Wiegman, Stanford Res. Inst. Tech. and Semiann. Rep. No. 1 (1973).
[25] M. F. Meier et al., Proc. 4th Symp. on Remote Sensing of Environment, 145 (1966).
[26] W. P. Waite et al., Remote Sensing of Environment I, 143 (1969—70).
[27] W. I. Linlor, Symposia on the Role of Snow and Ice in Hydrology, Banff (1972).
[28] J. C. Blinn III et al., J. Geophys. Res. **77**, 4366 (1972).
[29] A. T. Edgerton, A. Stogryn and G. Poe, Aerojet-General Corpn. Microwave Division Final Rep. 1285R-4, Contr. No. 14-08-001-11828 (1971).
[30] P. Gudmandsen, Proc. Int. Mtg on Radioglaciology, Lyngby, 124 (1970).
[31] S. Evans et al., J. Sci. Instrum. (J. Phys. E) **2** Ser. 2, 131 (1969).
[32] V. V. Bogorodskiy et al., Soviet Antarctic Expedition Bull. **6**, No. 65, 105 (1698).
[33] M. E. R. Walford, Nature **239**, 93 (1972).
[34] J. F. Nye et al., Nature **240**, 7 (1972).
[35] M. F. Meier et al., Proc. 7th Int. Symp. on Remote Sensing of Environment, Ann Arbor, Michigan, p. 1155 (1971).
[36] M. F. Meier, Int. Symposia on the Role of Snow and Ice in Hydrology, Banff (1972).
[37] C. Catoe, W. Nordberg, P. Thaddeus and G. Ling, NASA Rep. X-622-67-352 (1967).

Earth Survey Problems — Akademie-Verlag, Berlin 1974

# PROSPECTS FOR PHYSICAL OCEANOGRAPHY FROM SPACE

D. E. Cartwright

National Institute of Oceanography, Wormley, Surrey, UK

Apart from applications to marine biology, glaciology and near-shore processes, the principal areas of physical oceanographical research to which remote sensing from space are likely to contribute are: Lagrangian current tracking; global surface temperatures; wave height, roughness and wind parameters; altimetry of static and dynamic features of the ocean surface, including tides. Present and possible future requirements and progress in these topics are reviewed and discussed.

## 1. Introduction

The author is not associated with any specific project in remote sensing; this account is therefore an uncommitted, if largely personal, impression of the literature on remote sensing by a professional oceanographer.

One cannot expect remote sensing to solve all problems in physical oceanography, because (with the minor exception of data collection from sub-surface instruments) parameters of only the top few metres of the ocean at most can possibly be surveyed. Nevertheless, some surface parameters are so difficult to measure on a global, synoptic scale by shipboard techniques that their possible monitoring from a satellite offers new concepts of research in their respective fields. I shall enlarge on those fields in which I have most direct experience, rather than try to cover all possible applications at a superficial level. Aspects of physical oceanography which are more relevant to biology or glaciology are avoided, since these subjects are being treated by other authors, and coastal or near-shore processes such as shallow depth determination and tracing of river effluents are also avoided, since these are fine-scale effects better dealt with locally by airborne methods.

The areas to which satellite-based techniques seem to be most likely to contribute quantitatively are: Lagrangian current determination by float tracking; global mapping of surface temperature by infrared and microwave radiometers; ysnoptic mapping of wave height and surface roughness parameters by radar methods; and, most valuable because of the near impossibility of obtaining the data otherwise, altimetry of the static and dynamic features of the sea surface. These subjects are discussed in turn below. An important aspect of altimetry is the possibility of measuring ocean tides: this has received rather summary treatment in the general literature.

It should also be appreciated that the better known techniques for atmospheric sensing, especially the multi-band infrared methods from which air temperature and humidity can be derived, and to some extent cloud imagery, are themselves contributions to oceanography because they enable the ocean's infrared radiation budget to be computed.

## 2. Lagrangian Currents

Knowledge of ocean currents is fundamental to physical oceanography, or more particularly, if limited to the surface layers, to the study of the transfer of energy and momentum from the atmosphere. The Lagrangian motions (that is, the motions of "tagged" parcels of water) include dynamical features which are unobtainable from the well established Eulerian techniques, (which use moored sensors). Shipboard float-tracking is limited, and very costly in ship-days.

Tracking of drifting floats by satellite is in fact already established as one of the most straightforward and useful techniques of space oceanography, so there are few problems to discuss here. Most uses will continue to be detailed dynamic study of limited sea areas chosen by individual research groups, rather than any attempt at global coverage. Tracking of a batch of floats released simultaneously within a small space is probably the most economic and productive approach.

It is not possible to quote probable requirements in numbers, but there will certainly be a continual and expanding demand for data capacity. A desirable capacity of up to 500 floats, to be tracked for the order of a year has been suggested. Accuracy and frequency of fixing are perhaps the most important stipulations, even if these call for rather expensive surface equipment. The successful transponder system used in conjunction with the Eole craft, with fixing accuracy of a few hundred metres once or twice a day, is a good compromise: The related technique of data collection from fixed oceanic sensors is not discussed here as it does not differ essentially from land-based systems. It is widely use dat present, and future requirements depend on the unpredictable availability of costly arrays of moored buoys.

## 3. Surface Temperature

Global mapping of sea surface temperature to an accuracy of 1 °K at regular intervals of, say, 10 days, would add greatly to our knowledge of heat exchange and the generation of weather systems. With resolution of 20 km or better, it would also greatly facilitate the study of local "upwelling" of deep cold nutrient-rich water, the behaviour of major oceanic "fronts", and the boundaries of warm jet streams. The usual method of data acquisition, by collecting routine measurements from ships, is easy enough, but the concentration in commercial shipping an es is a severe restriction.

Again, one feasible technique, remote sensing by IR radiometry, has already been established by the Nimbus series and other satellites, but the advantage of high resolution (10 km) is offset by masking from clouds, often correlated with the sought oceanic feature, and the partial restriction to night viewing. Despite these disadvantages, however, useful mapping has been achieved by selecting distinctly clear-sky trajectories over a given area, especially where clouds are recognizable by their colder temperature. Here, the fact that surface temperatures do not change much in an interval of 10 days is advantageous. Even so, atmospheric absorption is not always known with much precision.

There is now a substantial literature on research into microwave radiometry for sensing sea temperature, although its validity is not strictly proven. This method would avoid the obvious disadvantages of infrared, but introduces new ones. The microwave radiation is more or less dependent on surface roughness, and emissivity is greatly enhanced by whitecaps and foam. It seems that the only way to reduce roughness-dependence is to increase the wavelength, but other factors limit this to 100 mm (3 GHz). However, the wider resolution, of order 100 km at the surface, is still distinctly useful, and if one has means of discriminating against severe conditions of foam and precipitation, large wavelength microwave radiometry may prove a viable competitor with infrared for oceanic purposes. As with the other techniques discussed below, efficiency of use would be much enhanced by complementary data from other sensors on the same satellite, enabling rejection of conditions in which a given sensor is known to be unreliable, and correction for moderate errors due to atmospheric conditions.

## 4. Waves and Wind

Research on ocean surface waves is mainly of two kinds. One kind of research studies the dynamic structure of the varying surface in great detail, and its relationship to similarly detailed fluctuations in the local air. Although airborne narrow beam radar altimeters have been used in such studies, it is unlikely that space methods could contribute usefully, even with laser beams. Such studies are best pursued with surface instruments in local sea areas. The other kind of research eschews all dynamical detail, and is content with one or two parameters to define the "sea state" adequately over a few hundred km$^2$ of surface and a few hours of time. The interest is in the global distribution of such parameters, either as real-time synopses, or for testing hindcasts, or for compilation of wave statistics. If surface "roughness" or rms slope is used as a parameter, this may reasonably be taken as an indicator of wind speed, with application to the generation of ocean current circulation and to atmospheric pressure gradients. The relevant techniques have been intensively studied, and there are high hopes that such parameters can be gathered from spacecraft by means of radar methods.

An advanced instrument package which embodies most requirements for wave/wind sensing is the S 193 at present being tested in Skylab. S 193 includes a scanning pencil-beam radar, a passive microwave radiometer, and a radar altimeter. The pencil beam scans a sub-satellite track 1200 km wide, every 120 km, with "footprints" of some 2 km. Allowing for the vertical angle, the scattering cross section of the surface can be interpreted in terms of "roughness", or wind speed. The radiometer gives a signal dependent partially on roughness of an area at some fixed inclination to nadir, partially on sea surface temperature, and also dependent on cloud cover and precipitation. It uses a higher frequency (about 10 GHz, 30 mm) than the radiometer systems aimed at temperature sensing (§ 3), and so the roughness dependence is supposed to be the dominant factor. Corrections for temperature can be derived from the infrared sensors included in the spacecraft. Its sensing of anomalies due to clouds, precipitation or foam can in turn be fed back either as a correction for atmospheric absorption in the radar signal or in extreme conditions as a warning to disregard the data. There is some advantage in sensing only the horizontally polarized components of the emission, since they have been shown to give a more uniform increase with surface roughness.

Roughness, however, is highly biased towards the very short wavelength components of the sea. The amplitudes of the longer waves, which in fact transmit most of the energy in the wave system, can only be inferred from roughness data by the results of other measurements and theory of the shape of the wave spectrum under steady wind conditions. The theory is somewhat speculative and it ignores swell produced by remote wind systems, so one needs an independent means of sensing a parameter which adequately describes wave height. This is provided by measuring the change in shape or "spread" in the return pulse from the radar altimeter. Besides giving a unique measure of wave amplitude, the spread is also useful as a correction factor for the more direct use of the altimeter signal to measuring mean surface altitude, discussed in § 5.

The S 193 package is still experimental, and a complete philosophy for its use or modification will only emerge from the analysis of the Skylab tests, in comparison with "ground truth" wave records. If it works as theory and measurement from aircraft indicate, it will show that an operational instrument should provide information that would be the equivalent of 40000 ship reports per day over the oceans.

## 5. Sea Surface Altimetry

The altitude of the sea surface above the earth's centre, viewed as a function of latitude, longitude and time, contains variations which are of great interest to oceanographers because, unlike all the other variables considered in this paper, direct measurements are relatively sparse and costly to obtain, and theoretical estimates are of dubious accuracy. The largest variations, of some tens of metres, are due to the shape of the geoid. The major features of the geoid are of course known from studies based on orbit perturbations, but there must remain many unresolved features of large wavenumber which could be sensed from space only by altimetry. These concern principally the geodesist, but oceanographers are interested in features associated with oceanic trenches and other geological anomalies.

The sea surface departs from the geoid by up to an order of a metre on account of the tilt induced by large current systems, and to a lesser extent by variations in density. Such departures are virtually static, but they could be disentangled from genuine geoidal features by independent measurements of the geoid (using say the shipborne GEON apparatus) in interesting localities such as the Gulf Stream. Study of current set-up by this means would be one of the few ways in which one could remotely sense the integrated internal motions of the sea.

### 5.1. Tides

The variations of surface altitude with time are of course independent of the geoid, and the most important of these are the tides, which are the most easily extracted from the noise background because of their high coherence with known astronomical variables. Amplitudes are of the order of a metre, but they have been shown to be extractable from noise amplitudes of the same order with only moderate numbers of data points because of this property of coherence. By the same token, tides are unique among geophysical time variables in that ground truth does not have to be supplied simultanesously with remote sensing; it can be supplied if necessary at the most tactical positions at any time in the future.

Remotely sensed tides are truly geocentric, unlike all other oceanic tidal measurements, which are perforce relative to the tidal motions of the earth's crust. The difference is interesting, but does not concern the present discussion. It consists of a direct crustal yielding to the gravitational forces, of order 0.2 m, which is easily calculable to more than sufficient accuracy, and an indirect effect of ocean loading which is not easy to calculate, but, being less than 0.1 m in most places, is within the expected limits of measuring accuracy. The field equations may be expressed in terms of either geocentric or crust-relative variables.

Ideally, one would wish to assign say four complex tidal "constants", that is two for the twice daily tides and two for the daily tides (unless one of these species is known to be weak) at every member of a grid of small areas, say 1° squares, over the deep ocean. In some circumstances, for example near an island where the tides are already known, one could reduce the requirement to two or even one tidal constant. Extraction of the constants with tolerable accuracy from a noise background of, say 0.5 m, would require some 100 independent corrected altimeter readings at each grid. Evidently, 100 traverses of a 1° square near the equator would in general take about 18000 orbits, that is a few years, although grid areas near the zenith latitudes would be traversed more frequently. Alternatively, one could consider much larger grid areas, perhaps 10°, and allow for linear spatial dependence of the tidal constants, thus increasing the effective number of constants per mesh from 4 to 12. One would then require perhaps 300 samples per mesh, but would obtain them ten times as fast. Or, one could apply the known local tidal dynamics as a "constraint" to reduce the number of arbitrary constants. There would probably be an advantage in choosing a grid pattern oriented along and normal to the sub-satellite tracks.

Given the data, then, an outstanding geophysical problem could be solved—the global tidal map. The feasibility depends on reducing the errors in altimeter performance and orbit tracking to less than a metre. It appears that altimeter design is already within this limit, provided on averages over a reasonable number of pulses and makes use of other remote sensors to correct for biases due to the atmosphere and surface waves. Tracking errors, for which one hears typical estimates of 1 to 5 metres, are therefore the main limitation. However, much of the total error in geocentric surface altitude may be greatly reduced by ground truth monitoring. For example, there are already several stretches of continental shelf zones along which the surface altitude at any time may be specified to a precision of about 0.2 metre (including a correction for atmospheric pressure). There are also lakes and seas where it could be specified even more precisely. On each oceanic trajectory between such truth zones, the tracking errors could thereby be reduced to within the desired limits.

Progress in the techniques of satellite altimetry is being made with Skylab and with the still more advanced equipment designed for GEOS-C. There is little doubt that a fair estimate of the major features of the geoid should emerge from these efforts. The next step should be to hindcast the expected tidal signal as well as possible from present surface knowledge, and test whether the residuals from the time-meaned altimeter signals are reasonably correlated with it. If so, the way may already be open for a serious independent tidal analysis as discussed above. At all events, the required precision will eventually be reached, given normal technological progress, and the value of all data sets will be cumulative.

## Acknowledgments

I thank J. F. R. Gower of Marine Science Directorate, Victoria BC, H. Charnock, Director of NIO, and Professor W. J. Pierson of New York University, for helpful talks and correspondence relevant to this survey. The many papers and articles used from the vast, expanding literature on geophysical remote sensing are too numerous to list in the context of this Symposium.

# ERTS 1 Results

Earth Survey Problems — Akademie-Verlag, Berlin 1974

# THE FIRST EARTH RESOURCES TECHNOLOGY SATELLITE (ERTS 1)

W. Nordberg

NASA/Goddard Space Flight Center, Greenbelt, Md, USA

The first Earth Resources Technology Satellite (ERTS 1) makes images of the earth's surface in four portions of the electromagnetic spectrum with sufficient spatial resolution and with a minimum of geometric distortions, so that these images may be applied experimentally to the study of geophysical processes relating to earth resources, to the exploration and conservation of these resources and to the assessments of environmental stresses. During the first six months of operation, ERTS 1 has imaged 6.5 million square kilometers of the earth's surface every day, covering most major land masses and coastal zones as well as both polar regions of this planet.

These images as well as the results of their analyses are available to all people throughout the world. Scientific investigators of all countries have been invited to participate in the utilization of ERTS 1 observations. Many of them have already demonstrated the great efficiency, economy and reliability of making earth surveys from space.

## 1. The ERTS 1 System

The ERTS 1 system consists of a spacecraft and a very complex ground facility, which, to a large extent, outweighs the spacecraft in complexity and significance. The spacecraft is an outgrowth of and very similar to the Nimbus meteorological satellites. Many of the sub-systems on the spacecraft are the same as those on Nimbus. An outline drawing of the ERTS 1 spacecraft and its payload is shown in Fig. 1. ERTS 1 differs from Nimbus mainly in the sensors which are carried by the spacecraft. They are: three Return Beam Vidicon (RBV) television cameras, and a Multi-Spectral Scanner (MSS). These two sensors are designed to measure sunlight reflected by the earth in four spectral bands, namely, in the green and yellow/red portions of the visible spectrum and in the infrared at a wavelength near 0.8 μm. The MSS alone is sensitive to a fourth band in the infrared, near a wavelength of 1.0 μm.

ERTS 1 also carries a Data Collection System which acquires transmissions from ground platforms distributed over North America. The satellite re-transmits the messages from these platforms to a central facility at the Goddard Space Flight Center. These messages consist of measurements such as water quality, rainfall, snow depth, and seismic activity. The measurements are made by the platforms directly and automatically at remote and inaccessible sites.

During daytime, the satellite passes from 80° N to 80° S in a south-southeasterly

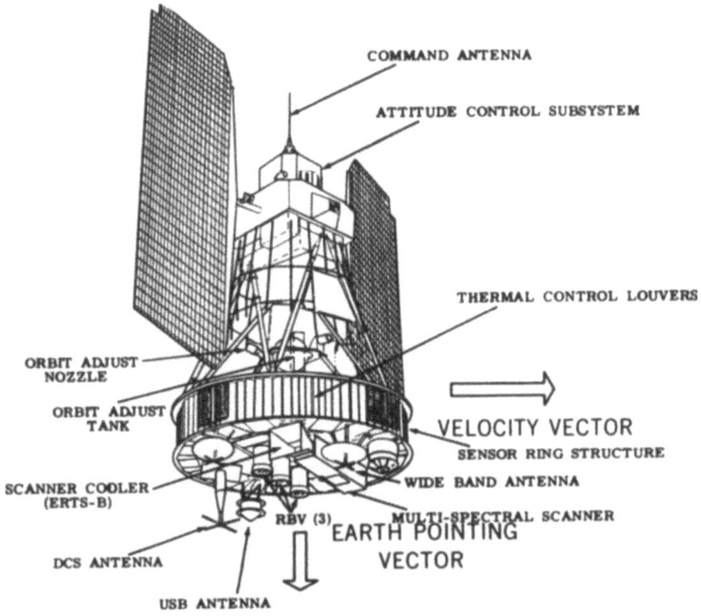


Fig. 1. Outline drawing of ERTS 1 spacecraft.

direction and transmits the optical measurements made by the return beam vidicon cameras and the multi-spectral scanner directly to receiving stations located at the Goddard Space Flight Center; Goldstone, California; Fairbanks, Alaska; and to a Canadian station in Saskatchewan. The government of Brazil has constructed a fifth station, which is expected to be fully operational during late 1973. During night-time, the satellite transmits the measurements recorded over other parts of the world to the three US stations only.

All satellite passes occur at approximately the same local time at each location, namely, between 0900 and 1000 in the morning every day. The fields of view of the two sensor systems are identical and cover a strip 185 km (100 nautical miles) wide directly underneath the spacecraft. Images are taken along selected portions of these orbital strips. At the equator, the centers of successive strips are spaced 2800 km apart, so that 14 such strips may be observed around the world every day. The orbit is adjusted so that a strip observed on one given day, in one given location, advances westward by about 170 km on the next day. In this fashion, as each strip advances from day to day contiguously, the entire world between 80°N and 80°S is viewed by these ERTS sensors once every 18 days. However, limited power, data transmission and, above all, data processing capacity, prevent acquisition of images over the entire world during every 18-day period. Thus, contiguous images are acquired every 18 days only over the North American continent. For the rest of the world, selected strips over portions of the orbit track are imaged. Every day, these strips amount to about 26000 km in length and 185 km in width. They are chosen on the basis of cloud cover forecasts and on the basis of investigators who have expressed interest in ERTS 1 coverage of such areas and have reached agreements with NASA in advance.

Every day about $10^{10}$ digital bits are transmitted electronically and are converted to about 1350 photographic images at the NASA Data Processing Facility at Goddard Space Flight Center. Approximately ten copies of each of these images are made and distributed to several US Government agencies: the US Army Corps of Engineers, the Department of Commerce, the Department of Interior, and the Department of Agriculture. Both the Department of Commerce and the Department of Interior archive all the photographs which they receive at their data centers, and ERTS 1 photographs may then be purchased from these data centers by the general public. In addition, NASA has reached agreements for analysis of ERTS images with about 320 investigators, of whom about 100 are from 36 countries outside the United States. These investigators attempt to apply the ERTS 1 observations of landforms and land use patterns to the assessment of mineral, agricultural and other land resources, the observations of snow, ice, lakes, and rivers to the assessment and prediction of water supply and quality and to the analyses of floods, and the observations of oceans and estuaries to the assessment of ocean food resources, sedimentation and erosion processes and the quality of the coastal environment.

## 2. Status of ERTS 1

ERTS 1 was launched on 23 July 1972. Picture taking was activated two days later. The spacecraft, its sensors and associated picture generation and distribution systems have been in operation since then. As of the end of June 1973, the system was still operational. ERTS 1 has met its primary and secondary objectives, which were to demonstrate the utility of multi-spectral images, acquired periodically over North America, to the survey and management of earth resources and the detection of environment stresses, and to acquire multi-spectral images of major land masses on a global scale, respectively. All design goals of the ERTS 1 system have also been largely fulfilled, namely, one year life of the spacecraft and its sensors and the generation of images from measurements over about 6.5 million square kilometers daily.

With the exception of three days during early August 1972, the ERTS 1 system has operated and acquired images on every single day since activation on 25 July 1972. During the first six months of operation, nearly 34000 scenes have been imaged throughout the entire world in each of the four spectral bands. A "scene" is an area 185 km wide and 185 km along the orbital path. This, on the average, amounts to 188 scenes per day.

The three return beam vidicon cameras and one of the tape recorders needed for acquiring images outside North America were deactivated shortly after launch because of two unrelated electrical malfunctions. However, the multi-spectral scanner and the other tape recorder have functioned normally until 29 March 1973. On that day, the second tape recorder malfunctioned and image coverage of the world outside North America had to be reduced to about 50 scenes or 9000 km long strips per day, since then. However, coverage of North America was not affected.

Most of the major land masses and coastal regions of the world and both polar caps have been mapped with the ERTS 1 system. Many of the investigators' test sites have been mapped periodically and most of North America has been mapped sixteen times since launch, as of mid-May 1973, cloud cover permitting.

## 3. Results from ERTS 1

Many characteristic features related to earth resources and environmental processes appear distinctly and separately in each of the four spectral bands of the ERTS 1 sensors. The most useful identifications and interpretations can be derived from the contrasts between the visible and infrared portions of the spectrum. This is illustrated in Figs. 2a and b which display southeastern England, with the city of London in the center, observed on 8 March 1973 in the red/yellow (2a) and the 1.0 μm infrared (2b) bands. The two spectral images which were taken simultaneously and are nearly in complete registry, show the drastically different characteristics of most features in the two spectral bands. A portion of the Channel, off the coast of Hastings, which appears relatively dark in both spectral bands, is seen in the lower right corner. The mouth of the Thames, however, shown in the center right, appears relatively bright in the yellow-red band, but very dark in the infrared. This is because in the infrared, any water, regardless of depth, composition or turbidity has extremely low reflectivity, while in the visible, shallow and turbid water is much brighter than clear and deep water. Thus, the very bright areas in Fig. 2a, north of the mouth of the Thames (center

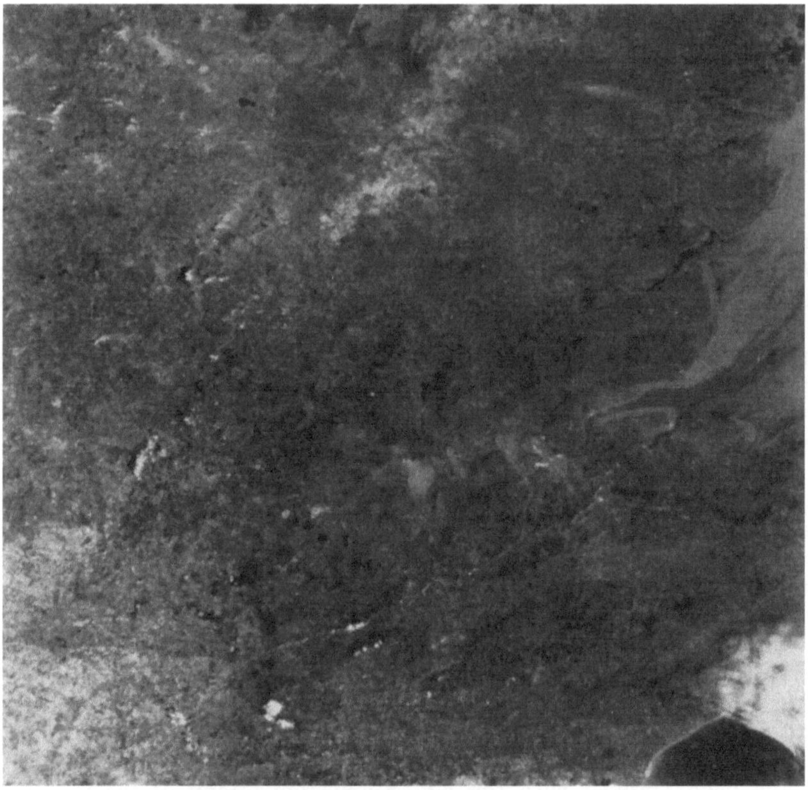

Fig. 2a. ERTS 1 image of southeastern England in yellow/red (0.6–0.7 μm) band obtained on 8 March 1973.

right) are shallow tidal mud flats; the portion of these mud flats which is exposed and above water can be clearly delineated in the infrared picture (Fig. 2b) which is indicated by the somewhat brighter strip, about 5 km wide, off the coastline near Southend-on-Sea. The deeper channels cut by the Thames and the tides, as well as the patterns of sedimentation and river effluents can be traced in the yellow/red band (Fig. 2a).

Cultural features also appear as nearly opposite brightness patterns in the two bands. For example, the densely built up area of the city of London and the radially spreading transportation arteries show very low brightness in the infrared while in the yellow/red band these areas are relatively bright. Also, the air pollution over the inner city and along the shore of the Thames, east of London, can be clearly identified by the white streaks, streaming off in a southeasterly direction in the yellow/red image (Fig. 2a). For reasons which are not fully known at this time but probably related to the composition of the plumes, these air pollution patterns are not apparent in the infrared image (Fig. 2b).

Both dry barren land and vigorous vegetation have a very high reflectance in the infrared. This is seen by the very light areas in the western and southern (left and lower) portions of Fig. 2a. A distinction between dry barren land and vegeta-

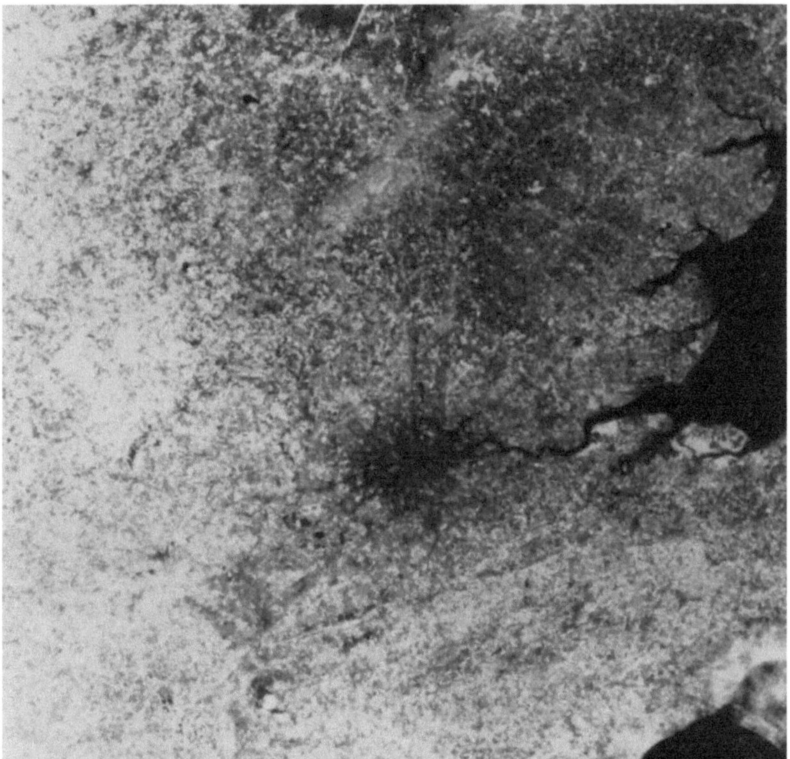

Fig. 2b. ERTS 1 image of southeastern England in infrared (0.8—1.1 μm) band obtained on 8 March 1973. The city of London is in the center.

tion can be made by comparing the infrared picture with the yellow/red one; in the latter, dry barren land appears bright, as seen in the lower left corner of Fig. 2a, while vegetation is generally dark as seen in much of the rest of Fig. 2a.

Unvegetated moist ground, especially when covered with deciduous forests whose leaves have fallen, appears very dark in both the infrared and visible bands; the general terrain shown in the upper right quadrant and at the very top of Figs. 2a and 2b belongs in this category.

These differing contrasts in each of the spectral bands can be displayed simultaneously in "false color" composites shown in Fig. 2c where the separate ERTS images in the green, red and infrared bands are superimposed on color film. In this particular composite, as well as in most ERTS color pictures, the infrared image was exposed to the red (magenta) dye of the film, the red image to the green dye and the green image to the blue (cyan) dye.

Perhaps the most significant result of the ERTS 1 mission is the realization that features related to land resources or to environmental processes can be mapped synoptically for very large areas on scales up to 1:250000 without

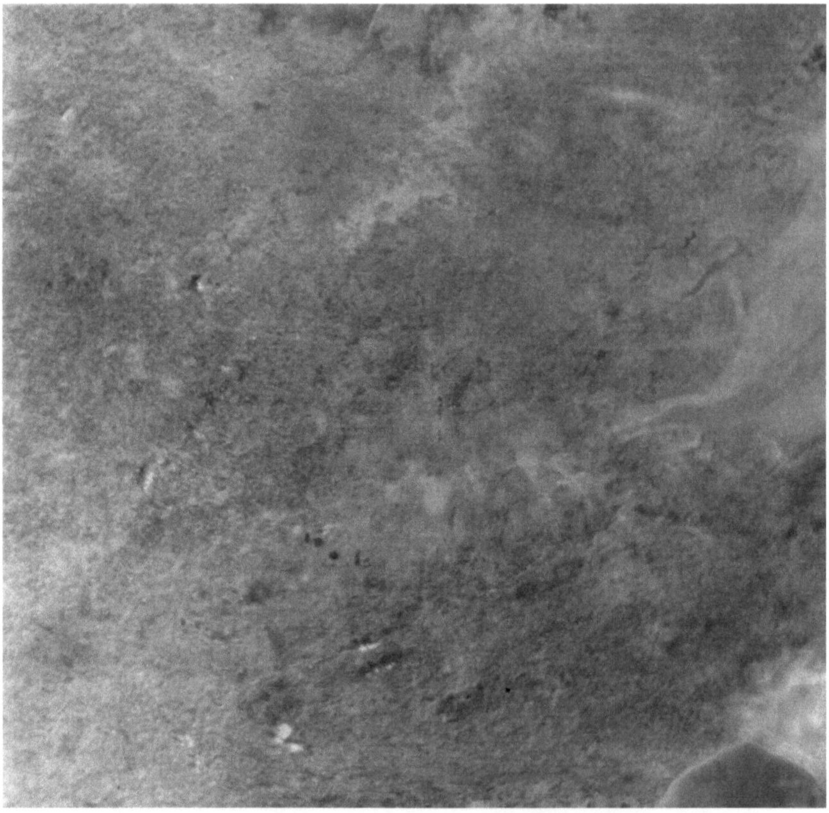

Fig. 2c. Color composite of green, yellow/red and infrared (0.8–1.1 μm) bands of ERTS 1 image obtained on 8 March 1973 of southeastern England.

Fig. 3. Montage of two color composite ERTS 1 images of north central Alps on 1:1000000 scale map. The right hand image was obtained on 6 October, the left hand on 7 October 1973. Lake Constance is seen in the upper left.

substantial loss of definition. This is illustrated in Fig. 3 where two ERTS views of the Alps showing the Inn Valley and Lake Constance were overlaid on a 1:1000000 scale map. The ERTS pictures produced from the transmitted signals are displayed on that scale and distortions are minimized so that these pictures serve practically as an instant map whose information content, economy and speed of production far exceed that of a conventionally produced map. These pictures are only "system" corrected, i.e. they are geographically referenced according to the known orbital position of the satellite at the time of observation and to the pointing of the sensor axis which depends on the known spacecraft attitude. No corrections are made for distortions due to non-linearities of the scanner. Yet, such system corrected pictures usually have no greater distortions than several hundred meters over the 185 km scene. These distortions can be further reduced to less than 100 meters by photo-processing on the ground but some of the thematic information content of such "scene corrected" images is sacrificed in the "precision" processing. Objects can be identified in the ERTS 1 pictures and put in the context of their overall environment even on a scale as

Fig. 4. ERTS 1 color composite image of San Francisco Bay (center and upper left), the San Joaquin Valley (upper right) and Monterey Bay (lower left) in California obtained on 4 April 1973.

large as 1:30000. Such objects include agricultural fields, snow and ice cover, rivers and lakes, tectonic structures, land slides, patterns of civil construction and of urban development. The color composite images of California shown in Figs. 4 and 5 are excellent examples for such mapping.

Delineating both the spectral characteristics and spatial or geometric patterns of features observed in ERTS 1 images permits a unique and extremely efficient classification of these features. In a typical agricultural scene such as shown for the Central California Valley (upper right, Fig. 4) or for the upper San Joaquin Valley (upper left, Fig. 5), 10—20 different categories of land use can be readily and automatically identified. When combined with existing knowledge of a particular area and with sporadic but strategically placed direct observations on the surface, such categorizations have resulted in the accurate identification of crops, measurement of acreages, and in the assessment of stresses on or damages to soils, crops, forests, rangelands or habitats by fires, insect infestations, human interaction, droughts, floods or topographic conditions. Many of these applications were summarzied by Poulton [1].

For example, based on the first ERTS 1 image of the agricultural area of San Joaquin county during July 1972, Colwell et al. [2] have been able to identify

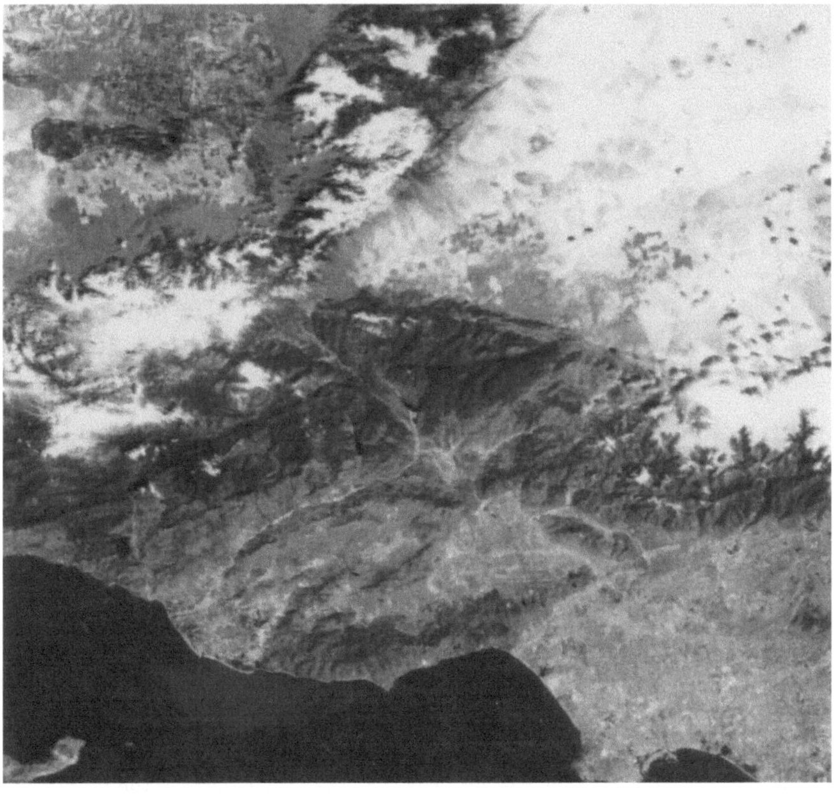

Fig. 5. ERTS 1 color composite image of the Los Angeles Basin (lower right), Mojave Desert (upper right) and mountain ranges (center) in southern California obtained on 14 March 1973.

17 land use and crop types in an analysis which required several orders of magnitude less effort than had this analysis been made by conventional means. That same area is shown eight months later in Fig. 4 (upper right). As expected, the agricultural patterns have changed drastically in that period. The very dark area in the upper right, at the eastern end of the San Francisco Bay estuary, had been identified in July as "asparagus, corn and alfalfa field crops" [2]. Now (Fig. 4), these fields have no vegetation and are very moist. On the other hand, the large round area in the upper center of Fig. 4, just north of the estuary and northwest of the dark region, is characterized by high infrared reflectance, indicating vigorous crops. In the July image that area had been identified as a mixture of fallow and burnt stubble fields. Clear delineations can be made in Fig. 4 between the moist, probably freshly plowed, uncultivated fields (very dark grey, regular field patterns), different types of crops (shades of red in regular field patterns), rangeland (light red, irregular patterns), redwood forests of differing density (dark red, irregular patterns) and mountain ranges covered with leafless trees (dark grey, irregular patterns). Some of these features can also be identified in Fig. 5

Fig. 6. ERTS 1 color composite images of St. Louis, Missouri: Mississippi, Missouri and Illinois Rivers during dry period on 2 October 1972 (left) and after major flooding on 31 March 1973 (right). Note that both images are reproduced on identical scales but immense flooding of entire river system makes right picture appear on much larger scale. Note also the area between Missouri River (lower left) and Mississippi River (center) northwest of St. Louis which has become almost an island in the right image but is shown as an integral part of the agricultural land to the west in the left image. The change of color tones of the terrain between the two images is because the high infrared reflectivity of the vigorous vegetation present in the October image appears red. This vegetation is absent in the March image causing the lack of red colors.

of the Los Angeles area. In addition, that area exhibits the various alluvial materials covering the Mojave Desert and the striking patterns of urbanization which extends over an 80 km × 50 km area in the lower right. The built-up center of the city appears bluish grey while the vegetation interspersed with residential areas makes such areas appear pink/grey. Similar patterns of urban development can be seen in Fig. 4 in the belt of cities surrounding the lower San Francisco Bay, ranging counter clockwise from San Francisco to San Jose to Oakland.

The very low reflectivity of water in the infrared is used extensively to map the amount of water covering large areas. This is especially significant to assess the damages caused by floods (Fig. 6), to plan and direct reclamation operations and to measure the amounts of water available in lakes, reservoirs, or other bodies of inland waters. The yellow/red band is used mainly to distinguish waters of differing quality, to map sedimentation and possibly to recognize biologically productive areas in ocean waters. This has resulted in the detection of natural or man-made processes affecting the quality of water, the quality of wetland areas, and the deterioration of coastlines. Such detection could form the basis of possible action which is required to conserve these resources. The green band is used primarily to analyse water depth because of the transparency of water in this spectral band. Navigational hazards, coral reefs and changes in underwater morphology can be detected and mapped in this fashion.

In the color composite of San Francisco, for example (Fig. 4), sedimentation and tidal flow patterns can be mapped in the San Francisco Bay and the outflow from the Bay through the Golden Gate and from various rivers in the Pacific Ocean can be detected.

The assessment and possible prediction of water run-off affecting very large regions can be made by analyzing intricate watersheds such as are displayed in Fig. 7 (right) for the entire Western Alps. Water run-off can be computed also from mapping the seasonal variation of snow cover displayed in sequential ERTS 1 images. Such applications have been described by Salomonson [3]. Fig. 7 also displays a significant difference in the spectral property of snow between the yellow/red (left) and infrared (right) bands. In the left picture the snow cover stands out by its much higher brightness compared with the surrounding dark terrain. In the right picture it is very difficult to delineate the snow cover, not only because the terrain has a greater reflectivity in the infrared, but also because snow is much darker in that spectral band. Furthermore, melting or old snow such as is seen mostly in this picture of the Alps, is much darker in the infrared than fresh, dry snow. This spectral characteristic of snow is used to map not only snow cover but also the state of snow melting.

The montage of ERTS 1 pictures of the Alps shown in Fig. 7 displays very clearly the relationship between river valleys and the tectonic structure. Note, for example, the virtual continuity of the line which extends from the lower Isere and upper Arly valleys (lower left), to the upper Rhone, just east of Lake Geneva (center), to the upper Rhine, south of Lake Constance and finally the Inn slicing through the northern edge of the Alps (upper right). A portion of this feature can be seen in greater detail in Fig. 3. Of course, many of these major features relating to geological structure have been mapped before especially in such well exploerd regions as the Alps. But some can be put into much better perspective using ERTS 1 images, even where previous explorations have been extensive. An example of this is shown in Fig. 5 where the well known and quite acteiv San Andreas fault can be seen extending from left to right in the center of

Fig. 7. Montage of ERTS 1 image of western Alps obtained along 5 orbitals passes in the yellow/red band (left) and 0.8 – 1.1 μm infrared band (right). Montages were made using images during 6 – 9 October and 18 – 20 December 1972.

the picture. Another fault stretches for about 150 km from the center to the lower left toward the coastline at the left edge of the picture. It has been hypothesized [4] that this fault continues northeastward through the desert along the line marked by the boundary between the blue/green and yellow/white tones in the upper right. Pease and Johnson [4] have related these previously unknown alignments to recent earthquake activities in southern California. Gawad and Silverstein [5] have also documented such relationships for other parts of the southwestern US. Geological interpretations of ERTS 1 pictures have also resulted in a rather fundamental revision of existing geological maps of Alaska which provide new insight into the potential mineral and petroleum deposits in that State [6]. They have also provided for an assessment of nickel and other mineral deposits in South Africa [7]. Many other similar applications were discussed by Lattman [8].

Generally, 25 to 50% new and previously unmapped geological structures are found in ERTS pictures even for relatively well explored areas. In lesser explored areas the new information content of ERTS pictures is even greater: for example, in the Wind River mountain range of Wyoming it took about five years to map 10—15% of the structural features known today. The remaining 85—90% were mapped from one ERTS picture in three hours [9].

Thus, lineaments expressed as depressions in the terrain or as color tone changes observed in barren regions have been shown to have a significant relationship to mineral deposits, findings of underground water and to the recognition of hazards to major civil construction projects. The derivation of these applications from the ERTS 1 observations is always based on a better understanding of the processes that are shaping the surface of the earth, gained from these observations.

Because of the vertical view of the earth afforded by ERTS and because of the relatively constant scene illumination, ERTS images can be composited and mosaiced easily, thus providing nearly instantaneous, synoptic views of entire states or countries. An example of this is shown in Fig. 8 where images obtained from three passes of ERTS 1 over the eastern United States, on 10, 11 and 12 October 1972, were montaged to form one contiguous, synoptic image of the whole area about 500 km wide and 500 km long.

This montage was made from 12 individual ERTS 1 images and was assembled with very little effort. It would have taken almost 50000 conventional aircraft photos and many man-years of labor to assemble a similar picture from aerial surveys.

The image is dominated by the water bodies of the Chesapeake and Delaware Bays and by their associated estuaries. The white sand covered barrier islands off the Atlantic coast and the marshes and tidelands between them and the mainland are synoptically surveyed. The interface between the silt laden waters of the James (lower left) and Potomac (center left) Rivers and the clearer waters of the Chesapeake Bay are delineated. The geological formations of the Appalachian mountain ridges can be seen in the upper left with the Delaware and Susquehanna Rivers breaking through gaps in these ridges. Land use patterns for this area in which more than 10% of the population of North America lives, are evidenced by varying shades of color. Dense forests are shown as dark red, agriculture and range lands are lighter red depending on moisture and crop type, fallow fields and construction sites are very light and swamplands are dark, as are densely urbanized areas; the cities of New York and Philadelphia can be seen in the upper right. Because the evidence of significant geophysical processes and formations is so clearly apparent, these images are also very valuable as an educational tool.

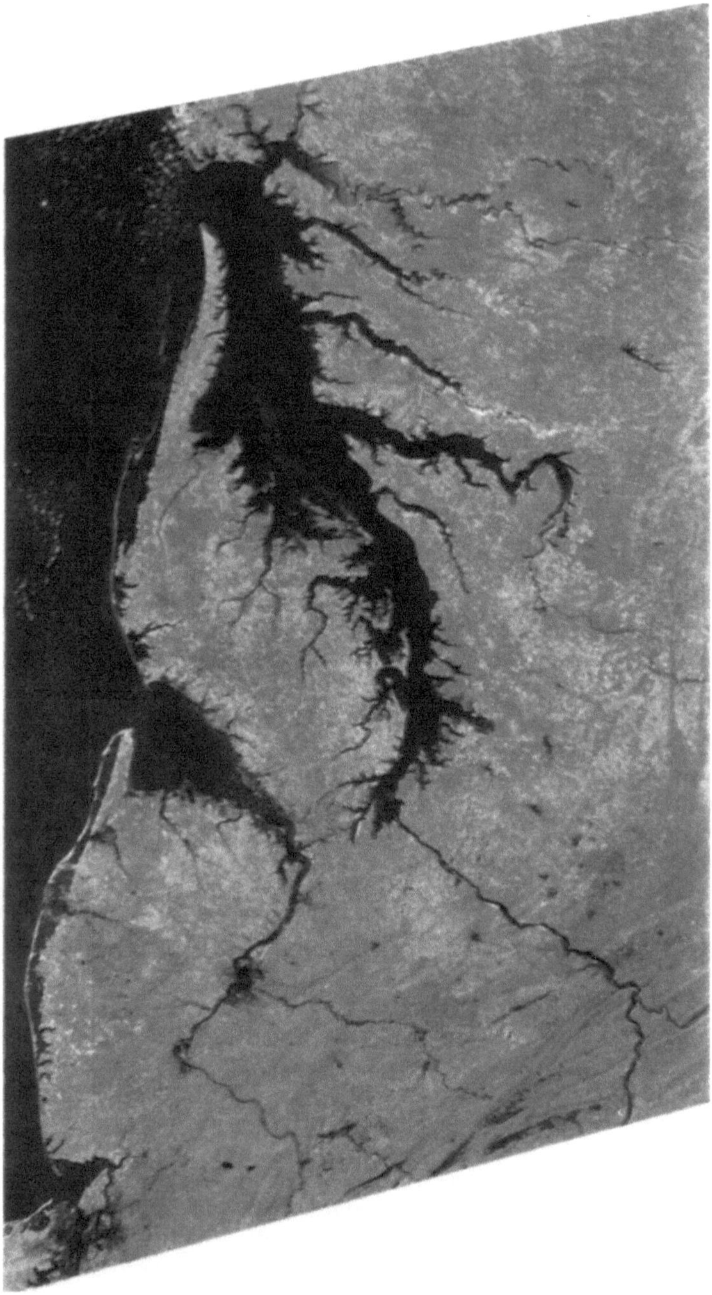

Fig. 8. Montage made from ERTS 1 color composite images of the eastern United States obtained during three consecutive passes on 10, 11 and 12 October 1972.

It should be borne in mind that the ERTS 1 system has never been intended to provide such applications on a routine and operational basis. Instead, the primary objective of ERTS 1 was to obtain multi-spectral images periodically over the United States, and to apply these images experimentally to investigations dealing with various parameters relating to earth resources. The fact that after six months of operation of ERTS 1 the use of these observations for purposes such as mentioned above has been taken for granted in many parts of the world is indicative of the phenomenal success of this first space mission dedicated to nationwide surveys of earth resources and intended to benefit all mankind.

## References

[1] C. E. Poulton, in: Symp. on Significant Results obtained from Earth Resources Technology Satellite 1, Volume II, Goddard Space Flight Center, Greenbelt, Md, May 1973.
[2] R. N. Colwell, in: An Integrated Study of Earth Resources in the State of California, Based on ERTS 1 and Supporting Aircraft Data, Progress Report, University of California, Berkeley, California, 31 January 1973.
[3] V. V. Salomonson and A. Rango, in: Symp. on Significant Results obtained from Earth Resources Technology Satellite 1, Volume II, Goddard Space Flight Center, Greenbelt, Md, May 1973.
[4] R. W. Pease and C. W. Johnson, in: Symp. on Significant Results obtained from Earth Resources Technology Satellite 1, Volume II, Goddard Space Flight Center, Greenbelt, Md, May 1973.
[5] M. A. Gawad and J. Silverstein, in: Symp. on Significant Results obtained from Earth Resources Technology Satellite 1, Volume I, Goddard Space Flight Center, Greenbelt, Md, May 1973.
[6] E. H. Lathram, I. L. Tailleur, W. W. Patton Jr, and W. A. Fischer, in: Symp. on Significant Results obtained from Earth Resources Technology Satellite 1, Volume II, Goddard Space Flight Center, Greenbelt, Md, May 1973.
[7] R. P. Viljoen and M. J. Viljoen, in: Symp. on Significant Results obtained from Earth Resources Technology Satellite 1, Volume 1, Goddard Space Flight Center, Greenbelt, Md, May 1973.
[8] L. H. Lattman, in: Symp. on Significant Results obtained from Earth Resources Technology Satellite 1, Volume II, Goddard Space Flight Center, Greenbelt, Md, May 1973.
[9] N. M. Short, in: Significant Results obtained from Earth Resources Technology Satellite 1, Volume III, Goddard Space Flight Center, Greenbelt, Md, May 1973.

Earth Survey Problems — Akademie-Verlag, Berlin 1974

# THE APPLICATION OF OBSERVATIONS MADE BY ERTS 1 TO THE MANAGEMENT OF AGRICULTURAL, FOREST AND RANGELAND RESOURCES

C. E. Poulton

Oregon State University, Corvallis, Oregon, USA
Author's Permanent address now: Earth Satellite Corparation, Berkeley, Calif., USA

Preliminary results in the evaluation of ERTS 1 data by the agricultural, forest, and rangeland resources disciplines are proving the unique suitability of the MSS system to the needs of these areas. Both human and machine interpretations have yielded significant problem-relevant information from data available only at the close of a crop year. This suggests that when temporal data for one full growing season can be included in the analyses, the full power of the system may well exceed all prelaunch expectations.

Crop and natural vegetation identification has been demonstrated at highly acceptable levels of accuracy by both human and human-machine interpretation systems. Most investigators reported accuracy levels clustering around 80% for information derived by interpretation of ERTS imagery. Accuracy ranges reported varied from 70% to slightly over 90%. The ability to derive problem oriented, useful information and to update many kinds of agricultural and natural vegetational statistics by simple, low cost human interpretation methods was clearly demonstrated. Tables 1, 2, and 3 summarize the breadth and nature of applications made by investigators reporting to the agriculture, forestry, and rangeland resources sessions during the March 1973 ERTS 1 Symposium held at the Goddard Space Flight Center, Greenbelt, Maryland, USA. Tables 1 and 2 confirm the wide range of interest and applicability of the ERTS system to important natural resources problems. Table 3 is especially significant because it shows that working both with ERTS alone and an ERTS 1/aircraft support system with very unsophisticated interpretation procedures, it is possible to derive highly useful information. While the full power of the ERTS 1 system is available only when the digital tapes are subjected to rather sophisticated computer analysis, it is especially

Table 1

Number of Investigators concerned with Broad Classes of Features

| Resource area | Recognition | Acreage estimate | Production | Monitoring | Insects and disease |
|---|---|---|---|---|---|
| Agricultural crops | 12 | 7 | 2 | | 1 |
| Forestry | 8 | 8 | 1 | 3 | 1 |
| Rangeland | 9 | 9 | | 5 | |

Table 2

Number of Investigators concerned with Specific Resource Features

| Soil mapping | Ecological resource inventory | Fire-related | Water resources | Erosion |
|---|---|---|---|---|
| 4 | 5 | 3 | 5 | 1 |

Table 3

System used for Interpretation by the Reporting Investigators

| ERTS alone | ERTS with aircraft | Human PI | Computer interpretation |
|---|---|---|---|
| 12 | 13 | 19 | 13 |

relevant to developing nations of the world that 19 investigators made very beneficial use of the imagery through conventional human image interpretation and by the use of long-standing photo interpretation techniques.

Preliminary cost effectiveness ratios for the interpretation of ERTS 1 data versus conventional survey and enumeration techniques were reported. These seem to cluster at about 9:1 in favor of the ERTS when expressed in terms of comparative man hour cost. Lauer and Krumpe [1] reported outstanding cost

Table 4

Interpretation Time and Costs associated with the Bucks Lake Study (50000 acres) (20250 hectares)

| Task | B/W photography | High-flight CIR photo | ERTS false-color image |
|---|---|---|---|
| Delineation of watershed boundary | 6 hours | 1 hour | 0.5 hour |
| Plotting effective areas on photos | 5 hours | 0 hours | 0 hours |
| Interpreter training and testing | 3 hours | 3 hours | 0.5 hour |
| Type delineation and classification | 30.5 hours | 17 hours | 1.25 hours |
| Total time required | 44.5 hours | 21 hours | 2.25 hours |
| Hourly wage | $ 7.00/hour | $ 7.00/hour | $ 7.00/hour |
| Total interpretation costs (time) | $ 311.50 | $ 147.00 | $ 15.75 |
| Total cost per acre (cents) (per hectare) | 0.622 (1.540) | 0.294 (0.726) | 0.032 (0.079) |
| Cost ratio | 20 | 9 | 1 |

effectiveness ratios for the use of both ERTS and high-flight aerial photography in forest land resource inventory. With each of these systems they produced a new inventory map shown in Fig. 1, both of which greatly exceeded the delineation density of the previously available resource inventory. Their cost effectivenes data are summarized in Table 4. Demonstrating the effectiveness of using ERTS imagery to update inventories in inacessible areas, Anderson, Shapiro and Belon [2]

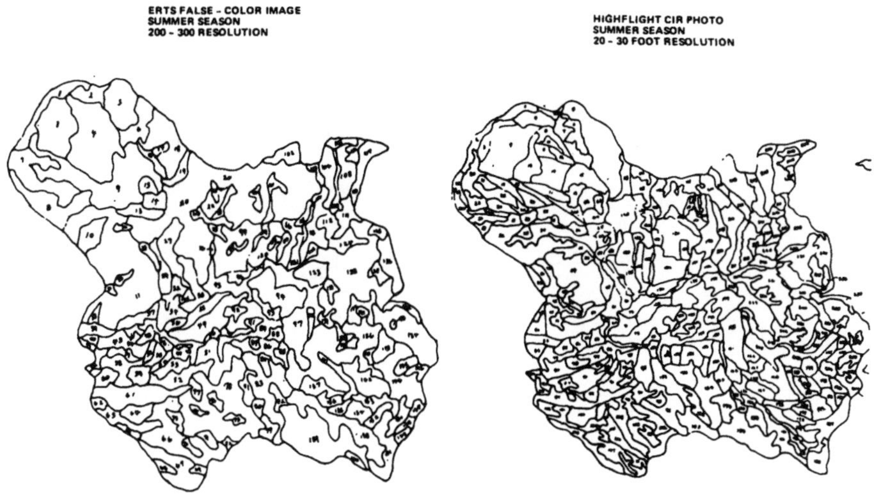


Fig. 1. State-of-the-art resource inventory map of the Feather River Watershed prepared from high flight (on the right) and from ERTS 1 (on the left). Compare the delineation density as an index with the amount of information developed by the various techniques.

were able, with only about 15 man hours of interpretation time, to replace a six-class inventory map of the Seward Peninsula, Alaska, with a twelve-class rendition of the vegetational resources of the study area (Fig. 2). For the entire State of Nevada, Tueller [3] derived the data shown in Table 5 at an image interpretation cost of 3 man days. This represents the first time a reasonably accurate proportioning of expenditures for rangeland rehabilitation between the public and

Table 5

Crested Wheatgrass Seedings: State of Nevada

| | Acres* | Total arid (%) |
|---|---|---|
| Public land | 743407 | 74.9 |
| Private land | 249579 | 25.1 |
| Total | 992986 | 100.0 |

* 1 acre = 0.405 hectare

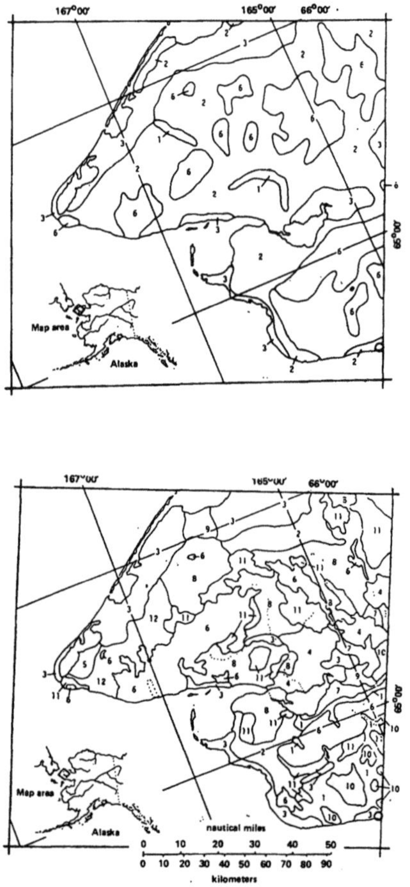


Fig. 2. Maps produced from an ERTS 1 image of the Seward Peninsula, Alaska, to improve resource definition for land use planning, resource allocation and resource development decisions. 1, shrub thicket; 2, upland tundra; 3, wet tundra; 4, fire scar; 5, senescent vegetation (?); 6, alpine barrens; 7, grassland tundra (?); 8, shrub thicket/upland tundra mosaic; 9, shrub thicket/wet tundra mosaic; 10, shrub thicket/alpine barrens mosaic; 11, shrub thicket/upland tundra/alpine barrens mosaic; 12, uplands tundra with some senescent vegetation.

the private sector has been possible. While the absolute accuracy of the data is not known, they are considered superior to any previous estimate. Comparable information could not possibly have been compiled with comparative effort and time.

Two studies were reported that could not have been attempted without the short cycle repetitive ERTS data. These studies involved variation in plant growth and development across the length and breadth of the United States. This kind of study requires short time lag between imaging of all study locations.

Table 6

Imperial Valley Crop Calendar for ERTS 1 Detected Field Conditions

| Crop type | Field condition code by date | | | |
|---|---|---|---|---|
| | 26 August | 1 October | 6 November | 12 December |
| *Field crops* | | | | |
| Barley (pasture | 4 | 2 | 1 | 1 |
| Barley (grain) | | 4 | 3 | 2 |
| Sorghum | 1 | 5 | 5 | |
| Oats | | 4 | 3 | 2 |
| Wheat | | 4 | 4 | 3 |
| Corn (grain) | | | 4 | 3 |
| Grass (Bermuda and Alicia) | 2 | 2/1 | 1 | 1/5 |
| Grass (rye) | 4/3 | 3/2 | 2 | 1 |
| Grass (Sudan) | 1/5 | | | |
| Alfalfa | 1/4 | 1/2 | 1/2 | 1/2 |
| Sugar beets | 3 | 2/1 | 2/1 | 1 |
| Cotton | 1 | 1 | 5/4 | 5/4 |
| Rape | 3 | 2 | 1 | 1 |
| *Vegetable crops* | | | | |
| Lettuce | 3 | 2 | 2/1 | 1 |
| Mustard | 3 | 2 | 1 | 1 |
| Cabbage | 4 | 2 | 1 | 1 |
| Cantaloupe (fall) | 1 | 1/5 | | |
| Cantaloupe (spring) | | 4 | 3 | 2 |
| Cucumbers (fall) | 2 | 1/5 | | |
| Cucumbers (spring) | | 4 | 3 | 2 |
| Watermelon | | | 4 | 3 |
| Tomatoes | | | 4 | 3 |
| Carrots | 3 | 2 | 2/1 | 1/2/5 |
| Onions (bunching) | 2 | 1 | 1 | 1/5 |
| Onions (dry) | 4 | 4 | 2 | 2 |
| Squash (summer) | 2 | 1/5 | | |

Field condition code legend: 1 - (Red) growing crops; 2 - (Purple) wet soil seeded crops; 3 - (Lavender) plowed soil; 4 - (White) bare soil; 5 - (Yellow) harvested stubble.

The reports presented at te ERTS 1 Symposium in March 1973 covered agricultural crop identification and acreage determination and an unexpected inventory of a newly introduced crop (Table 6). By matching image condition to crop development calendar, Johnson and Coleman [4] were able to identify and inventory all of the crops listed in this table. Documentation of the rate at which urbanization is reducing prime agricultural crop acreage was demonstrated by Draeger [5] by interpreting urban area extent from Apollo 9 and ERTS 1 imagery (Table 7). Monitoring data such as these may eventually convince some of the affluent nations of the importance of guiding urban expansion into areas not valuable for human food production. Other investigators reported on the monitoring and mapping of large wild fires in forest and rangeland environments with more

Table 7

Urban Change in Phoenix Area

| Town or city | 1969 Urbanized area (acres) | 1972 Urbanized area (acres) | Amount of agricultural land (acres) lost |
|---|---|---|---|
| Chandler | 1920 | 2880 | 960 |
| Glendale | 4000 | 6240 | 2240 |
| Mesa | 8320 | 13440 | 5120 |
| Phoenix* | 56680 | 64680 | 8000 |
| Scottsdale* | 6400 | 9600 | 3200 |
| Tempe | 7200 | 15840 | 8640 |

* 1972 urbanized area totals do not include sparse residential developments in wildland areas.

accurate impact and acreage determination than by conventional methods, monitoring of potentially catastrophic frost kill and of forest and crop land insect problems, quick and effective update of statistics on rangeland types and rehabilitation acreages, and rapid production of a reconnaissance soil survey in a developing nation.

The potentiality for estimation of important agricultural food crops is indicated by the work of Thomson [6] by his determination of rice acreages in a California study area (Table 8). Two different attempts at the determination of rice acreage are shown in columns 3 and 4. The close agreement with the second attempt and the true acreage is suggestive of the power of the digital ERTS system for the detailed analysis of agricultural landscapes, since it is possible to consider the unique spectral signature of each 0.4 hectare ground resolution area. Thinking of the importance of being able to extrapolate interpretable images of both natural vegetation and crop analogues among regions and, hopefully, globally, Poulton and Welch [7] have shown reasonable success in demonstrating repeatable images over a wide expanse intraregionally for selected natural vegetations and the

Table 8

Comparison of Rice Field Acreages estimated by Recognition and by Recognition plus Convex Mixtures

| Rice field | True acreage | Recognition estimate | Recognition plus conmixed |
|---|---|---|---|
| 1 | 150 | 119 | 144 |
| 2 | 212 | 174 | 205 |
| 3 | 106 | 71 | 95 |
| 4 | 159 | 130 | 163 |
| 5 | 176 | 149 | 179 |
| 6 | 194 | 173 | 196 |
| 7 | 224 | 214 | 236 |
| Total | 1221 | 1030 | 1218 |

important world food crop, rice. Treating the interrelationship of agricultural, forestry, and rangeland resource problems in the land use planning context, a multi-discipline team is demonstrating the feasibility of ecological resource inventory from both ERTS and high-flight imagery that draws together considerations of landforms, geology, natural vegetation, and soils as an information base for land use planning at county and state or regional level [8].

These preliminary results suggest that full use of the ERTS system may be one of the most significant developments of this century in helping man to feed and clothe himself and to relate compatibly with his environment.

A record of the present paper, with color illustrations, covering applications in the agriculture, forestry, and rangeland resources areas, is published in [9]. This publication will shortly be available in limited supply through the NASA Goddard Space Flight Center, Greenbelt, Maryland, 20771, U.S.A.

The full papers of the various contributing scientists, with color illustrations, have been published in: Symposium on Significant Results obtained from the Earth Resources Technology Satellite 1, Vol. I, Technical Presentations, Sections A and B, National Aeronautics and Space Administration, Scientific and Technical Information Office. Washington, D. C. 1973. (NASA Publ. No. Sp-327). The full text of the following citations [1—9] will be found on the pages indicated in this publication.

## References

[1] D. T. Lauer and P. F. Krumpe, Symp. on Significant Results obtained from ERTS 1, Goddard Space Flight Center 1973 (p. 97).
[2] J. H. Anderson, L. Shapiro and A. E. Belon, Symp. on Significant Results obtained from ERTS 1, Goddard Space Flight Center 1973 (p. 67).
[3] P. T. Tueller, Symp. on Significant Results obtained from ERTS 1, Goddard Space Flight Center 1973 (p. 77).
[4] C. W. Johnson and V. B. Coleman, Symp. on Significant Results obtained from ERTS 1, Goddard Space Flight Center 1973 (p. 19).
[5] W. C. Draeger, Symp. on Significant Results obtained from ERTS 1, Goddard Space Flight Center 1973 (p. 197).
[6] F. J. Thomson, Symp. on Significant Results obtained from ERTS 1, Goddard Space Flight Center 1973 (p. 181).
[7] C. E. Poulton and R. I. Welch, Symp. on Significant Results obtained from ERTS 1, Goddard Space Flight Center 1973 (p. 53).
[8] G. H. Simonson, D. P. Paine, C. E. Poulton, R. D. Laurence, J. H. Herzog and R. J. Murray, Symp. on Significant Results obtained from ERTS 1, Goddard Space Flight Center 1973 (p. 1451).
[9] C. E. Poulton, Symp. on Significant Results obtained from ERTS 1, Goddard Space Flight Center 1973, Vol. II, Summary of Results, NASA Goddard Space Flight Center, Md, USA, Publ. No. X-650-73-127.

Earth Survey Problems — Akademie-Verlag, Berlin 1974

# ERTS 1 MSS IMAGERY: A TOOL FOR IDENTIFYING SOIL ASSOCIATIONS

F. C. Westin

Plant Science Department and Remote Sensing Institute,
South Dakota State University, Brookings, South Dakota, USA

**Soil association maps show the spatial relationships of land units developed in unique climatic, geologic, and topographic environments, and having characteristic slopes, soil depths, textures, available water capacities, permeabilities, and so on. From these characteristics of the soil broad interpretations can be made such as how the soil is suited for various agronomic and engineering uses. ERTS 1 imagery was found to be a useful tool in the identification of soil associations since it provides a synoptic view of an 8 million acre (3.2 million hectare) scene, which is large enough for the effect to be seen on soils of climate, topography, and geology. A regional view also allows soil associations to be observed over most, if not all, of their extent. This aids in selecting typical sampling sites and provides a check on the homogeneity of the associations. ERTS 1 MSS imagery also provides four spectral bands taken every 18 days which give data on relief, hydrology and vegetation, all of which bear on the delineation and interpretation of soil associations. Enlarged prints derived from the individual spectral bands and shown in gray tones were useful for identifying soil associations.**

## 1. Introduction

Soil maps are an integral part of an effective agricultural research and advisory program. Soil maps are used for farm and ranch planning, for crop and grass yield estimates, for land use planning, for irrigation planning, for drainage planning, for assessing potentialities for special crops, for rural zoning, and for forest management.

Soil maps are of different scales ranging usually from 1:15840 to 1:7000000. The large scale maps necessary for detailed land planning show the extent of individual soils, and are made by boring holes and walking over the land so that delineations are observed over their entire extent. These soil survey maps are expensive to make and publish.

Small scale soil maps, called soil association maps, are geographic associations of one or several soils and are usually at scales of 1:500000 to 1:1000000. Simonson [1] states that each soil association consists of a set of geographic bodies that are segments of the soil mantle covering the land surface. Soil associations may include like or unlike soils linked together by mode of occurrence rather than degree of similarity. Field checking of soil association maps is done at infrequent intervals depending upon the scale and use of the map. Although soil association maps of small scale are not as precise for interpretations as detailed soil maps of large scale, they cost much less to make and they do have use for broad planning purposes and for education.

It should be emphasized that the soils shown on soil maps are not defined in terms of profiles alone. According to the soil survey staff of the USDA [2], each soil unit is a particular kind of landscape and it is these landscapes of soils that are classified and shown on soil association maps. Although soil profiles cannot be seen on air or satellite imagery, soil landscapes are visible. In this regard it should be stated that soil survey experience is necessary if maximum use is to be made of ERTS 1 imagery for identifying soil associations. Soil landscapes exhibit a characteristic surface geometry such as relative frequency of streams, and a characteristic surface composition such as the percentage of bare soil areas. Other features used in differentiating soil landscapes include vegetation and hydrology.

Aerial photographs have been widely used to differentiate soil association landscapes for 40 years. The bulk of these photographs have been black and white panchromatic prints although color and infrared materials have had some use. The aircraft has been the primary platform for obtaining these images. Now, since the summer of 1972, ERTS 1 MSS satellite imagery available in four spectral bands and at 18 day intervals brings a new tool to the study of soil association landscapes. It is the purpose of this paper to report on some results using ERTS 1 MSS imagery to identify soil associations.

## 2. Procedure

The ERTS 1 multi-spectral scanner records the electromagnetic energy that comes from features on the earth's surface. The radiation is in four bands as follows: band 4, 0.5—0.6 μm; band 5, 0.6—0.7 μm; band 6, 0.7—0.8 μm; and band 7, 0.8—1.1 μm. The spectral energy coming from a soil or crop or other feature depends on its molecular composition. Consequently the various earth's features exhibit unique tonal signatures on the different bands. Tonal signatures can result from multiband viewing or from monochromatic prints derived from one spectral band and shown in gray tones.

The ERTS 1 imagery used for this study consisted of 9 inch transparencies at a scale of 1:1000000. From these transparencies three kinds of images were prepared for study: (i) color transparency composites at 1:1000000 scale made from bands 4, 5 and 7; these were viewed over a light table with about 3 power magnification; (ii) monochrome transparencies of individual bands from which enlargement negative prints of scale 1:500000, 1:250000, 1:100000 and 1:60000 were prepared; these are the prints used in all figures of this paper except Figs. 7 and 8; (iii) diazo transparencies of each of the four bands were prepared in different colors and also each of them was prepared in black diazo. Black diazo transparency enlargements were used in Figs. 7 and 8.

In this study an effort was made to study ERTS 1 imagery using equipment usually available to a soil survey office. This consists mainly of access to a darkroom with a good enlarger having a non-diffuse light source. A diffuse light enlarger tends to blend and smooth out boundaries on the images, which is not desired. A densitometer was used to quantify some data for Fig. 8 but the reflectance differences of the bands used are apparent to the eye and use of the densitometer is not required to differentiate the crops shown in the figure.

### 3. Results

The results of this study are shown in nine figures. The imagery shown in these figures is all monochromatic derived from the individual spectral bands and shown in gray scale. These monochrome images were sharp and were enlarged to scales of 1:500000 and 1:250000 and even to 1:100000 without deteriorating to any great extent. One enlargement from a diazo transparency to 1:60000 is shown in Fig. 7. Colvocoresses and McEwen [3] indicate that MSS bulk images have good spectral consistency and good object detectability at a maximum printing scale of 1:250000.

The ERTS frames each cover over 8 million acres* and an advantage of this for use in soil association identification is that it encourages a regional view. Soil associations can be observed over much or all of their area of extent on a scene of this magnitude. Since soil associations occur in repeating patterns, the delineations in general can be observed to see if a change in land use, slope, drainage pattern, hydrology or other feature has changed. A regional view is useful in another way to the soil scientiest in that it encourages a regional approach to soil problems. This may ease state line soil correlation problems.

The earth's surface is curved and cannot be shown in a flat map without some distortion. Colvocoresses and McEwen [3] state, however, that the MSS ray is always within 5.78° of the nominal vertical and the curvature of the earth across one MSS scene is less than 2° so that the image approximates an orthographic view. This means that the reflection on the imagery of the soils, vegetation and other features has a minimum distortion and reflectance measurements can be reasonably quantitative.

ERTS images afford a near real-time view of the scene so that the current use of soil associations can be studied. Comparisons of reflectances of the different bands keyed to ground truth identification provide a means for study of the use of the soil associations.

Climate is one of the factors of soil formation. Fig. 1 shows the appearance of soil associations on two bands over a climatic range.** The soils on these three scenes are classified as follows: Hapludolls and Haplaquolls dominate in Iowa; in Central South Dakota the fields in cultivation are Haplustolls while the rangeland soils generally have more slope and are dominantly Camborthids; in Western South Dakota the soils are mostly Haplustolls on the flat areas and Torriorthents in the slopes. The ERTS scenes shown illustrate these broad soil patterns in two ways, field size and land use. In Iowa field size is relatively small—40 to 80 acres, principally. In Central South Dakota field size is 80 to 160 acres or even larger. In Western South Dakota field size is large but the pattern may be one of alternate strips of wheat and fallow. In Iowa the dark tones of band 7 indicate growing crops over most of the scene at this late summer date. These crops are corn and soybeans primarily with some alfalfa and pasture. This field size and crop pattern indicates a humid climate. In Central South Dakota the white squares on band 7 are fallow fields (indicating a semi-arid climate) while the dark fields are wheat or milo. Also in this area only about half of the soils are under cultivation. Band 7 for the western scene shows fallow fields, wheat fields and alternate wheat fallow

---

* 1 acre = 0.405 hectare.

** The scales given refer to the original pictures. The present figures are reduced as follows: Figs. 1, 6—9 to 65%, Figs. 2—5 to 75% of the original size.

Figure 1 Appearance of soil associations over a climatic range on MSS-5 and 7 of ERTS 1 Negative prints Scale 1:500,000

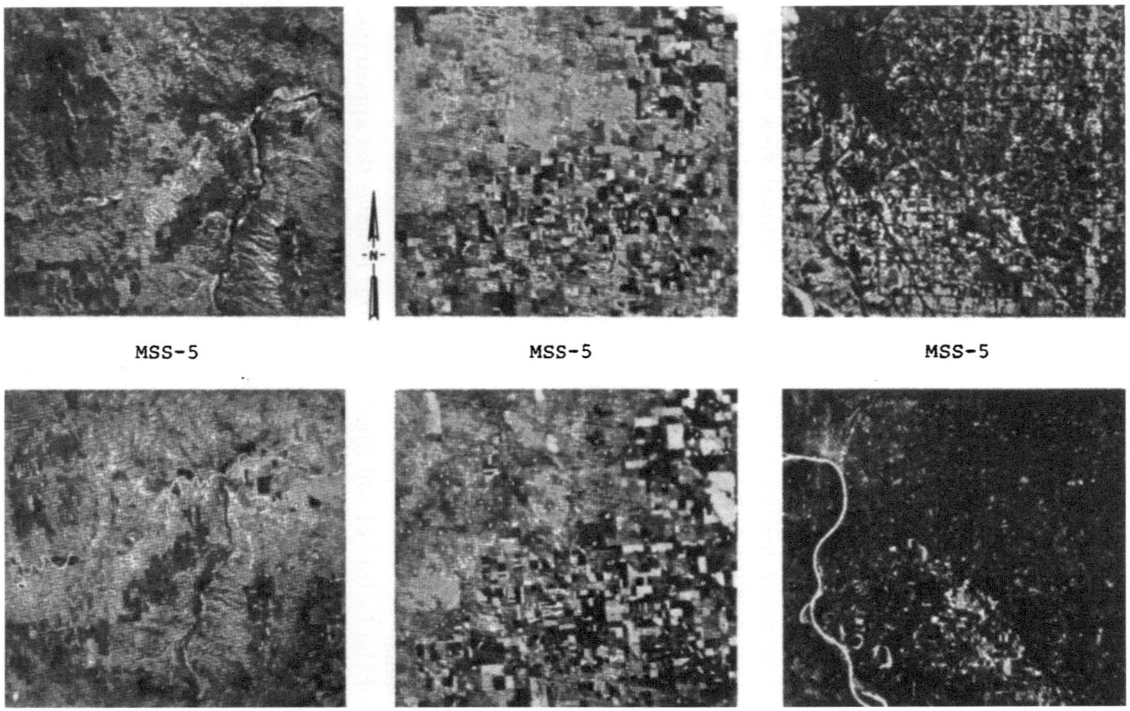


Western South Dakota-355 mm. Ann. Precipitation; 7.2°C Ann. Temperature; mostly rangeland, a few fields with wheat fallow; stripcropping

Central South Dakota-432 mm. Ann. Precipitation; 7.2°C Ann. Temperature; about half rangeland half cultivated; large fields; winter wheat fallow

Northwest Iowa-737 mm. Ann. Precipitation; 8.8°C Ann. Temperature; intensively cultivated; small fields; corn, soybeans, hay, pasture

Figure 2 Appearance of Steeply Sloping Area in Northeastern South Dakota
Bands 5 and 7 of ERTS 1 (Negative Prints)

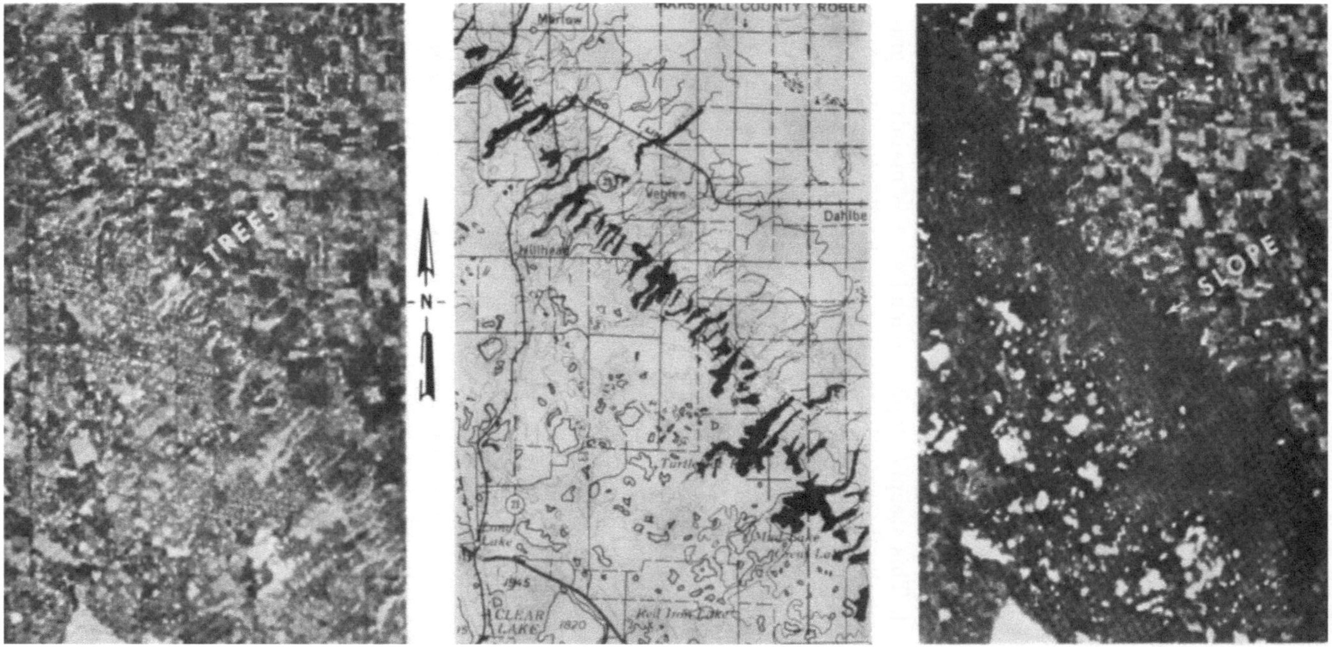


MSS-5 29 July 72 16484
Scale 1:250,000

USGS Topographic Map
Scale 1:250,000

MSS-7 29 July 72 16484
Scale 1:250,000

Two kinds of vegetation (grass and deciduous trees) are present on the steeply sloping area passing diagonally across this scene. On band 5 the reflectance of the trees is less than that of the grass or the adjacent crop areas and so they appear lighter on the print. On band 7 both trees and grass on this steep slope have high reflectance compared to the adjacent land and lakes and so the slope stands out as a dark, wide band on the print.

strips where the soils are cultivated. However, in this area most of the soils are in rangeland indicating an arid climate. The band 5 print for each of the three climatic scenes provides additional information to the soil scientist. Roads and towns stand out on this band. Also more information on crops can be deduced by comparing bands 5 and 7. For example, ground truth studies have shown that on band 7 both wheat and milo have high reflectance in the Central South Dakota scene. By comparing the high reflectance fields (dark tones on these negative prints) of bands 7 and 5, the milo fields can be separated from the wheat fields since milo has low reflectance and wheat high reflectance on the 5 band. This point is discussed again and illustrated in Fig. 8. Thus, in an area where agriculture has been carried on for some time the field size and crop pattern appear to give reliable correlations with broad soil associations, and these relationships are visible on ERTS 1 images.

In Fig. 2 two bands of an ERTS 1 scene are compared with the United States Geological Survey topographic map. The intent here was to determine if a prominent relief feature (in this case the steep-sided northeast-facing Prairie Coteau) could be differentiated. This steeply sloping glacial area is mainly in grass with deciduous trees, principally basswood, growing in the deeply incised valleys. Fig. 2 shows that in band 5 the reflectance of the trees is less than that of the adjacent grasslands and so the areas in trees are clearly delineated as lighter-toned patches. However the reflectance of the grass vegetation on the steep slope is not delineated. On band 7 the reflectances of the trees and grass are similar and the trees and grass together have a lower reflectance than the adjacent farmland and so this area stands out as a dark, wide band. The soils on this steep slope are thinner, stonier versions of the soil being farmed on leveler slopes and are classed as Entic Haploborolls compared to Typic Haploborolls and Typic Argiborolls on the adjacent areas. These thin, steep, stony soils are in grass and trees because this is the most intensive use possible. Although the soil profiles cannot be seen on the ERTS 1 images, deductions concerning their characteristics can be made.

In Figs. 3, 4 and 5 two bands of an ERTS 1 image are compared with the South Dakota Soil Association Map of Westin and Bannister [4]. Fig. 3 is of the western South Dakota Cretaceous plain and is typical of the drier parts of the Great Plains. In Table 1 the characteristics and interpretations of some of the soils shown on Fig. 3 are listed. This soil information is from the South Dakota Cooperative Soil Survey data store and illustrates the kind of information that can be prepared for small scale soil association maps.

A comparison of the soil map with the ERTS 1 prints in Fig. 3 illustrates that the steep land, No. 50, Samsil association (thin, shale-derived Torriorthents), can be accurately delineated. Also the bottomland, No. 62, Glenberg association (alluvium-derived Torrifluvents), is easily visible. The ERTS prints also show that the Glenberg soil delineation is actually narrower than it is shown to be on the soil map. The other two main soil associations, the upland and terraces, are not as easy to differentiate from each other although they stand out from the Samsil and Glenberg associations. No. 32, the Ralph association (soft silty shale-derived Argiustolls) has a pattern indicating some cropping and thus is usually distinguishable from No. 44, the Pierre association (clayey, shale-derived Camborthids) which is usually in grass. There are two terrace soils shown, No. 36, the Caputa association (clay loam alluvium-derived Argiustolls) and No. 39 the Satanta association (loamy, alluvium-derived Argiustolls). These two soils apparently cannot be distinguished from each other on ERTS 1 MSS images. Both soils also are

Figure 3 Comparison of Soil Association Map with MSS-5 and 7 of ERTS 1
Scale 1:500,000 Negative Prints Western South Dakota
12 Oct 72 17064 Argiustolls Torriorthents Camborthids

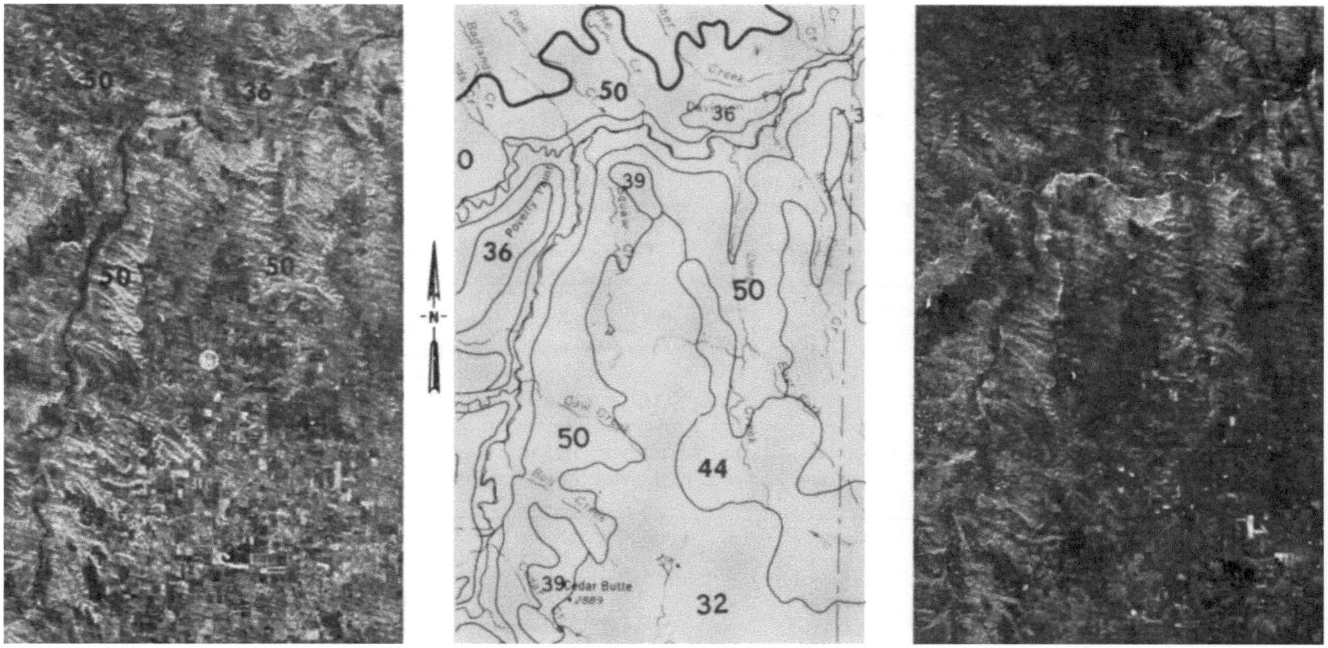


MSS-5

Soil Association Map
Scale 1:500,000
AES Info Series No. 3
SDAES, SDSU Brookings, SD

MSS-7

Table 1

Characteristics and Qualities of Soils shown on Fig. 3
(from the Soil Survey Data Store)

| Map No. | Slope | Principal soil series | Classification | USDA texture | Capability subclass | Winter wheat yield Bu/A | Dominant rangesite |
|---|---|---|---|---|---|---|---|
| 32 | undulating | Ralph | Aridic Argiboroll, fine-silty, mixed | loam | III c | 20 | Silty |
| 36 | gently undulating | Caputa | Aridic Argiustoll, fine, mixed, mesic | clay loam | III c | 24 | Clayey |
| 39 | gently undulating | Satanta | Aridic Argiustoll, fine-loamy, mixed | loam | III c | 22 | Silty |
| 44 | rolling | Pierre | Ustertic Camborthids, very fine, montmorillonitic, mesic | clay | III s | 18 | Clayey |
| 50 | steep | Samsil | Ustic Torriorthent, clayey montmorillonitic | clay | VI s | not adapted | Shallow |

| | Estimated Engineering Properties* | | | | | Degree of Limitation | | | |
|---|---|---|---|---|---|---|---|---|---|
| Map No. | Liquid limit | Plasticity index | Permea-bility (in/hr) | Avail. water capac. (in/in) | Shrink swell potential | Septic tank filter field | Sewage lagoons | Corrosivity uncoated steel | Corrosivity concrete |
| 32 | 15—38 | 5—19 | 1.2—2.0 | 0.18—0.20 | low | severe | mod. | high | mod. |
| 36 | 25—40 | 11—35 | 0.6—1.2 | 0.19—0.22 | mod. | severe | mod. | high | mod. |
| 39 | 22—36 | 2—15 | 0.6—2.0 | 0.16—0.22 | low | slight | mod. | mod. | mod. |
| 44 | 50—82 | 30—50 | < 0.06 | 0.08—0.12 | high | severe | severe | high | high |
| 50 | 50—100 | 25—65 | 0.06—0.2 | 0.08—0.12 | high | severe | severe | high | high |

* Explanation of the column headings for engineering properties can be found in "Guide for Interpreting Engineering Uses of Soils", USDA Soil Conservation Service, issued Nov. 1971. Washington, D.C. USGPO Stock, No. 0107-0332.

Figure 4 Comparison of Soil Association Map with MSS-5 and 7 of ERTS 1
Scale 1:500,000 Negative Prints Southwest South Dakota
19 Aug 72 17065 Argiustolls Haplustolls

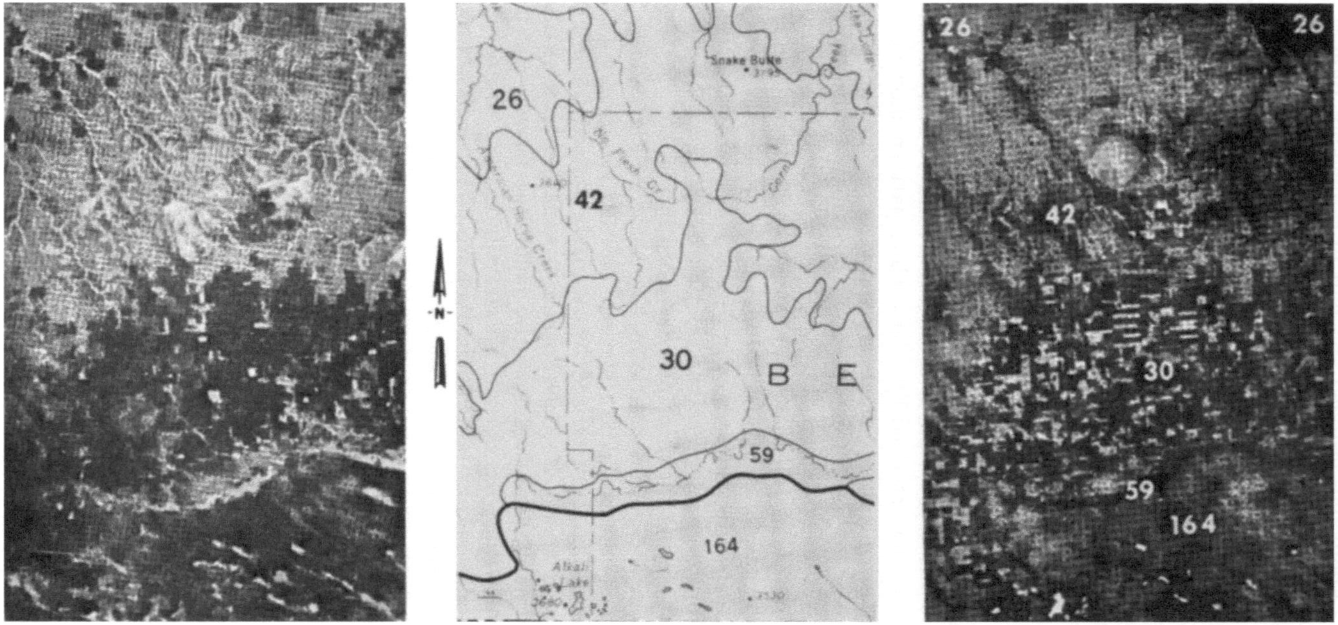


MSS-5

Soil Association Map
Scale 1:500,000
AES Info Series No. 3
SDAES, SDSU Brookings, SD

MSS-7

difficult to distinguish from No. 32 except that they usually are intensely cropped while No. 32 is about half cropped and half in rangeland. Band 7 gives a better definition of the steeply sloping soil association (50) than does band 5. This is thought to be due to the fact that since a smaller mass of vegetation is produced in this area of dominantly shallow soils, these areas would have a lower reflectance on band 7. Thus the tone on band 7 of these steeply sloping soils would be lighter than that of the adjacent soil associations.

Fig. 4 is from Southwest South Dakota and on it two ERTS 1 MSS bands are compared with the soil association map of a portion of the Tertiary high plains area. The main soil associations here include the Nebraska sandhills No. 164, Valentine association (Ustipsamments, in grass), No. 59, Minatare association (Natrargids, in grass), No. 30, Keith association (loess-derived Argiustolls in winter wheat or fallow), No. 42, Oglala association (loamy, sandstone-derived Haplustolls, mainly in grass), and No. 26 Kadoka association (silt-stone-derived Argiustolls, mainly in grass). Both bands 5 and 7 show all of these soil associations except perhaps the separation of soil Nos. 26 and 42 is somewhat vague. The latter however, is characterized by a lacy white pattern shown especially well on band 5. This pattern is due to the evergreen trees of the Pine Ridge escarpment. The trees have a low reflectance on band 5 compared with the adjacent grasslands. The reflectance differential of the trees and grass is considerably less on the 7 band. Band 7 is useful, however, to show the stockponds and lakes especially prevalent in the sandhills. Also the tonal differences seen in the sandhills soil association are caused by differences in vegetative cover of the rangesites present. Finally, the crops growing on soil association No. 30, and the others as well, can be catalogued by using the principles described in Fig. 8.

Fig. 5 is from the glacial prairie in Northeast South Dakota. The soil associations apparent on this ERTS scene are: No. 193 Kranzburg association (loess-derived Haploborolls characterized by regularly shaped fields having no lakes or poorly drained areas), No. 199 Forman association (glacial till-derived Argiborolls and Argiaquolls in a hummocky landscape having marshes and lakes), No. 201 Forman-Buse association (glacial till-derived Argiborolls and Haploborolls in a choppy end moraine landscape), No. 205 Renshaw association (glacial outwash-derived Haploborolls on irregular relief), No. 200 Forman soils on a sloping rather than undulating topography, No. 212 Peever association (clay loam glacial till-derived Argiborolls on a nearly level topography), and No. 219 Lamoure association (Haplaquolls derived from alluvium). As was discussed earlier for Fig. 2 the steeply sloping soils in this area stand out as a dark strip on band 7 while the area in trees can be distinguished on band 5. Also band 7 gives superior definition to all water which aids in differentiating soil associations in glacial landscapes since the presence of lakes usually means there are poorly drained soils in the landscape.

In Fig. 6 the appearance of eroded black and light colored soils and sediments is shown on the four MSS bands of ERTS 1. The dark colored soils are derived from shale high in manganese which outcrops adjacent to the Missouri River reservoir. These dark-colored soils are largely barren of vegetation and can be distinguished on negative prints of bands 6 and 7 as a light colored fringe. These soils have a low reflectance on these bands compared with the adjacent areas in grass. On bands 4 and 5 they have the same reflectance as adjacent areas and hence are not apparent.

Light colored sediments, such as those found in the badlands of South Dakota

Figure 5 Comparison of Soil Association Map with MSS-5 and 7 of ERTS 1
Scale 1:500,000 Negative Prints Northeast South Dakota
29 July 72 16484 Argiborolls Haploborolls

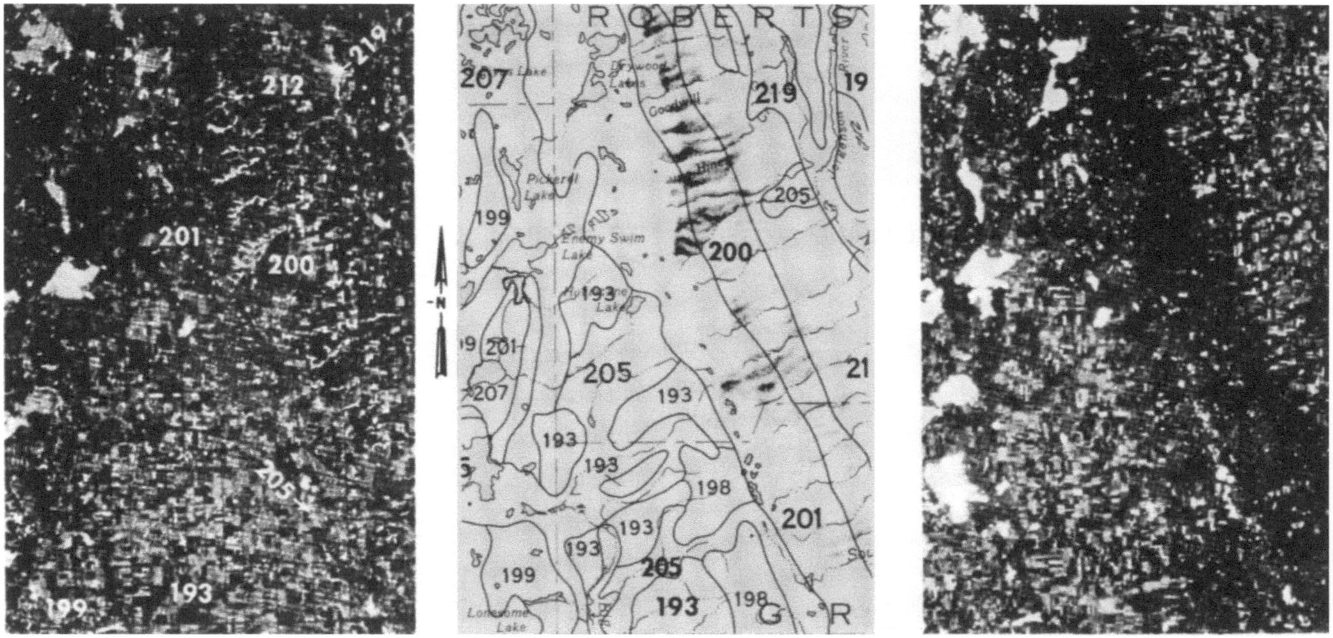


MSS-5

Soil Association Map
Scale 1:500,000
AES Info Series No. 3
SDAES, SDSU Brookings, SD

MSS-7

Figure 6 Appearance of erosion of black and light-colored soils and sediments on ERTS 1 imagery Negative prints Scale 1:250,000

(Black eroded soils have low reflectance on IR bands 6 and 7 compared to adjacent vegetated areas and stand out as a white fringe. Light colored eroded soils have higher reflectances and thus appear darker on all bands than adjacent vegetated areas).

Erosion of black, shale-derived soils above Missouri River Reservoir ERTS 1 17 Aug 72 16551

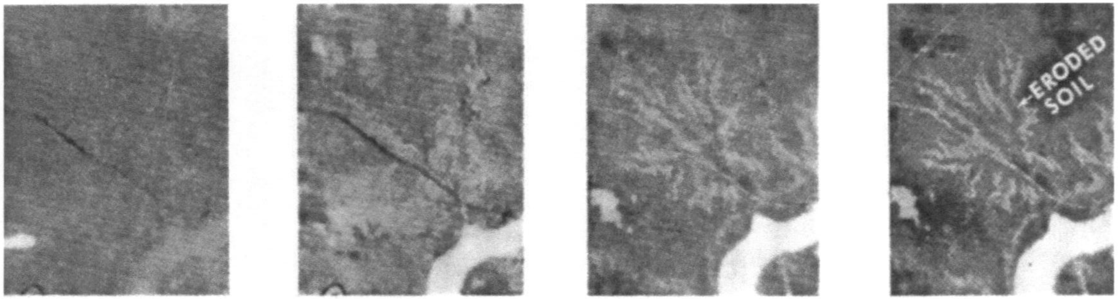


MSS-4 MSS-5 MSS-6 MSS-7

Erosion of light-colored soils in Badlands ERTS 1 19 Aug 72 17065

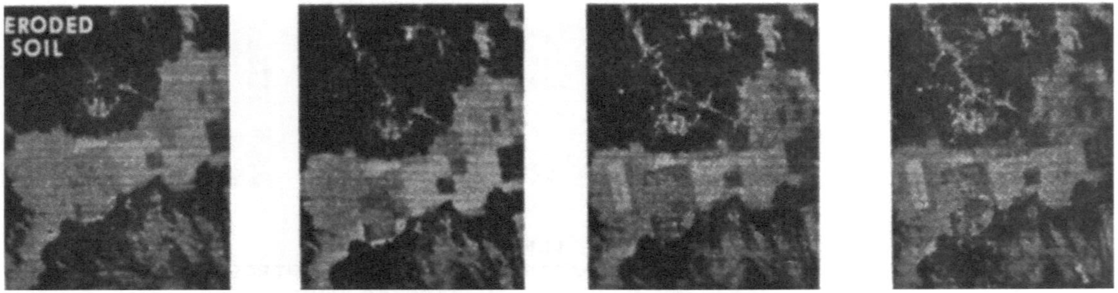


MSS-4 MSS-5 MSS-6 MSS-7

Figure 7 Enlargements of Band 7 of an ERTS 1 Image

Utilization of soils can be determined by comparing reflectances of the MSS bands. These enlargements show that individual fields can be identified for purposes of determining land use.

Negative Prints made from Band 7 of a black diazo transparency 17 Aug 72 16551-7

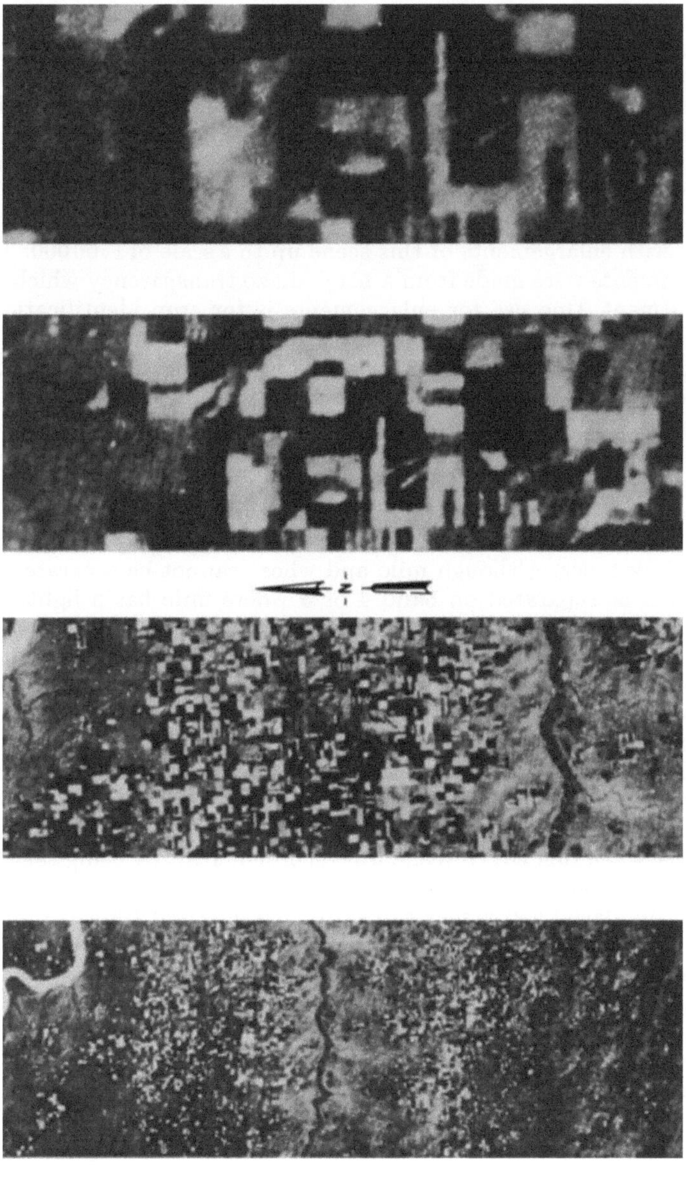


Scale 1:1,000,000 Scale 1:500,000 Scale 1:100,000 Scale 1:60,000

which are derived from the White River formation, have high reflectances compared with adjacent grass areas. They show as very dark gray or black tones which reflect somewhat differently from the grass areas on all four bands. The reflectance differential of white sediments and grass is greater on bands 4 and 5, however, than on bands 6 and 7.

Monitoring of erosion of soils and sediments by ERTS 1 adds a new dimension to the study of soils on small scale soil association maps. Bare soil areas on slopes contribute sediment to lower lying areas which is a serious pollution problem. The situation causes more concern if a reservoir receives this sediment since this affects the useful life of the reservoir.

Shown in Fig. 7 is a portion of band 7 of an MSS image of scale 1:1000000 along with enlargements of this scene up to a scale of 1:60000. The image and the enlargements were made from a black diazo transparency which produces a sharp enlargement. One use for enlargements is for crop identification since on them field boundaries can be located and ground truth obtained.

Fig. 8 illustrates how the crop use being made of soil associations can be determined by comparing the reflectances of crops on the four MSS bands. Enlargements of 1:100000 appear to work well for this purpose. The land sections are marked out on the enlargements of the bands and comparisons made of the reflectances of the various crops or land uses on the MSS bands. For example on the negative prints shown fallow fields have low reflectance on band 7 and thus appear in light tones. Native grass has a medium gray tone and wheat and milo have dark tones. Although milo and wheat cannot be separated on bands 6 or 7, they can be separated on band 4 or 5 where milo has a light tone and wheat a dark tone. These examples show that in this area enlargement prints coupled with ground truth data can give very satisfactory preliminary figures on the crop use being made of soil associations. Reports on crop identification involving the computer are given in [5, 6].

Fig. 9 shows how four lakes in a glacial landscape appear on the four ERTS MSS bands. Water has a low reflectance on bands 7 and 6 and all water on these scenes will be detected. On the other hand, marsh vegetation and water have about the same reflectance on bands 4 and 5. Thus, it is seen that two of these lakes were marshes when this ERTS scene was recorded. Thus by using the MSS bands marsh and open water can be separated.

Certain glacial soil associations are characterized by lakes, marshes and other poorly drained areas, while others have mature landscapes with well developed surface drains and no poorly drained areas. Haplaquolls occur in the former landscape while they are largely absent in the latter except along the stream courses. Thus knowledge of the hydrology furnished by ERTS 1 images is an aid in identifying soil associations.

A question that suggests itself when viewing ERTS images is "How accurate are they?" Colvocoresses and McEwen [3] fitted early MSS images to ground control and found rms distortions of up to 1000 m, always larger than the 300 m required by the US National Map Accuracy Standards. Later, however, the images improved and their best results for a frame had an rms of 192 m. Apparently changes in mirror speed and spacecraft attitude can be made further to improve the MSS geometric quality.

In summary, the ERTS 1 imagery is a useful tool for the soil surveyor in the preparation of soil association maps. One image covers so large a scene that regional relationships become apparent. This aids in assessing the effect of soil forming

Figure 8

Identification of crops grown on Soil Association Opal-Promise (Vertic Haplustolls, very fine, montmorillonitic, mesic) on 4 bands of ERTS 1. Negative prints made from black diazo transparencies. Sections 1 through 5 of T106N, R77W, Lyman County, S.D. 17 Aug 72 16551 Scale 1:100,000

Ground Truth

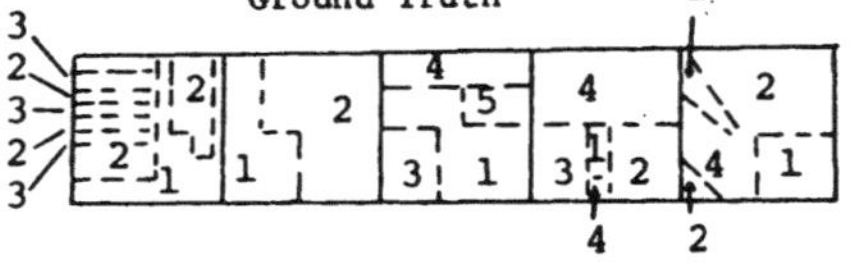

1. Fallow
2. Wheat
3. Milo
4. Native grass
5. Tame grass

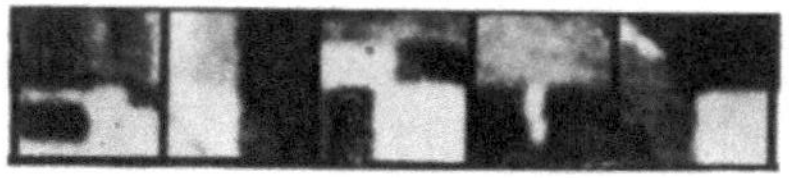

MSS-7

MSS-6

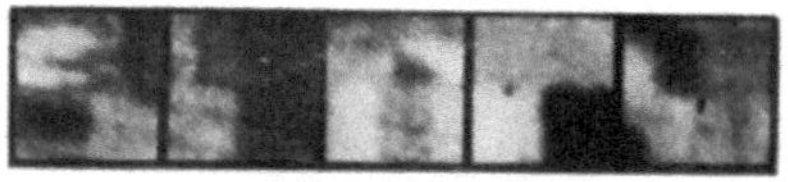

MSS-5

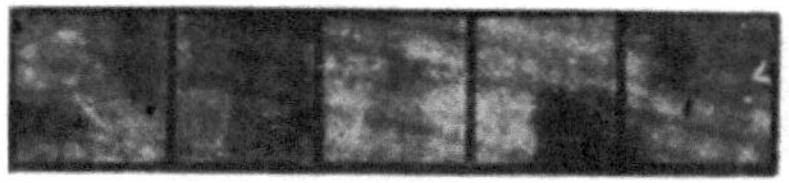

MSS-4

Reflectances on MSS bands 4,5,6 and 7 for crops measured on Macbeth Reflection Densitometer Model RD 219

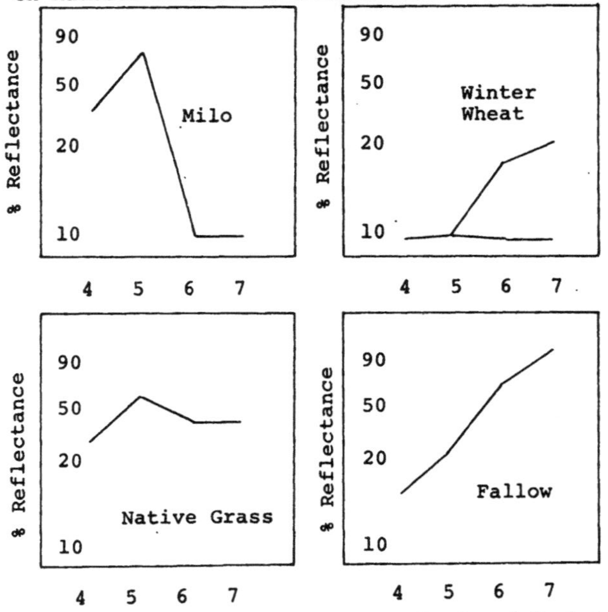


Wheat and milo can be separated on bands 4 and 5, but not on bands 6 or 7. Fallow and native grass can be separated on bands 6 and 7, but not clearly on bands 4 or 5. Thus, utilizing band 6 or 7, fallow and native grass can be distinguished from each other and from wheat and milo. Then, using band 4 or 5, milo can be distinguished from wheat. Note that on these negative prints the low reflectances have a light tone and high reflectances a dark tone.

Figure 9 Appearance of lakes and marshes on 4 MSS bands of ERTS 1
Eastern South Dakota glacial soil areas
21 Sept 72 16491 Scale 1:250,000 Negative Prints

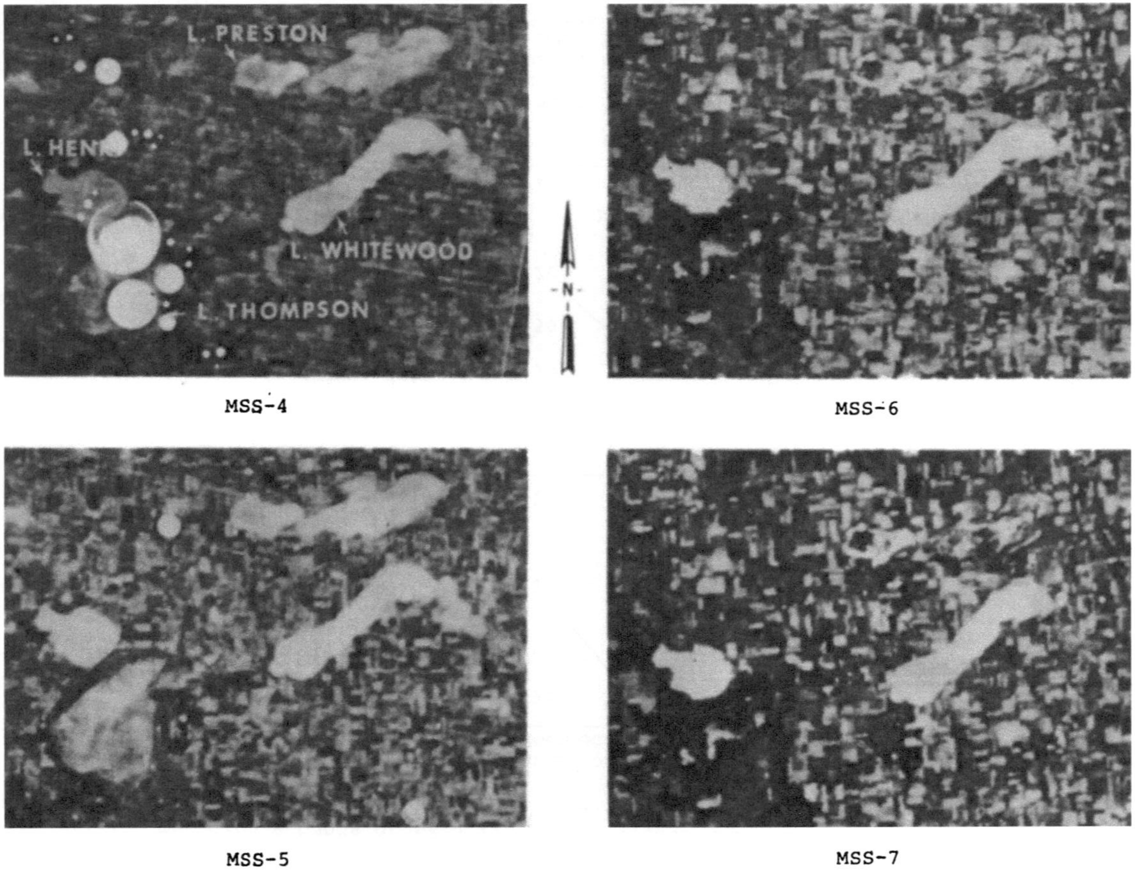

factors as well as providing a means of checking the patterns of soil associations over most or all of their area of extent. Each image comes in four spectral bands taken every 18 days. Each band gives specific data on relief, hydrology, and vegetation all of which bear on soil association patterns. The bands can be prepared in color composites or photographic or diazo prints, thus allowing a wide choice of images for study and publication.

## Acknowledgments

Supported in part by contract No. NAS-5-21774. Authorized for publication as Journal Series 1189, South Dakota Agricultural Experiment Station, South Dakota State University, Brookings, South Dakota.

## References

[1] R. W. Simonson, Soil Sci. Soc. Amer. Proc. **35**, 959 (1971).
[2] Soil Survey Manual, ARS, USDA Washington, D.C. 1951 (p. 131).
[3] A. P. Colvocoresses and R. B. McEwen, in: Proc. NASA Symp. Goddard Space Center NASA Sp-327, Symp. on Significant Results obtained from the ERTS 1, March 1973, New Carrolton, Md. (1973) Vol. I, Sec. B, p. 887.
[4] F. C. Westin and D. L. Bannister, AES Info. Series No. 3, S. D. Agr. Expt. Sta., Brookings, S. D. (1971).
[5] M. L. Horton and J. L. Heilman, in: Proc. NASA Symp. Goddard Space Center NASA Sp-327, Symp. on Significant Results obtained from the ERTS 1, March 1973, New Carrolton, Md. (1973) Vol. I, Sec. A, p. 27.
[6] M. F. Baumgardner, S. J. Kristof and J. A. Henderson, in: Proc. NASA Symp. Goddard Space Center NASA Sp-327, Symp. on Significant Results obtained from the ERTS 1, March 1973, New Carrolton, Md. (1973) Vol. I, Sec. A, p. 213.

Earth Survey Problems — Akademie-Verlag, Berlin 1974

# THE DEFINITION OF THE GEOTECTONIC DOMAINS OF THE SOUTHERN AFRICAN CRYSTALLINE SHIELD BY ERTS 1 IMAGERY AND ITS ECONOMIC IMPORTANCE

M. J. Viljoen and R. P. Viljoen

Geological Research Department, Johannesburg Consolidated Investment Co. Ltd., Randfontein, Republic of South Africa

The ancient granite-greenstone cratons and the younger metamorphic mobile belts constitute the two major geotectonic units of the Southern African crystalline shield. Each has its own distinctive structural style. The cratons are characterized by arcuate vulcano-sedimentary belts encircling or partially encircling diapirically intrusive gneissic tonalitic batholiths. These rocks, with their well-defined structural imprint, are sharply transgressed by younger, more potash-rich granites. The metamorphic mobile belts, on the other hand, are characterized by para-gneisses and other high grade metamorphites. Plastic flowage folding phenomena involving complex refolding and the development of basin and dome structures are the most diagnostic features.

Portions of these two geotectonic environments have been studied on ERTS 1 imagery. It was found that the broad synoptic view provided by this imagery is ideally suited to a study of the diagnostic macro-structures, and that the different geotectonic styles are clearly recorded. ERTS 1 imagery thus allows a more accurate definition than exists at present of the contact zones and internal structures of the two domains. This could be of considerable economic significance in view of the fact that along closely analogous contact regions of the ancient cratonic Superior Province of the Canadian Shield, basic/ultrabasic bodies containing some of the world's largest nickel deposits have been emplaced.

## 1. Introduction

The main objective of the present study has been the identification and documentation of the important large scale features which characterize and therefore identify the major geotectonic environments of the Southern African crystalline shield. The two main geological domains recognized are the ancient, so-called granite-greenstone cratonic nuclei and the younger encircling metamorphic mobile belts. The study to date has involved the examination of ERTS 1 colour composite imagery covering portions of these environments (Fig. 1). It is hoped at a later stage to extend the study so as to define the full extent of the important contact zone between the cratonic nuclei and the metamorphic mobile belts as well as to study the internal structures and lithologies of these two environments in more detail. It is envisaged that the above study could provide data of far-reaching significance as far as nickel mineralization is concerned.

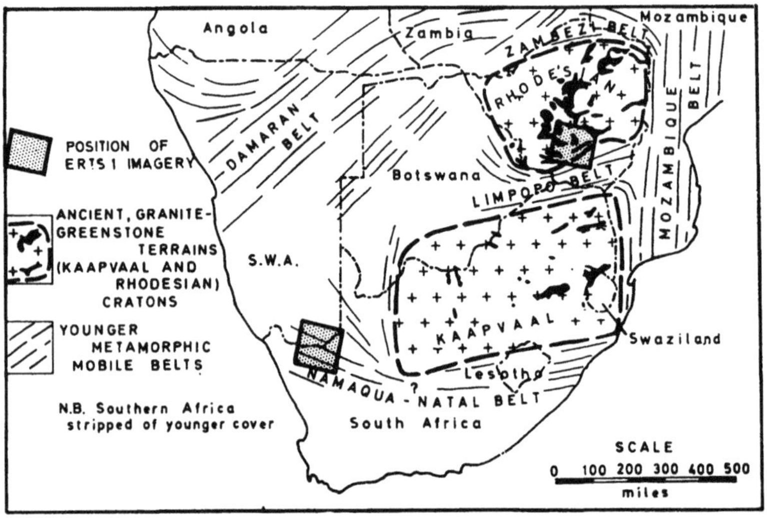


Fig. 1. The Southern African crystalline shield showing the main geotectonic domains and the areas covered by the ERTS 1 images discussed in this paper.

## 2. The Granite-Greenstone Cratons

Archaean granite-greenstone terrains often constitute the most ancient (pre $2.5 \times 10^9$ years) and stable portions of the continents and appear to be remarkably similar in all of the better known shield areas of the world. Vulcano-sedimentary sequences known as greenstone belts are the most distinctive features of these terrains and represent the remnants of previously more extensive accumulations of volcanic and sedimentary rocks. The greenstone belts in their present form appear to "float" in the form of deformed synclinal "keels" or "rafts" in a complex granitic terrain.

Despite their great age such belts have not been subjected to high grades of thermal metamorphism and primary volcanic and sedimentary structures are commonly encountered. Most belts have been subjected to strong structural deformation which is responsible for their characteristic form. Narrow arcuate schist belts trending at various angles from the main belt are particularly important as are large-scale isoclinal folds with axes more or less parallel to the regional grain of the belt. The typical granite greenstone pattern showing in particular the arcuate synclinal schist slivers is depicted in Fig. 2.

The lowermost volcanic assemblage of most greenstone belts commences with a suite of ultramafic and related rocks, intermediate to acid tuffs and magnesium-rich basalts. Some of the ultramafics represent subaqueous peridotite lavas while others form part of layered intrusive ultramafic complexes. The lowermost volcanic succession is overlain largely by metatholeiitic basalts (greenstones) with narrow interlayers of acid lava and chert. Shales, greywackes and quartzites form the uppermost stratigraphic assemblages of most belts.

Several different types of granitic rocks each with its own tectonic style, petrology and chemistry are generally present. The oldest are invariably gneisses of tonalitic composition, and often intrude the lowermost volcanics in the form of diapiric domes. The arcuate schist belts tend to drape around these domes and are

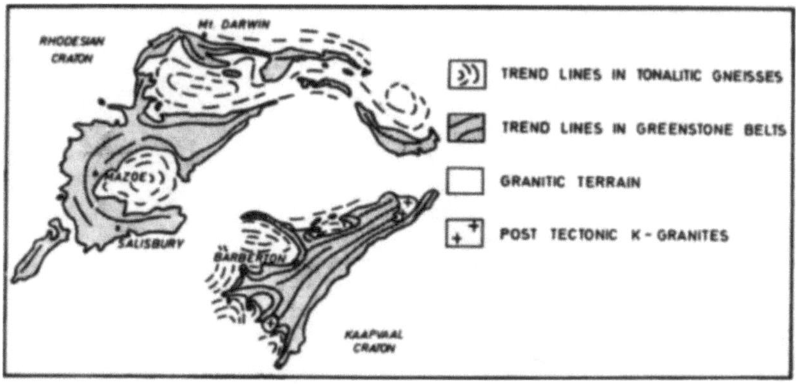


Fig. 2. Typical granite-greenstone pattern showing arcuate greenstones draped around tonalitic granite domes. Note also younger cross-cutting potassium-rich granites.

responsible for the distinctive greenstone pattern noted in Fig. 2. The tonalitic gneisses often give rise to low-lying featureless country. Younger more potassium-rich granite plutons cause abrupt truncation of earlier formed structures and fabrics both in the older gneisses and in the greenstone belts. They are generally more resistant to erosion and give rise to high-lying ground or boulder-strewn hills.

## 3. The Metamorphic Mobile Belts

Surrounding and dividing the cratonic nuclei are a series of metamorphic mobile belts. Such belts are essentially linear zones containing high grade, granitized metamorphic tectonites which show abundant evidence of rheid flow and plastic flowage folding. This can lead to the production of a complex pattern of re-folded folds in a continuous sequence of deformation as shown in Fig. 3 which depicts an area from the Namaqualand mobile belt of the Cape Province of South Africa. Major linear zones of transcurrent dislocation are common, particularly along the marginal zones of the mobile belts and there is usually evidence of repeated fault movement. From the above and from a comparison of Figs. 2 and 3 it is clear that the major structures typical of mobile belts are very different from those encountered in greenstone belts and they reflect fundamental differences in the geotectonic setting of these two environments.

## 4. ERTS 1 Imagery as an aid to the Identification of the Distinctive Patterns of the Crystalline Shield

### 4.1. The Granite-Greenstone Environment

ERTS 1 image No. 1103-07291 covers part of the southern sector of the Rhodesian craton and portion of the linear north marginal zone of the Limpopo mobile belt (Fig. 1). Fig. 4 shows the ERTS image (colour composite) of this area and the geological interpretation is shown in Fig. 5. The Great Dyke traverses

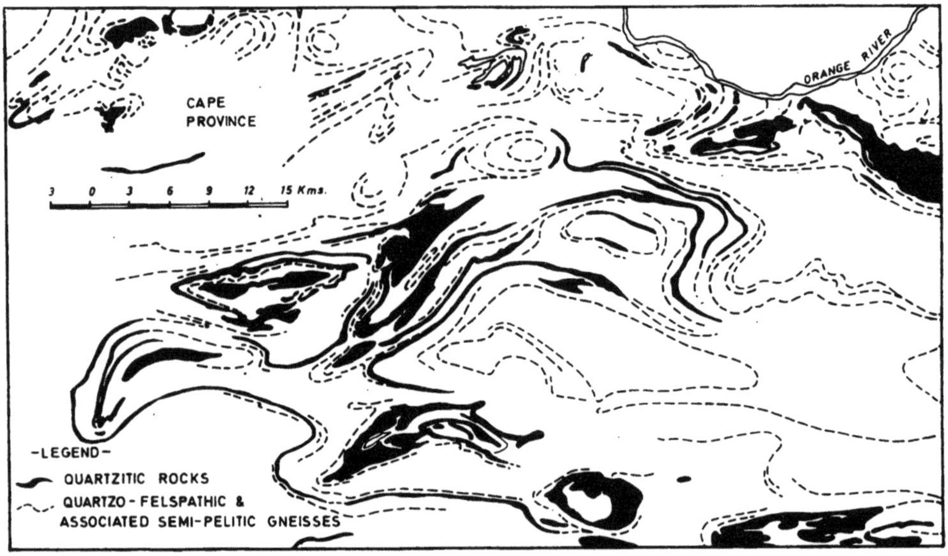


Fig. 3. Typical metamorphic mobile belt pattern showing complex flowage folding phenomena of a high grade metamorphic assemblage.

the central part of the image and is the most conspicuous and clearly defined feature. The bulk of the Dyke is ultramafic in nature (grey blue on colour composite and dark grey on black and white prints) but contains very well defined (heavily vegetated) layers and canoe shaped remnants of gabbro which give an exceptionally clear view of the internal structure of the body.

All the features of the granite-greenstone environment are clearly visible. These include the arcuate schist belt slivers which are striking features in the case of the Buhwa, Gwanda and Filabusi belts. Certain of the stratigraphic components of the greenstone belts are also distinguishable, particularly in the case of the Shangani belt where the economically important lowermost stratigraphic assemblage comprised of resistant layered ultramafic bodies and some banded ironstones forms a resistant zone flanking the eastern margin of the belt. This zone, clearly distinguished by dark grey colour tones, can be traced southwards where it appears to split, one branch trending westwards and the other southeastwards where it constitutes the northern flank of the Filabusi belt. Nickel mineralization has been encountered at two places along this clearly discernible zone. The Shabani and Mashaba ultramafic complexes are also clearly visible and the waste dumps from asbestos mining operations are conspicuous. The younger intrusive nature of the Mashaba ultramafic body is well demonstrated by the western limb which is superimposed on a north-south stratigraphy consisting of greenstone belt slivers lying within tonalitic gneisses.

In all cases the internal structure (mainly folding) and the gross stratigraphy of the greenstone belts is discernible. In addition all the major mines in the area are clearly recognized by their waste dumps.

The distinctive geotectonic styles of the various cratonic granites are also to be seen. The generally featureless terrain which gives rise to lighter colour tones is underlain by the tonalitic gneiss assemblage. Arcuate greenstone-belt tongues

Fig. 4. Colour composite print of ERTS image No. 1103-07291, Rhodesia.

protrude into this terrain and probably define or partly define individual diapiric tonalitic batholiths. Numerous cross-cutting potassium-rich granite bodies occur in the area and as a result of their somewhat more resistant and jointed nature have greater water holding properties. This has resulted in more luxuriant plant growth which clearly defines such bodies, particularly on colour composite prints.

The most well defined of these bodies intrudes the Shangani belt causing abrupt truncation of all the stratigraphic trends. In addition several other similar but somewhat smaller bodies are also discernible in the western sector of the image. An extensive and as yet unmapped elongated zone of potassium-rich granites has recently been identified by the Rhodesian Geological Survey in the zone between the Shabani and Buhwa greenstone belts. The contacts of this granite

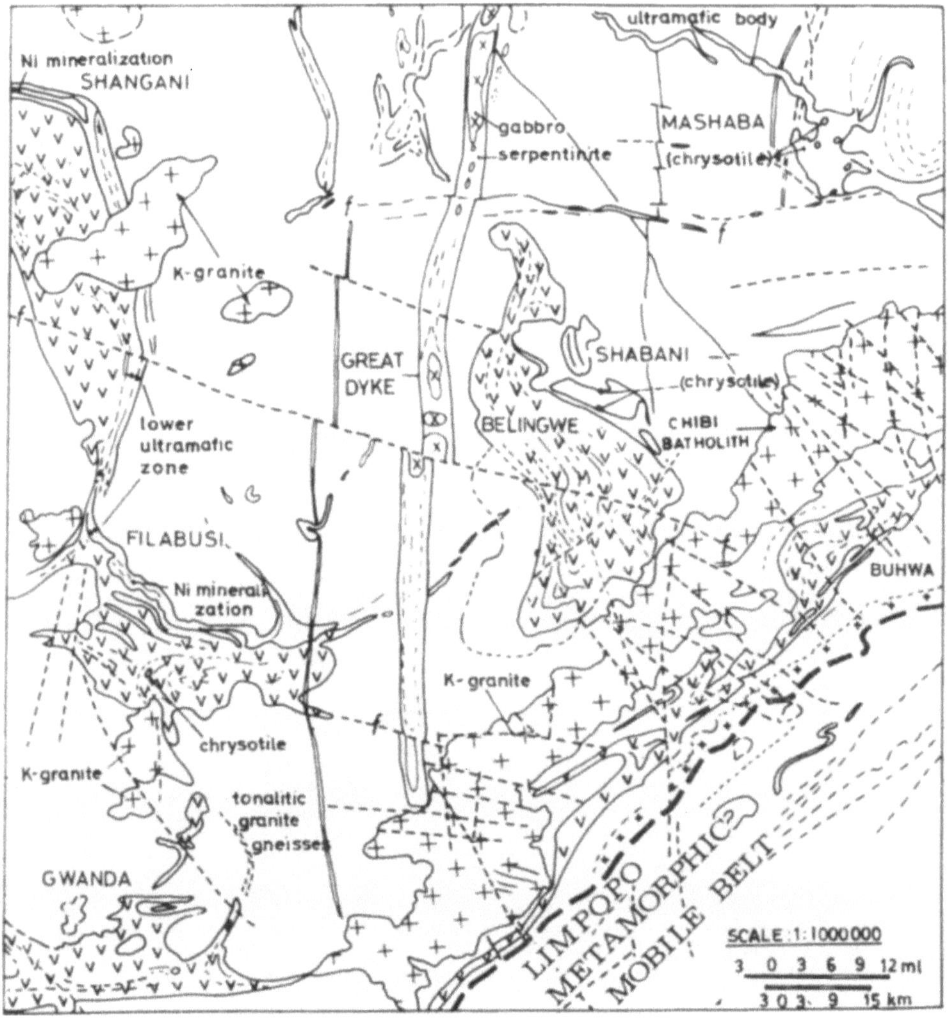


Fig. 5. Photogeological interpretation of ERTS image No. 1103-07291 showing part of the southern region of the Rhodesian craton and the north marginal zone of the Limpopo belt.

(termed the Chibi Batholith), which might be a marginal manifestation of the Limpopo metamorphic mobile belt, can be remarkably accurately mapped from the image. The body is characterized by a spectacular joint and/or fault system normal to the contact of the mobile belt. The jointing has lead to the development of numerous component granite domes which are surrounded by thick vegetation.

Besides the Great Dyke two narrow but well defined parallel satellite dykes are also conspicuous features of the image. These together with a number of other dykes which are also clearly seen give a good indication of the type and extent of faulting which has affected the entire area. Three major east-west trending faults with right lateral movement have displaced the Dyke and greenstone belts.

## 4.2. The Mobile Metamorphic Belt Environment

Part of the north marginal zone of the Limpopo mobile belt occupies the south eastern part of the image No. 1103-07291. Although lacking the large scale flowage folding features, present only in the central zone of the belt, the distinctive

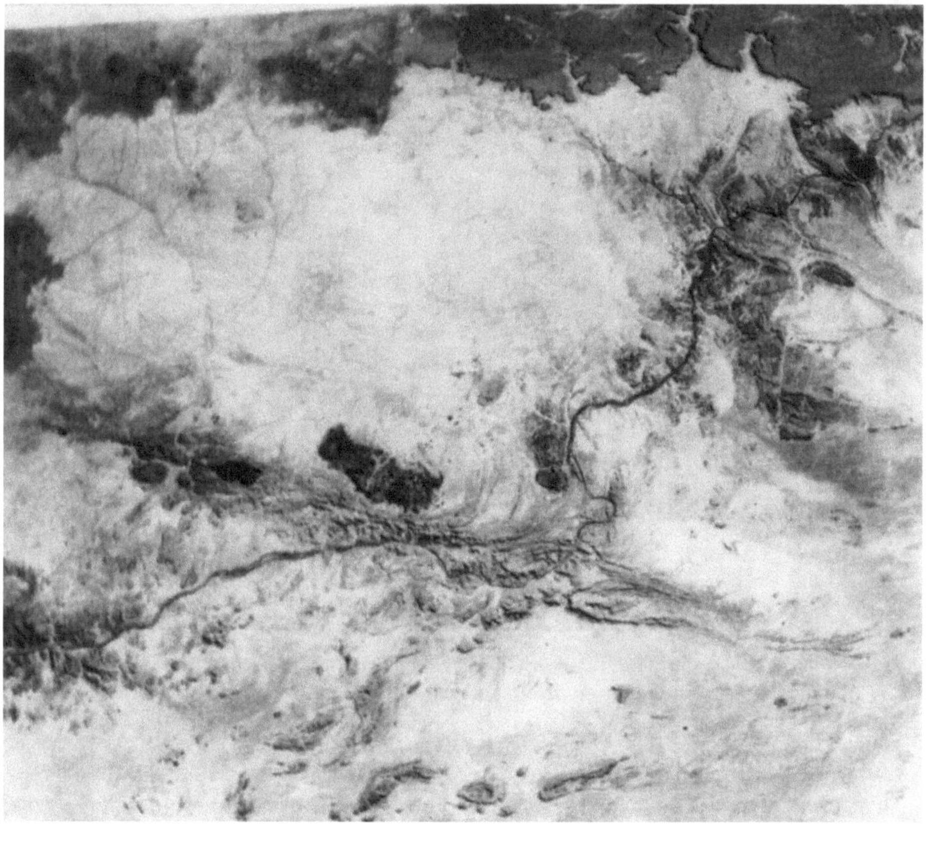

Fig. 6. Colour composite print of ERTS image No. 1055-08053 of the N.W. Cape Province of South Africa.

linear grain, tonal differences, and absence of arcuate forms readily distinguishes this terrain from the granite-greenstone terrain to the north.

The strong linearity of trends along the north maring is due to tectonic overprinting in the form of shearing and major transcurrent dislocation, probably just after the formation of syntectonic eutectoid granites one of which can be defined on the image.

As an example of the more typical flowage folding pattern of the mobile belts, ERTS image No. 1055-08053 (Fig. 6) covering part of the Namaqualand belt along the Orange River in the N.W. Cape Province and southern S.W. Africa has

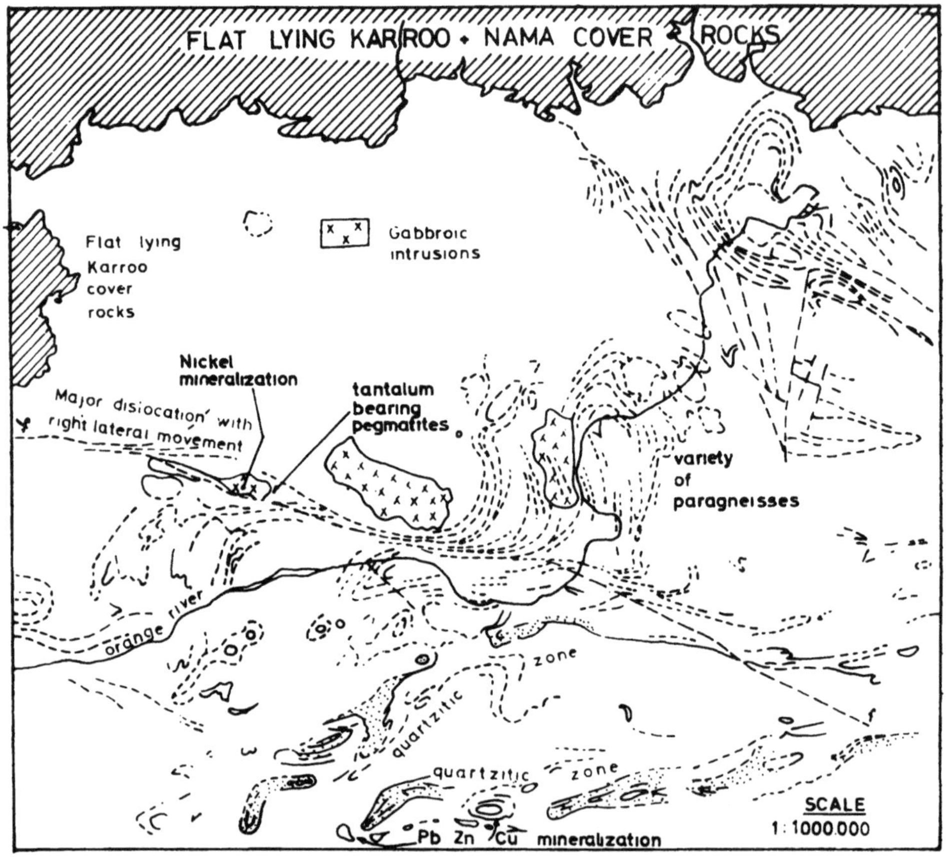


Fig. 7. Photogeological interpretation of ERTS image No. 1055-08053 showing part of the mobile metamorphic belt environment of Namaqualand, N.W. Cape Province.

been studied as part of this investigation. Only a few unpublished maps of a small part of this area are available and the interpretation presented (Fig. 7) is thas based largely on the ERTS image.

The typical flowage folding pattern with the development of intricate fold structures in the form of cross folds, "eye" folds etc., is clearly a widespread phenomenon of the area studied. Most of the rocks are paragneisses with amphibolitic material becoming important towards the northeast. A quartzitic assemblage which splits into two ranges occurs in the southern portion of the area and, although not defined before, is readily apparent from the ERTS imagery due to its resistant nature. In addition a major zone of dislocation with right lateral movement and not known to exist previously, is a dominant feature of the central portion of the image. Three gabbroic intrusions appear to be closely associated with this zone and nickel/copper mineralization has been located in one of these. Tantalumbearing pegmatites are also present within the same zone which therefore appears to have acted as the locus for emplacement of a variety of mineral deposits. A low escarpment formed of flat-lying Nama and Karroo sedimentary strata forms a conspicuous feature along the northern part of the image.

## 5. Summary and Conclusions

It is the writer's opinion that the broad synoptic view provided by the imagery is ideally suited to a study of the diagnostic macro-structures of the two geotectonic domains considered in this paper. ERTS 1 imagery should allow for a better understanding and more accurate definition than exists at present of the contact zones of these two geologic environments. As noted earlier, this may be of considerable economic importance, since basic/ultrabasic rocks containing some of the world's largest nickel deposits have been emplaced along similar contact regions within the Canadian Shield. These include the Thompson belt deposits at the junction of the Superior craton with the Churchill mobile belt and the Sudbury nickel irruption lying within the contact region of the Grenville front (mobile belt) and the Superior granite-greenstone province.

## Acknowledgments

The writers would like to thank the management of the Johannesburg Consolidated Investment Co., for affording them the opportunity of undertaking the investigation on the ERTS 1 imagery and for providing the funds for one of us to attend the symposium. Spectral Africa Ltd., are gratefully acknowledged for their co-operation and for the superb printing of the ERTS imagery which greatly aided in the geological interpretation.

Earth Survey Problems — Akademie-Verlag, Berlin 1974

# DYNAMICS AND MORPHOLOGY OF BEAUFORT SEA ICE DETERMINED FROM SATELLITES, AIRCRAFT, AND DRIFTING STATIONS

W. J. CAMPBELL[a], P. GLOERSEN[b], W. NORDBERG[b] and T. T. WILHEIT[b]

[a]US Geological Survey, University of Puget Sound, Takoma, Washington, USA
[b]NASA/Goddard Space Flight Center, Greenbelt, Md, USA

A series of measurements from drifting stations, aircraft, the ERTS 1, Nimbus 4 and Nimbus 5 satellites have jointly provided a new description of the dynamics and morphology of the ice cover of the Beaufort Sea. The combined analysis of these data shows that the eastern Beaufort Sea ice cover is made up of large multi-year floes while the western part is made up of small, predominantly first-year floes. The analysis suggests that this distribution might be quasi-steady-state and that the dynamics and thermodynamics of the region are more complex than hiherto known. The measurements consist of: (i) high resolution ERTS 1 imagery which is used to describe floe size and shape distribution, short-term floe dynamics, and lead and polynya dynamics; (ii) tracking by Nimbus 4 of IRLS drifting buoys to provide ice drift information which enhances the interpretation of the ERTS 1 imagery; (iii) Nimbus 5 microwave (1.55 cm wavelength) imagery which provides synoptic, sequential maps on the distribution of multi-year and first-year ice types; (iv) airborne microwave surveys and surface-based observations made during 1971 and 1972 in conjunction with the AIDJEX (Arctic Ice Dynamics Joint Experiment) program.

## 1. Introduction

Strong interest in sea ice was generated by the results of recent climate models [1—5] indicating that because of the effect of sea ice on the heat budget of the Arctic the distribution and amount of sea ice is an important and possibly a major influence in the climate of the Northern Hemisphere. Fletcher [4] points out that sea ice distribution in both the Arctic and Antarctic should be monitored as an indicator of changes in oceanic heat exchange.

During the 1960's it became obvious that observations from the traditional single drifting stations were intrinsically inadequate in providing sufficient data on the complex air-ice-ocean interaction of the Arctic Ocean with which to test the developing large scale numerical models of the ice canopy. Several investigators [6—8] developed mathematical models which indicated that a force neglected in earlier models, the internal ice force generated by momentum exchange between the ice floes, was a dominant one and must be included in any realistic model for sea ice dynamics. Recently, two new observing techniques, both derived from the space program, made possible the inference of the rheological properties of sea ice over vast areas, namely simultaneous measurements of ice, water, and air motion from an array of drifting stations relayed by satellites, and the satellite acquisition of sequential, synoptic images of radiation emitted and reflected by sea ice which relate to the dynamics of floes, leads, polynyas and ridges.

Sunlight and infrared observations of the Arctic ice canopy obtained by early meteorological satellites [9—12] revealed that it undergoes large scale space variations at small time scales and is the most variable solid feature of the earth's surface. However, these early observations were of low resolution; thus, individual ice floe trajectories and the dynamics of leads and polynyas could not be studied with them. Because the Arctic is dark for many months and often cloud covered when sun lit, it was not possible to obtain an adequate sequential view using visible and infrared images. Furthermore, these early data were not coupled with detailed surface (ground truth) observations.

In 1970 an international and interdisciplinary Arctic Ice Dynamics Joint Experiment (AIDJEX) was formed with the objective of reaching, through coordinated field experiments and theoretical analysis, a fundamental understanding of the dynamic and thermodynamic interaction between arctic sea ice and its environment. AIDJEX launched three spring pilot experiments in the Beaufort Sea in 1970, 1971 and 1972, in which surface-based measurements were combined with photographic, infrared and microwave observations made during over-flights with the NASA Convair 990 aircraft. Some of these observations are described in [13—15]. Subsequent to these experiments, which combined very effectively airborne remote sensing with extensive surface measurements, very detailed images of Beaufort Sea ice from the ERTS 1, synoptic microwave pictures of the entire Arctic from Nimbus 5, and tracking of automatic platforms on the ice from Nimbus 4 became available. In this paper, we wish to show how the combination of data from surface experiments, aircraft, and these three satellites has given us a new understanding of the sea ice of the Beaufort Sea.

## 2. Aircraft Microwave and Surface Observations of Sea Ice

Remote sensing from the NASA CV-990 aircraft flown in conjunction with the AIDJEX 1970, 1971 and 1972 pilot experiments involved 14 over-flights of AIDJEX areas, numerous flights over the transition zone extending from the north coast of Alaska out to the active sea ice, and one extending from 75°N to the conjectured steady-state center of the Beaufort Sea Gyre (Fig. 1).

The 1970 experiment showed that by mapping microwave emission at 1.55 cm wavelength it was possible to distinguish ice from liquid water under all meteorological conditions during all seasons [13]. This finding was of great importance from both dynamic and thermodynamic viewpoints. Strong microwave emissivity differences amounting to about 20°K in brightness temperature between two types of ice were also apparent in these observations. The 1971 AIDJEX-NASA experiment was designed to shed some light on the phenomenon of these varying microwave emissivities. High altitude flights were conducted to obtain images of the largest possible area of sea ice by flying overlapping parallel flight lines, and two microwave images of a 93 km by 93 km area of ice were obtained under clear and cloudy conditions. Gloersen et al. [16] showed that the varying ice emissivities were due to varying ice types, with the cold brightness temperatures corresponding to multi-year ice (thickness of two or more meters, low salinity, many empty brine inclusions near the surface, re-frozen crystals) and the warmer brightness temperatures corresponding to first-year ice (less than two meters thick, high salinity, most brine inclusions full, crystals of the original freezing). This demonstrated an all-weather capability of distinguishing multi-year and first-year ice over a large area of the Arctic.

One of the microwave mosaic images from this study is shown in Fig. 2. The AIDJEX camp, where most of the surface-based measurements were made, is

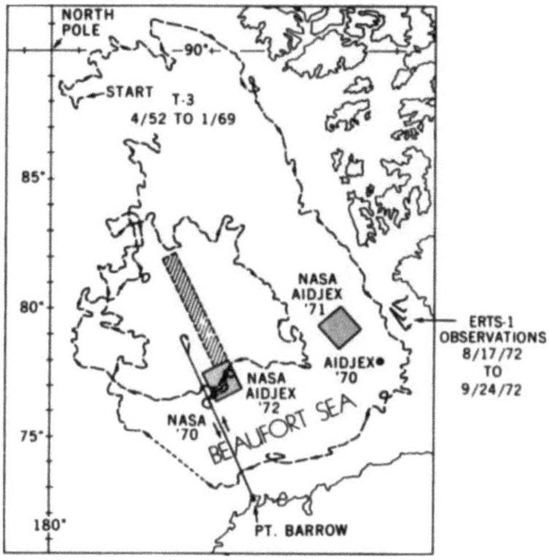


Fig. 1. Positions of NASA/AIDJEX experiments, showing also the drift track of Fletcher's Ice Island (T—3).

shown by a white circle. It was located on the north edge of a very large multi-year ice floe in the eastern Beaufort Sea (Fig. 1). This large floe as well as numerous smaller ones appear in the image as having a low brightness temperature (blue, dark green), whereas the first-year ice has a high brightness temperature (yellow, orange). The white triangle to the northeast of the camp indicates the position of the CRREL tellurometer strain array [17] which was situated entirely on first-year ice with a thickness of from 1.3 to 2.0 meters, having only small ridges and leads at the time of observation. The extensive ground truth and microwave measurements during the AIDJEX-NASA 1971 experiment served to correlate definitely the signatures to varying ice types.

The 1972 AIDJEX Pilot Experiment was considerably more extensive than the 1971 experiment. The area of study was several hundred kilometers further west of the 1971 site (Fig. 1). Three manned stations were occupied for two months and even unmanned AIDJEX-IRLS drifting buoys, left in place at the termination of the main experiment, were placed at ranges of about 400 km around the manned array (positions and drift trajectories are shown in Fig. 6). High level microwave mosaic images of the area encompassing the three manned stations [16] were obtained; part of one is shown in Fig. 2, in comparison with an image of an area of equal size obtained a year earlier in the eastern Beaufort Sea.

The most startling difference between these images of a large area of sea ice in the western Beaufort Sea and that of an equal area in the eastern Beaufort

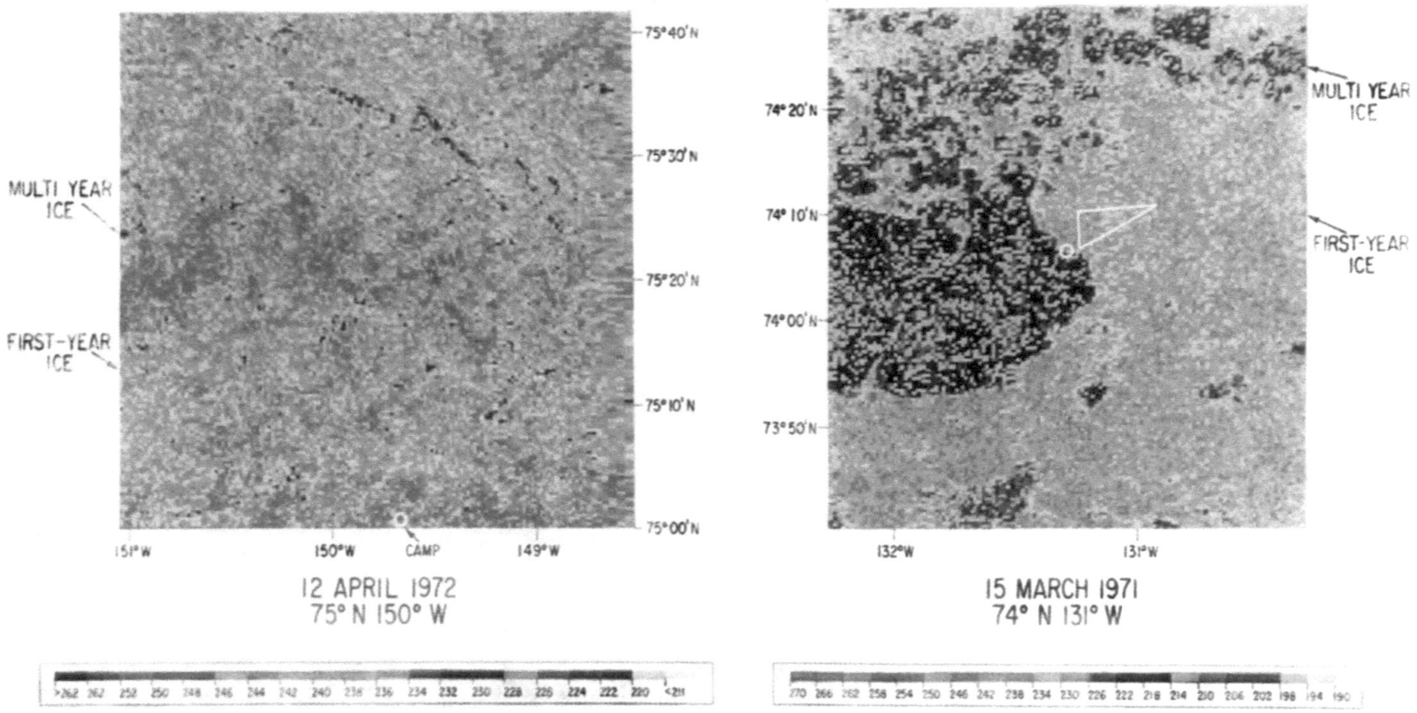
MULTI YEAR ICE
FIRST-YEAR ICE
75°40'N
75°30'N
75°20'N
75°10'N
75°00'N
151°W
150°W
CAMP
149°W
12 APRIL 1972
75° N 150° W
>262 262 252 250 248 246 244 242 240 238 236 234 232 230 228 226 224 222 220 <211
74°20'N
74°10'N
74°00'N
73°50'N
132°W
131°W
MULTI YEAR ICE
FIRST-YEAR ICE
15 MARCH 1971
74° N 131° W
270 266 262 258 254 250 246 242 238 234 230 226 222 218 214 210 206 202 198 194 190

Fig. 2.

Sea obtained a year earlier is that there are no large multi-year ice floes in the western sector as there were earlier in the eastern sector at the same latitude. Surface measurements obtained by participants in the 1972 experiment [18] showed that the main camp was located on a floe made up of cemented pieces of first-year and multi-year ice, but the amount of multi-year ice was apparently less than that of first-year ice; therefore the integrated signature over the area resolved by the radiometer (500 m $\times$ 500 m) appears radiometrically to be first-year ice. An area of $1.1 \times 10^4$ $m^2$ of sea ice adjacent to the main camp containing both ice types was studied in great detail [19] by measuring the surface microwave emissivities at 13.4 GHz in a grid of forty-four points along with ice crystallographic sampling, and the average brightness temperature of those points was found to be that of first-year ice, in agreement with observation of that area obtained from the aircraft. The microwave images show that there was significantly more first-year than multi-year ice and that there definitely were no large multi-year floes. According to the brightness temperatures observed, there could not have been more than 10% multi-year ice in the area, if the emissivity of first-year ice is 0.95.

This was surprising because it was generally believed that the ratio of old and new ice was fairly homogeneous throughout the Arctic Basin, with the exception of coastal areas. To find such a large area of predominantly first-year ice so deep within the ice pack contradicted long-held thermodynamic and dynamic beliefs. Therefore, after these findings had been obtained from four successive flights over the AIDJEX array, it was decided to fly a mission covering an area extending from the array to as deep within the Beaufort Sea Gyre as logistical constraints permitted. The microwave image obtained is shown in Fig. 3, and the position of the area is shown in Fig. 1. The image shows that just north of the AIDJEX array there were small multi-year floes (low brightness temperatures) with larger ones starting at 76°45′ N latitude and continuing to 79° N. North of 79° N the pack was essentially solid multi-year ice. The lineaments of high brightness temperatures in this area are due to new ice covering recently re-frozen leads and polynyas. No large multi-year floes similar to those observed a year earlier in the eastern Beaufort Sea were observed along this entire pass.

Fig. 4 shows that microwave images can be used to study ice dynamics. The dark green linear features are new leads, some being narrower than 500 meters and therefore not fully resolved by the radiometer, having a preferred SE—NW orientation in the left image. The right image shows the same area three days later, during which period a strong cyclone passed south of the AIDJEX array, and the lead orientation has switched almost orthogonally to a SW—NE direction.

←

Fig. 2. Comparison of ice morphology of western and eastern areas of the Beaufort Sea ice as observed with an imaging microwave radiometer operating at a wavelength of 1.55 cm. The locations of main AIDJEX camps are indicated by the white circles. The white triangle indicates the position of the ice strain array during 1971 experiment.

→

Fig. 3. Transect of the Beaufort Sea Ice Gyre (location shown in Fig. 1) as observed with the 1.55 cm microwave imager on 18 April 1972. The main AIDJEX camp is indicated by a white circle.

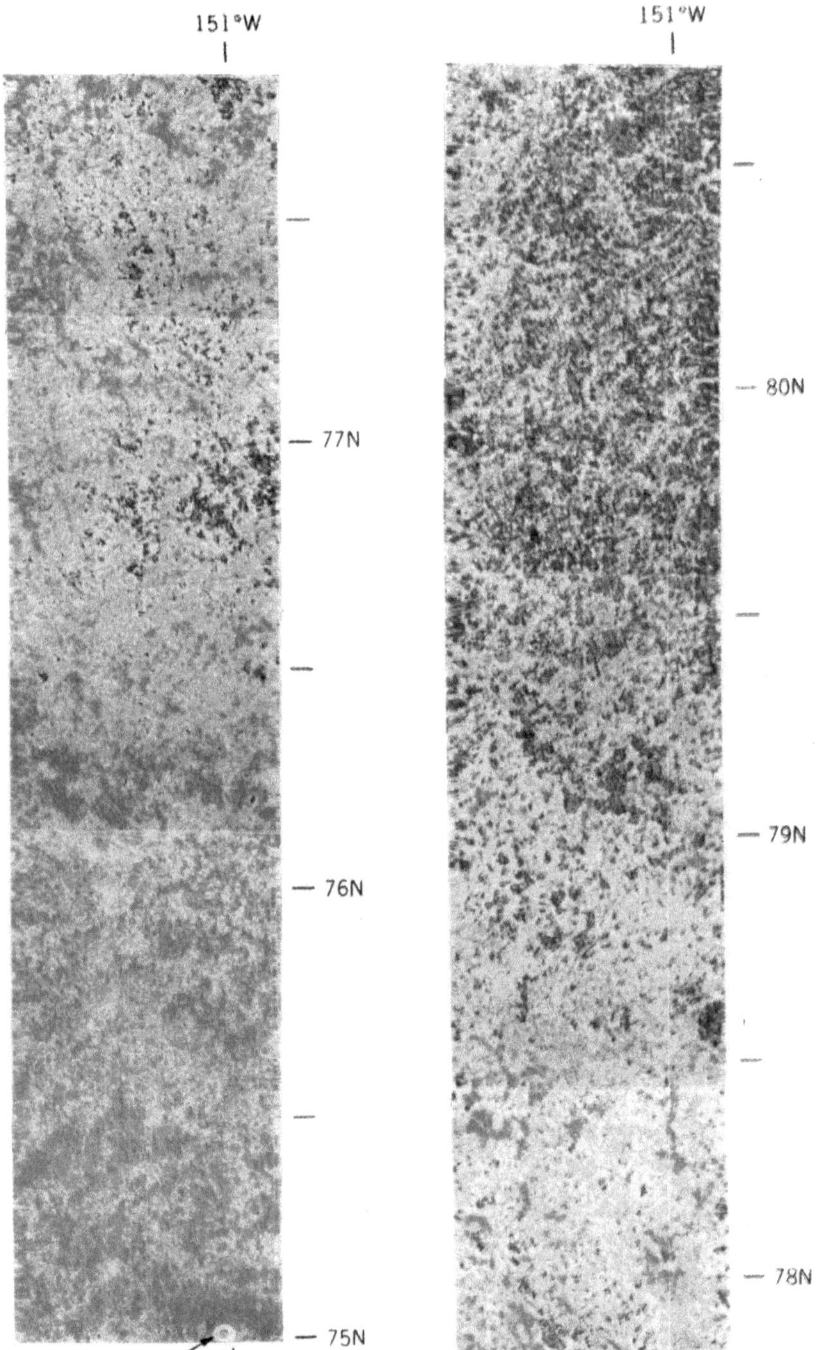


Fig. 3.

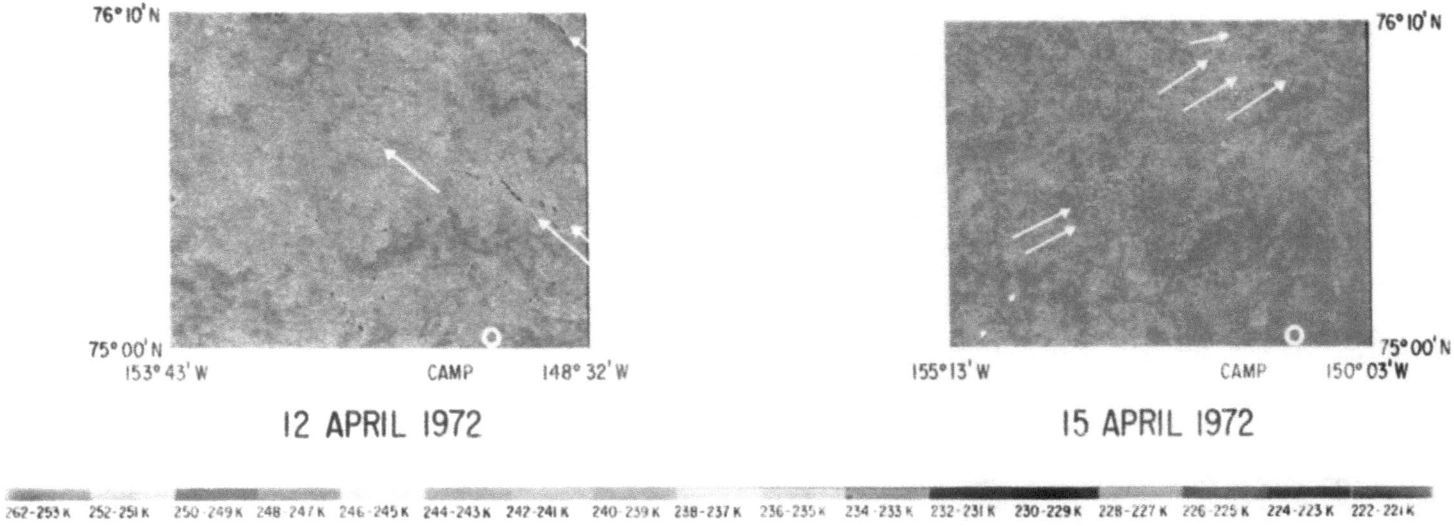


Fig. 4. Changes in the lead structure shown in microwave images of sea ice. The orientation of the leads is indicated by the white arrows. The white circles again indicate the location of the main AIDJEX camp.

## 3. ERTS 1 Observations of Beaufort Sea Ice Morphology

Although important studies of sea ice were made using data from meteorological satellites [9, 11, 12] it was not until the launching of ERTS 1 that imagery of sufficiently high resolution was available to allow the delineation of individual floes, leads and polynyas. A limited number of images of the Beaufort Sea from ERTS 1 was available for the period from late July until mid-October 1972. On 22 August 1972, ERTS 1 obtained images from both the eastern and western Beaufort Sea, roughly in the area where the 1971 and 1972 over-flights had been conducted. Fig. 5c shows the ERTS 1 image strip (160 km × 480 km) of the eastern Beaufort Sea. Very large, round shaped ice floes cover most of the area, some as large as 55 km across and ten more than 35 km across. These same floes can be identified in images (not shown) obtained on 2 August; in the three-week summer period between these observations they underwent very little weathering while moving towards the southwest, with only a few showing any discernible change of shape. The longest time sequence of ERTS 1 images covering one set of flows extended from 17 August to 24 September 1972 and consists of five passes over an area just east of the center of the 1971 AIDJEX grid. Two large floes (13 km × 28 km and 16 km × 27 km) could be tracked during the five and one-half week period and one showed no signs of weathering. The drift trajectories of these two floes as derived from ERTS 1 images are shown in Figs. 1 and 6. A later image obtained on 12 October in the same area again shows numerous large round floes.

The shape and size of the large floes observed in the eastern Beaufort Sea resemble those of the multi-year floes shown in microwave images obtained in the same area in 1971 (Fig. 2). Although we cannot positively say that these large floes were indeed multi-year floes, the fact that they endured a summer melt season and much dynamic activity while undergoing only minor weathering strongly suggests that they were. Also on 22 August 1972, 100 minutes after the images comprising Fig. 5c were obtained, ERTS 1 overflew the western Beaufort Sea, and obtained the image strip shown in Fig. 5a in approximately the same geographic area where the 1972 experiment took place. The most striking difference between this strip and the one shown in Fig. 5c is the complete absence of any large floes. The mean size of the floes in the western area is more than an order of magnitude smaller than those in the eastern area, having typical dimensions of 1—2 km across, and they have an angular, fragmented shape. It is important to bear in mind that these images (Fig. 5a and c) show ice at the same latitude and are essentially synoptic. ERTS 1 images obtained on 1 August, 5 September and 8 September (Fig. 5b) in approximately the same area as shown in Fig. 5a also show no large floes but numerous small floes that appear to have undergone recent strong deformation. The general picture of this area given by the August and September ERTS 1 images is one of relatively small, angular ice floes undergoing strong dynamic activity with large strains occurring frequently. We cannot tell the age of the numerous small floes observed in the western Beaufort Sea solely with the use of the ERTS 1 data since it is possible that many of the floes are fragmented multi-year floes. However, when we compare these images with the 1972 microwave images (Figs. 3, 4 and 5) of the same area, we are tempted to conclude that the area in the summer and fall was largely made up of first-year ice, as it was during the spring.

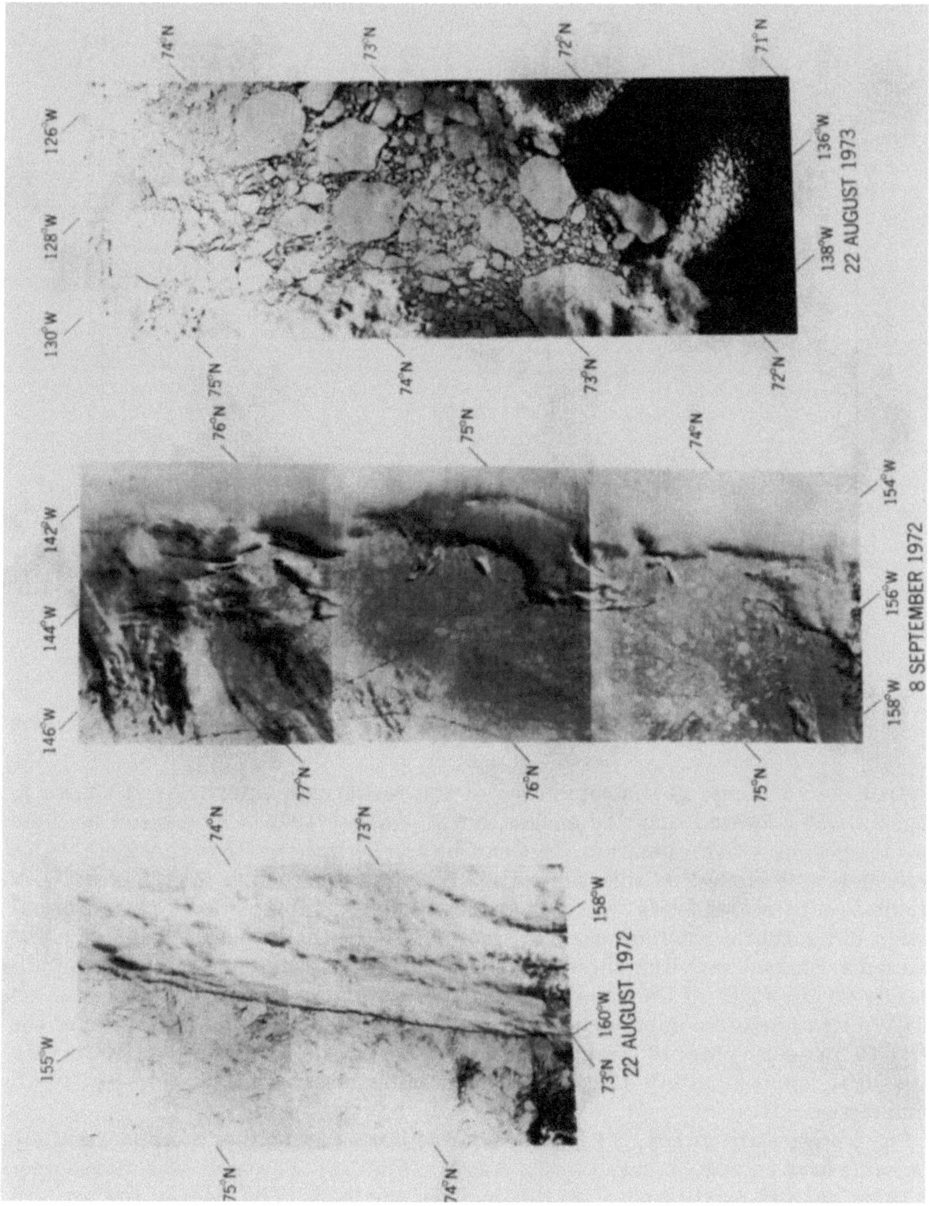


Fig. 5. ERTS 1 composite images of Beaufort Sea ice obtained in the approximate positions of the 1971 and 1972 AIDJEX experiments.

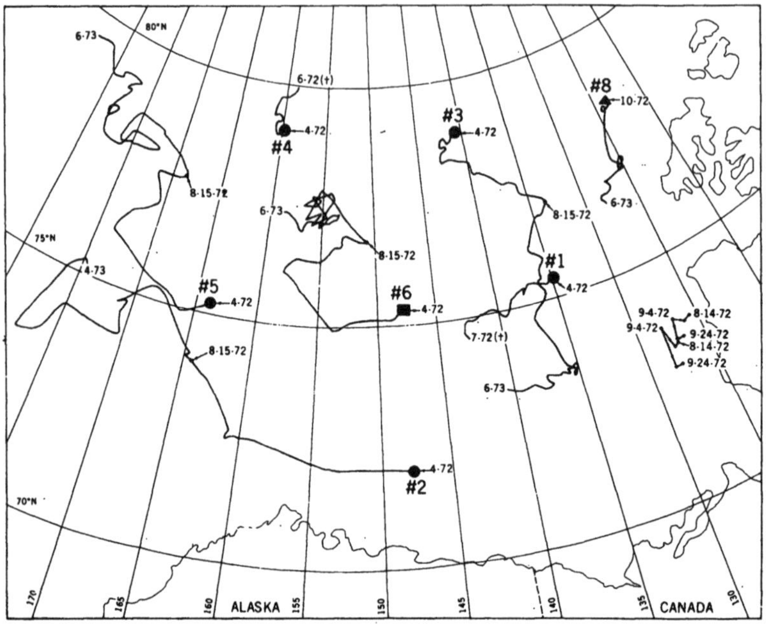


Fig. 6. Drift tracks of IRLS buoys in the Beaufort Sea. Also shown are the drift tracks of two large ice floes off Banks Island derived from ERTS 1 imagery.

## 4. Ice Drift by means of ERTS 1 Imagery and Nimbus 4—IRLS Platforms

An ERTS 1 image of the same area of the western Beaufort Sea as shown in Fig. 5a was acquired one day earlier, on 21 August 1972. Comparison of these two images, in which numerous floes can be tracked, shows that considerable ice deformation occurred at this time. On 21 August there was a preferred N—S orientation of major leads, whereas on 22 August there was no preferred orientation but rather a random array of small leads. In the interval of one day, the triangular-shaped polynya shown in the southwest corner of Fig. 5a, approximately 45 km north of the amorphous edge of the pack, had been stretched in an E—W direction and compressed in the N—S direction. The high resolution of the ERTS 1 imagery permits strains to be estimated, which in this case were very large with approximately 10% E—W stretching and approximately 6% N—S compression.

Fig. 5b is a strip of ERTS 1 images obtained on 8 September of approximately the same area as was observed on 22 August (Fig. 5a). The melt season was over at this time and newly frozen ice fills most of the leads. The floes in the western part of this mosaic resemble the numerous small floes visible in Fig. 5a, but in the south-center section a N—S band of larger, rounded floes is seen. The origin of these large floes cannot be determined from the ERTS 1 pictures because sequential images of a given area can, at best, be obtained only on three successive days at 18-day intervals. It is, therefore, highly desirable to have additional data on the ice dynamics which should be obtained between the periods of ERTS 1 observations. As mentioned earlier, seven IRLS buoys were positioned around the manned drift station array of the 1972 experiment and left in place when the manned stations were vacated. These buoys continued to operate, giving daily coordinates and surface atmospheric pressure, throughout the following summer; five of them are still operating as of June 1973. The general drift trajectories of these buoys is shown in Fig. 6. The data from these buoys are stored in the AIDJEX data bank, and a report on the Beaufort Sea ice dynamics as deduced from IRLS data is under preparation by AIDJEX scientists (P. Martin, personal communication).

The IRLS ice drift data complement the ERTS 1 data in that they allow us to infer the ice dynamics that took place between successive satellite images. As a first example, let us return to Figs. 5a and b. At the time these images were made, buoy No. 6 (the buoy left at the vacated Main Camp) was located at the northern edge of each image. Its respective positions on 22 August and 8 September were 76°58′N 151°56′W and 77°47′N 154°29′W. Thus it moved 96 km to the NW during that period. Another buoy, No. 2, which was drifting approximately 200 km to the west of the area of each image, experienced about the same drift toward the NW. Since we may assume that the differential motions within the area are less than the translation of the area itself, it appears that the larger floes shown in the center of Fig. 5b were advected into the area from the east, and that at the time of the earlier image (Fig. 5a) they were just east of the image area.

Another good example of the advantage of complementary ERTS 1 sequential images with buoy tracking can be found in the case of the two large floes tracked from 17 August to 24 September in the eastern Beaufort Sea (Fig. 6). The floes had similar trajectories, and considerable relative strains occurred. During the 19 August—4 September drift, the floes converged with an average strain rate of 2.3% per day, and during the 4—21 September drift they diverged with an average strain rate of 1.5% per day. These floes were stituated at the edge of the ice pack and had much open water around them, thus deformation could easily occur. While these are large strains, they are less than those observed at approximately the same time in the western Beaufort Sea as discussed earlier. During the drift period of these two floes another buoy, No. 3, was located approximately 360 km to the NW. During the period 17 August—4 September, the buoy drifted SSW for 44 km while the tracked floes moved WNW, thus the area between them converged in the NW—SE direction and moved west. On the other hand, during the 4—22 September period, the buoy moved SSW for 75 km and the tracked floes moved south about equally which implies that a large area of the eastern Beaufort Sea ice pack had a uniform translation. Thus, we have demonstrated how the presence and absence of ice deformations can be derived from data obtained simultaneously by two differnt satellite techniques in adjacent areas.

Correlations between the drift of buoys and floes seen by ERTS 1 in the same area under conditions of pronounced ice drift have been established for short periods (2—3 days). The ERTS 1 imagery and the buoy data agree that during the month of August the entire southern Beaufort Sea ice pack drifted mainly west, with the eastern area drifting toward the SW and the western area drifting toward the NW. An ERTS 1 image of an area of the Beaufort Sea located between the areas shown in Figs. 5a and b (western sector) and 5c (eastern sector) was obtained on 27 August. It clearly shows numerous floes of an intermediate size between the large ones observed in the east and the small ones observed in the west, with an average size in the order of 10 km across. Of all ERTS 1 images analyzed for the summer 1972 over the consolidated pack ice this image shows the largest amount of open water, estimated at 5% of the total area.

Considering these data, we find the ice of the southern Beaufort Sea to be a non-homogeneous mixture of floe sizes, very large and rounded in the east and smaller but rounded in the center. Even smaller but angular floes appear in the western sector. The general direction of ice drift in the southern Beaufort Sea is to the west during this period. The 1971—1972 experiment tempts us to conclude that the large eastern floes are multi-year and the small western ones are predominantly first-year floes.

## 5. Nimbus 5 Microwave Synoptic Observations of Arctic Sea Ice Canopy

The Electronically Scanning Microwave Radiometer (ESMR) mounted on the Nimbus 5 satellite is providing synoptic images of microwave emission at 1.55 cm wavelength over the entire globe, including the Arctic. Two of these images are shown in Fig. 7.

The image for 16 December 1972 shows low brightness temperatures throughout most of the Arctic Ocean with the exception of a large warm zone north of the coast of Alaska, a smaller one extending from north of the McKenzie Delta into the Amundsen Gulf, large warmer zones in the Laptev Sea and East Siberian Sea, with the largest and warmest zone in the Kara Sea. Based on the results of the AIDJEX experiments, the areas of low brightness temperatures are believed to be covered by multi-year ice and areas of high brightness temperatures by first-year ice or a mixture of the two that is predominantly first-year. This assumption breaks down only where there is a mixture of high emissivity new-year ice with extremely low emissivity open water, where the ice covered areas are smaller than the 30 km $\times$ 30 km resolution of the radiometer. In this case, the mixed composition of the area will produce a brightness temperature much higher than that of open water but lower than that of first year ice, namely corresponding to multi-year ice. The Bering Sea and Greenland Sea are typical for such areas in Fig. 7. However, such open water areas which are partially covered with new-year ice can easily be identified in sequential images where the overall brightness temperature will increase rapidly as the area becomes more and more covered with first-year ice. Based on this information we have identified various ice types in the two microwave images (Fig. 7), as shown in Fig. 8.

In the image obtained on 16 December 1972 (Fig. 7), a broad tongue of multi-year ice extends south from the central Arctic Ocean into the eastern Beaufort Sea as far south as 72° N, adjacent to Banks Island and north of the McKenzie Delta. First-year ice, or a mixture of predominantly first-year and multi-year (which we will hereafter call first-year), extends into the western Beaufort Sea in a broad triangular zone with its base running along the coast from the McKenzie Delta to the Bering Strait, and its apex at approximately 78° N 160° W. This distribution of ice types now inferred on a synoptic scale for the entire Beaufort Sea bears out the inferences made from the transect microwave image obtained

Fig. 7. Nimbus 5 microwave ($\lambda = 1.55$ cm) images of the Arctic. →

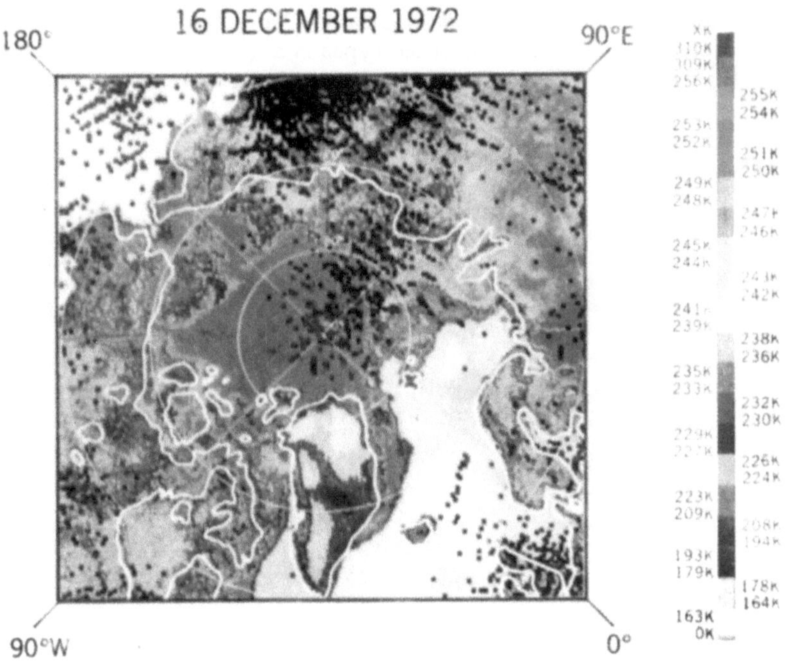
16 DECEMBER 1972
180°
90°E
90°W
0°

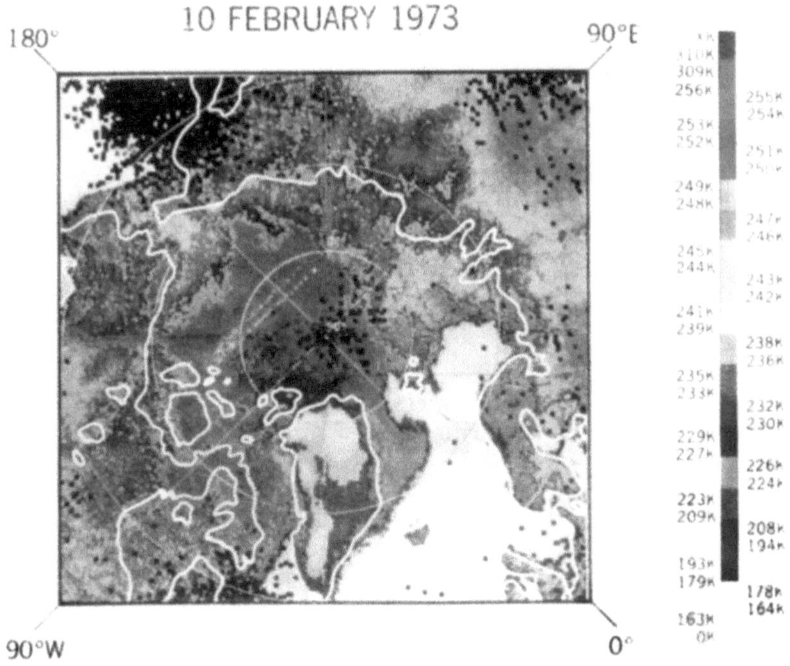
10 FEBRUARY 1973
180°
90°E
90°W
0°

during the 1972 experiment (Fig. 3). The transition from first-year to multi-year ice at 160° W occurs at 78° N in the satellite image and at 79° N in the aircraft transect at approximately the same longitude. Combined aircraft and satellite data suggest that this distribution of ice types is a quasi-steady-state feature of

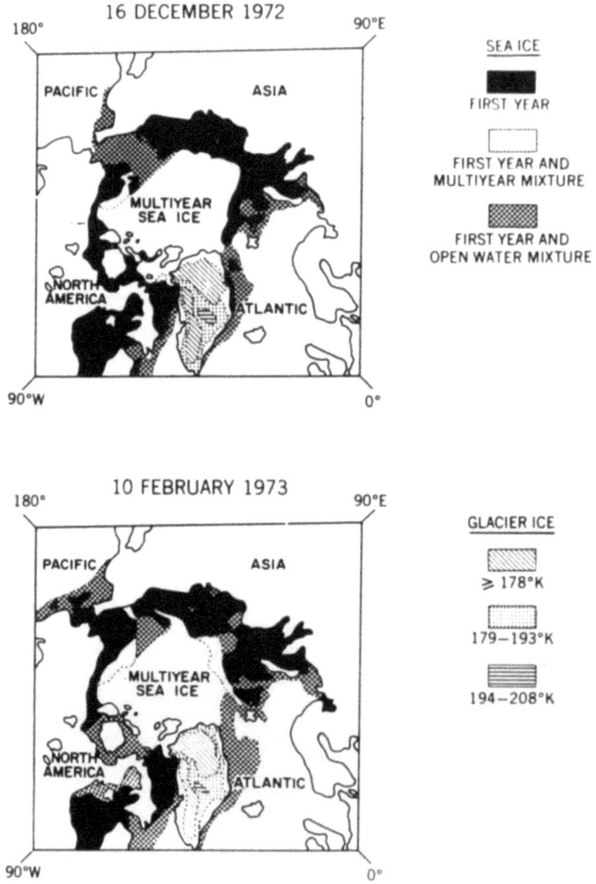


Fig. 8. Analysis of images shown in Fig. 7.

the Beaufort Sea, at least for the winters 1971/1972 and 1972/1973. This inference is further supported by another Nimbus 5 microwave image acquired on 10 February 1972 (Fig. 7). Noting the Beaufort Sea in this image and its analysis in Fig. 8, we can see that the same broad tongue of multi-year ice extends into the eastern Beaufort Sea from the central Arctic, and slightly cooler brightness temperatures are found between Greenland and the Pole. The triangular zone of first-year ice in the western Beaufort Sea seen earlier is still there but it has broadened and shifted slightly westward. Thus, the non-homogeneous distribution of ice types seems to persist throughout the winter 1972/1973, with slight fluctuations of the first-year ice zone of the western Beaufort region.

Analyzing the microwave images (Fig. 7), we see that as the winter progressed the first-year ice in the East Siberian and Laptev Seas occupies the same area whereas the first-year ice zone in the Kara Sea has either expanded further north between Severnaya Zemlya and Franz Josef Land or the ice in that area was not consolidated earlier. Of all the Arctic Basin first-year ice zones observed in these images, the one in the western Beaufort Sea is the only one not bordered by islands and not located over shelves in shallow seas where lower surface water salinities exist at the time of freeze up. Therefore, the first-year zones north of Asia may result predominantly because they form in enclosed, shallow structured seas having large influxes of fresh water and hence low surface salinities, whereas the one north of Alaska may be the product of ice dynamics related to the behavior of the central Arctic Ocean ice. Another pronounced variation of microwave emittance is observed over the Greenland ice cap (Fig. 7). Brightness temperature variations of 50 °C occur across the ice cap with the highest emittance corresponding roughly to the highest elevations of the central cap. Although we do not as yet fully understand the cause of these brightness temperature variations occurring on glacier ice, we are certain that they are not due to variations in the physical temperature of the ice but rather are emissivity variations probably induced by variations of the small scale ice stands. We believe that these may be analogous to the emissivity variations between first year and multi-year sea ice. In that case the lower emissivity of multi-year ice is believed to be partly caused by the recrystallization due to seasonal thawing and re-freezing. In the case of the glacier ice such crystal metamorphosis could be caused not only by temperature variations but also by pressure gradients within the ice.

## 6. Conclusions

This study illustrates how a variety of surface based, aircraft and satellite observations can be combined and jointly analyzed to yield new findings which would not have resulted if the data were treated separately. We have found that ERTS 1 imagery can be used to study ice floe morphology and dynamics at time scales of several days to months and lead and polynya dynamics at scales of one to several days. Nimbus 4—IRLS drifting buoys have permitted the mapping of gross ice drift of the Beaufort Sea. They have also provided the continuity necessary for the derivation of pack ice motion occurring between intermittent ERTS 1 observations of a given area. The combining of buoy and imaged ice floe trajectories yielded information on gross ice deformation. Microwave (1.55 cm wavelength) images obtained during 1971 and 1972 experiments combined with those from Nimbus 5, permitted the delineation of first-year and multi-year ice types at several scales over the entire Arctic.

During the 1971—1973 period, the Beaufort Sea ice canopy consisted of non-homogeneously distributed floe sizes and types, with the eastern sector composed of large, rounded multi-year floes and the western sector composed of small, angular predominantly first-year floes. It appears that the strong dynamical activity of the general westward ice flow through the Beaufort Sea results in pronounced fracturing of the large multi-year floes that are advected into the eastern Beaufort Sea from the north. The non-homogeneous distribution of ice types, floe sizes and deformational strains derived from this ensemble of observations implies that the western Beaufort Sea must be a zone of net annual ice divergence, from which the first-year ice that survives the summer melt would flow into the transpolar drift stream and/or the Beaufort Sea Gyre. In either case this ice would be transported northward from this area. In order to prevent the net influx of multi-year ice during the summer season from the eastern Beaufort Sea which would occur under the presently held concept of a steady-state Gyre, the position and/or structure of the Gyre must undergo seasonal variations.

## In Memoriam

The data collected by the crew of the NASA CV-990 "Galileo" airborne observatory during their many Arctic Ocean missions were of key importance to this and other investigations of Arctic Sea ice. The crew's dedication and skilled labors have opened new frontiers of knowledge. We mourn the loss of some of these fine men as a result of the tragic crash of that aircraft on 12 April 1973.

## Acknowledgment

One of us (W. Campbell) was supported in this endeavor by the NOAA/NESS Spacecraft Oceanography Center under Contract No. NA-916-73.

## References

[1] M. I. BUDYKO, Meteorol. Gidrol. **3, 11** (1968).
[2] M. I. BUDYKO, Climate Changes, Hydrometeorological Publishing House, Leningrad 1969 (in Russian).
[3] J. O. FLETCHER, Rand Corporation Rep. No. R-444-PR, 1965.
[4] J. O. FLETCHER, in: Proc. Conf. on Long-Range Polar Objectives, US Dept. of Transportation, Washington, D.C. 1968.
[5] Y. MINTZ, WMO Tech. Notes, No. 66 (1965).
[6] W. J. CAMPBELL, J. Geophys. Res. **70**, 3279 (1965).
[7] D. L. LAIKHTMAN, Trans. Leningrad Hydrometeorol. Inst. No. 7 (1958) (in Russian).
[8] M. I. RUZIN, Trans. Arctic and Antarctic Inst. and Main Geophysical Obs. **226**, 123, (1959) (in Russian).
[9] H. P. NELSON, S. NEEDHAM and T. D. ROBERTS, Sea Ice Reconnaissance by Satellite Imagery, Final Rep. to NASA 1970.
[10] W. J. CAMPBELL, in: Proc. 7th Int. Symp. on Remote Sensing of Environment, Willow Run Labs. 1971 (p. 937).
[11] E. P. MCCLAIN and M. D. BALILES, Mariners Weather Log, **15** (1), 1 (1971).
[12] J. C. BARNES, D. T. CHANG and J. H. WILLAND, J. Geophys. Res. **77**, 453 (1972).
[13] T. T. WILHEIT, W. NORDBERG, J. BLINN, W. J. CAMPBELL and A. EDGERTON, Remote Sensing of Environment, **2**, 129 (1972).
[14] W. J. CAMPBELL, in: Proc. Symp. on Earth Resources Technology Satellite 1, 1972, Goddard Space Flight Center 1973 (p. 129).

[15] P. Gloersen, W. Nordberg, T. J. Schmugge, T. T. Wilheit and W. J. Campbell, J. Geophys. Res. **78**, 3564 (1973).
[16] P. Gloersen, W. Nordberg, T. C. Chang, T. J. Schmugge, W. J. Webster, T. T. Wilheit and W. J. Campbell, paper presented at Symp. on Sea-Air Interactions in Polar Regions, AGU 1972.
[17] W. D. Hibler III, W. F. Weeks, S. Ackley, A. Kovacs and W. J. Campbell, J. Glaciol. **12**, 187 (1973).
[18] W. D. Hibler III, W. F. Weeks, A. Kovacs and S. Ackley, to be published in J. Glaciol. (1974).
[19] D. Meeks, R. O. Ramseier, G. Poe, W. J. Campbell and A. Edgerton, EOS Trans. AGU **53**, 1017 (1972).

Earth Survey Problems — Akademie-Verlag, Berlin 1974

# APPLICATIONS OF ERTS IMAGERY TO SNOW AND GLACIER HYDROLOGY

M. F. Meier

US Geological Survey, Tacoma, Washington, USA

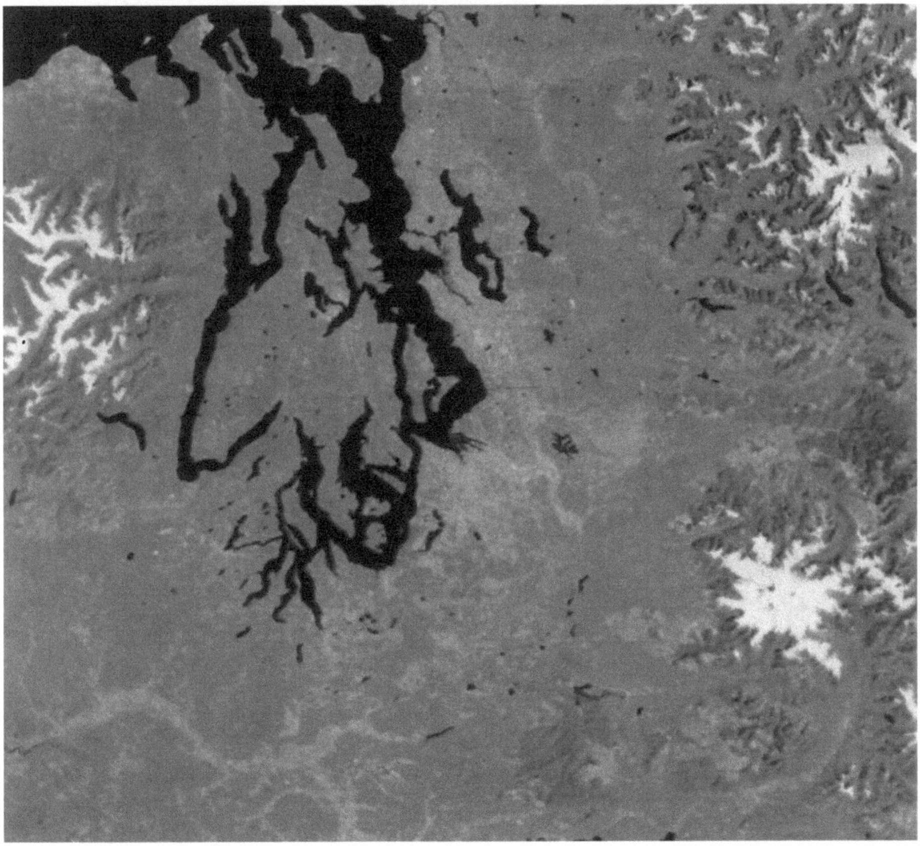

Fig. 1. ERTS image of part of Western Washington showing the eastern Olympic Mountains (left) Puget Sound (center), the southwestern North Cascades (upper right), and Mount Rainier (lower right). Alder Reservoir (to the left and below Mt. Rainier) and Commencement Bay (center of image) show a blue color due to glacial sediment. NASA image 1006-18313, 29 July 1972.

This paper explores the usefulness of ERTS-type snowcover data to scientific and operational aspects of the management of water resources and to the study of glaciers. The conclusions presented here are based on a limited sample of data and are preliminary.

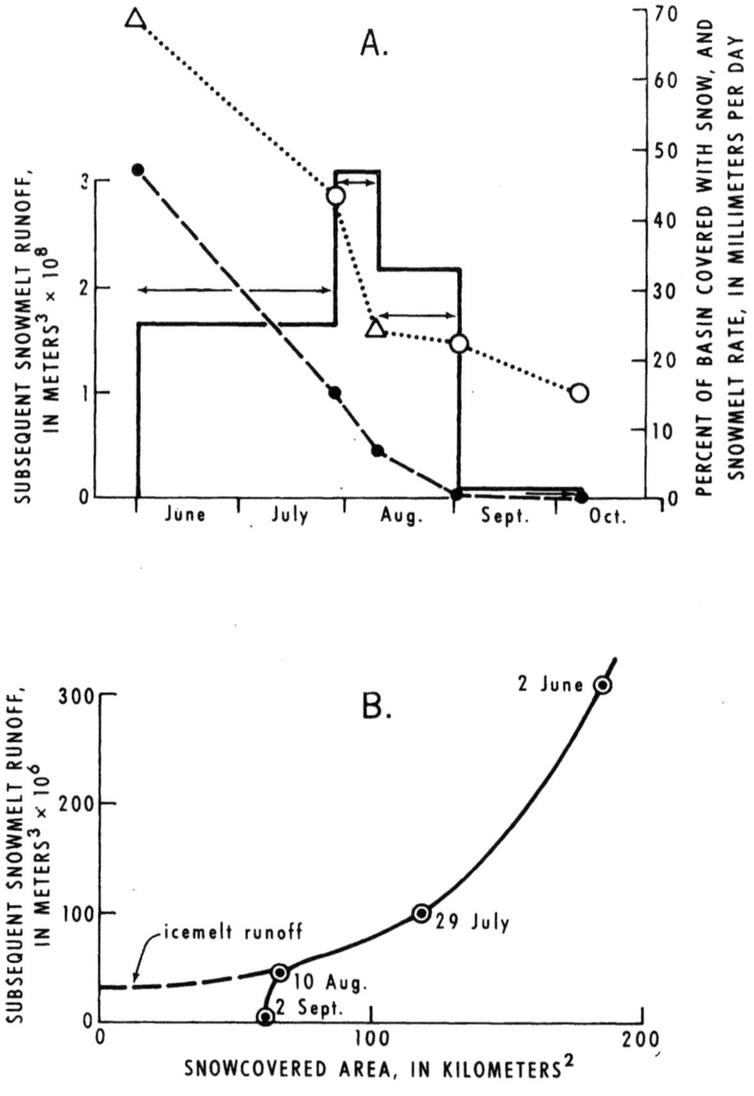


Fig. 2. A. Snowcovered area (dotted line), subsequent snowmelt runoff volume (dashed line), and rate of snowmelt (heavy line) in the Thunder Creek drainage basin; thin line is average observed snowmelt rate. Triangles are data from 19.8 km altitude aerial photographs; circles are data from ERTS 1.

B. Subsequent snowmelt runoff as a function of snowcovered area, Thunder Creek drainage basin.

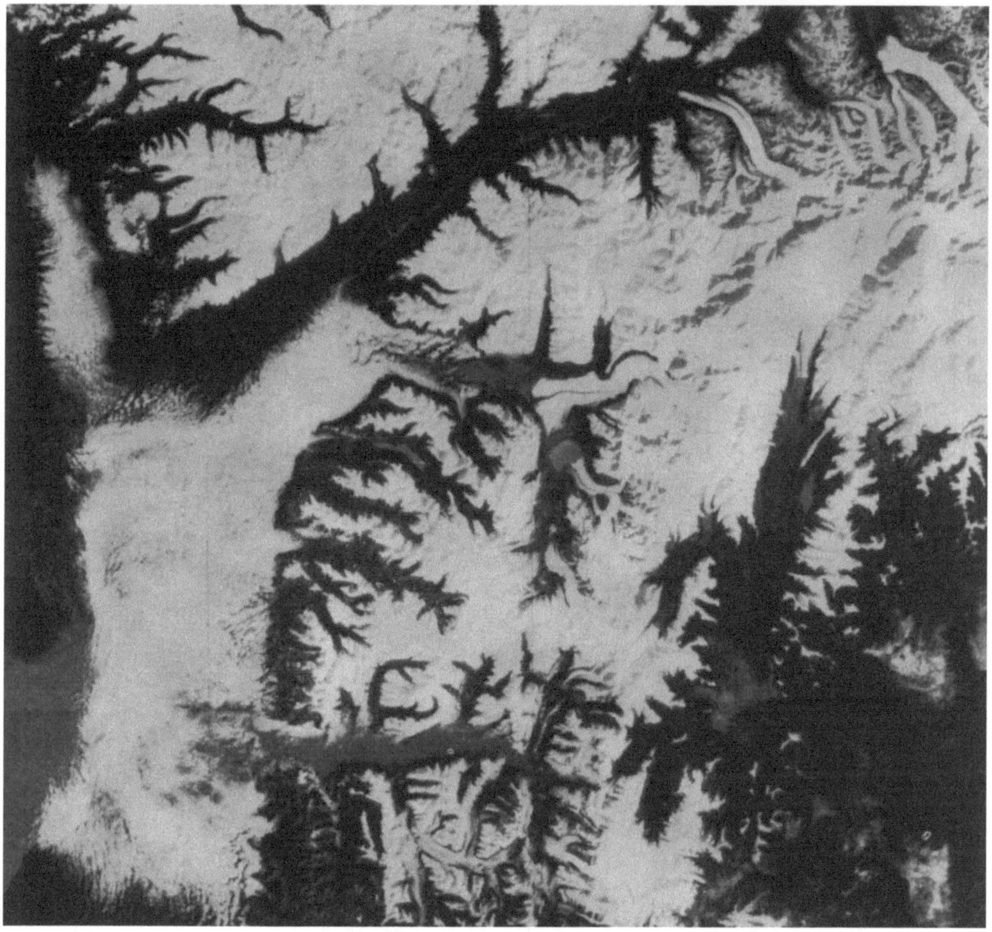

Fig. 3. ERTS image of Anchorage and the eastern Chugach Mountains showing an unusually sharp snowline. NASA image 1066-20451-4, 27 September 1972.

The percentage of snowcover area on specific drainage basins was measured from ERTS imagery by video density slicing using the Stanford Research Institute Electronic Satellite Image Analysis Console (ESIAC). Results were repeatable to 4% of the snowcovered area. Data from ERTS images of the melt season snowcover in the Thunder Creek drainage basin in the North Cascades, Washington (Fig. 1) were combined with existing hydrological and meteorological observations to enable calculation of the time distribution of the water stored in this mountain snowpack (Fig. 2). Similar data could be used for frequent updating of expected inflow to reservoirs. Equivalent snowline altitudes were determined from area measurements. Variations in snowline altitudes were determined by combining enlarged ERTS images with maps; an accuracy of about 60 m was attained under favorable conditions (Figs. 3, 4). Ability to map snowcover or to determine snowline altitude depends primarily on cloud cover and vegetation and secondarily on

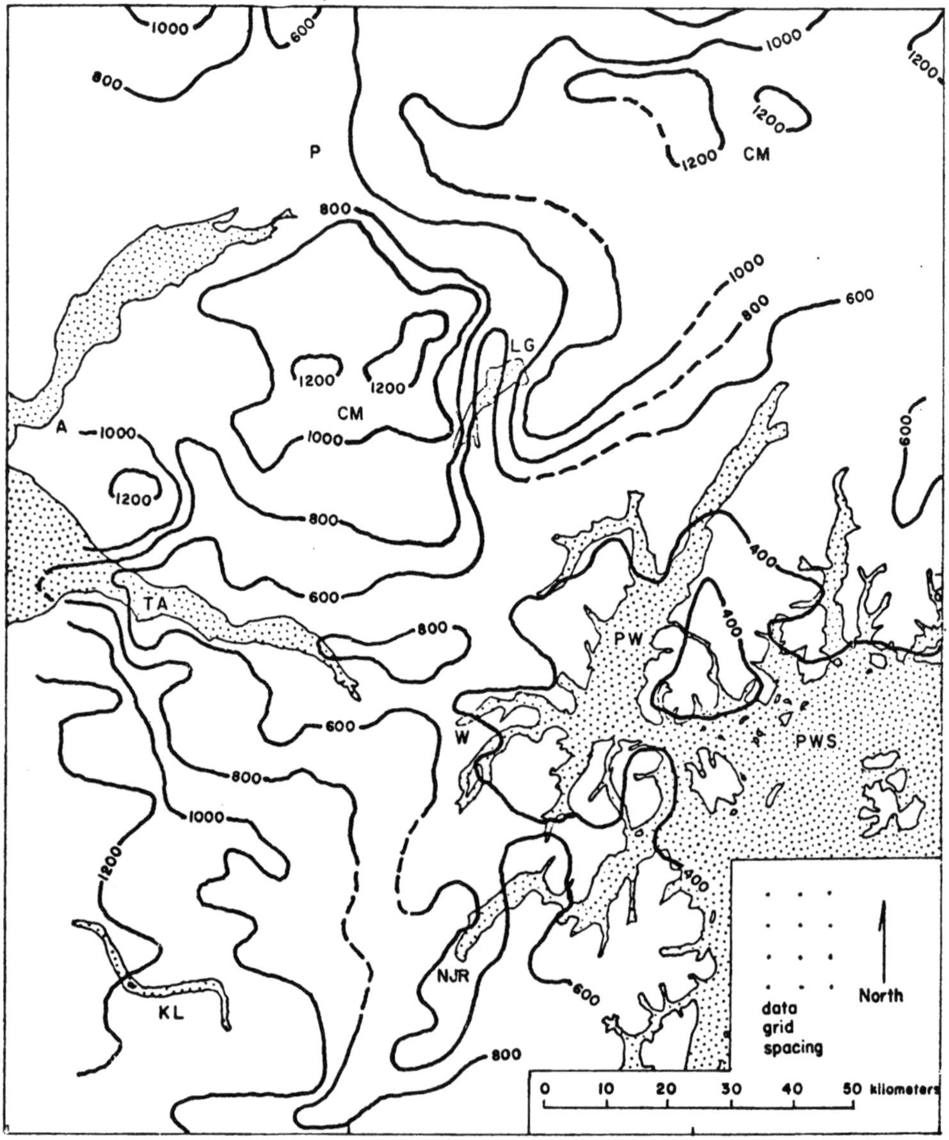


Fig. 4. Map showing the altitude of the snowline in the Anchorage area, Alaska, 27 September 1972. Data derived from ERTS images 1077-20451 and 1066-20453. Locations indicated as follows: A Anchorage, P Palmer, W Whittier, TA Turnagain Arm, CM Chugach Mountains, KL Kenai Lake, NJR Nellie Juan River, LG Lake George, PW Port Wells, PWS Prince William Sound.

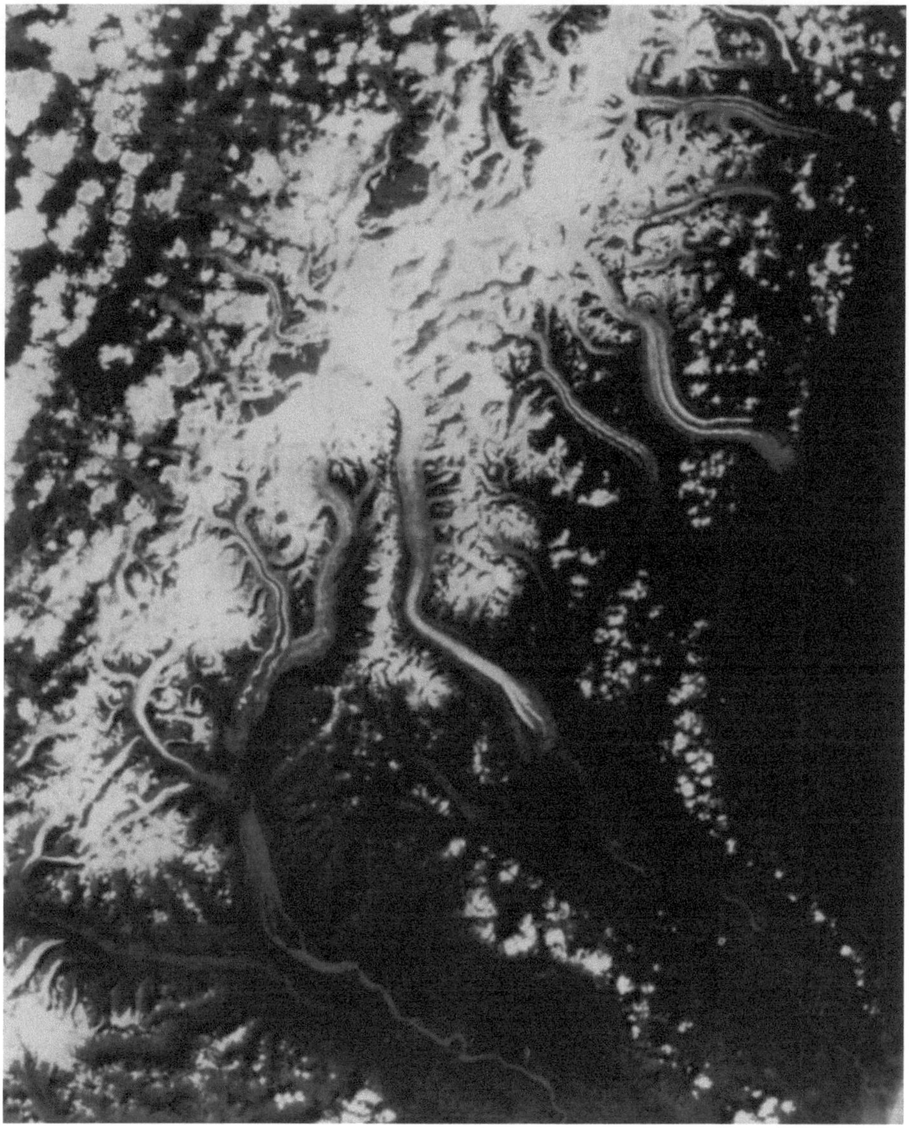

Fig. 5. ERTS image of the Alaska Range near Mt. McKinley, Alaska. The glacier with a prominent dark, wiggly medial moraine in the left center is Yentna Glacier, which is undergoing a surge. NASA image 1033-21020-4, 25 August 1972.

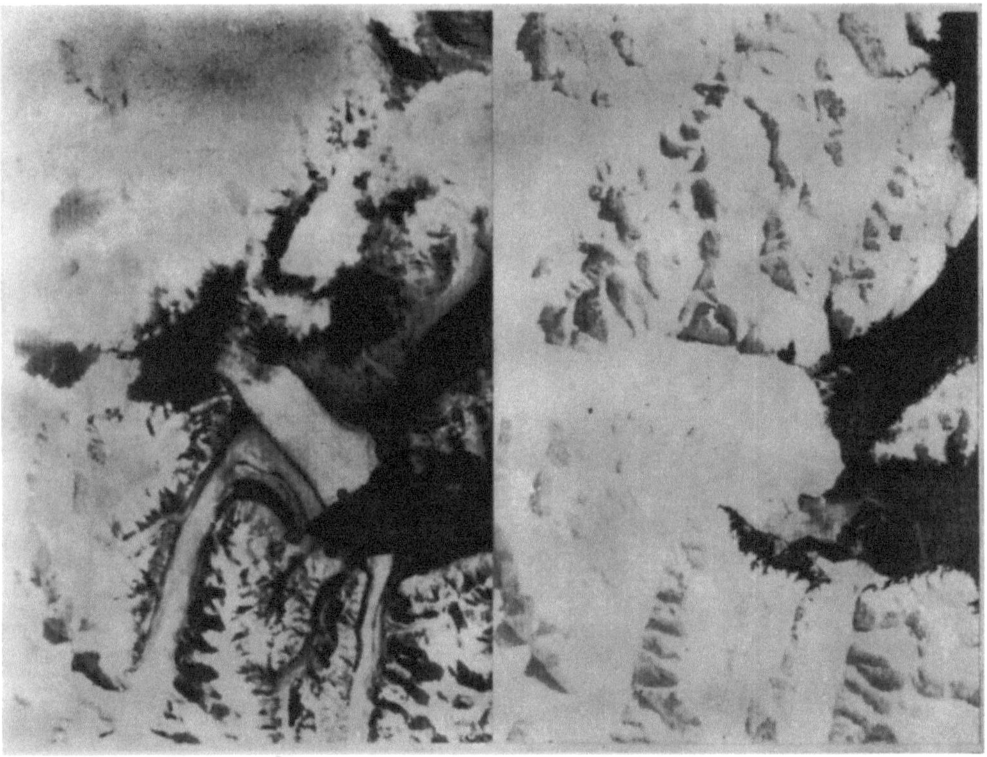

Fig. 6. Changes in terminus of Hubbard Glacier from 20 September (left) to 8 October (right). Terminus is 11 km wide. Image 1059-20052-6 (left) and 1077-21020-4.

slope, terrain roughness, sun angle, radiometric fidelity, and amount of spectral information available.

ERTS imagery was also used to measure glacier accumulation area ratios, detect subtle flow structures on glaciers, identify surging glaciers (Fig. 5) and monitor changes in tidal glacier termini (Fig. 6).

A more complete version of this paper is to be published in the *Proceedings of the ERTS-1 Symposium, Greenbelt, Maryland, 5—8 March 1973*, under the title "Evaluation of ERTS Imagery for Mapping of Snowcover on Land and on Glaciers".

Earth Survey Problems — Akademie-Verlag, Berlin 1974

# APPLICATIONS OF ERTS DATA TO OCEANOGRAPHY AND THE MARINE ENVIRONMENT

G. A. MAUL

National Oceanic and Atmospheric Administration,
Atlantic Oceanographic and Meteorological Laboratories, Miami, Fla, USA

The multi-spectral scanner (MSS) on board the Earth Resources Technology Satellite (ERTS) has returned some remarkable imagery with applications to oceanography. Theoretical background for understanding the penetration and reflectance of light from the sea is briefly given for the MSS bands. Examples from the recent symposium on significant results include studies of sea ice, fisheries, charting, beach erosion, limnology, estuarine circulation, ocean features in the lee of islands, ocean currents, and internal waves. Several enhancement techniques are discussed and exemplified in the context of its application.

## 1. Introduction

The attenuation coefficient $\alpha$ for light traveling through water is dependent on its wavelength $\lambda$. The attenuation coefficient is defined as the sum of the absorption coefficient $a$ and the scattering coefficient $b$. A general minimum occurs near 0.5 μm with several orders of magnitude increase near 1.1. μm. These wavelengths represent the lower cutoff of MSS 4 (0.5—0.6 μm) and upper cutoff of MSS 7 (0.8—1.1 μm) of the multi-spectral scanner (MSS); MSS 5 (0.6—0.7 μm) and MSS 6 (0.7—0.8 μm) are intermediately situated. The underwater irradiance $H$ in each of these bands is given by

$$H(\phi, Z) = \int_0^\infty \phi(\lambda) H_0(\lambda) \exp \left[ -\int_0^z \alpha(Z, \lambda)\, \alpha Z \right] d\lambda,$$

where $\phi$ is the response of the bandpass filter, $H_0$ is the irradiance at the sea surface, and $Z$ is depth.

Calculations for the ERTS bands, normalized, and expressed in per cent, are graphically depicted in Fig. 1 [1]. Of the energy that penetrates pure water, MSS 7 has an inverse e-folding depth of much less than 1 m, whereas MSS 4 decays to 1/e of its surface value at approximately 20 m. These are optimistic by a factor of at least 2, for even the cleanest seawater has $\alpha$ values double that of pure water. Practically, water information in MSS 6 and 7 are constrained to be surface features, whereas MSS 4 and 5 can offer information at depth.

Chandrasekhar [2] formulated an expression for the reflectance $R(Z, -)$ at the sea surface from the theory of radiative transfer. Reflectance is defined as

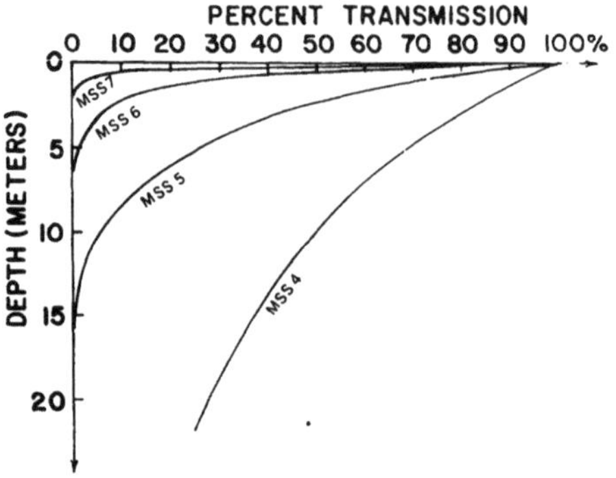


Fig. 1. Percentage of the light that penetrates a column of pure water reaching a given depth for the four MSS bands (after Charnell and Maul [1]).

the ratio of the upwelling (−) to the downwelling (+) irradiance. At the sea surface, $Z = 0$, and for the simple case of isotropic scattering,

$$R(Z, -) = 1 - H(\mu, \omega_0)\,(1 - \omega_0)^{1/2},$$

where $\omega_0$, the scattering albedo, is the ratio of the scattering coefficient to the attenuation coefficient, $H(\mu, \omega_0)$ are the $H$-functions tabulated by Chandrasekhar, and $\mu$ is the cosine of the zenith angle. The significant variable is $\omega_0$. Examining the two limiting cases, note that $R$ approaches 1 when $\omega_0$ approaches 1 (scattering dominant) and as $\omega_0$ approaches 0 (absorption dominant), $R$ approaches 0 because $H(\mu, 0) = 1$. Thus if all the light that enters the ocean is scattered, reflectance is 100% and the ocean appears white, and if all the light is absorbed, reflectance is zero and the ocean is black.

The upwelling irradiance above the sea surface is the sum of $R(Z, -)$ plus the reflection from the surface of direct and diffuse sunlight. Three spectra taken with near equal solar zenith angles, differing chlorophyll-$a$ concentration (affecting absorption, $a$) and differing volume scattering coefficients $\beta_{45}$ (affecting scattering, $b$) are given in Fig. 2. These are irradiance measurements taken from shipboard at 4 m above the surface in the Straits of Florida; water depth was in excess of 100 m. These measurements are in qualitative agreement with calculations made from isotropic reflectance theory given above, and serve to give relative energy levels for MSS 4, 5, and 6 for different oceanic waters. Clearly ERTS bands have information in them on the biomass of the ocean.

The radiant intensity $N(\lambda, \theta)$ from a nadir angle $\theta$ received at the satellite is attenuated by absorption and scattering in the atmosphere. For Rayleigh scattering, Beer's law gives an exponential solution with a scattering coefficient

$$S(\lambda) = \frac{32\pi^3\big(n(\lambda) - 1\big)^2 \varrho}{3\lambda^4 N \varrho_0},$$

where $N$ is the number of molecules at STP, $n$ is the index of refraction at STP, $\varrho_0$ and $\varrho$ are the standard and actual densities of dry air. This brings out the well-known inverse fourth powr law. Integration over the spectral response of the MSS filters would give approximate scattering ratios of 9:5:3:1 for the MSS 4,

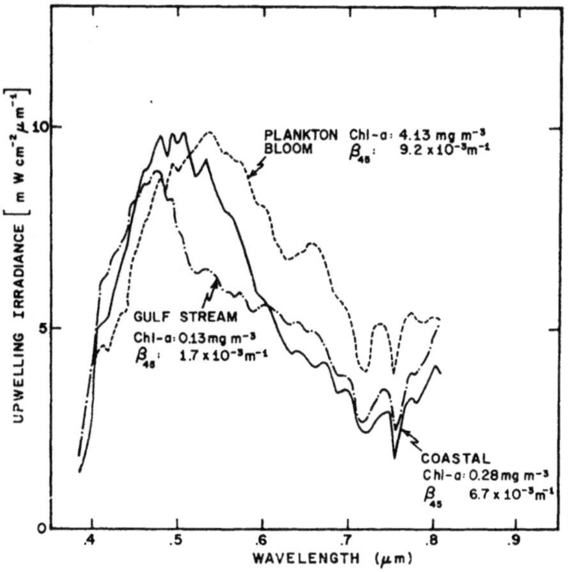


Fig. 2. Spectra of upwelling irradiance taken from 4 m above the sea surface in the Straits of Florida during 1—4 November 1972. The solar zenith angle was about 30°; sea state, wind speed, percent cloudiness, and downwelling skylight were approximately the same. The volume scattering function $\beta_{45}$ is for blue (0.436 μm) light.

5, 6, and 7 respectively. For the same surface signal, contrast is reduced by a factor of almost 2 when viewing in MSS 4 as compared with 5. Many authors report most satisfactory oceanographic results using MSS 5 for this reason.

## 2. Significant Results

In the sections that follow, the work of several contributors to the Symposium on Significant Results obtained from ERTS 1 [3] are summarized. The figures were given by the authors as representative of their work. Where results are not specifically from the symposium, the proper reference is given at the conclusion of the paper.

### 2.1. Sea Ice

Barnes reports that sea ice is detectable in all of the MSS bands and can be distinguished from clouds through a number of interpretive keys. Overall, MSS 4 and 5 appear to be better for mapping the ice edge, whereas MSS 7 provides greater detail in the ice features. Fig. 3 is imagery showing sea ice along the east coast of Greenland on 25 September 1972. The left-hand image is MSS 4 and the right-hand image is MSS 7. Differences in the reflectance in the two spectral bands are believed to be associated with different ice types. Areas of probable brash or rotten ice have a considerably higher reflectance in MSS 4 than in MSS 7; however, ice floes that are difficult to distinguish in MSS 4 have a significantly higher

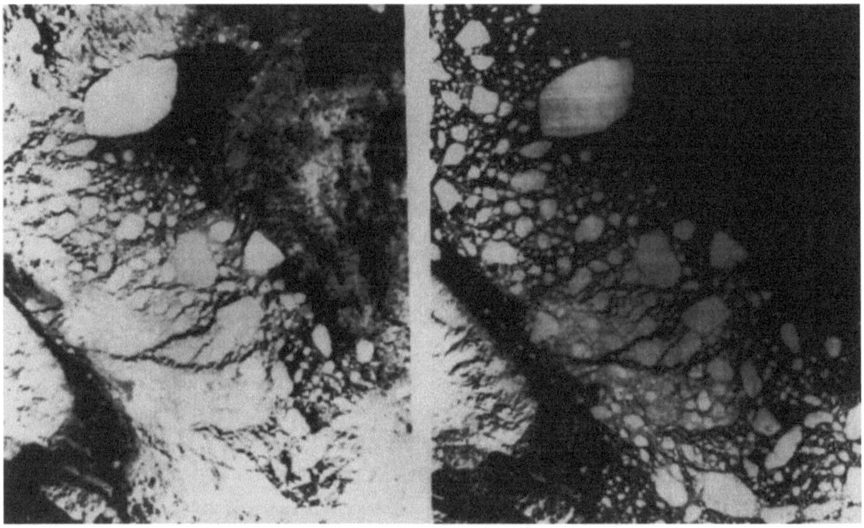

Fig. 3. Sea ice detected in MSS 4 (left) and MSS 7 (right). The larger floes in the image are 20—30 km across (after Barnes [3]).

reflectance than the surrounding brash ice in MSS 7. Many surface features on ice floes can be seen in MSS 7; this is caused by the high absorption of liquid water in the near infrared (MSS 7).

### 2.2. Fisheries

Kemmerer and Benigno studied the application of spacecraft data to the menhaden fisheries in the Mississippi Sound region of the northern Gulf of Mexico. The goal here is to improve fisheries by locating areas of potentially high yield. MSS 5 proved to be most useful in these efforts. Multiple regression analysis of oceanographic variables collected during satellite transit showed that water depth and turbidity had statistical significance with fish catch data; this is confirmed by Maughan, Marmelstein, and Temple. Fig. 4 is a false color density slice of MSS 5, taken on 5 August 1972. In this technique a color is assigned to each of several discrete photographic densities (gray shades). The black dots are areas of fish school locations observed during transit. Each color represents a narrow density range within the range containing the fish school locations; the fish

→

Fig. 4. Location of menhaden schools in the Mississippi Sound are shown by black dots. The false color density slice of MSS 5 shows that the fish locations are just in, or adjacent to, regions of high turbidity (orange). Horizontal scale is approximately 50 km (after Kemmerer and Benigno [3]).

Fig. 4

schools fall either adjacent to or within this narrow range. Szekielda and Curran report correlation with ocean chlorophyll measurements of the ratio MSS 4/MSS 5; the higher the ratio the more chlorophyll. This is not supported by the data of Fig. 2 ∅.

## 2.3. Charting

Shoreline information needed to update nautical charts and maps is emphasized on MSS 7, whereas shallow bathymetric features are available in MSS 4 and 5. Fig. 5 is from Williams and shows images in MSS 4, 5, and 7 and a conventional map of the Cape Cod area on 1 September 1972. There are many inaccuracies in the portrayal of the coastline and inland lakes when compared with the imagery. The discrepancies are caused by cartographic generalization from larger scale maps and, in some instances, actual changes in landforms. The imagery provides a very cost-effective method for increasing map accuracy, portraying additional environmental information, and for frequent updating at scales of 1:250000 or smaller. Polcyn has obtained correlations over the Bahamas banks of depth measurements at 2, 3 and 8 m; a mathematical model for depth measurements using ratios of MSS 4 and 5 has been developed. Ross found that where reflective bottom and clear water are found, MSS 4 images can be used with density contouring for estimating depths in 2, 5 and 10 m steps. These findings may provide significant data for assisting in the verification of navigation hazards in remote areas as well as updating other chart products.

## 2.4. Beach Erosion

Slaughter has been applying ERTS aircraft photographs to determine the relationship of existing groins. This is being done to determine if further groin construction is warranted. Kerhin has been using similar imagery for recognition of beach ridges and sand waves. The ERTS imagery has marginally useful resolution for the features themselves; however, indirect information, such as sediment transport patterns inferred from the imagery (see below) will prove useful in this regard.

## 2.5. Limnology

Strong made an observation of an algae bloom on Utah Lake on 12 September 1972. Fig. 6 shows an anticyclonic gyre in this shallow (ca. 3 m) and highly entrophic lake. The imagery is an example of color infrared image reconstruction. MSS 4, 5 and 6 were registered and projected on the same screen through blue, green and red filters respectively. The algae are highly reflective in MSS 6 and appear red on the color composite. Utah Lake is known for its high turbidity, and during summer months algae blooms frequently follow calm periods. Brigham Young University conducted a limnology survey on the day of satellite transit and found an extensive algae cover, mostly *Anabeana*, that was several centimeters thick.

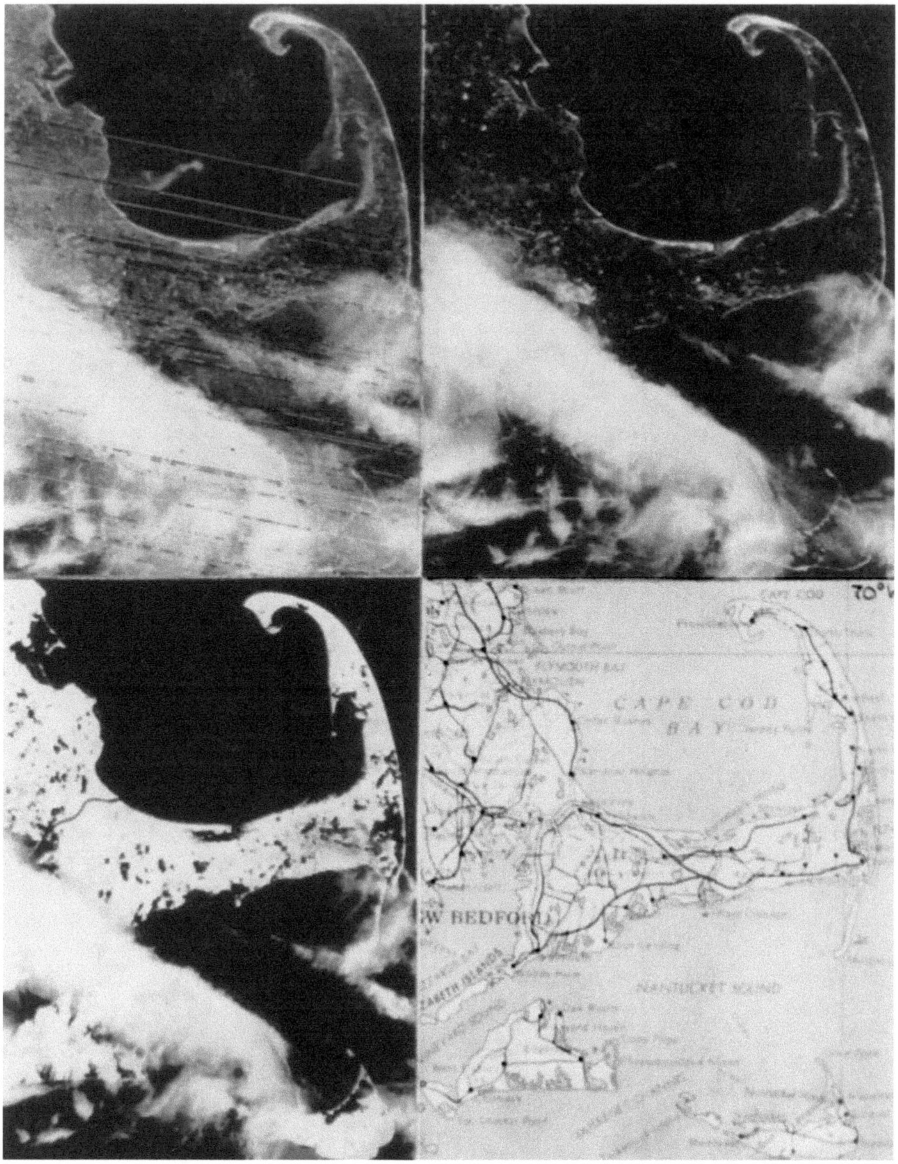


Fig. 5. MSS 4 (upper left), MSS 5 (upper right), MSS 7 (lower left) and a conventional map of the Cape Cod area. Horizontal scale on the images is approximately 80 km (after Williams, EROS Program, USGS).

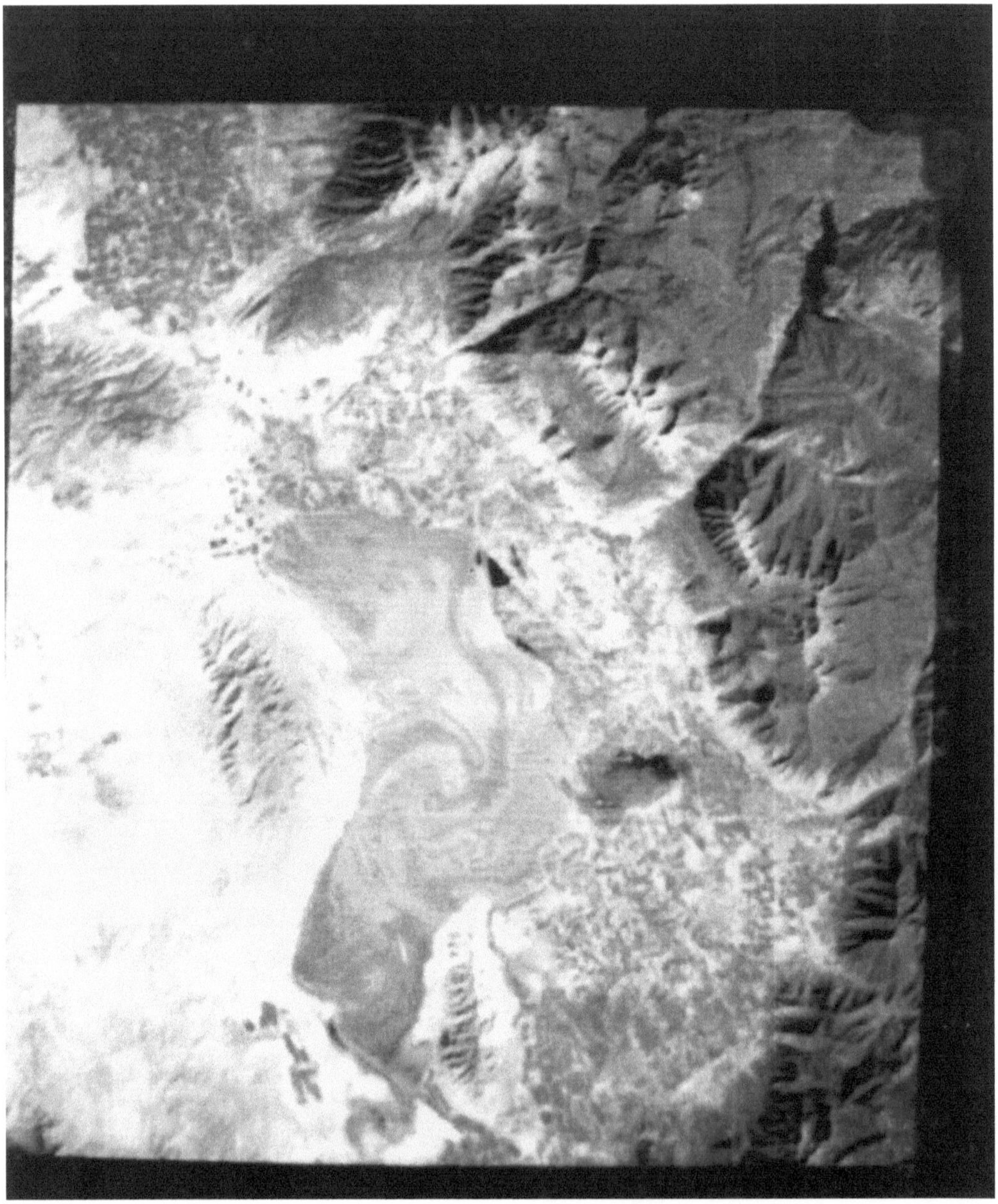

Fig. 6. Color infrared recomposition of MSS 4 (blue), MSS 5 (green), and MSS 6 (red) over Utah Lake. The area of the lake is about 375 $km^2$ (after Strong [3]).

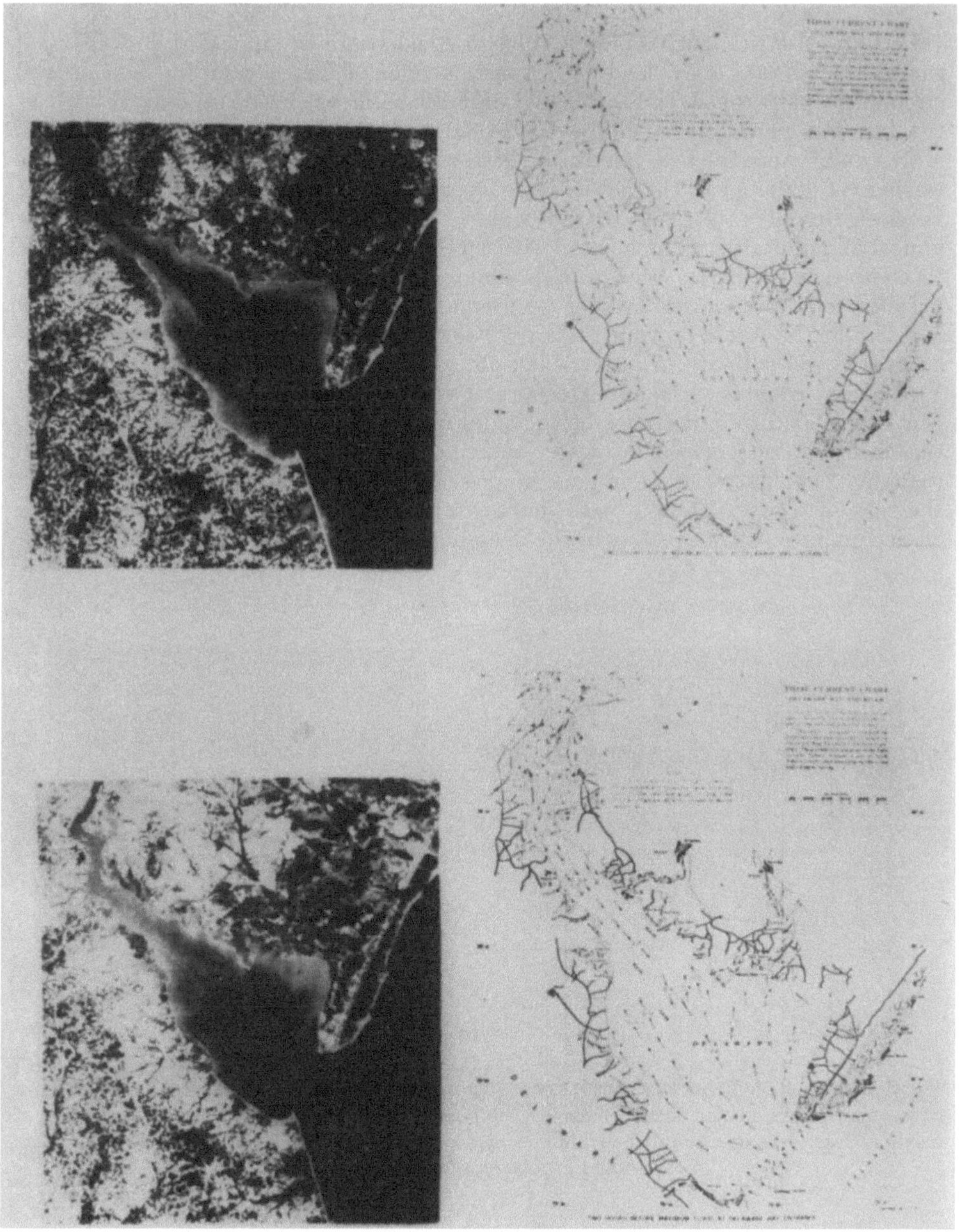

Fig. 7. Two views of Delaware Bay at different stages of the tidal current cycle: two hours before maximum flood at the bay entrance (upper) and two hours after maximum flood (lower). These are compared with conventional National Ocean Survey (NOAA) tidal current charts (after Klemas, Srna, and Treasure [3]).

## 2.6. Estuarine Circulation

Certain estuaries, such as Cook Inlet in Alaska, are so large that conventional survey techniques may lead to erroneous conclusions concerning the overall circulation patterns. Wright, Sharma and Burbank used natural tracers such as the fine sediment (glacial flour) from deglaciation to study water movements. MSS 4 and 5 were the best bands for these waters where suspended sediment loads range over 400 mg $l^{-1}$. The ERTS imagery was useful in identifying the boundary between fluvial waters and marine waters. This boundary is important because almost all pollutants are associated with turbid waters; further, location is valuable to commercial fisheries because fish congregate there and for navigation safety because logs and other flotsam accumulate at such boundaries. Several new features of the water movements have been identified from ERTS imagery with surface vessel support. Another example of application to estuaries is given in Fig. 7 from Klemas, Srna and Treasure. Two views of Delaware Bay are given in this figure at different stages of the tidal cycle. The images are MSS 5 showing the clear (darker) near-shore water entering the bay; ebbing waters have high turbidity. The patterns are in general agreement with the tidal current charts for the time of transit. Thus, in this shallow estuary natural tracers can be used for environmental studies just as in the deep regime of the fjord-like Cook Inlet.

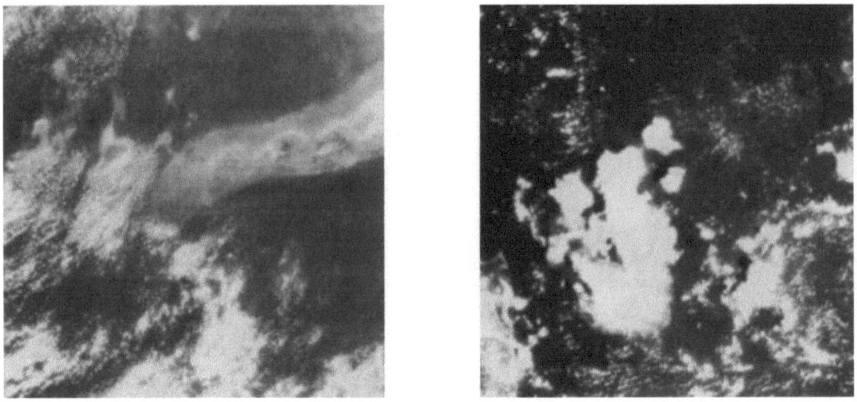

Fig. 8. The cyclonic boundary of the Loop Current in the Gulf of Mexico seen in MSS 4 as a consequence of the color change between the coastal water and the current (left), and as a result of higher sea state and hence higher reflectance (right) as seen in MSS 5. Horizontal scale is 180 km in each scene (after Maul [3]).

→

Fig. 9. Low reflectance features in the lee of the Antilles appear brighter on this negative print mosaic. Sketch to the right outlines the features in an area of known eddies (after Hanson [4]).

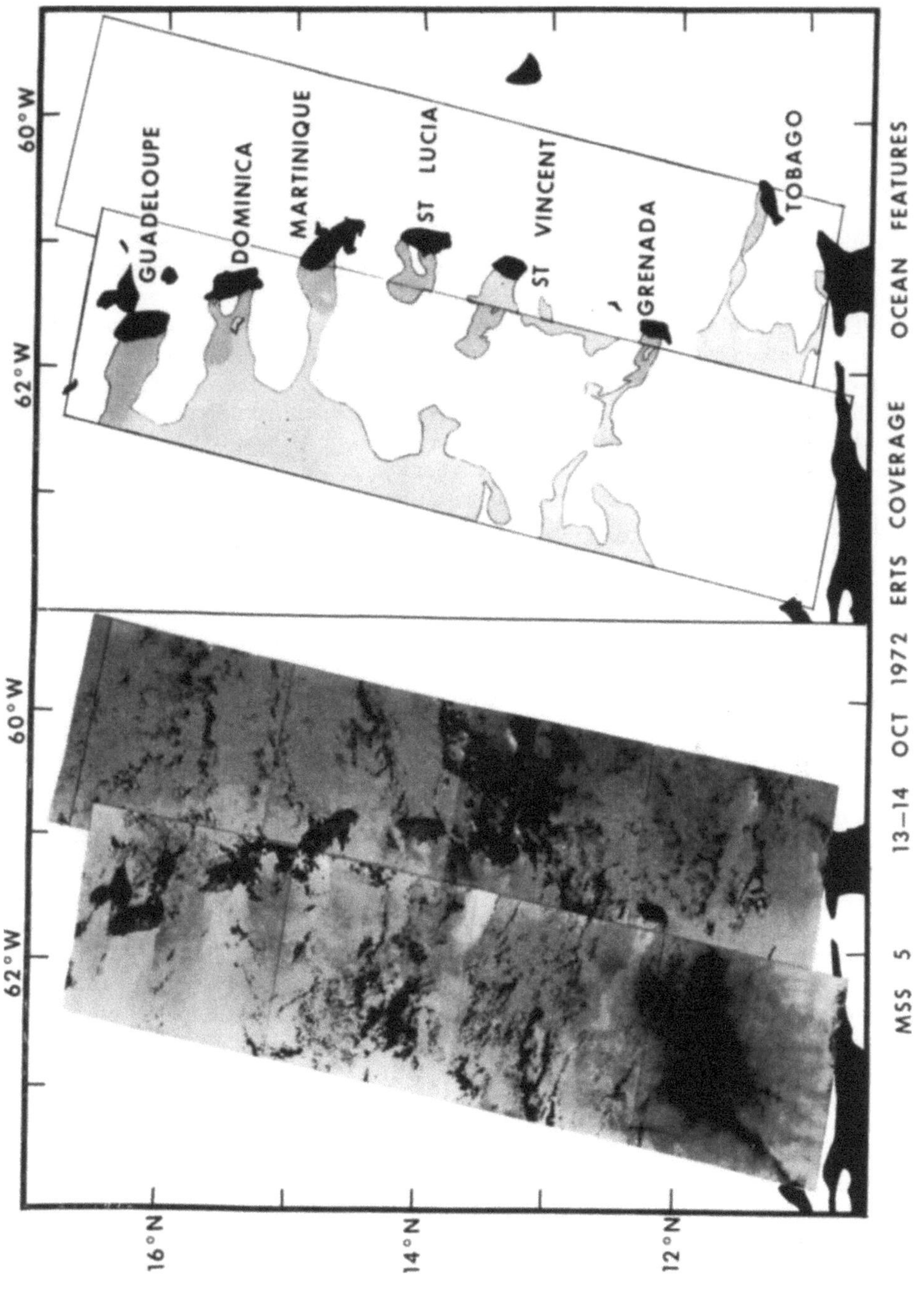
60°W
62°W
60°W
62°W
GUADELOUPE
DOMINICA
MARTINIQUE
ST LUCIA
ST VINCENT
GRENADA
TOBAGO
16°N
14°N
12°N
MSS 5 13–14 OCT 1972 ERTS COVERAGE OCEAN FEATURES

## 2.7. Ocean Currents

Detection of baroclinic flows in the subtropics is not possible on a year-round basis by infrared techniques because the sea surface becomes seasonally isothermal in many areas. The location of water masses in hurricane-prone areas such as the Gulf of Mexico is important to transportation, fisheries, and weather forecasting. Maul is making cruises every 36 days into the Gulf to locate the Loop Current by ship and satellite. The findings to date have been several-fold: The edge of the current is an accumulation zone for surface algae and phytoplankton which frequently delineate the boundary by high reflectance in MSS 6. In addition a change of sea state frequently occurs, steeper waves and more whitecaps occurring in the current. Fig. 8 shows how the cyclonic edge of the current has been detected by ERTS supported by concurrent surface observations. The left-hand panel (MSS 4) shows the greener (lighter shade) water of Florida Bay delineating the edge of the bluer (darker shade) Loop Current. This is an expected consequence of analysis of the data in Fig. 2. The right-hand panel (MSS 5) shows the current appearing brighter than the surrounding waters. Maul interprets this as the overriding influence of whitecaps and foam in the higher sea state of the current and suggests that a sea state correction measurement is necessary in order to quantify any ocean color measurements. Alternately, the possibility exists that the MSS observations could be used to estimate sea state.

## 2.8. Ocean Features in Island Lees

Hanson, Hebard and Cram are studying the application of ERTS data to locating eddies in the lee of the Antilles. These eddies are known zones of high productivity and possibly good tuna fisheries. Fig. 9 is a mosaic of images from two consecutive days. These are negative prints so that the lighter appearing features to the west of the islands are regions of low reflectance. Note also the rapid change in these features, particularly in the lee of St. Vincent where the one-day delay in adjacent transits produces a remarkable change in the imagery. Hanson [4] has shown that these features align with the local wind and may be caused by a modification of the sea and/or swell in a wind shadow of the island. The features are seen in all bands and hence are surface phenomena but have not been confirmed in ground truth cruises to date. A firm conclusion as to the cause of these features has not been reached nor has the impact of their existence on the environment been analyzed.

## 2.9. Internal Waves

Apel [5] has interpreted the wavelike features of Fig. 10 as surface manifestations of internal waves. The patterns on this image are in agreement with refraction patterns computed for internal waves of 1000 m wavelength for the bathymetry in the southeastern New York Bight. The left-hand panel is a contrast stretched computer enhancement of the original scene (right-hand panel) by Maul et al. [6]. The contrast stretching technique is to assign all values above a certain cutoff on the MSS digital tapes to the highest exposure and all those below another threshold to the lowest exposure. Thus, only the radiance steps of interest are

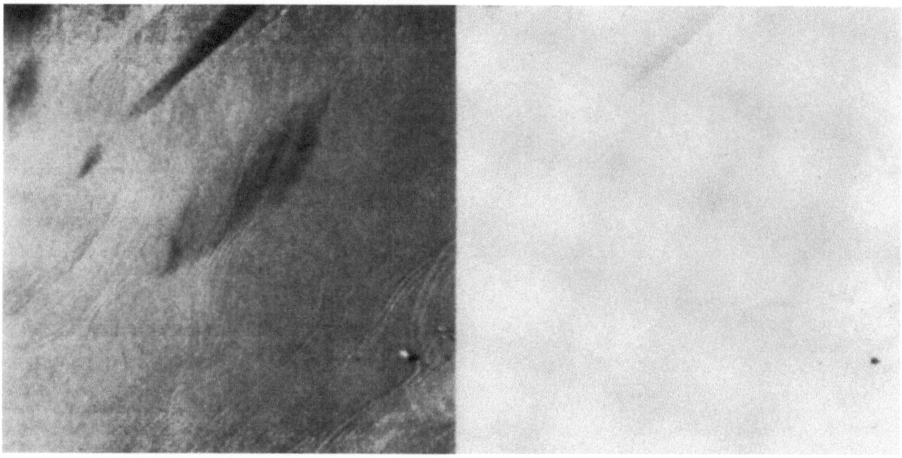

Fig. 10. Contrast stretched computer enhancement (left) of patterns interpreted by Apel [5] to be surface manifestations of internal waves impinging on the continental shelf off New York. Horizontal scale in the enhancement by Maul et al. [6] is about 60 km.

assigned the intermediate gray levels, and many significant portions of the ocean scene are revealed. In this case the surface reflection patterns generated by the slick water/rough water patterns, typical of these waves, are markedly enhanced.

## 3. Conclusions

ERTS has proven itself to be a valuable tool to oceanographers in many disciplines. In several cases it has provided the first repetitive synoptic view with high spatial resolution ever obtained. Observations have been made of the effect man has on his environment [1], and in this way the data have focused attention on new or emerging problems. The problem of scientifically correlating these data with additional surface measurements and optimizing multi-spectral pattern recognition lies ahead. Indeed the labor and the reward of ERTS are just beginning to surface.

## References

[1] R. L. Charnell and G. A. Maul, Nature **242**, 451 (1973).
[2] S. Chandrasekhar, Radiative Transfer, Oxford University Press, London 1950.
[3] NASA SP-327, Symposium on Significant Results obtained from ERTS 1 Vol. I (B), 1973.
[4] K. J. Hanson, personal communication 1973.
[5] J. R. Apel, personal communication 1973.
[6] G. A. Maul, R. L. Charnell and R. H. Qualset, paper presented at Amer. Soc. Photogrammetry Fall Convention, Orlando, Fla, October 1973.

# Contributed Papers

Earth Survey Problems — Akademie-Verlag, Berlin 1974

# MAPPING OF SNOW COVER IN THE SWISS ALPS FROM ERTS 1 IMAGERY

H. Haefner[a], R. Gfeller[a] and K. Seidel[b]

[a]Department of Geography, University of Zürich, Zürich, Switzerland
[b]Department of Photography, Swiss Federal Institute of Technology, Zürich, Switzerland

## Abstract

A semi-automated mapping system was developed by combining the possibilities of a Quantimet 720 for the individual discrimination of the critical density level in each image with the advantage of high resolution photographic techniques for the execution of density slicing. The visually determined density level on the monitor is used as reference time to expose a lith-film. The snow cover will be transferred onto a special topographical map 1:500000, but for this procedure SCC imagery is necessary. Investigations were undertaken on the accuracy of the areal measurements in the different MSS bands. The results show that image-series with an old and melting snow cover differ greatly in the reproduced areas, whilst image-series taken shortly after a new snowfall produce more or less the same results. This opens a possibility for a further characterization of the surface of the snow cover by using all four bands in combination. For areal measurements of the snow, band 5 is suggested. Additional investigations are necessary to achieve a sufficient accuracy and to overcome the problems of geometric as well as photometric distortions.

## 1. Problem

The development of an automated mapping system of the snow cover in high mountain areas and the estimation of its areal extent and changes from satellite imagery is handicapped by the following problems: separation of clouds from snow, influence of mountain shadows (differing in each image-series), influence of various exposures and steepness of slopes, radiometric variations in different image-series, variations in the quality and the characteristics of the snow cover, snow cover in forested areas. Because of these many changing variables it is extremely difficult to develop an automated mapping system based only on density measurements and to achieve measuring results of sufficient accuracy. We agree with Meier [1, p. 2] that for the time being the interpretation still depends mainly on the human eye and brain control, but can be rationalized and improved tremendously by using specific instruments.

It is the purpose of this paper to report on the first activities and results with ERTS 1 imagery to develop such a semi-automated mapping system and on investigations regarding the accuracy of the areal measurements using images of different bands.

## 2. Measuring Device

For a quick but controllable separation of snow-covered areas by means of density slicing of a part or an entire ERTS image a Quantimet 720 (QTM 720) was used. With this instrument an electronic separation of the grey tones is possible and areas within an adjustable density range can be measured and registered in digital form. The ERTS image is displayed on a monitor via a high resolution TV camera (720 lines). Parallel to this display each of the 500000 picture points is classified according to its grey tone and in respect to the adjusted tone level. The results are also visible on the monitor. All picture points below the tone level appear in "white", the ones above in "black" (or vice versa). The "white" picture-points are counted and the ones within a given frame are indicated in digital form. The output of a picture is reproduced in 0.1 sec. The tone level separating the "black" and the "white" picture points can be varied continuously with a potentiometer, and is calibrated in optical densities with a standardized wedge.

Each ERTS image is reproduced on the monitor and the tone level adjusted by visual control so that it corresponds best with the snow/no-snow boundary. With this method it is possible to overcome tonal differences of the snow from one image to the other. By using masking techniques selected areas such as a watershed of a river, a mountain valley etc. can be isolated.

## 3. Photographic Techniques

The 720 lines of the QTM 720 do not provide a sufficient resolution to analyse an entire ERTS image of 2340 scan-lines with the desired accuracy. For an exact analysis it would be necessary to divide the picture in at least 9 or even 16 equal squares and to reproduce one after the other on the monitor. To avoid this procedure and to reach a separation of the snow cover on an entire ERTS image with sufficient accuracy, the visual controlled discrimination with the QTM 720 is combined with photographic techniques. The tone level is adjusted to the snow/no-snow boundary as described for each image individually. The calibrated position of the potentiometer then provides the corresponding density value, which is used as reference time to expose a lith-film. The developed high resolution film contains only two density steps in black and white, representing the snow- and ice-covered and the snow-free areas of an entire ERTS image in all its detail (Fig. 1).

From this clear tonal separation into "black" and "white" it is easy to establish the areal extent of the snow cover and the snow line altitudes by an exact measurement of the snow cover, if necessary for especially defined regions and a transference into a topographic map 1:500000 with contour lines to determine the exact position of the snow line altitudes for different regions, exposures etc. In addition, the different photographic tone separations of all four MSS bands can be processed further additively or subtractively for a better understanding of the differences as received in the four bands (§ 4).

The combined electronic-photographic method can be summarized by the following steps: (i) presentation on the QTM 720 monitor and tonal discrimination of the snow for each image individually by visual control; (ii) improvement of the accuracy by taking the calibrated potentiometer position as reference for the exposure time of a high resolution lith-film; (iii) photographic density slicing on the lith-film of the snow-covered from the snow-free areas; (iv) transfer on to specific topographic maps 1:500000.

## 4. Measurements in Different MSS Bands

Careful measurements of the snow cover in all four MSS bands of the same image-series with the method described gave some interesting results: Sometimes the measured areas of the same image-series differ rather substantially from each other. Band 4 then always produces the largest snow cover, slightly larger than band 5, and there is a continuous decrease in bands 6 and 7. A characteristic

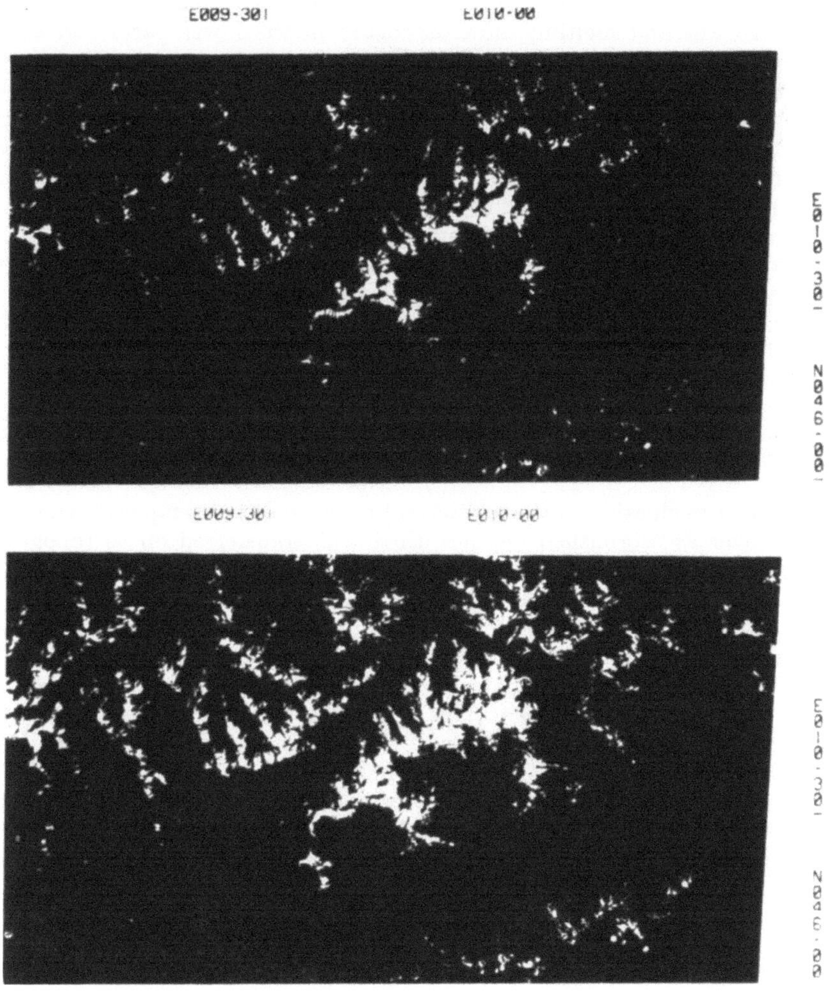

Fig. 1. Discrimination of snow cover from ERTS-1076-09442 of 7 Oct. 1972, on lith-film in MSS band 5 (below) and 7 (above).

example is the ERTS 1 image 1076-09442 of 7 October 1972, of which the same sections from band 5 and band 7, reproduced on lith-film, are demonstrated in Fig. 1. The differences in the determined snow covers are obvious. On the other hand there are imageseries which do not show this phenomenon, e.g. image 1093-09384 of 24 October 1972.

A comparison with weather maps and ground observation of the snow cover indicates a strong correlation with the date of the latest snowfall. Image-series shortly after a new snowfall do not have the differences in the areal extent between the four bands whilst series with extreme differences always occur after a longer period of fair weather. The respective dates for the two examples given are: for image 1076-09442 of 7 October 1972 the preceding snowfall was on 25 September 1972 and for image 1093-09384 of 24 October 1972 on 23 October 1972.

It has to be concluded that we have to distinguish between an "old snow cover" with relative wet and melting snow especially in the lower parts and a "new snow cover" of relatively dry snow. Further investigations along this line are needed in combination with spectrophotometric measurements and measurements of the temperature gradient on the ground before final conclusions may be drawn as to whether this phenomenon can be related to such criteria as snow types, snow depth, surface temperature, water content etc. There seems to be possibility of obtaining additional information on the characteristics of the surface of the snow by using all four bands in combination.

For mapping the areal extent of the snow cover band 5 produces the best results even if the surface is sometimes a little smaller than in band 4. But discrimination in band 5 in general is more accurate and easier to accomplish because the snow boundary is better visible, whilst in band 4 haze often disturbs a clear tonal separation.

## 5. Determination of the Position of Snow Line Altitude

For an exact evaluation of the altitudinal position of the temporary snow lines the snow cover as established on the lith-film is transferred on to topographic maps with contour lines. For this purpose special maps were produced. The official topographic map 1:500000 has only contour intervals of 200 meters and does not show the hydrographic features in sufficient detail. Therefore a special map was constructed from the official topographic map 1:100000 containing contour intervals of 50 meters and the hydrography in more detail as the only features, and then was reduced to 1:500000. This again allows mapping of the distribution of the snow with the desired accuracy needed for good resolution of the ERTS system. Experiments with SYC imagery showed that only rather limited areas could be transferred with sufficient accuracy and that it was difficult to eliminate distortions. SCC images will be of great advantage for this purpose and therefore we decided to not proceed but to wait with this procedure until the geometrically corrected imagery is available.

## 6. Conclusions

Mapping the snow cover with ERTS imagery still cannot be done on an automated basis but depends mainly on the visual control. With a combination of a Quantimet and a photographic technique the accuracy as well as the rapidity of the mapping procedure could be improved substantially. Nevertheless there are still many problems to be solved and the results achieved do not satisfy us regarding the following:

*Geometrical accuracy:* Even if the snow cover can be separated clearly, the accuracy is not sufficient for high mountain terrain with rough relief, steep slopes etc. Instead of the measurement on a map in an orthographic projection we would prefer to know the extent of the real surface of the terrain.

*Interfering objects:* Shadows, clouds etc. and other interfering objects can be eliminated when obvious, but it is rather difficult and very time consuming to reach a total elimination and especially to include in a measurement all snow-covered areas which are hidden by mountain shadows. In addition the tonal presentation of snow in sun-exposed positions and in north-oriented positions varies greatly. Even when using masking techniques and stepwise discrimination the measuring results still lack the desired accuracy.

The method developed is therefore appropriate for an immediate and quick look investigation fo the changes of the snow cover but does not solve the problem of an automated mapping system in the long run. This has to be developed on a digital basis considering a digital terrain model, variation of illuminance, sun-angle, shadow-effect etc. to correct the photometric and geometric differences.

## 7. Outlook

The method described will be applied to map and inventory the seasonal fluctuations of the snow cover in the Alps and Pre-alps of Switzerland during a yearly cycle and to establish the position and variation of the temporary snow line altitude for various climatic regions, exposures etc. The data will then be correlated with ground measurements of snow depth, surface runoff and meteorological data to develop a model for surface runoff prediction.

## Acknowledgment

Appreciation is expressed to NASA Goddard Space Flight Center, Greenbelt, Maryland, USA, for enabling us to participate in the ERTS 1 program; to Prof. Dr. W. F. Berg, Head, Department of Photography, Swiss Federal Institute of Technology, Zürich, for his great support, and to "Jubiläumsspende für die Universität Zürich" for funding the measurements with the Quantimet.

## Reference

[1] M. F. Meier, in: Symp. on Significant Results obtained from ERTS 1, NASA Goddard Space Flight Center, Md, Paper W 19 (1973).

Earth Survey Problems — Akademie-Verlag, Berlin 1974

# REMOTE SENSING TECHNIQUES FOR SUPPORT OF COASTAL ZONE RESOURCE MANAGEMENT

R. O. Piland and Staff of Earth Resources Laboratory

NASA Manned Spacecraft Center, Houston, Texas, USA

Many natural resources exist in coastal zones, including waterways for commerce, waters for fisheries, marshes for marine-life breeding and wildlife habitat, and beaches and water for recreational purposes. These resources lead to economic development and population pressures which can result in damage or destruction of the natural resources unless careful management is exercised. The planning and management process requires a variety of information, and remote sensing techniques, involving aircraft and satellites, can contribute. In support of land use planning a simple and rapid procedure has been developed for the preparation of 1:24000-scale land use maps and associated inventories using high altitude aircraft imagery. A comparative analysis has been conducted utilizing ERTS 1 imagery. An automated system for classification of vegetation in the nearly inaccessible marshes has been demonstrated and applied using aircraft multi-spectral scanner data and pattern recognition techniques. The critical water parameter, salinity, has been measured remotely using L-band passive microwave radiometry. Earlier infrared techniques have been extended to provide automatically prepared surface temperature maps with atmospherically corrected quantitative values derived from calibrated thermal scanner measurements.

## 1. Introduction

The coastal areas of the world represent a significant natural resource. They provide waterways for commerce, water bodies for fisheries, marshes for marine life breeding and wildlife habitat, and beaches and waters for recreational purposes. These characteristics lead to economic development and increases in population whose requirements may result in damage or destruction of the natural resources unless careful management is exercised. For example, land with suitable characteristics for industry, businesses, and residences is limited by the water boundary on one side and on the other by the requirement to be reasonably close to the water. Consequently, land in the coastal areas is generally in short supply. The land shortage often leads to the draining and filling of marshland in order to provide for the needs listed above. This reduction in marshland potentially reduces our marine resources since it is a prime breeding area for such resources. Man-made changes such as dredging or channelization can adversely affect the environmental conditions, e.g. salinity, necessary for certain marine resource well being. Untreated effluents from industrial and residential communities can obviously have detrimental effects.

Consequently there is a tension between the need for development and the need for preservation of natural conditions. It is critical therefore that the planning and management process be handled skillfully in coastal areas to effect an optimized balance between the forces of development and preservation.

The planning and management process requires a variety of information on which to base its actions. It is reasonable to assume that remote sensors located in aircraft and satellites could provide a portion of this information. Consequently a number of remote sensing studies have been conducted to develop and/or demonstrate techniques for use in land use inventory, marsh vegetation classification, and water characteristics surveys. The remainder of this paper summarizes the results of these studies. Additional detail is provided in the reference material listed in the bibliography.

## 2. Geographical Test Area

The remote sensing techniques discussed herein are intended to be applicable over a wide range of coastal areas. The specific studies were conducted in an area known as the Mississippi Gulf Coast, which is about eighty miles in length and extends from the City of Mobile on the east to New Orleans on the west. The coastal political units include Hancock, Harrison, and Jackson Counties of the State of Mississippi. They encompass approximately 1800 square miles (4670 $km^2$) and contain a population of 200000 people.

The westernmost County of Hancock is least developed and populated with only 10% of the 200000 coastal population. Harrison, in the center of the three counties, is developing rapidly based on tourism, shipping and light industry. Jackson, the easternmost county, is also in a period of growth with a concentration of shipbuilding and petro-chemical industries. All three counties derive significant income from fishing and the associated processing industry. A goodly portion of the eighty miles of coast consists of sand beach, much of which has been developed by pumping sand from offshore. The coastal area includes 100 square miles of salt marshes with their relevance to marine resources. The coastal water body is the Mississippi Sound which extends from Mobile Bay in the east to Lake Pontchartrain in the west. It is bounded by the coast and a series of barrier islands located some ten miles offshore. The passes between these islands allow the interchange of waters between the Sound and the Gulf of Mexico. The Sound is relatively shallow, averaging 12 ft (3.66 m) in depth. It is fed by a number of rivers including the Pearl, Pascagoula, and Biloxi. Its dynamics, however, are controlled primarily by the tidal conditions of the Gulf of Mexico.

The land use studies have been conducted in the three coastal counties and illustrative results will be presented for the Harrison County area. The marsh vegetation studies were conducted in a marsh at the western end of the test area, and the water characteristics studies have been conducted in the Mississippi Sound and adjacent Biloxi Bay which empties into the Sound.

## 3. Land Use

### 3.1. High Altitude Imagery

In order to plan the use of the land it is necessary to have a knowledge of how the land is presently being utilized. Consequently, a basic tool of the planning agency is a land use inventory. For a number of years planning agencies have made use of aerial photography as an aid to the preparation of land use maps and inventories. This photography was generally black and white and obtained at altitudes of typically 12000 ft. Land use maps of 1:24000 scale were then derived from the photography and associated surface measurements. While the use of aerial photography represented a considerable improvement over the use of surface information alone, significant effort was expended in the matching and mosaicing of the large number of photographs required. For example, a photograph taken from 12000 ft with a 6 inch focal length camera would encompass a field of view or area of 3 miles by 3 mile, or 9 square miles. Photography of the three coastal counties encompassing some 1800 square miles would require 300 to 400 photographs which would have to be mosaiced, or at least located on a base map, in order to utilize the information contained on the photographs. A further limiting consideration was the fact that each frame was taken under a different illumination condition resulting in different shading of the images. This low altitude also resulted in significant geometric distortion away from the center of the image.

A study was conducted to demonstrate the value of utilizing high altitude imagery (60000 ft) and color infrared film in the preparation of land use maps as might be required by coastal regional planning agencies. It was intended further to develop and document a detailed procedure for the preparation of land use maps and inventory information using existing techniques such that the task could be accomplished with a minimum of time, skill, equipment and cost.

The imagery utilized in the demonstration was obtained from an altitude of 60000 ft utilizing an RC-8 aerial mapping camera mounted in an RB57F aircraft. A typical frame of this imagery is presented in Fig. 1. The area represents a 15 mile by 15 mile square in the vicinity of the City of Gulfport in Harrison County. The large area (225 square miles per frame) coverage obtained from the high altitude imagery results in a requirement for only 15—20 frames to cover three counties as compared with 300—400 frames for the 12000 ft altitude condition discussed earlier. The scale of the high altitude photography is 1:120000.

A number of steps are required to obtain the land use information from the imagery. These steps are presented in detail by Vegas in [1] and only a few of the basic considerations will be treated here. For legal and related purposes most of the United States is divided into 6 mile by 6 mile townships. Consequently it was decided to use the township as the basic area for the investigation. Township lines may be located relatively easily on the imagery since their boundaries are often roads or are coincident with crop boundaries. After a given township is located within a frame it is marked off and blown up five times to achieve the desired 1:24000 scale. The geometry—range and section lines—are then placed on the enlarged photo by transference from existing topographic maps. A degree of inaccuracy results but this is not considered critical for land use planning. Conventional photo interpretation techniques are then applied to the original color infrared frames of photography using magnifying eyepieces or a 10 × stereo viewer and the resulting interpretations are noted directly on the enlarged black and white photo. A typical land use map, developed in this manner, is presented in Fig. 2. It was determined that 18—22 categories of land use, as listed on the figure, can be derived from imagery with this resolution. A map is developed for each township and combined into a county atlas. An additional step can provide an acreage compilation for each of the 22 categories for each township. The use of a simple random dot-grid counting technique produces results of acceptable

Fig. 1. High altitude (60000 ft) imagery of Mississippi Gulf Coast.

accuracy. While such a technique is not automated, its simplicity allows it to be done with personnel with a minimum of training.

A major consideration in land use mapping is the currency of the product. Too often land use maps, particularly in dynamic coastal regions, are outdated before completed. The present procedure allows an area of approximately 1800 square miles to be completed in two months by six workers with moderate training and skill.

A second major consideration is cost. Map production is costly and since the maps must be frequently updated they must be relatively cheap to develop and reproduce. The cost of developing the land use maps and associated inventory statistics for the three counties was approximately \$ 18000. The cost of developing similar products using low altitude imagery was estimated to be \$ 72000. The

Fig. 2. Township land use map developed from high altitude imagery.

cost of acquiring the imagery is not included in this comparison. Copies of the maps can be reproduced using Ozalid process copiers with a material cost of approximately 10 cents per sheet.

Updating the land use maps as new imagery becomes available can be accomplished at a small fraction of the initial cost.

### 3.2. Satellite Imagery

The study described in the preceding paragraphs has demonstrated the effectiveness of high altitude imagery for land use inventory. This effectiveness results from the combination of relatively large field of view and adequate resolution to classify those categories of land use pertinent to land use planning. It would appear reasonable to examine how imagery obtained at still higher altitudes from satellites with a correspondingly larger field of view may contribute to land use inventory.

Fig. 3 presents an image of the Mississippi Gulf Coast area as obtained with the ERTS satellite multi-spectral scanner on 19 January 1973. This image encompasses 10000 square miles of area and is approximately 100 miles by 100 miles

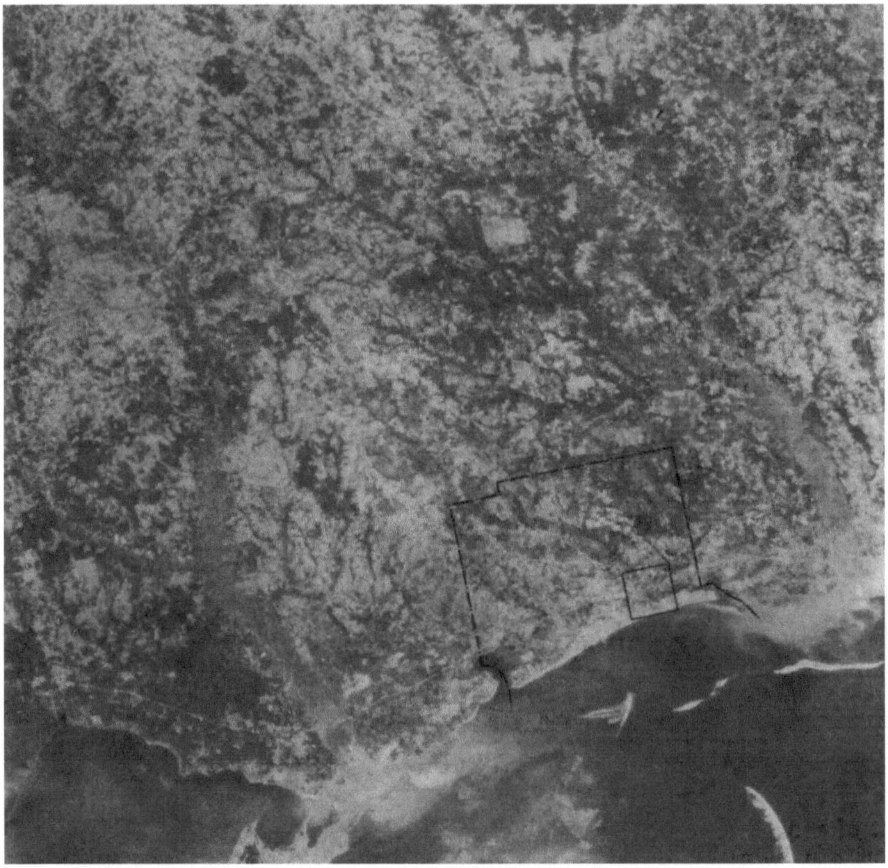

Fig. 3. ERTS imagery of Mississippi Gulf Coast.

on a side. This compares with the 225 square miles in the aircraft imagery previously described.

For reference, the township (6 miles by 6 miles) of Fig. 2 and Harrison County (700 square miles) are outlined on this frame. As stated above, the imagery from 60000 ft provided adequate resolution for land use inventory at the township level. While the satellite imagery increases the field of view, i.e. total area of interest contained in one frame, the resolution is decreased with a resulting decrease in the number of categories that can be identified and classified. To maximize our capability of photo interpretation the area of the frame encompassing Harrison County was enlarged. It was found that the image could be enlarged four to eight times to a scale of 1:250000 to 1:120000 with beneficial results. Photo interpretation techniques were applied to the enlarged imagery and it was determined that between 5 and 10 categories of land use could be delineated. It was further determined that the limited size of a township and the limited number of categories interpretable made it an unlikely size area for use with ERTS data. A land use map of Harrison County at a scale of 1:250000 is shown in Fig. 4 with a list of the categories that were determined in the initial investigation. This number will vary with the skill of the interpreter, his familiarity with the area, the character of the area, the degradation of the imagery due to clouds and haze, the season of the year that the imagery was obtained, and the processes by which the imagery was derived from the initial data transmission. Subsequent investigations have shown considerable increase in the number of categories that can be delineated. It is believed that the results presented here represent a baseline and that a greater number of categories can be obtained by optimizing the conditions listed above. Two particularly promising areas include the use of near "first generation" imagery and the use of two or several sets of sequential seasonal imagery over the same area.

The photo interpretation techniques discussed are available to a large community of users since there is no particular requirement for expensive or sophisticated equipments.

Considerable work has been done in the past decade directed toward automated classification of surface characteristics, e.g. land use, utilizing multi-spectral scanners and computer-based pattern recognition techniques. A number of such investigations are now underway utilizing the ERTS data. These techniques will be valuable to process the huge quantities of data that will be obtained from multi-spectral scanners operating over large areas in satellites or high altitude aircraft. The automatic classification techniques, utilizing aircraft mounted scanners, will also be particularly useful in smaller areas which have difficult access problems. Coastal marshes are an example of such a condition.

### 3.3. Marshland Ecology

Marshes are known to represent a valuable link in the ecological chain, with particular importance to the development of living marine resources. Man-made and natural effects such as draining and filling, and hurricanes and erosion obviously can affect the condition of the marsh. A change in salinity conditions is a typical result of such activities. To measure these changes directly with remote sensing techniques is difficult. However, the use of inference techniques offers considerable promise. The most visible part of the marsh is its vegetation. The vegetation conditions can provide the marsh ecologist with a variety of information about the marsh. Consequently it would appear that a most profitable approach for remote sensing would be to apply automated techniques to the identification and mapping of marsh vegetation.

A marsh at the western end of the Mississippi Gulf Coast, adjacent to Lake Pontchartrain, was utilized in this investigation [2]. A photograph of the marsh area and its location in shown in Fig. 5. Photo interpretation techniques would allow a degree of separation of the various species of vegetation; however, the use of pattern recognition techniques results in the classification map shown in

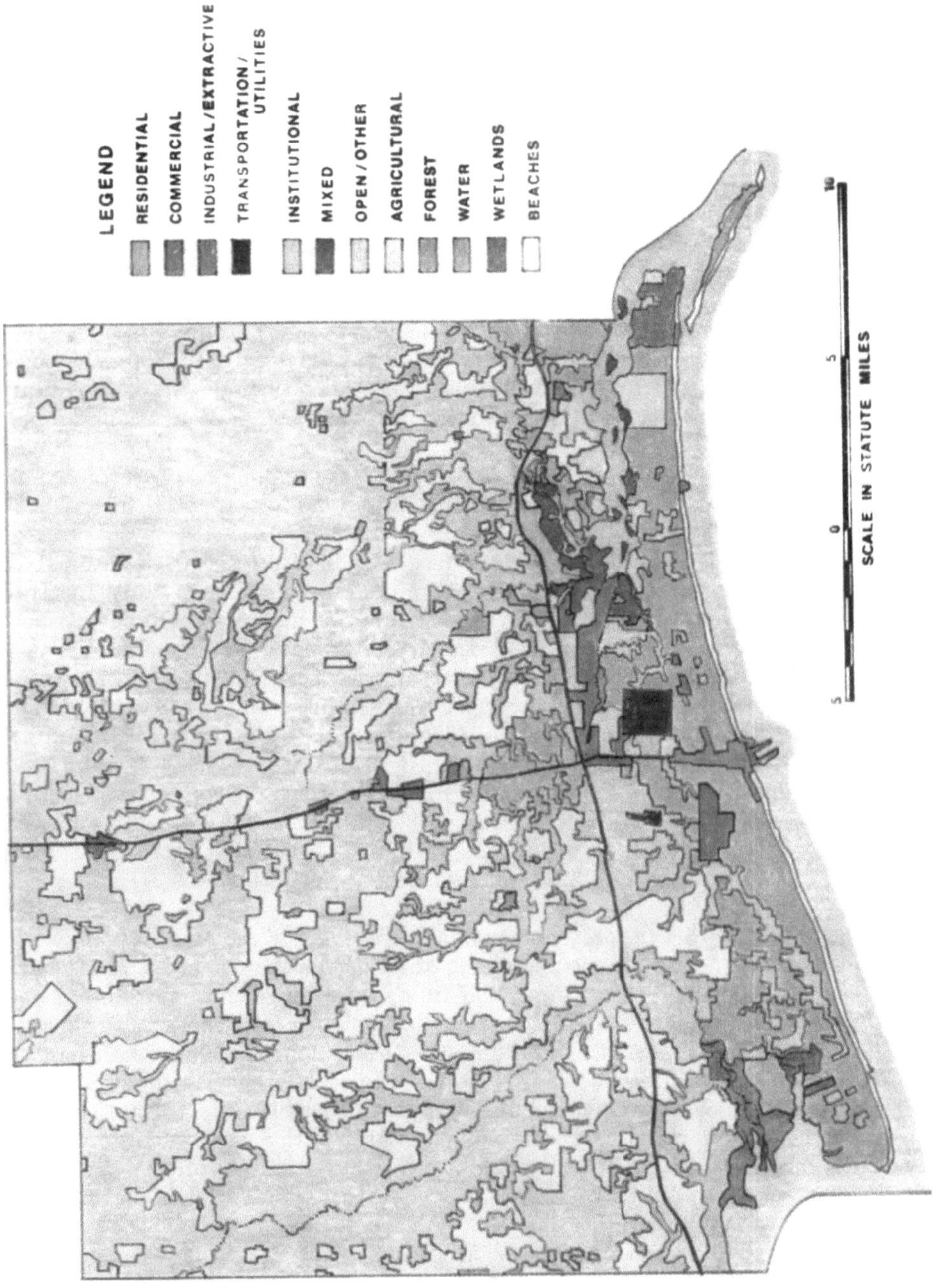


Fig. 4. Harrison County land use map developed from ERTS imagery.

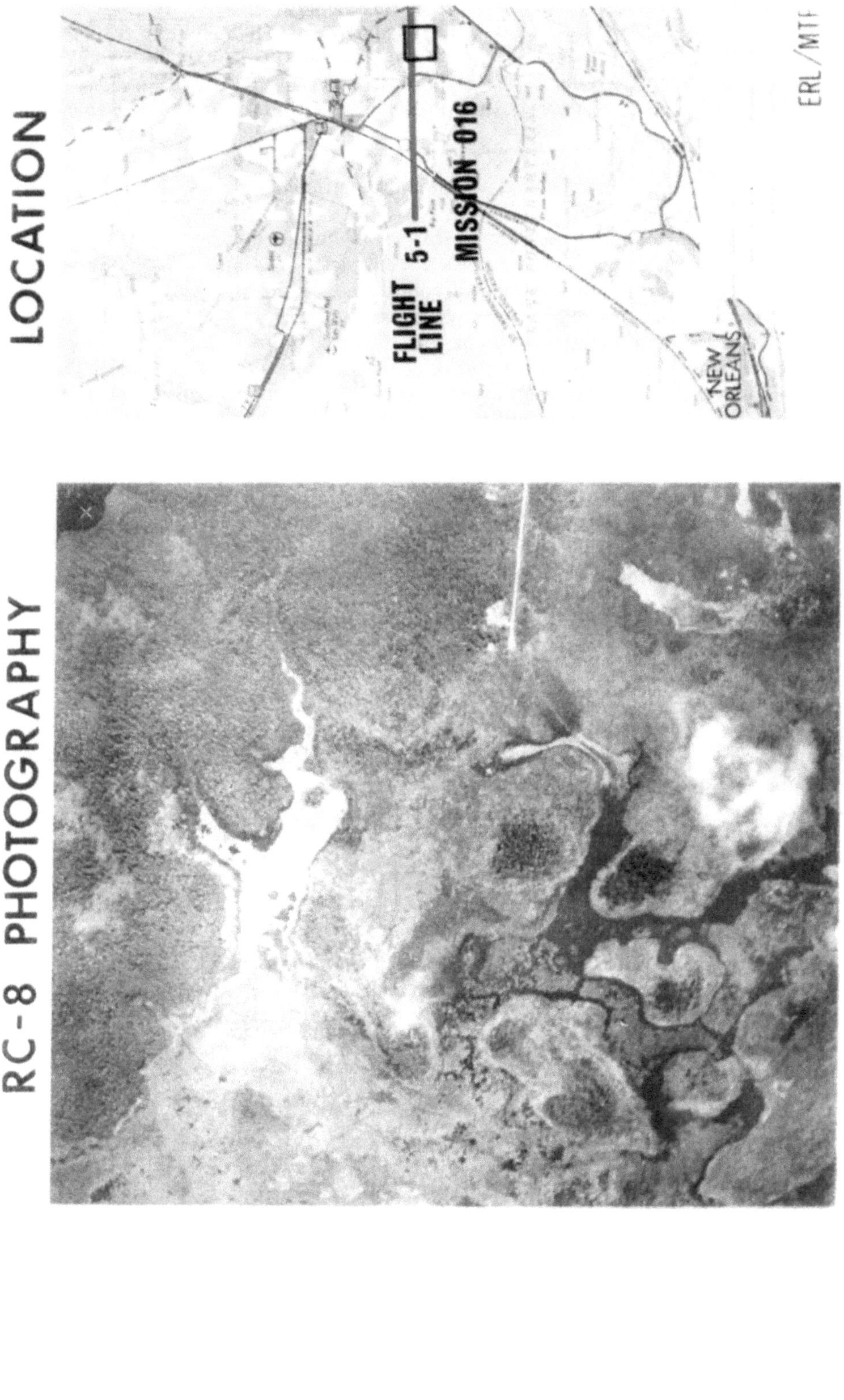


Fig. 5. Marsh ecological studies, photograph and location of test area.

Fig. 6. Seven categories of vegetation and water conditions are indicated. By inference we know that the area classified as spartina patinas is likely to be wet with shallow water of a brackish condition. Annual classification of the vegetation in the marsh thereby would indicate the extent of salinity intrusion resulting from activities in the coastal areas and indicate where there is a need for remedial action.

### 3.4. Coastal Water Characteristics

The third major element of this coastal area consists of the adjacent water bodies. Like the marsh these water areas are subject to changes which may work adversely to the normal environmental conditions. While the relative inaccessibility of the marsh makes remote surveys desirable, the highly dynamic nature of the water bodies due to tidal conditions, runoff, river inflows, and meteorological conditions makes remote, rapid surveys of coastal waters equally desirable. These variables, coupled with relatively large areal extent, result in considerable difficulties in surveying and understanding the water conditions. An example of this variability is shown in Fig. 7 where the surface temperature, salinity, chlorophyll and turbidity of the Mississippi Sound is shown for three different seasonal (July, November, January) and tidal conditions. These contour maps were derived from measurements made from a fleet of over 50 surface vessels. The use of such a number of surface vessels is impractical for operational use. Remote sensing from aircraft and satellites offers promise as a technique for measuring certain water parameters of interest. Initial results of one such investigation in the Mississippi Sound is presented by Atwell [3].

The measurement of water temperature utilizing thermal infrared radiometers and more complex scanners has received considerable study and practical use. The objective of the current study was to extend these techniques to provide accurate, quantitative temperature surveys. This has involved the use of a calibrated instrument, the application of atmospheric corrections to the measured data, and the development of particular processing and formating techniques to provide a useful end product. Fig. 8 presents an example of the results. The upper portion of the figure shows the test area of Biloxi Bay which empties into the Mississippi Sound. The imagery in the center of the figure is the photographic output of the infrared system. The lighter shading indicates warmer water and the darker shading indicates cooler water. These data contain no corrections for atmospheric effects and provide no indication of the absolute values or differences of temperature. The lower portion of this figure presents the corrected and processed data. Details of the atmospheric corrections are presented in [4] and [5]. The final format includes grey shades representative of the temperature variations. These shadings represent the corrected or actual temperatures. In addition, the formating techniques result in the actual temperature being printed on the figure at selected intervals. Examination of these data would indicate a maximum variation in temperature over the total area of 7° centigrade. Either of the two formats would indicate the warm water plume at the left of the figure which is the effluent from a nearby utility plant. For certain applications the imagery is adequate; for others, however, there will be a requirement for the quantitative results. For example, the location of effluents being discharged into the water body may be accomplished with the imagery alone. The assessment of shrimp breeding conditions over a several month period, on the other hand, would require

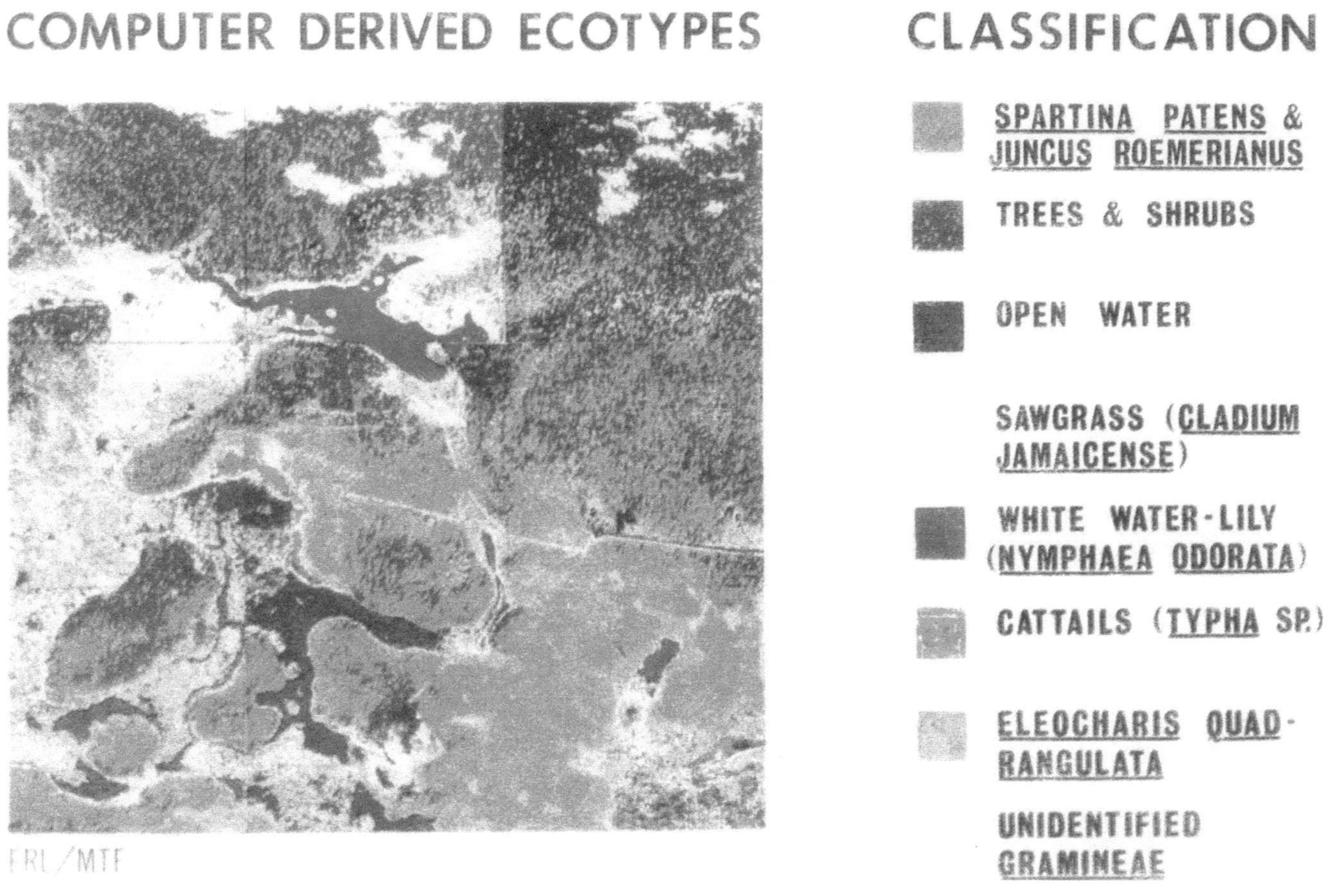


Fig. 6. Marsh ecological studies, computer derived ecotypes.

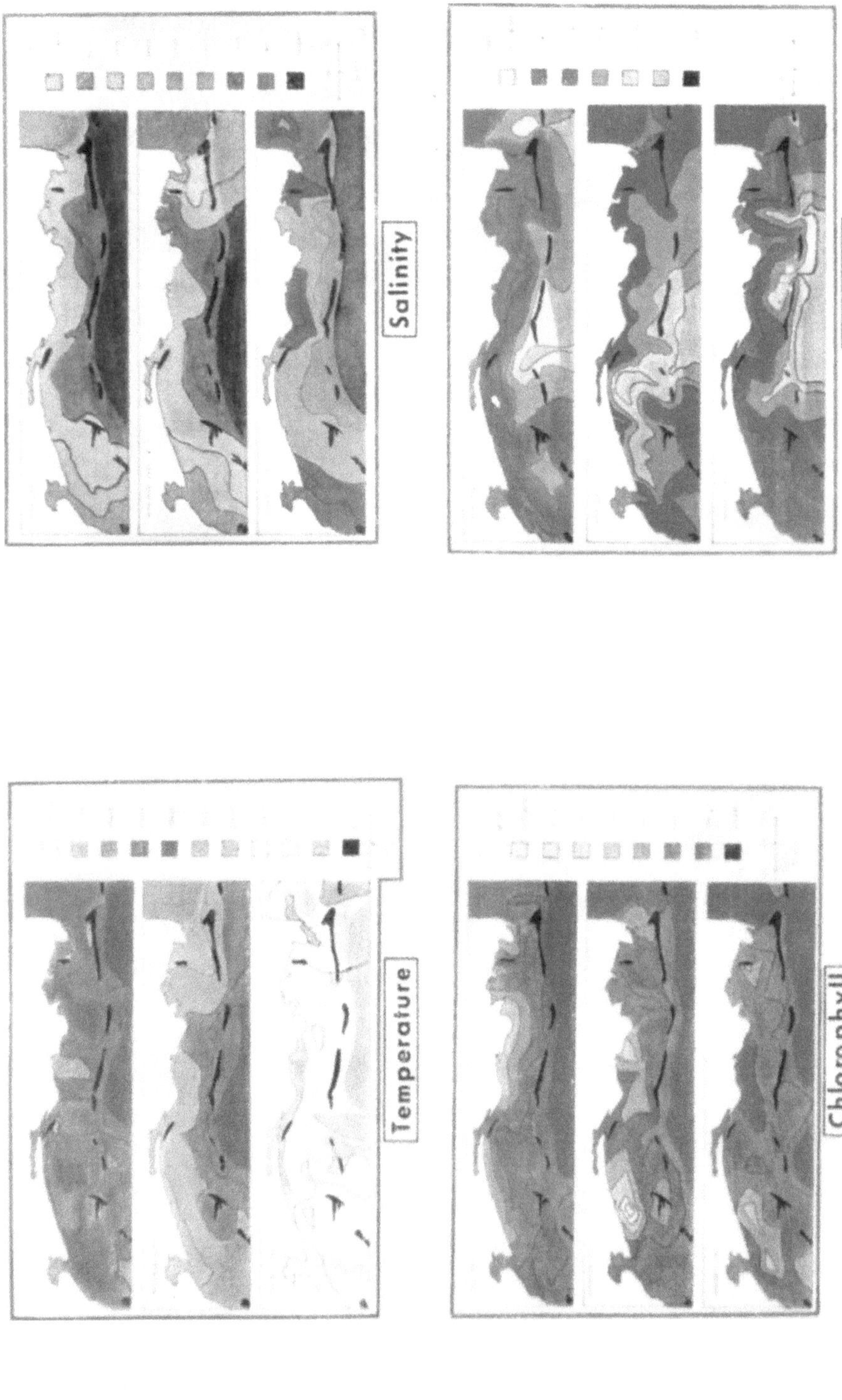


Fig. 7. Surface characteristics of Mississippi Sound.

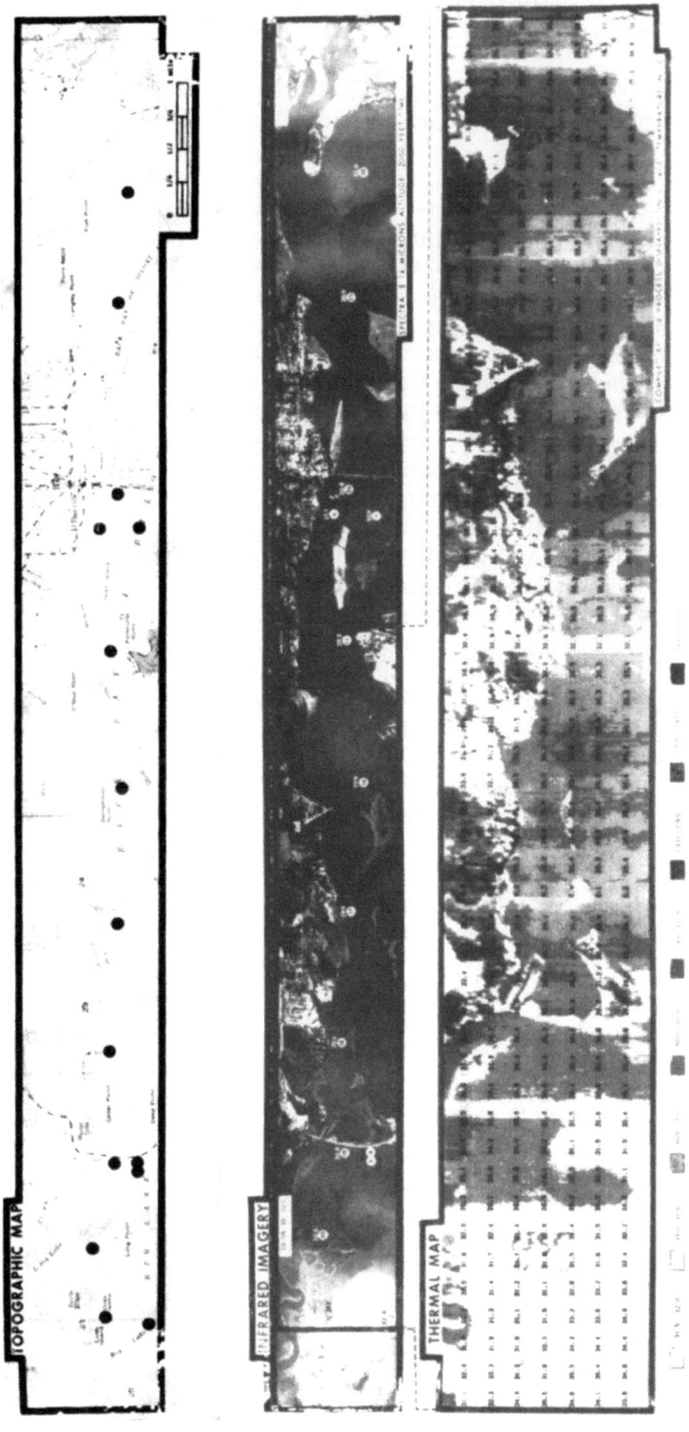


Fig. 8. Automated airborne infrared data processing, Biloxi Bay.

the quantitative data. Utilizing light aircraft, relatively low-cost sensors, and one surface reference measurement, accuracies within one degree centigrade appear to be attainable.

Salinity conditions are closely related with the well being or development of many types of marine organisms and adjacent marsh vegetation. For example, the oyster thrives in salinity conditions below 15 parts per thousand. If these conditions are increased by 1 or 2 parts per thousand the conditions become conducive to the growth of the oyster drill with its resulting depredations. Such salinity changes can come about quickly through dredging, channelization or the development of jetties. The measurement of salinity conditions with remote sensors is not as developed as the measurement of temperature with infrared sensors. A promising approach involves the use of a passive L-band (1.4 GHz) microwave radiometer. The measurement made by the radiometer is related to the dielectric constant of the water which, in turn, is related to the salinity content of the water. Thomann [6] presents the results of an investigation of this technique conducted in the Mississippi Sound. Example results are presented in Fig. 9 where salinity derived from the radiometer measurements is compared with salinity content determined from water samples collected at the same time as the remote measurements. The results of this investigation indicate a present capability to measure salinity to an accuracy of 3—5 part per thousand. Improved instrumentation should improve the accuracy considerably. A factor to be considered in salinity measurements is antenna size as limited by the aircraft or satellite and the associated surface footprint or resolution. The Skylab L-band radiometer with a $1\,\mathrm{m} \times 1\,\mathrm{m}$ antenna had a footprint or resolution element of 60 square miles. Such a resolution may be inadequate for coastal surveys particularly in mixed land-water areas. Aircraft-mounted radiometers operating at low altitudes result in a much improved situation. However, the physics of this situation indicates a need for careful planning in relation to the use of this technique.

Other water parameters of interest that appear feasible with remote sensing techniques include chlorophyll and turbidity. A number of studies have been conducted or are underway to develop the relationship between these parameters and spectrometer measurements in the visible and very near infrared regions of the spectrum. As these studies are completed they will provide additional valuable tools for the monitoring of coastal water conditions.

## 4. Concluding Remarks

A number of investigations have been conducted to develop and/or demonstrate techniques for the measurement of parameters of importance to coastal zone monitoring and management. These include photo interpretation techniques and procedures for the development of land use information from high altitude aircraft and satellite imagery; computer based pattern recognition techniques utilizing multi-spectral scanner data for marsh vegetation classification; and infrared and microwave techniques for the monitoring and surveying of coastal water temperature and salinity characteristics.

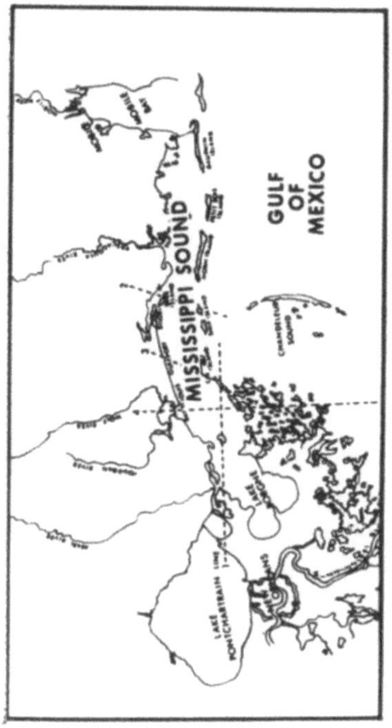


THE MISSISSIPPI SOUND

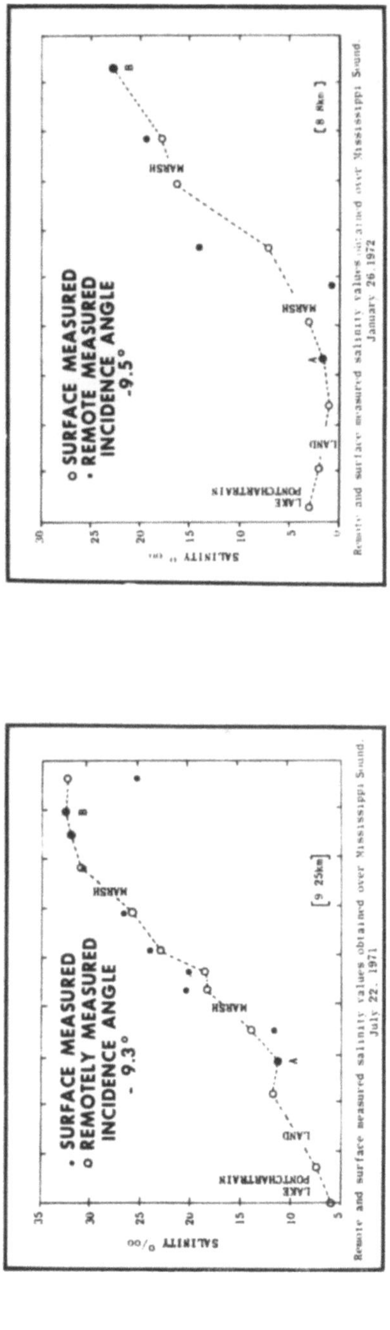


Fig. 9. Salinity measurements using L-band passive microwave radiometry.

## References

[1] P. L. VEGAS, ERL Rep. No. 031, NASA/ERL, Bay St. Louis, Mississippi, October 1972.
[2] W. G. CIBULA, ERL Rep. No. 020, NASA/ERL, Bay St. Louis, Mississippi, July 1972.
[3] B. H. ATWELL, ERL Rep. No. 048, NASA/ERL, Bay St. Louis, Mississippi, April 1973.
[4] R. D. BOUDREAU, ERL Rep. No. 014, NASA/ERL, Bay St. Louis, Mississippi, April 1972.
[5] R. D. BOUDREAU, ERL Rep. No. 029, NASA/ERL, Bay St. Louis, Mississippi, September 1972.
[6] G. C. THOMANN, ERL Rep. No. 036, NASA/ERL, Bay St. Louis, Mississippi, November 1972.

Earth Survey Problems — Akademie-Verlag, Berlin 1974

# THE CARTOGRAPHIC AND SCIENTIFIC APPLICATION OF ERTS 1 IMAGERY IN POLAR REGIONS

R. B. SOUTHARD and W. R. MACDONALD

Topographic Division, US Geological Survey, Washington, D.C., USA

The first Earth Resources Technology Satellite (ERTS 1), launched by the National Aeronautics and Space Administration in July 1972, is providing valuable data for investigations of the most inaccessible and hostile regions of the earth—the Arctic and Antarctic. ERTS images and map products derived from them offer a whole new dimension in source material for multidiscipline investigations in the earth sciences. For the first time scientists can view synoptic, repetitive scenes of the polar regions in four spectral bands. Ongoing experiments funded by NASA and conducted in the US Geological Survey have demonstrated the feasibility of revising coastlines on maps of Antarctica, detected gross changes in the northern limits of the three largest ice shelves in the world, and led to the discovery of uncharted mountain ranges.

## 1. Introduction

The launching of the first Earth Resources Technology Satellite (ERTS 1) and the successful operation of the onboard imaging systems introduced a whole new approach in monitoring and mapping the polar regions of the earth. Anticipating the potential usefulness of ERTS imagery as source material for cartographic products, the US Geological Survey proposed a series of experiments, which were approved and funded by the National Aeronautics and Space Administration. The approved proposal, "The Cartographic Applications of ERTS/RBV Imagery in Polar Regions" (designated SR-194), set in motion a number of investigations to determine the feasibility of expediting the ongoing USGS mapping operations that support the immediate and long-range goals of the United States Arctic and Antarctic Research Programs, which are administered and funded by the National Science Foundation. Other experiments in the proposal were designed to implement recommendations of the Committee on Polar Research (CPR) of the National Academy of Sciences, National Research Council [1] and the Working Group on Geodesy and Cartography, Scientific Committee on Antarctic Research (SCAR) [2, 3].

The feasibility and economy of using ERTS imagery is being investigated for compiling maps of unmapped areas at scales of 1:250000 to 1:1000000, for supplementing planimetric information such as crevasse fields and glacier flow lines, for preparing photomaps at various scales, and for map revision. Other experiments will include delineation and change detection of gross ice features, measuring seasonal variations of sea-ice boundaries, and mapping regional areas at 1:10000000 scale.

## 2. Status of ERTS 1

The Earth Resources Technology Satellite (ERTS 1), launched 23 July 1972 by the National Aeronautics and Space Administration (NASA), is in a circular, sun-synchronous, near-polar orbit at an altitude of 900 km. It circles the earth

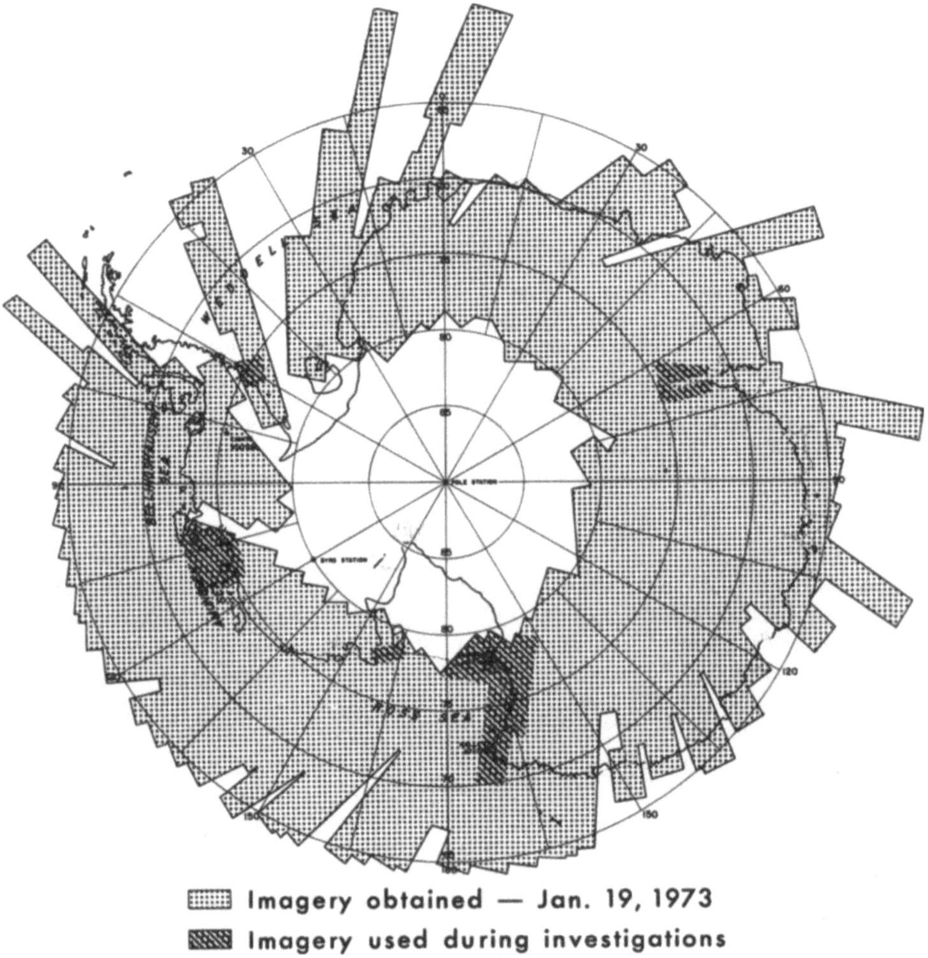


Fig. 1. ERTS 1 space imagery (MSS), Antarctica index.

every 103 minutes (14 orbits a day) and repeats a given orbital track every 18 days. The science package includes three boresighted return-beam vidicon (RBV) cameras, each recording imagery in a discrete spectral band, and a multi-spectral scanner (MSS), which records in four spectral bands including near infrared [4].

The US Geological Survey planned to use imagery from the RBV cameras for cartographic experiments because of its favorable geometric characteristics. Shortly after the satellite achieved the desired orbit, however, the switches that activate the RBV cameras malfunctioned and one of the two onboard tape recorders became inoperable. It was therefore necessary to substitute MSS imagery as source material for the experiments. Although the geometric characteristics of MSS are not as desirable as those of RBV, the distortions of MSS are systematic

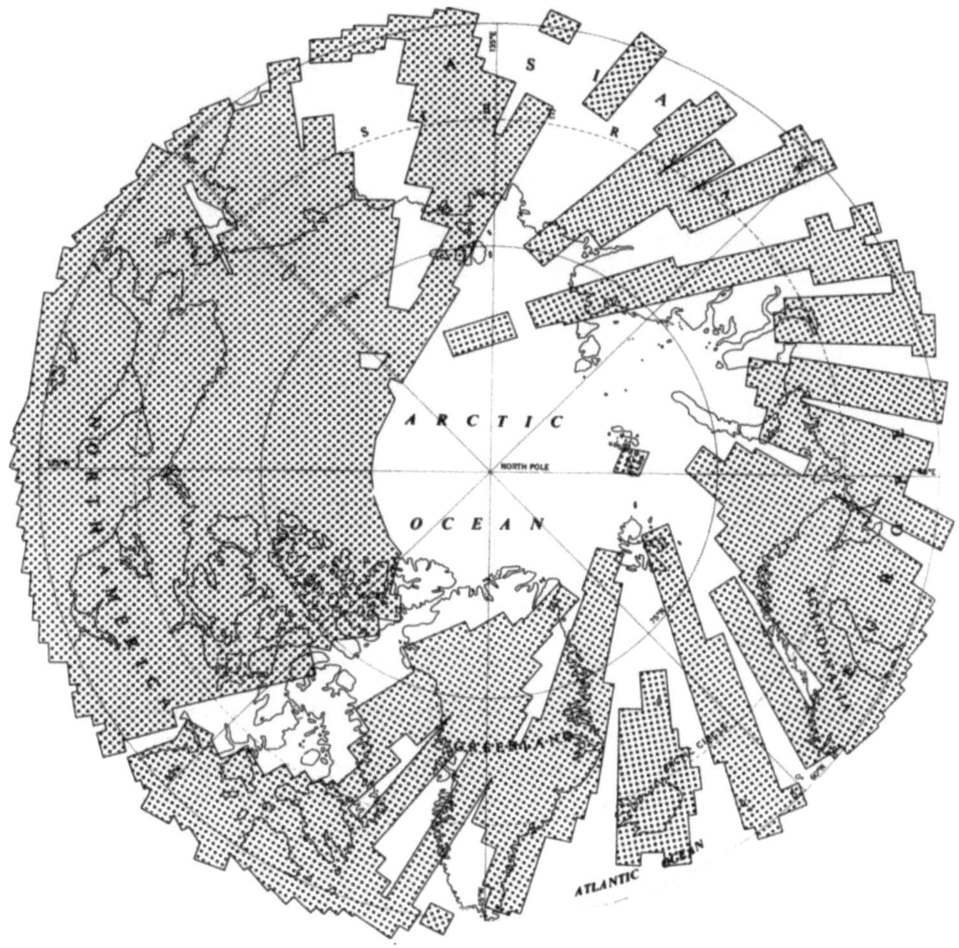


Fig. 2. ERTS 1 space imagery (MSS), Arctic index.

and the imagery has good spatial and spectral resolution. (Perceptional and geometric image qualities of MSS and RBV are described in [5].)

Widely scattered MSS imagery has been received over both polar regions (Figs. 1 and 2). Imagery beyond the range of the receiving stations in the United States and Canada was stored on a tape recorder and later transmitted on command to the receiving stations. Unfortunately, in early April, sporadic noise was interjected into the second recorder, which degraded the images, and it was turned off by NASA in April 1973. The tape recorder functioned well until this event and actually had exceeded its 500-hour designed lifetime by about 10%. Without the recorder, the ERTS 1 satellite has very limited capability of providing any further imagery of the polar regions. No further imagery can be obtained of areas beyond the range of the few existing ground stations.

## 3. Imagery Requirements

For the preparation of maps in support of the US Antarctic Research Program, the United States has obtained aerial photography over an area of about 3250000 km². This effort has cost many years and many millions of dollars. ERTS 1 now provides the capability of producing a single frame of imagery covering an area of about 34000 km² (185 km by 185 km). About 100 frames of ERTS imagery would cover the same area now covered by over 100000 aerial photographs. Thus we have seen the development of a system that produces highly useful and readily available synoptic imagery for meeting many of the mapping requirements of scientific investigations of the polar regions.

For use in the ERTS experiments, imagery with no more than 10% cloud cover was specified. Only about 5% of the imagery received so far met this specification (Fig. 1). All available coverage is now being examined to find additional usable MSS imagery (50% cloud cover or less) of the polar regions.

NASA delivered the first MSS scenes of the Arctic region in October 1972, and of the Antarctic in December 1972. Of the four spectral bands, the near infrared (band 7) appears to offer the best data for cartographic applications and image interpretation in the polar regions.

## 4. Investigations

### 4.1. Strip Mosaic and Change Detection

An important objective now is to compile 1:1000000-scale photoimage mosaics along the coastal areas of Western Antarctica and, eventually, along the coastal areas of all Antarctica. These imagery products will enable the US Geological Survey to build a historical record which, when compared with existing maps and sequential ERTS coverage, will show changes in size, shape, and position of features such as ice shelves, glaciers, and ice tongues (Fig. 3). A strip photomosaic, comprising portions of seven ERTS 1 images, was compiled at 1:1000000 scale along the Victoria Land Coast of Antarctica between Cape Adare and Harbord Glacier. This mosaic, covering an area of 185 km by 644 km, depicts 45 glaciers and ice tongues, numerous ice shelves, and the northern extent of the Transantarctic Mountains. This experiment is a pilot effort, and other 1:1000000-scale photoimage mosaics along the coastal areas of West Antarctica between the Ross Ice Shelf north and westward to 180° east longitude will be compiled under this proposal.

Glaciers, ice tongues, and ice shelves were clearly identifiable on the 1:1000000-scale ERTS imagery. This pilot project indicated that the imagery was of sufficient resolution to be used as a source for photoimage revision and for glaciological change detection. Further investigations proved that ERTS imagery can be used effectively for planimetric revision of small-scale maps, and this technique is being applied to the six 1:250000-scale topographic maps that fall in the area of

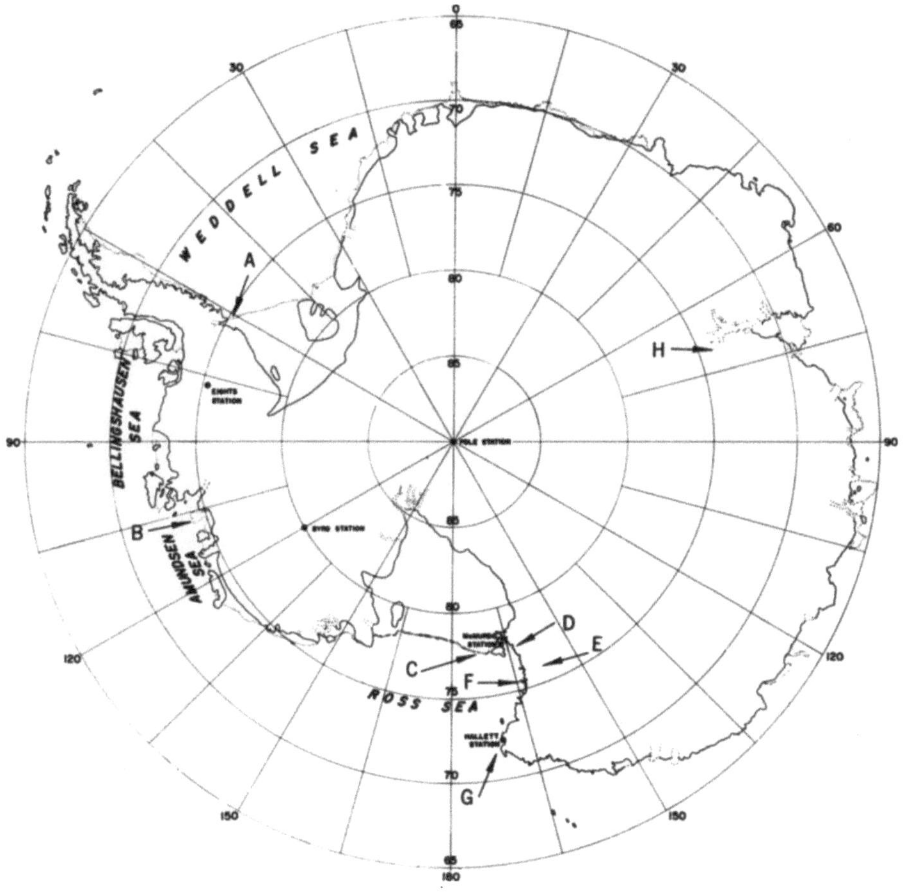


Fig. 3. Locations of significant changes evident from ERTS imagery (Dec. 1972—Feb. 1973):

A. Ronne Ice Front
B. Thwaites Ice Tongue
C. Ross Ice Shelf
D. McMurdo
E. New features on IMW
F. Drygalski Ice Tongue
G. Hallett (fast ice)
H. Lambert Glacier (new features)

the Victoria Land strip mosaic. A detailed cartographic analysis has not been completed over the entire area of the mosaic, but the coastline was carefully compared with existing maps, and several major changes in coastal features were found. The comparison showed that the new satellite imagery can greatly facilitate meaningful revision of the existing maps (Figs. 4 and 5).

Because the 1:250000-scale source maps were compiled from multiyear aerial photography, it would be erroneous to use the map publication date as a benchmark for determining ice movement. Accordingly, in our analysis, it was necessary to determine the date of aerial photography used as source for each feature on the existing map. In the future, scientists need only refer to the taking date of the ERTS imagery.

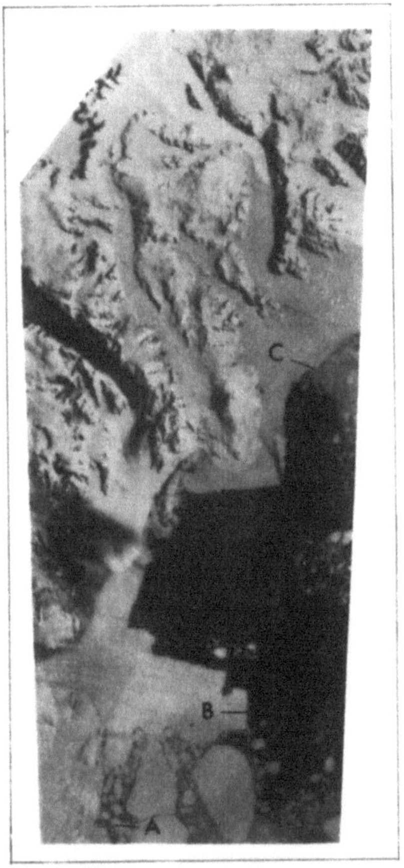


Fig. 4. Drygalski Ice Tongue, Victoria Land Coast Area. Left, three 1:250000-scale US Geological Survey topographic maps compiled from source data, 1955–1964. Annotated revisions based on ERTS 1 imagery. Right, ERTS photoimagery mosaic showing significant changes:
A. Harbord Glacier B. Drygalski Ice Tongue C. fast ice

## 4.2. Map Revision

Fig. 6 illustrates the application of ERTS imagery for evaluating and revising published maps. The graphic on the bottom is a composite of two 1:500000-scale sketch maps compiled from conventional photographs taken during the austral summer of 1965–1966. The graphic on the top is a mosaic of parts of seven scenes of MSS bulk imagery. Area coverage is about 117000 km$^2$. The two triangles represent geodetic positions used to fit the imagery mosaic to the map base. The ERTS imagery greatly improves the absolute and relative positioning of shoreline configurations, ice tongues, and other map features. Because of the limited amount of control and the large number of conventional photographs used to compile the sketch maps, it was not possible to maintain scale and position throughout the map compilation, especially in areas devoid of readily identifiable

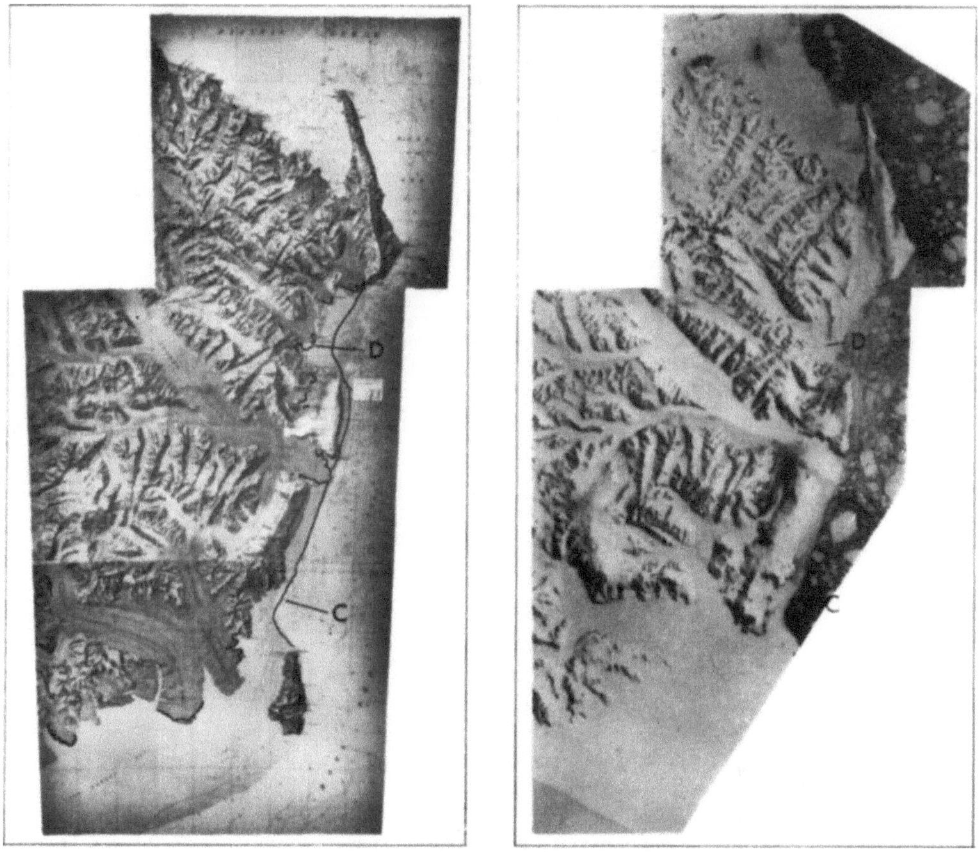


Fig. 5. Cape Adare, Victoria Land Coast Area. Left, three 1:250000-scale US Geological Survey topographic maps compiled from source data, 1961–1964. Annotated revisions based on ERTS 1 imagery. Right, ERTS photoimagery mosaic showing significant changes: C, boundary of fast ice and bay ice has changed; D, the shape of Honeycomb and Ironside Glaciers has changed and their tongues have advanced about 3.2 km.

features that make good pass points. A single ERTS scene covers the same area as 1320 1:40000-scale photographs. Comparing the two parts of Fig. 6, note the change in the configuration of the coast, as indicated by numbers 1, 3, and 4, and in the position of Burke Island, number 2. Also note the change in size and position of the Thwaites Iceberg Tongue (number 5). Area has increased from 44200 km$^2$ (map) to 71500 km$^2$ (image), and the position has shifted about 8 km.

### 4.3. Small-Scale Mapping

The main immediate application of ERTS 1 imagery in the polar regions, particularly in Antarctica, is for compiling 1:1000000-scale maps and photoimage mosaics. The need for million-scale coverage of Antarctica has been recognized by several organizations, including SCAR [1–3]. Because of the virtual impossi-

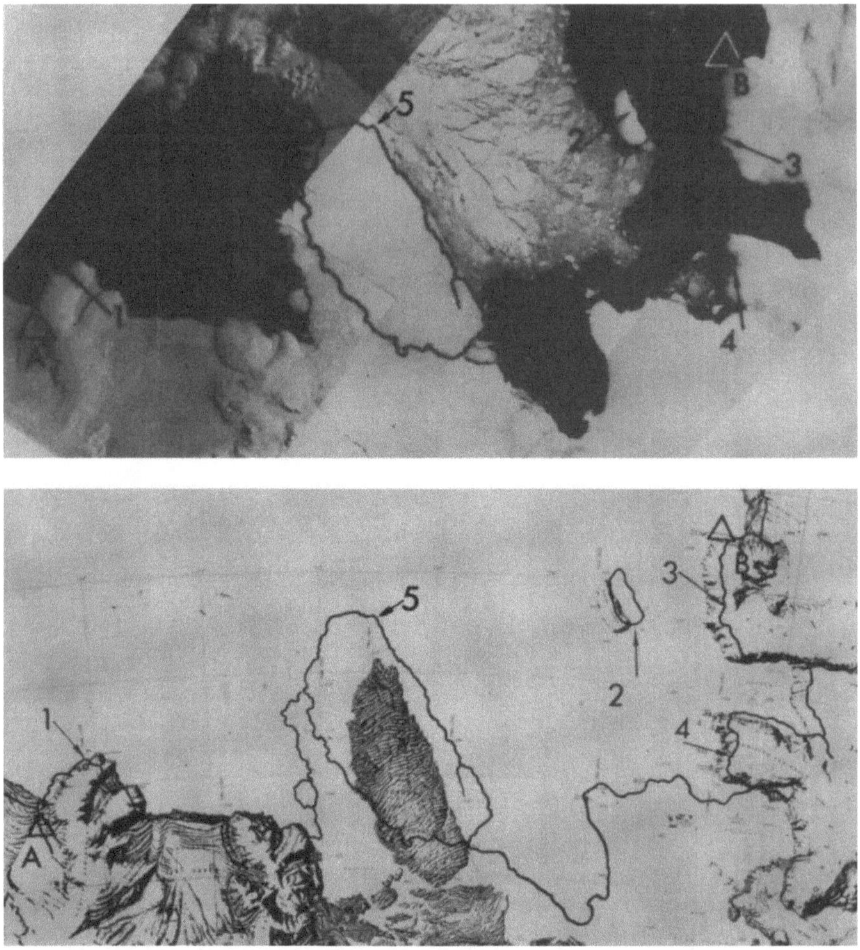


Fig. 6. Thwaites Iceberg Tongue. Annotated mosaic of ERTS 1 imagery (top) and corresponding sketch map (bottom). Annotations on the map show significant changes to be the coastline and position changes to map features that can be made with ERTS 1 imagery. Geodetic control points are shown at A and B.

bility of taking photographs with normal aircraft at altitudes high enough for efficient million-scale mapping, efforts before the launching of ERTS 1 were directed primarily toward mapping the coastal and mountain areas at 1:250000 scale. ERTS imagery, however, meets the needs for million-scale photomapping which is expected to prove most beneficial to the international scientific community. To cover an IMW map area, cartographers will only have to assemble 15 to 20 ERTS scenes rather than 12000 conventional photographs. Production cost will decrease as production rate increases. The user will have at his disposal a visual representation of vast areas that have never been mapped. Moreover, he will not have to wait years before the image maps are available.

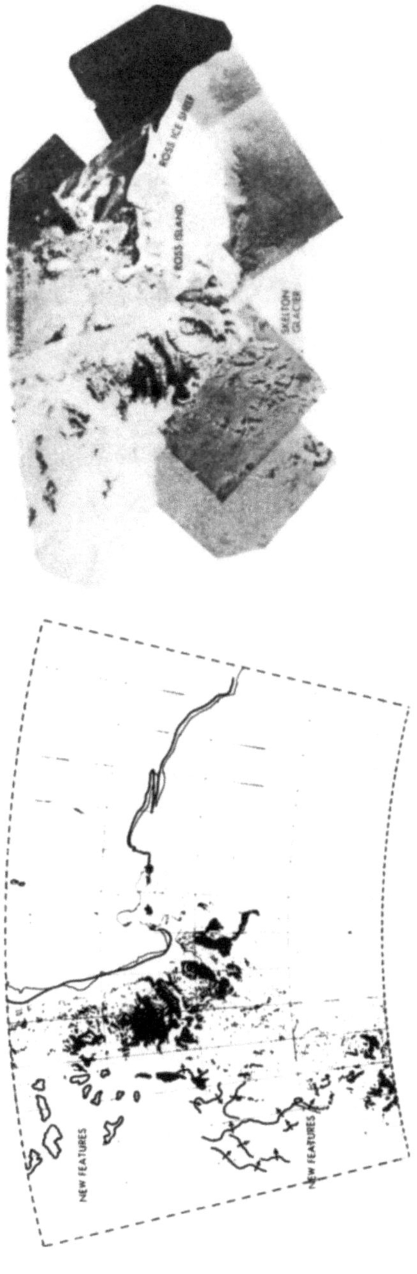


Fig. 7. McMurdo 1:1000000-scale map (IMW series)/ERTS imagery. With the aid of the ERTS imagery mosaic, newly discovered mountains, land features, and coastal changes will be depicted on the USGS 1:1000000-scale manuscript prior to publication.

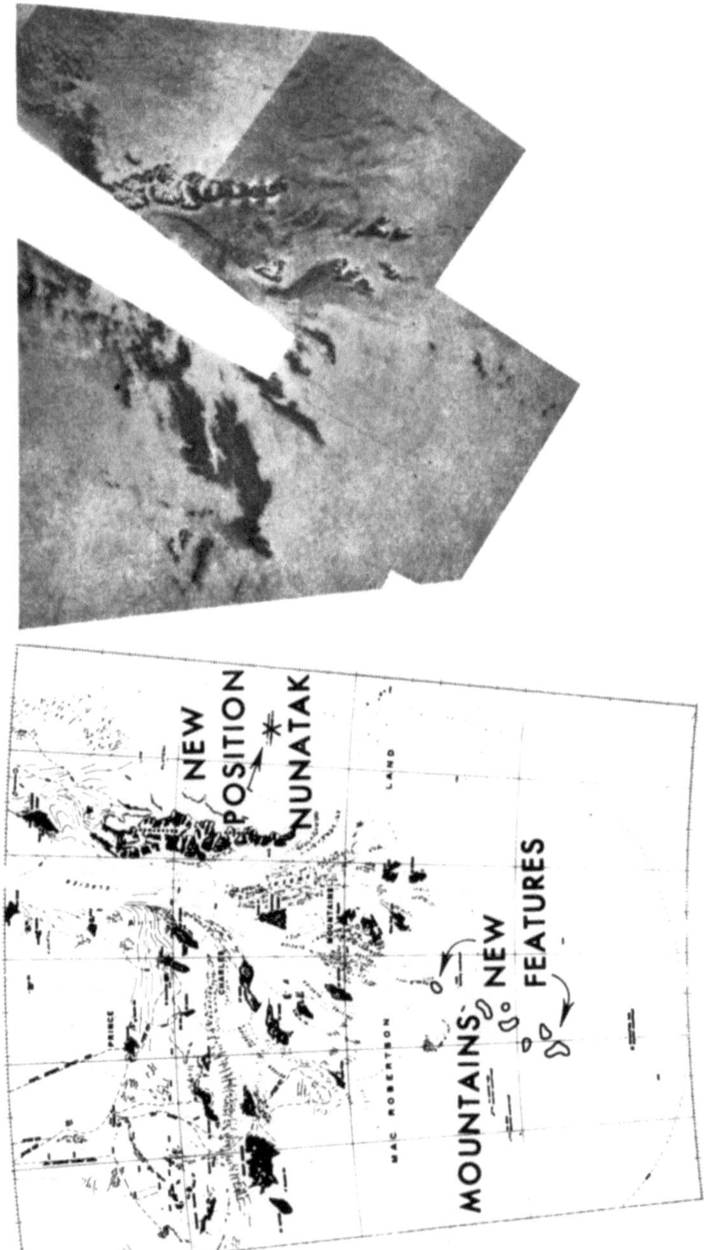


Fig. 8. Australian 1:1 000 000-scale map (IMW series)/ERTS imagery. With the aid of ERTS imagery, it is possible to analyze and evaluate existing source maps. The imagery reveals new mountains and other significant features which can be added to the Australian 1:1 000 000-scale map.

Comparisons carried out on the USGS-compiled McMurdo Sound Region, IMW Sheet ST 57—60 (Fig. 7), and on the Australian-compiled IMW Sheet SS 40—42 (Fig. 8) clearly demonstrate the application. Revisions and additions to the two sheets are readily apparent. The most obvious changes are the large block

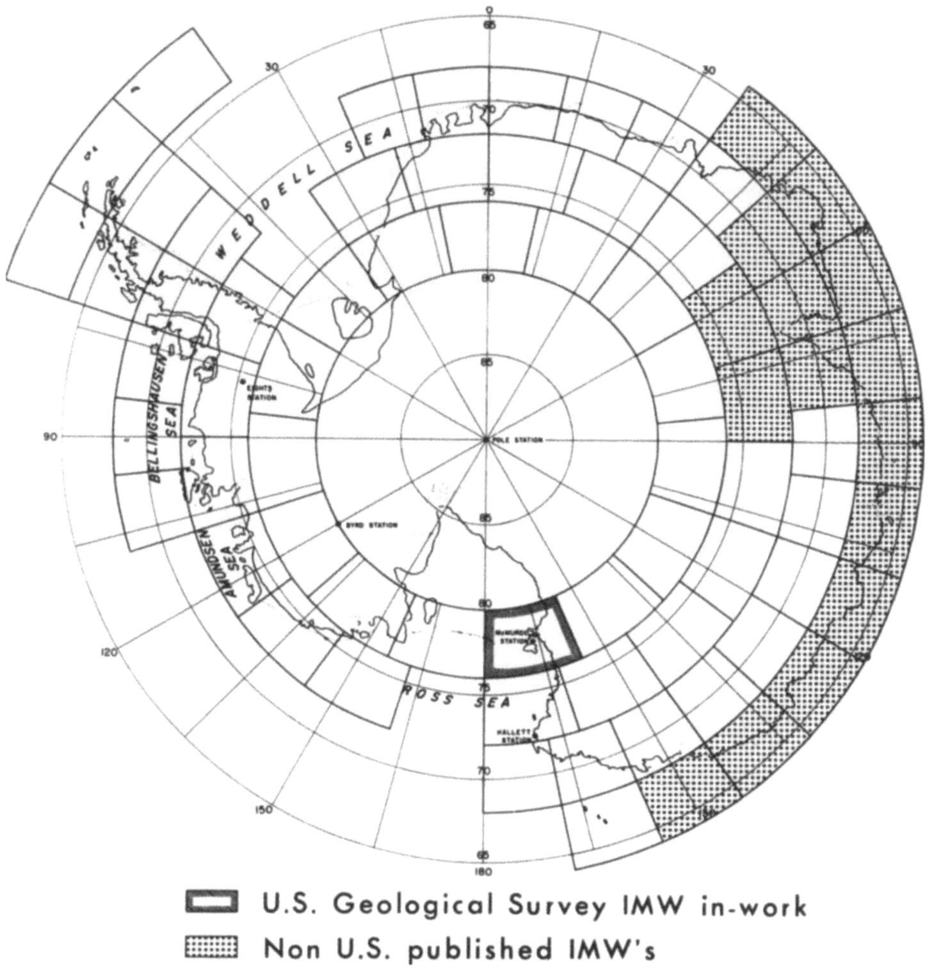


Fig. 9. Antarctica IMW 1:1000000-scale index.

of new and unmapped geographical features revealed by the ERTS imagery. Noteworthy revisions include the repositioning of the Ross Ice Shelf Front (about 6.4 km north) and Franklin Island (7.2 km south). The position of Franklin Island has been in contention for many years and has been continually reported in error by US ships.

Reference to the IMW 1:1000000-scale index of Antarctica (Fig. 9) and the space-imagery index (Fig. 1) indicates the value of ERTS 1 imagery to the IMW mapping program. Because of the 99° orbital inclination (ascending node) of the satellite, imagery cannot be obtained between 82° S latitude and the pole. Accordingly, if all the available imagery is of satisfactory quality, which has yet to be determined, 71 maps can be compiled.

## 5. Scientific Application

Comparison of ERTS scene 1154—19322, dated 24 December 1972, with the published Ross Island 1:250000-scale map disclosed a unique change in the Erebus Glacier Tongue (Fig. 10). Further examination of photographs and histor-

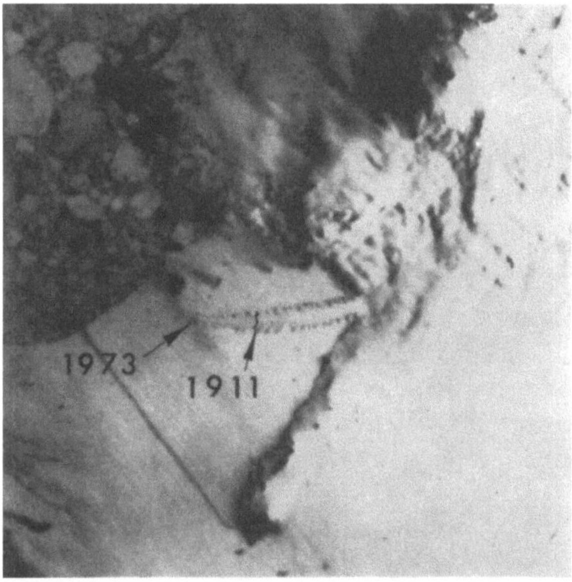


Fig. 10. ERTS imagery of Erebus Ice Tongue. Analysis of ERTS images and library sources determined that the Erebus Glacier Tongue has advanced 9.6 km since 1947. Place where the tongue broke off in 1911 is annotated.

ical maps indicated that the present position of the tongue is about the same as it was in 1910 [6] and that the tongue has advanced about 9.6 km since 1947 and 4.8 km since 1962. A lateral shift or curving of the leading front toward the mainland seems to have occurred since 1970. Sources are not adequate to determine whether the movement occurred gradually or over a very short period. Some evidence does seem to indicate that the tongue may have gone through a surging period. Perhaps it has completed a growth cycle and will again break off as it did in 1911. If ground investigation proves it to be a surging glacier, it will be the first one found in Antarctica and therefore of keen interest to glaciologists.

Further indication that ERTS imagery is useful for detecting glaciological changes is given in Fig. 11. ERTS scene 1212—11183, dated 20 Feb. 1973, was compared with the published 1:500000-scale sketch map of Ellsworth Land and Palmer Land and with aerial photographs to determine that the Ronne Ice Shelf has advanced about 16 km since January 1966.

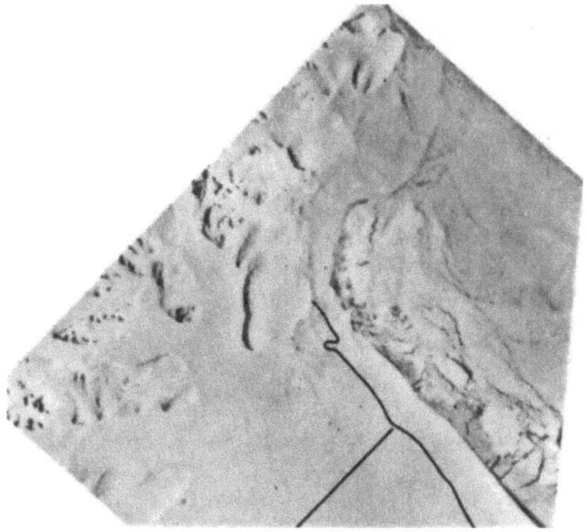

MAP POSITION
BASED ON 1966 SOURCES

Fig. 11. ERTS imagery of Ronne Ice Shelf. The edge of the shelf has advanced about 16 km since 1966.

## 6. Future Plans

In addition to experiments discussed in this paper, plans include investigating the use of MSS imagery:

(i) to compile a small-scale (1:10000000) photomap of the entire continent of Antarctica
(ii) to compile thematic maps of sea-ice conditions
(iii) to begin an analysis of the variations of the sea-ice boundary and rate of deformation of pack ice
(iv) to compile physical maps of the Arctic region
(v) to compile 1:1000000-scale photoimage mosaics of Alaska as a reference for evaluating ecological effects of North Slope and pipeline development and for revision of published maps.

With the ERTS imagery as a current guide, Doppler satellite-tracking equipment will be used to establish geodetic control for Antarctic mapping. Survey teams will be sent to the field during the Antarctic austral season to establish geodetic positions ($x$, $y$ and $z$) on pre-selected points identifiable on ERTS scenes.

## 7. Conclusion

Though we have had only a few months of experience in using ERTS imagery for investigations of the polar regions, we are sure that it can form the basis of a quality map product that will meet most users' needs. We are also sure that

ERTS imagery can be used as source for revision of published small-scale and medium-scale maps of coastal areas in the polar regions.

The US Geological Survey is responsible for maintaining the United States SCAR Library, distributing copies of all US-completed cartographic products of Antarctica to the eleven other Antarctic Treaty nations. We therefore already have an established system which can be expanded to distribute ERTS imagery on request to member nations of SCAR.

## References

[1] Polar Research—A Study published by the National Academy of Sciences (NAS), Mar. 1970 (Chap. 5).
[2] SCAR Bulletin No. 3, Sept. 1959 (p. 591).
[3] SCAR Bulletin No. 26, May 1967, International Activities (p. 579).
[4] Data Users Handbook, National Aeronautics and Space Administration, Goddard Space Flight Center, Document No. 71SD4249, Appendix A.
[5] A. P. COLVOCORESSES and R. B. MCEWEN, Progress in Cartography EROS Program, US Geological Survey, paper presented to XII Pan American Consultation on Cartography, Pan American Institute of Geography and History, Panama 1973.
[6] Map of Ross Island: West Coast, scale 1:100000, Dec. 1912.

Earth Survey Problems — Akademie-Verlag, Berlin 1974

# REMOTE SENSING AND SPATIAL INFORMATION SYSTEMS

D. Steiner

Department of Geography, Faculty of Environmental Studies, University of Waterloo, Waterloo, Ont., Canada

Researchers, planners and administrators are facing the ever increasing problem of arriving at rational conclusions and decisions concerning resources, people and environment in a short time. There is a growing need for useful information and, consequently, for collecting, processing, displaying and disseminating data in an efficient manner. As an answer to this problem computerized spatial information systems have been or are being established in many parts of the world at the local, regional, national and, in some cases, the international level. The Commission on Geographical Data Sensing and Processing of the International Geographical Union (IGU) has been instrumental, through two international symposia organized in Canada, in bringing together experts in the information systems field, assessing the present state-of-the-art and making the findings known to a wider segment of the community concerned with spatial data and their use. The present paper is essentially a review of the results of these two IGU meetings. In particular, however, it looks at the possibilities and problems of integrating remote sensing data into spatial information systems. To this end, the components of and processes in information systems in general are first discussed, ranging from data specification, data acquisition, data input, data storage and data processing to data output. Based on this background information the second part of the paper then examines the special problems arising from the use of remote sensing data. The emphasis is on image data and semi-automatic and automatic techniques. Therefore, the topics reviewed include image data storage and image data processing (geometrical transformations, radiometric transformations, pattern recognition).

## 1. Introduction

A system is a set of objects characterized by a number of attributes tied together into a whole through relationships between the objects and between their attributes [1]. It can be described in terms of the seven elements: function (objective which the system attempts to achieve), inputs (materials, energy, etc., which are processed by the system), outputs (desired and undesired products), sequence (interrelationships among component processes and order in which they operate), environment (surroundings within which the system operates), physical agents (material, i.e. non-human, resources which operate in the system sequence), and human agents (human resources operating in the system sequence) [2]. General systems theory has developed in response to the growing needs for problem solving in a world of increasing complexity and rapid change. It provides the necessary platform for pooling and integrating knowledge from various disciplines [3].

Information systems are systems which collect, store, process, display and disseminate information for users engaged in decision making in research, planning and management. The core of the system (storage and processing with associated inputs and outputs) is usually in computerized form. The above mentioned seven elements can be readily translated into the case of an information system. In particular, the function of such a system is to convert data to useful information. Spatial information systems deal with data on phenomena distributed in geographical space. They differ therefore from other information systems in that locational identifiers are attached to the data. Often the term "geographical information system" is used. Rather than identifying geography as a discipline, "geographical" here simply means "spatial". Since spatial data are conventionally shown on maps, we can define a spatial information system as a system that handles map data. We make a distinction between base (planimetric, topographic, hydrographic) maps and thematical maps, and, consequently, we can also use this criterion to characterize spatial information systems. An example for a map compilation system of the first type is the automatic system for hydrographic charting of the Canadian Hydrographic Service [4]. Thematical systems can range in topic from geology [5], over hydrology [6], glaciology [7], air pollution [8], land use and natural resources [9], population and real estate [10] to transportation [11], to give only a few examples. Furthermore, such systems may be established at different levels in terms of geographical area covered. Examples are the following: local: Système d'Information Urbaine et Régionale of the Quebec Urban Agglomeration [12]; regional: Land Use and Natural Resources Information System of New York State [13]; national: Canadian Geographic Information System based on the Canada Land Inventory [14]; international: World Weather Watch [15].

A spatial information system is an open system, i.e. it can be regarded as a subsystem of a larger real world system [16]. It is the result of the desire and need to exercise control over environmental, economic and social processes. Clearly the complexity of these processes in a real world man-environment system requires their modeling at a certain level of selection, simplification, generalization and abstraction. A spatial information system takes on the minimum function of providing the data needed to work with such models. At a higher level of sophistication we can visualize monitoring functions being built directly into such a system. However, while a considerable degree of automation has already been achieved in the monitoring of industrial activities [17], the complexity and our lack of understanding of the man-environment system rules out this level for spatial information systems at the present time [18].

An assessment of the present state-of-the art in the spatial information systems field has been the concern of the Commission on Geographical Data Sensing and Processing (Chairman: R. F. Tomlinson, Ottawa, Canada) of the International Geographical Union for the last few years. With financial support from UNESCO two symposia were held in Ottawa, Canada, in 1970 and 1972 respectively, and the results have been published [19, 20]. It is the purpose of this paper to give a summary of the Commission's findings. We will do this by paying special attention to the problem of integrating remote sensing data into spatial information systems: Section 2 looks at components and processes in spatial information systems in general; Section 3 is devoted to issues that are specifically related to the handling of remote sensing data.

## 2. Principles of Spatial Information Systems

It is common to distinguish five main phases or components which must be considered in the conceptualization, design, development, implementation and maintenance of a spatial information system [21]: (1) data specification (the act of defining the sets of data needed for certain programs and projects); (2) data acquisition (the recording of measurements that report on the state of the features chosen); (3) data input, storage, retrieval, processing and output (manual and/or automated data handling operations); (4) data dissemination (the delivery of data to users by man or machine); (5) data application (activities of the users carried out upon receipt of the data). The five phases comprise an iterative chain of events, and, obviously, interrelationships exist between them. The development of a balanced system requires the simultaneous consideration of all phases and interrelationships. In the following sections we will discuss briefly some of the main considerations in each phase up to phase 3 (compare with Fig. 1). The exclusion of phases 4 and 5 from this discussion is justified by the fact that they become user-specific and they cannot possibly be dealt with in the context of this contribution.

### 2.1. Data Specification

One would generally assume that the specification of data to be collected is uniquely governed by the need for certain types of information. This process is by no means a perfected science. From a scientific point-of-view, the theories and models used to describe and analyze the real world man-environment system decide upon which data should be collected. In the absence of a well established body of theories and models, however, the data need is not so clearly defined. A continuous feedback procedure must be pursued: Initial models can be built with available data, whereupon a more elaborate data base can be acquired for the purpose of refining the models [22, 23]. Secondly, deviations from an ideal situation occur simply because, with the involvement of governments, data specification becomes invariably part of the political process, and data collection is subject to budget constraints [24]. There may be other factors involved. Depending on the data base already available and on prospects for returns in a development program, economic considerations may decide on the level of detail at which data are collected (reconnaissance, semi-detailed, detailed). Finally, institutional factors such as availability of trained manpower and data analysis capabilities may play a part, particularly in developing countries [25].

The information content of data in spatial information systems can be regarded as three-dimensional: A thematical value with a spatial and a temporal location [22]. Data specification must therefore cover these three dimensions. In terms of theme or topic, i.e. with respect to the types of variables to be collected, a functional approach which makes use of the frequent overlap in subject matter concern between different users is most desirable [26]. With respect to the space dimension, spatial extent (area to be covered), spatial resolution and spatial frequency (complete versus sampling coverage) have to be specified. Similarly, for the time dimension, temporal extent (period of coverage), temporal resolution and temporal frequency (rate of repetition) have to be decided upon. The latter is important for change detection in dynamic situations. For an appropriate specification of this parameter one must have an understanding of the processes that cause change [27]. A further consideration is related to data accuracy. For example, one has to decide which scale of data measurement, nominal, ordinal, interval or ratio [28], is adequate. The use of a nominal scale is premised on the existence of a classification. A classification has value in that it facilitates the communication of complex information. On the other hand, it can never be a substitute for original data at a higher level of measurement. Some advocate a parametric approach to data gathering where phenomena such as soils or land uses are characterized by parameters on a continuous scale rather than by type within a preconceived classification [31].

### 2.2. Data Acquisition

Once the need for certain types of data has been defined the next step will be to determine whether they are available from existing sources or whether they have to be collected for the program in question. Two broad groups of data sources can be distinguished: (i) administrative data: these are data produced as by-products through the record keeping functions of operating agencies; typical examples are tax assessment records, building permits, vehicle registration and utility billing [29]; (ii) survey data: these are data collected for specific purposes or programs, either as a one-time effort or at regular intervals (such as censuses); we can distinguish between the mapping of characteristics of the land and the environment (land use, land capability, resources) and the acquisition of data on population characteristics by means of questionnaires and interviews.

Data collection procedures can consist of contact or non-contact (remote sensing) methods. Contact methods may involve direct recording, such as in interviews or in visual and instrumental field observations on the ground, or remote recording. An example for the latter case are the data platforms used within the ERTS Data Collection System (DCS) which measure various ground phenomena and transmit the measurements via the satellite as a relay station [30]. Data source formats include written documents, such as field notes, forms and tables, graphical documents, in particular maps of various kinds, and, in the case of remote sensing, pictorial documents, i.e. imagery in actual or latent image form. More advanced data collection procedures may involve the recording of data directly in machine readable format (for example, mark sense sheets, punch tape, magnetic tape). Whatever the original format, what one needs for a spatial information system eventually is, of course, data which are in two-dimensional mappable form.

A further consideration in the data acquisition phase relates to the spatial and temporal specifications indicated in the previous section, i.e. data have to be collected within a particular spatial and temporal framework. Almost all spatial information systems are based on areal units of some kind. Because of their overriding importance let us have a brief look at the different solutions. Areal collection units can be of the natural, the institutional or the arbitrary type. "Natural" implies a breakdown of the study area into subareas such that discontinuities coincide with unit boundaries. A classification based on single or on multiple features must exist which permits to delineate natural units. The easiest is the delimitation of single-feature areas corresponding, for example, to different soil or land use types. Such areas form the basis of the Canadian Geographical Information System [14]. The identification of more complex multi-feature areal units has been achieved by some form of an integral or landscape approach. Each unit represents a typical combination of physical features with certain relationships between them. This approach has gained some popularity particularly for vast, relatively unexplored, regions and is heavily practised, mainly on the basis of aerial photographs, by such institutions as CSIRO in Australia and the Soil Science Laboratory of the University of Oxford in England [31, 32]. Institutional units are areas of a political, administrative or legal nature. Census information is the most common example of data based on an institutional framework. Information systems based on a cadastral parcel framework are particularly attractive because many data may be generated by operating agencies at this level already. In large areas of North America systematic land surveys have led to a pattern of parcels approaching the regularity of a square grid. An example of an information system based on such parcels is the Minnesota Land Management Information System (MLMIS) [33]. Drawbacks of unequal sizes and possibly changing boundaries attached to natural and institutional areal units can be avoided by the use of arbitrary (usually square) grids. The Land Use and Natural Recources Inventory (LUNR) of New York State is based on a 1 km square grid [13]. The choice of one or the other type of spatial framework depends on the purpose of a system, the degree of detail and flexibility required and last, but not least, on the state of development of the region concerned. It may also happen that different sets of data are based on different frameworks, in which case the problem of integration on the basis of some common denominator occurs [34, 35].

### 2.3. Data Input, Storage, Retrieval, Processing and Output

Preparing spatial data for input to a computer accessible data store involves attaching locational identifiers, an operation commonly known as geocoding. The simplest type of locational identifier is an external index, such as a geographical name, and address or a postal code. On a stand-alone basis, however, such indices preclude a processing system from doing any real spatial analysis or mapping. It must therefore be possible to convert external codes to coordinate information. For example, address conversion systems have been developed for urban areas, where data are available frequently on the basis of street addresses [36]. Topological principles of geocoding have been employed for describing lines as being connected to other lines or areas being adjacent to other areas, ect. Such a scheme is useful for redundant encoding together with coordinate information for error checking. This is the case, for example, in the Dual Independent Map Encoding (DIME) system developed for the US census [37]. A true spatial information system must be based on a coordinate system. The choice is between terrestrial coordinates (latitudes and longitudes in a spherical or geodetic system) and plane coordinates based on some map projection. Of course, a conversion is always possible, but it is important that a common language is agreed upon for communication [38]. Coordinate encoding can be done explicitly in a line format (lines described as strings of point coordinates) or implicitly in a raster format (a matrix of numbers representing data values). In both cases compact encoding schemes can be employed to alleviate storage problems [39, 40].

Assuming that spatial information systems are normally based on digital principles, data have to be converted for input into a machine-readable form by digitization. Map data consist of points, lines and areas and are digitized by semi-automatic point digitizers and line tracers, automatic line followers or automatic raster scanners [41]. The latter may be of the optical drum scanner type [14] or the electronic image tube type (for example, TV cameras). Data in a character (letters and numbers) format (for example, on tables) are conventionally digitized by keypunching. More recently, some success has been attained in automatic character recognition [42].

For data storage the physical as well as the logical organization of data has to be considered. The physical organization is dictated by available memory devices. The memory of present computer systems is by necessity hierarchically organized since there is a tradeoff between access speed and capacity. Therefore, there is a range from low capacity, fast and direct (random) access devices, on which data are stored temporarily to be available as operands for processing, to large slow and serial access memories for the permanent storage of large amounts of data [43]. Ideally, one would like to have spatial data available in two-dimensional map replica form at all times, but this, for the above reason, is possible temporarily only. The recent development of optical memory techniques, particularly the ones associated with lasers and holographic principles, however, may provide a breakthrough in the near future and make possible the replacement of the present memory hierarchy with the cumbersome transfer of data by a single large direct access device [44]. The logical organization of data is known as the problem of file structures. The permanent storage of data is sequential in one way or another. However, the use of pointers makes it possible to divorce the logical from the physical data organization. A pointer behind a data value tells where the next logical entry is located [45]. Such file structures may be very flexible and facilitate updating. Of further interest to spatial information systems is the possibility of employing memories of the associative (or content addressable) type, where data can be fetched in terms of content rather than memory location. The integrated circuit (semiconductor) technology, which is presently being extended into the memory field, provides such a capability [46].

Present-day computers are essentially linear, or sequential, machines and therefore relatively slow. Often, a capability for carrying out computations in parallel would be desirable. This has been achieved in special computers such as the Illiac IV, which has 64 parallel processor units, each with its own memory [47]. While such machines exist as prototypes only, a certain degree of parallelism can be obtained in regular computers with bit plane storage schemes and word parallel processing [48]. Furthermore, speed can be increased by having certain operations interpreted and executed at machine level through non-programmable wired-in logic. Such techniques are known as microprogramming [49].

Batch processing can be contrasted with interactive processing. The latter becomes increasingly popular because of the possibility of combining the best abilities of machines and man in real time. It has become largely possible through the advent of minicomputers which enable a user to set up a dedicated system at relatively low cost [50]. Interactive systems make use usually of a cathode-ray display tube as a terminal, and input devices such as light pens, tablets, joysticks or mice [51]. Interactive approaches have been used particularly at the input end of information systems, for example, for on-line error correction of map digitization and name placement in map compilation [41]. They are also attractive, though more complex to implement, for retrieving data in a query mode [52].

Software considerations relate to systems software and to applications software. The former, consisting of such components as assemblers, macro processors, loaders, compilers, and operating systems, are normally provided by the computer manufacturers, but are also becoming available from other sources, particularly universities [53]. For applications software the level of language (0 level: machine language; 1st level: assembly language; 2nd level: procedure-oriented language; 3rd level: problem oriented language) has to be decided upon [54]. Most programs have been written in a 2nd level language such as FORTRAN. Special languages are available for interactive work, for example, BASIC [55]. Third level languages have been designed for the query of stored data. An example is TARELA developed for the Geographically Referenced Data Storage and Retrieval System (GRDSR) associated with the census of Canada [56]. Some type of two-dimensional language specifically designed for the description and analysis of spatial data would be highly desirable, but there has been only sporadic progress in this area [57].

To review possible types of processing operations on spatial data in any detail is obviously beyond the scope of this paper. However, we will briefly mention major categories of operations. We can distinguish between geometrical operations and thematical operations. The former operate on the geometrical configuration, the latter on the thematical content of spatial data. Furthermore, it is convenient to make a distinction between transformations and descriptions. A transformation turns a given map into a new map. It can be of the point, local or global variety. In a point operation, the result at an output point is dependent only on the corresponding input point, in a local operation it depends on a neighborhood area around the input point, and in a global operation it is affected by the entire map. These definitions are largely used in picture processing [40], but are obviously readily applicable to maps as well. Descriptions are applied locally or globally and provide some sort of summary measure for the map portion in question. It is worthwhile to note that for many purposes and under many circumstances the thematical values of spatial data can be regarded as constituting a surface. Therefore surface processing concepts are of special interest [39].

Data output is in the form of tables or, more commonly, maps. Character information is output on a typewriter terminal, a high speed printer or a display tube with built-in character generators. Optical projection of templates (on a plotter) [41] and, more recently, holographic character generation [58] are other possibilities. Data in map form can be output in the line or in the raster mode. For line work a variety of drum and flatbed pen plotters are on the market [41]. Some plotters are equipped with an optical writing system where a small spot of light is used to draw on a photosensitive surface [59]. Line data can also be generated on cathode-ray tubes of the calligraphic type [52]. Microfilm plotters are devices that transfer CRT output through a camera onto photographic film. A newer technique is the one of electron beam recording, whereby the beam exposes film directly in the tube vacuum [60]. Output of map data in the raster mode can be achieved by a variety of devices. The most popular, because of its general availablity, is the high speed printer. Others include electrolytic plotters (commonly used for facsimile transmission), electrostatic and ink jet plotters (61], as well as cathode-ray tubes operating in a TV-like mode [62] and various plotters using a film exposing light beam (the counterpart to the raster scanner on the input side) [63, 64].

## 3. Remote Sensing Subsystems

Having now some understanding of the principles of a spatial information system in general we can look at the special problems we must face when trying to integrate remote sensing data into such a system. The components and processes involved from the collection of remote sensing data to the input to the

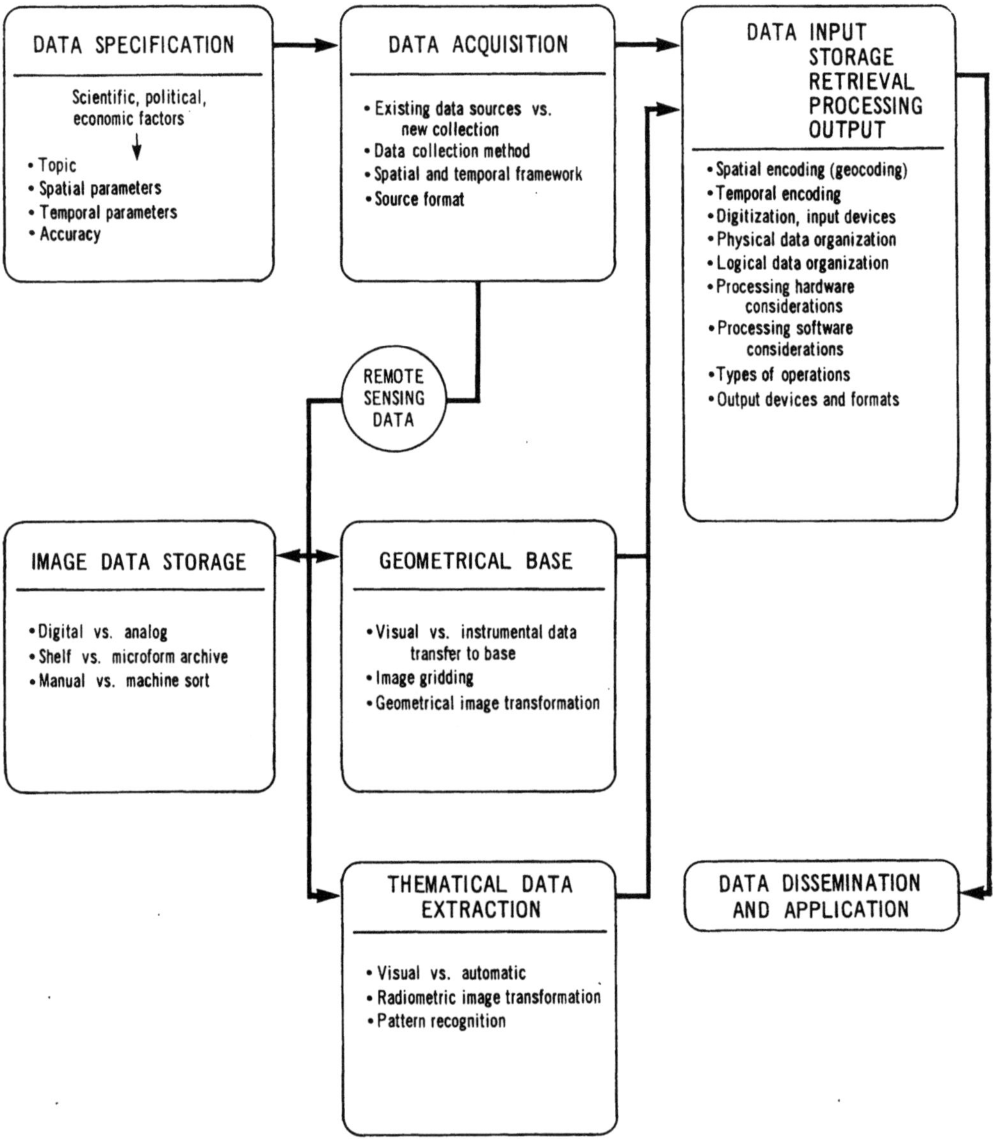


Fig. 1 : Major considerations in a spatial information system with a remote sensing subsystem information system can be thought of as forming a remote sensing subsystem (see Fig. 1). Remote sensing data are available in actual or latent image form. One immediate problem is the fact that remote sensing data are usually non-selective. The wealth of data collected has to be sorted and analyzed into selected useful information in a map format, whereupon it can be input to the information system in a normal fashion. The conventional procedure is visual analysis and interpretation. However, if we ever wish to make full use of the data providing potential of remote sensing some means of automating the data extraction process are needed [65]. We will refer to this question in the following as the thematical data extraction problem. A second obstacle is the usual occurrence of geometrical distortions in remote sensing imagery. One must be able to reference extracted data spatially to standard map coordinates. Solutions to this problem will be briefly reviewed under the heading of "the geometrical base problem". From the point-of-view of the overall information system then both geometrical and thematical image data handling can be regarded as preprocessing operations. A third problem we will examine below is the one dealing with image data storage. The conventional solution is some kind of manually sortable archive. It is obvious that, with the present state-of-the-art of computer memory technology, permanent storage of image data directly on-line to a digital computer is out of the question. However, we can think of having a machine sortable analog image data archive where selected images are transported to a read station. The storage of image data in analog, particularly optical analog, form is not only of interest from the point-of-view of compactness but also because of the possibility of direct input to optical processing systems. Such systems are capable of performing important image data handling operations as we will see below.

### 3.1. Image Data Storage Problem

The technology of semi-automatically or automatically sortable document retrieval systems has advanced to a point where we can seriously consider organizing a collection of images on this basis. Probably the greatest problem is the necessity of converting images to a microform. It is true that micro image storage and retrieval systems have been developed primarily for documents containing text or graphics. In this case, because of the binary nature of the "image", resolution is no problem. For remote sensing data gray level information must be available. Technical improvements, however, seem to make the miniaturization of continuous tone images feasible without too much loss of information. For example, weather satellite images are now converted to microforms on a routine basis by using electronic equipment specifically designed for high speed and high quality copying [66]. On the other hand, we should not forget that there is an inherent limitation in photographic recording in that there is a tradeoff between spatial resolution and gray tone resolution [67].

A variety of document retrieval systems has been designed and many of them are commercially available. In microfilm systems images are reproduced on 16 or 35 mm roll film. The pictorial records must be widely spaced to permit the insertion of codes for sorting. Newer systems make use of film cassettes. The user requests information via a keyboard and the computer supplies a cassette and a document number. The cassette has to be mounted in a film reader manually, but then the computer conducts the further search and displays the desired image [68]. In film card systems several pictorial records are printed on a film sheet (microfiche principle). A sorting of microfiche cards is achieved by means of magnetic codes [69] or with holes for a computer controlled needle sort technique [70]. Similar solutions can be used for aperture card systems which use opaque cards with the document mounted in a window [71]. Other developments are in the field of magnetic tape systems as an adaptation of video tape recording technique used for television [72].

### 3.2. Geometrical Base Problem

For input to a spatial information system data must be referenced to a coordinate grid based on some map projection. Remote sensing imagery does not normally have map-like quality and, consequently, spatial data referencing cannot be done without some intermediate operations. It is often found that, once a picture has been interpreted, the simplest and fastest approach to solving this problem is visual transfer of data to a base map without any instrumental aids. This is obviously not of very high, but often sufficient, accuracy. The comparison of images and maps can be facilitated by superimposing a distortion grid (plus possible additional information such as rivers or coast lines) on an image. This, however, requires knowledge of the image distortion (and therefore appropriate geometrical calibration) and the calculation of the grid. Image gridding has been applied to weather satellite imagery for quite some time now [66]. More accurate results can be obtained by the instrumental transfer of image locations to a map. Traditional analog photogrammetric methods are well established [73]. Of more recent advent are digital transfer methods. The location of image points is digitized, and a computer calculates the corresponding map positions. If distortions due to terrain elevation are negligible the measurements can be carried out on mono-imagery by using a conventional map digitizer [74]. Otherwise digitization must be done on a stereo model. Various digital readout devices have been developed which can be attached to conventional photogrammetric instruments [75]. Another solution is provided by analytical stereoplotters [74]. Some special devices have been built to handle more exotic remote sensing products such as line scanner and radar imagery [76, 77].

The possibly most interesting, but also the most elaborate, solution to the geometrical base problem is the one of geometrical image transformation. According to whether distortions induced by terrain height differences are considered or not one can distinguish between 3-D and 2-D transformations. The latter can be of the global or local type. Some simple global transformations can be achieved easily by analog optical means [78]. A well known example is tilt removal by projective transformation in conventional photogrammetry. Considerable research has been done on digital image transformations and, in fact, some procedures have been in operational use for some years. The geometrical processing of weather satellite imagery is an example [79]. Usually digital processing is based on local transformations, i.e. a picture is segmented into a number of mutually exclusive regions and a separate operation is carried out for each region. However, it is frequently useful to do an overall global correction first. It may then be possible to eliminate residual errors with simple low order local transformations [80]. A picture segmentation may be based on arbitrary polygons [81] or regular grids [79]. A combination of digital and hybrid techniques is employed in line scan modulation systems such as the ones used presently for the geometrical correction of ERTS imagery. The calculations are done by a digital computer, but the actual correction is implemented by scan pattern shaping either on the input or output side [82].

3-D image transformations are generally known under the name of "orthophotography". They consist, by necessity, of local operations. Orthophoto production comprises two basic components, image registration in 3-D (or terrain model profiling) and image mapping. The former consists of a correlation of corresponding points on the left and the right hand picture and the measurement of local $x$-parallaxes. The latter makes use of this information and converts local areas around these points to an output image with the desired orthographic property. Stereo image registration is commonly obtained by having an operator follow manually the terrain model with a scaning slit or by automatic electronic correlation with flying spot scanners [83]. Solutions for the image transfer include direct projection with variable magnification achieved by controlling the distance from the projector to the recording film plane according to terrain profile information, optical printing involving zoom lenses and rotation devices and electronic scan pattern shaping [84]. The latter two implementations have the advantage that they can improve the geometrical image quality by taking slope within local areas into consideration. A relatively recent innovation is the concept of stereo orthophotography. One image of a stereo pair is converted to a regular orthophoto. The opposite member is used to create a stereomate, which has horizontal parallaxes introduced as a function of terrain' height. Stereo orthophotography provides a means of combining accurate photogrammetric mapping with detailed photo interpretation in an ideal fashion [85].

### 3.3. Thematical Data Extraction Problem

While we are interested in automated data handling techniques for the purpose of thematical data extraction, we should not overlook the fact that the human eye-brain system is still the most effective image processor [65]. Certain simple human functions can, however, be duplicated by machines at higher speeds. Moreover, machines can handle multi-dimensional (for example, multi-spectral) data without problems. In contrast, the human capability for comparing several pictures simultaneously is definitely limited. We can distinguish broadly three groups of image data handling operations which are amenable to automatic implementations. The first group comprises radiometric image transformations. Such transformations, in contrast to geometrical operations, leave the geometry of a picture intact and operate on the radiometric content. Decision-making as to the content of the picture is not involved, Instead, the goal is to transform a picture into a new derived picture in such a way that it displays certain desirable properties which aid a subsequent human or machine interpretation. The second group comprises techniques which, on the basis of measurements taken on an image, do involve decisions with respect to phenomena shown on that image. In other words, they make use of taxonomic models. They can be summarized under the heading of pattern recognition. A third category, which we can call parametric mapping, has to do with predictive models (regression techniques): Some measurable image variable is related directly or indirectly to a ground parameter on a continuous scale. Examples are the prediction of plant productivity on the basis of radiometric levels [86] and the prediction of population size from dwelling unit counts [87]. In the following, we take a somewhat closer look at radiometric image transformations and pattern recognition.

Image transformations can consist of point or local operations and we can separate image enhancement from image restoration techniques [40]. The first category attempts to make certain aspects of the radiometric content of a picture more apparent, the second category tries to compensate for factors which have previously degraded the picture quality. These operations can be implemented in digital or analog (primarily electronic and optical) form. Digital calculation has the great advantage of complete flexibility and high accuracy, but it has to pay a price in terms of computing time if implemented on a conventional serial computer. In contrast, many analog solutions achieve partial or complete parallelism, which increases the speed, but the range of possible operations is normally limited and the accuracy of the results is lower [65].

A simple image enhancing point operation is density slicing (equidensitometry). It is of use if one may assume that certain density levels are associated with certain ground phenomena. Density slicing can be implemented digitally [88], analog photographically [89, 90] and analog electronically [91]. Increasingly popular is also the translation of density slices into color tones [92]. Another point operation, involving two or more pictures, is image addition or subtraction. The usual purpose is to enhance differences between different spectral bands or between images taken at different times. The results are normally shown in color. Common implementations include photographic superimposition [93], multiple optical projection and electronic combination [94].

Local image enhancing operations are image smoothing and image sharpening. The former suppresses the high-frequency content of a picture, the usual purpose being noise elimination, whereas the latter conversely suppresses low frequencies, the result being some form of edge enhancement. Such operations can be implemented in the spatial domain by convolving the filter function with the original picture, using digital [95], non-coherent optical [96, 97], photographic [98], electro-optical [99], optical-electronic [100] and electronic analog [101] solutions. Spatial filtering operations can also be carried out in the frequency domain. This requires transforming a picture to its frequency spectrum (Fourier transform), suppressing parts of this spectrum and reconstructing an image from what remains. The usual implementation is by means of a coherent optical processor [67, 102], but the development of fast algorithms makes the digital calculation more practical [103]. The concept used in image smoothing and image sharpening can be readily generalized to any type of bandpass filtering, including intermediate frequencies. Moreover, directional filtering can be applied as well. It is then possible to enhance or suppress features in an image that are associated with a certain frequency and/or direction [104].

Image restoring point operations try to remove radiometric distortions. If these distortions are systematic and the factors causing them are known (for example, look angle effects, lens shading) they can be eliminated by digital [80] or analog [105] correction. More difficult to handle is the occurrence of random variations. If multispectral data are available some form of channel ratioing may be helpful in this case [106]. Local operations are involved in procedures which restore the high-frequency content of a picture, which was lost previously as a result of, for example, image motion (modulation transfer correction) [80], and which remove coherent noise, a frequent degrading factor in the transmission of images from spacecraft [88].

Pattern recognition is concerned with the classification of sets of measurements into classes of interest. In the case of remote sensing, measurements are derived from imagery in actual or latent form. A set of measurements on one object is called a pattern. Most work has been done on multispectral scanner data [107]. More recently, experiments have been reported using digitized aerial or space photographs [108]. In some cases multi-temporal [109] and textural information, the latter derived either from the spatial or the frequency domain of a picture [110, 111], have been employed for recognition. Pattern recognition can also be based on shapes. In this case, an often appropriate implementation is by means of a local convolution operation, whereby a template matching the object being sought is scanned over the picture [40].

The concept of pattern recognition can be best understood by visualizing the process geometrically as a partition of the measurement space by decision boundaries into mutually exclusive regions, one for each category. Two groups of procedures can be distinguished: discriminant techniques and clustering techniques. In the first case, *a priori* information in the form of selected ground truth is available. This information is used as a training sample to establish some type of discriminant functions on the basis of which the unknown patterns can be classified (often called supervised learning). In the second case such information is not given. One tries to find structures in the measurement space in the form of pattern clusters and to identify the nature of these clusters afterwards (unsupervised learning). In both cases some measure of similarity, usually related to a distance or correlation concept, forms the basis of the process [107]. Most pattern recognition experiments have been done on digital computers, but in some cases faster electronic analog machines have been used as well [112].

Discriminant techniques can be of the statistical or deterministic type. Statistical methods are based on probability distributions [113], whereas deterministic solutions do not involve probability concepts. Furthermore, statistical approaches can be divided into a parametric and a non-parametric category. Parametric solutions make use of probability distributions of which the form is assumed to be known and analytically describable. The problem is to determine the parameters of the distribution. The best known example is the one which involves normal (Gaussian) distributions. The assumption of equal covariance in all groups leads to a linear discriminant analysis, its relaxation to allowing unequal covariances to a quadratic solution [114]. A typical example for a non-parametric approach is the histogram method: the measurement space is quantized into small regions, and decisions are taken on the basis of the frequencies with which patterns occurring in these regions belong to certain classes [115]. Deterministic methods are commonly based on some minimum distance concept, for example, distance to class centroids or distance to nearest neighbors [114].

Many different procedures have been proposed for pattern recognition by clustering. They are all essentially sequential. One either looks at the similarity of all possible pairs of patterns and lumps them together to successively larger groups, or one examines the total collection of given patterns and partitions the measurement space into successively smaller regions employing a certain discontinuity criterion. Examples of clustering methods include minimum distance techniques [116], minimum variation techniques [117], and dense space seeking techniques [107, 115]. The formidable computation problem in digital clustering can be alleviated by making use of the fact that neighboring image points are likely to belong to the same class. The introduction of a contiguity constraint reduces general clustering in the measurement space to spatial clustering (i.e. clustering in geographical space) [118].

Pattern recognition applied to remote sensing data has dealt primarily with the recognition of agricultural crops [119], but experiments with soil [120], rock [121] and natural vegetation [122] types have also been carried out. Good recognition accuracies of 90% and higher have been reported frequently, but this applies usually to relatively small areas only. Signature extension, i.e. wide area interpolation or extrapolation remains a major problem.

## 4. Conclusion

If we look at presently existing spatial information systems we find that, in general, remote sensing data are not as yet used to an extent which would reflect their information-providing potential. There are a number of reasons for this situation which we cannot review here. From a technical point-of-view the integration of remote sensing data into information systems is certainly possible. Operational solutions exist for the geometrical base problem. A safe way to overcome the thematical data extraction problem is still conventional visual image interpretation, except that this requires skilled interpreters, who are in short supply. The increasing need for data, and also the increasing potential of such programs as ERTS for providing data, make a certain degree of automation mandatory. We do not expect that fully automatic data extraction techniques will be completely operational in the near future. Once more, the best hope in the meantime seems to lie in man-machine interactive solutions.

## References

[1] A. D. Hall and R. E. Fagen, in: Modern Systems Research for the Behavioral Scientist, Aldine, Chicago 1968.
[2] F. J. Endelman et al., J. Env. Systems **2**, 3 (1972).
[3] R. McDaniel and M. E. E. Hurst, Comm. on College Geography Publ. No. 8, Assoc. Amer. Geogr., Washington 1968.
[4] A. R. Boyle, in: Geographical Data Handling Vol. 2, IGU Comm. on Geogr. Data Sensing and Processing, Ottawa 1972 (p. 1125).
[5] J. W. Harbaugh and D. F. Merriam, Computer Applications in Stratigraphic Analysis, J. Wiley, New York 1968.
[6] J. A. Gilliland and A. Treichel, Canad. J. Earth Sci., **5**, 1518 (1968).
[7] C. S. L. Ommanney and D. Gagnon, in: International Geography 1972 Vol. 2, Univ. of Toronto Press, Toronto 1972 (p. 980).
[8] G. B. Morgan et al., Science **170**, 289 (1970).
[9] S. A. Denenberg et al., NARIS, Center for Adv. Computation, Univ. of Illinois, Urbana 1971.
[10] B. Alfredsson et al., FRIS A:1, CFD Central Board for Real Estate Data, Sundbyberg, Sweden 1970.
[11] H. Haack, in: Papers 8th Ann. Conf. of URISA, Kent State Univ., Kent 1970 (p. 139).
[12] S. Filion, in: Les Méthodes de la Cartographie Urbaine, Dépt. de Géographie, Univ. de Sherbrooke, Sherbrooke 1972 (p. 81).
[13] D. J. Belcher et al., LUNR Final Rep. Vol. 1, Center for Aerial Photogr. Studies, Cornell Univ., Ithaca 1971.
[14] R. F. Tomlinson, in: Land Evaluation, Macmillan of Australia, South Melbourne 1968 (p. 200).
[15] B. Zavos, in: Proc. 7th Int. Symp. on Remote Sensing of Environment, Vol. 1, Willow Run Labs., Univ. of Michigan, Ann Arbor 1971 (p. 207).

[16] T. Hermansen, Reg. Sci. Assoc. Papers **22**, 107 (1969).
[17] C. L. Smith, Computing Surveys **2**, 211 (1970).
[18] J. S. Gardner, ETL-CR-72-1, US Army Engin. Topogr. Labs., Ft. Belvoir, Va. 1972.
[19] R. F. Tomlinson, Environment Information Systems, Univ. of Saskatchewan, Saskatoon 1970.
[20] R. F. Tomlinson, Geographical Data Handling, 2 Vols., IGU Comm. Geogr. Data Sensing and Processing, Ottawa 1972.
[21] B. S. Wellar, 54th Ann. Conf. of AIP, Paper No. 06, Amer. Inst. of Planners, Washington 1971.
[22] T. Hermansen, Lund Studies in Geogr. Ser. B, Human Geogr. No. 37, 1 (1971).
[23] P. Carrere et al., in: Data Banks for Development, Observatoire Economique Méditerranéen, Marseille 1971 (p. 135).
[24] R. L. Lowry, in: Proc. Business and Econ. Statistics Section, Amer. Statist. Assoc., 86 (1962).
[25] K. P. Rodgers, in: Earth Resources Survey Systems Vol. 1, US Govt. Printing Office, Washington 1972 (p. 179).
[26] G. M. Gaitz, Planning Techniques Paper No. 1, Dept. of the Environment, London (1971).
[27] R. H. Alexander, in: Earth Resources Survey Systems Vol. 1, US Govt. Printing Office, Washington 1972 (p. 327).
[28] D. Harvey, Explanation in Geography, E. Arnold, London 1969.
[29] J. Salmona, in: Data Banks for Development, Observatoire Economique Méditerranéen, Marseille 1971 (p. 33).
[30] J. P. Eaton and P. L. Ward, in: 4th Annual Earth Resources Program Review Vol. 3, NASA Manned Spacecraft Center, Houston 1972 (p. 56-1).
[31] J. A. Mabbutt, in: Land Evaluation, Macmillan of Australia, South Melbourne 1968 (p. 11).
[32] A. B. Brink et al., MEXE Rep. No. 940, Milit. Engng. Exp. Establishment, Christchurch, England 1966.
[33] G. W. Orning and L. Maki, Minnesota Land Management Information System Study Rep. No. 1, Center for Urban and Regional Affairs, Univ. of Minnesota, Minneapolis 1972.
[34] A. C. Hamilton, The Canad. Surveyor **23**, 12 (1969).
[35] D. A. Nichols et al., Proj. THEMIS Tech. Rep. T-71-4, Dept. of Geography, Univ. of California, Riverside (1971).
[36] P. J. Hubert, The Canad. Surveyor **23**, 151 (1969).
[37] US Bureau of the Census, Census Use Study Rep. No. 4, Washington 1970.
[38] L. A. Gale, The Canad. Surveyor **23**, 40 (1969).
[39] T. K. Peucker, Comm. on College Geography Resource Paper No. 17, Assoc. Amer. Geogr., Washington 1972.
[40] A. Rosenfeld, Picture Processing by Computer, Academic Press, New York 1969.
[41] A. R. Boyle, in: Geographical Data Handling Vol. 2, IGU Comm. Geogr. Data Sensing and Processing, Ottawa 1972 (p. 634).
[42] L. D. Harmon, Proc. IEEE **60**, 1165 (1972).
[43] E. C. Joseph, in: Computer Science, J. Wiley, New York 1972 (p. 103).
[44] J. A. Rajchman, RCA Optics, 67, RCA Labs., Princeton, N.J. 1971.
[45] G. G. Dodd, Computing Surveys **1**, 117 (1969).
[46] D. F. Calhoun, in: Computer Science, J. Wiley, New York 1972 (p. 1).
[47] D. L. Slotnick, Sci. American **226**, 76 (1972).
[48] E. G. Johnston, in: Picture Processing and Psychopictorics, Academic Press, New York 1970 (p. 427).
[49] S. S. Husson, Microprogramming: Principles and Practices, Prentice-Hall, Englewood Cliffs, N.J. 1970.
[50] B. Soucek, Minicomputers in Data Processing and Simulation, J. Wiley, New York 1972.

[51] R. J. Pankhurst, in: Advances in Information Systems Science Vol. 3, Plenum Press, New York 1970 (p. 215).
[52] C. T. Meadow, Man-Machine Communication, J. Wiley, New York 1970.
[53] J. J. Donovan, Systems Programming, McGraw Hill, New York 1972.
[54] A. F. Cardenas, in: Computer Science, J. Wiley, New York 1972 (p. 315).
[55] J. E. Sammet, Programming Languages: History and Fundamentals, Prentice-Hall, Englewood Cliffs, N.J. 1969.
[56] R. J. Ion et al., GRDSR, Dominion Bureau of Statistics, Ottawa 1970.
[57] M. F. Dacey, Ontario Geography **5**, 71 (1970).
[58] D. Meyerhofer, RCA Optics, 53, RCA Labs., Princeton, N.J. 1971.
[59] Experimental Cartography Unit, Automatic Cartography and Planning, Architectural Press, London 1971.
[60] D. M. Avedon, Info. Display **8**, 15 (1971).
[61] J. Peters et al., Printer/Plotter Considerations, Varian Data Machines, Palo Alto, Calif. 1972.
[62] A. M. Noll, Communications ACM **14**, 143 (1971).
[63] P. J. Min and D. R. Thompson, in: Papers 37th Ann. ASP Meeting, Amer. Soc. Photogrammetry, Washington 1971 (p. 720).
[64] M. Bryant, Photomation, Optronics Int., Chelmsford, Mass. 1973.
[65] J. K. Hawkins, in: Automatic Interpretation and Classification of Images, Academic Press, New York 1969 (p. 199).
[66] J. A. Leese et al., ESSA Tech. Rep. NESC 53, Nat. Env. Satellite Center, Washington 1970.
[67] K. Preston, Coherent Optical Computers, McGraw Hill, New York 1972.
[68] A. Teplitz, Annual Review of Info. Sci. and Technology **5**, 87 (1970).
[69] B. A. Lipetz, Sci. American **215**, 224 (1966).
[70] S. R. Ray, Rep. No. 224, Dept. of Computer Sci., Univ. of Illinois, Urbana 1967.
[71] D. D. Moyer, The Canad. Surveyor **23**, 132 (1969).
[72] C. A. Steinberg, Info. Display **7**, 25 (1970).
[73] H. G. Jerie, Int. Symp. on Photointerpretation ISP Comm. VII, Intercomm. Working Group IV/VII, ITC, Delft 1970.
[74] J. Diello, Paper for Int. Cartogr. Assoc. Conf., Rome Air Development Center, Rome, N.Y. 1972.
[75] G. Petrie, The Cartogr. J. **3**, 60 (1966)
[76] R. B. Forrest and D. P. Hattaway, Bendix Tech. J. **1**, 81 (1968).
[77] M. V. Jonah and C. Dicarlo, in: Papers ACSM-ASP Techn. Fall Conference, Amer. Soc. Photogrammetry, Washington 1970 (p. 207).
[78] J. W. Hardy, Paper for Int. Congr. for Photogrammetry, Itek Corp., Lexington, Mass. 1968.
[79] C. L. Bristor, Techn. Memo NESCTM-3, Nat. Env. Satellite Center, Washington 1970.
[80] P. Will et al., IBM Research RC 3027, IBM T. J. Watson Research Center, Yorktown Heights, N.Y. 1970.
[81] C. Linstedt, IBM Research RC 3314, IBM T. J. Watson Research Center, Yorktown Heights, N.Y. 1971.
[82] Bendix Aerospace Systems Division, BSR 2855, Ann Arbor, Mich. 1969.
[83] D. C. Kowalski, Bendix Tech. J. **1**, 63 (1968).
[84] J. J. Edmond, Bendix Tech. J. **1**, 32 (1968).
[85] S. H. Collins, The Canad. Surveyor **22**, 167 (1968).
[86] R. S. Driscoll et al., in: Papers 38th Annual Meeting of ASP, Amer. Soc. Photogrammetry, Washington 1972 (p. 358).
[87] R. Binsell, Rep. USGS Contract No. 14-08-0001-10654, Dept. of Geography, Northwestern Univ., Evanston, Ill. 1967.
[88] F. C. Billingsley, in: Proc. Conf. on Parallel Image Processing, NASA Goddard Space Flight Center, Greenbelt, Md. 1972 (p. 41).
[89] A. Reinhold, Albrecht-Thaer Archiv **11**, 203 (1967).

[90] E. Ranz and S. Schneider, Bildmessung u. Luftbildwesen **38**, 123 (1970).
[91] R. R. McHail, in: Papers 37th Ann. Meeting of ASP, Amer. Soc. Photogrammetry, Washington 1971 (p. 290).
[92] D. Rutland, Info. Display **7**, 20 (1970).
[93] C. E. Molineux, Photogramm. Engng **31**, 131 (1965).
[94] J. D. Lent, in: Monitoring Earth Resources from Aircraft and Spacecraft, NASA Sci. and Tech. Info. Office, Washington 1971 (p. 9).
[95] A. Rosenfeld et al., in: Picture Processing and Psychopictorics, Academic Press, New York 1970 (p. 381).
[96] E. A. Trabka and P. G. Roetling, J. Opt. Soc. Am. **54**, 1242 (1964).
[97] J. K. Hawkins and C. J. Munsey, J. Opt. Soc. Am. **54**, 998 (1964).
[98] D. S. Ross, in: Papers 35th Ann. Meeting of ASP, Amer. Soc. Photogrammetry, Washington 1969 (p. 301).
[99] J. K. Hawkins, in: Pictorial Pattern Recognition, Thompson, Washington 1968 (p. 373).
[100] W. Swindell, Appl. Opt. **9**, 2459 (1970).
[101] M. D. Freedman, Bendix Tech. J. **2**, 86 (1969).
[102] A. R. Shulman, Optical Data Processing, J. Wiley, New York 1970.
[103] H. C. Andrews, Computer Techniques in Image Processing, Academic Press, New York 1970.
[104] R. Chevallier et al., Photogrammetria **26**, 17 (1970).
[105] D. T. Edson, in: Papers ASP-ACSM Fall Convention, Amer. Soc. Photogrammetry, Washington 1971 (p. 554).
[106] F. J. Kriegler et al., in: Proc. 6th Int. Symposium on Remote Sensing of Environment Vol. 1, Willow Run Labs., Univ. of Michigan, Ann Arbor 1969 (p. 97).
[107] K. S. Fu, Tech. Rep. EE 71—13, School of Elect. Engng, Purdue Univ., Lafayette 1971.
[108] P. E. Anuta and R. B. MacDonald, Remote Sensing of Environment **2**, 53 (1971).
[109] D. Steiner, Photogramm. Engng. **36**, 187 (1970).
[110] R. M. Haralick, in: 4th Ann. Earth Resources Program Review Vol. 2, NASA Manned Spacecraft Center, Houston 1972 (p. 38-19).
[111] G. C. Lendaris and G. L. Stanley, Proc. IEEE **58**, 198 (1970).
[112] R. F. Marshall and F. J. Kriegler, in: Proc. 7th Int. Symp. on Remote Sensing of Environment Vol. 3, Willow Run Labs., Univ. of Michigan, Ann Arbor 1971 (p. 2169).
[113] P. W. Cooper, in: Automatic Interpretation and Classification of Images, Academic Press, New York 1969 (p. 97).
[114] D. Steiner, Geoforum **2**, 75 (1970).
[115] R. M. Haralick, CRES Tech. Rep. No. 133-11, Univ. of Kansas, Lawrence 1969.
[116] G. P. Cardillo and D. Landgrebe, LARS Info. Note 101866, Purdue Univ., Lafayette 1968.
[117] D. Steiner et al., in: Proc. 6th Int. Symp. on Remote Sensing of Environment Vol. 2, Willow Run Labs., Univ. of Michigan, Ann Arbor 1969 (p. 895).
[118] G. Nagy and J. Tolaba, IBM J. Research and Development **16**, 138 (1972).
[119] Laboratory for Agricultural Remote Sensing, Remote Multispectral Sensing in Agriculture Vol. 4, Purdue Univ., Lafayette 1970.
[120] M. G. Tanguay et al., in: Proc. 6th Int. Symp. on Remote Sensing of Environment Vol. 1, Willow Run Labs., Univ. of Michigan, Ann Arbor 1969 (p. 33).
[121] R. D. Watson and L. C. Rowan, in: Proc. 7th Int. Symp. on Remote Sensing of Environment Vol. 3, Willow Run Labs., Univ. of Michigan, Ann Arbor 1971 (p. 2043).
[122] H. W. Smedes, in: Earth Resources Survey Systems Vol. 2, US Govt. Printing Office, Washington 1971 (p. 345).

Earth Survey Problems — Akademie-Verlag, Berlin 1974

# A LOW-COST SYSTEM FOR REPRODUCING ERTS IMAGERY

J. S. MacDonald[a], D. S. Sloan[a], R. E. Barrington[b] and W. Rolfe[b]

[a]MacDonald, Dettwiler and Associates Ltd., Vancouver, British Columbia, Canada
[b]Communications Research Center, Ottawa, Ontario, Canada

A system is described which is capable of reproducing the scanner (MSS) or vidicon (RBV) imagery from ERTS at full resolution and relatively moderate cost. This system, which uses a high resolution cathode-ray tube system and photographic film to display and record the data, has demonstrated its ability to produce high-quality imagery at both the Canadian and Brazilian ERTS ground stations. The system, in its simplest form, reproduces an unannotated image from one MSS or RBV spectral band at a time. The MSS images are corrected for line length, sinusoidal scan and S-bend distortions. This system has proven very useful as a check-out of the total ground station and of some of its components, such as tape recorders, and provides a simple yet effective method for quality control of the tapes shipped from the ground station. More important, it provides the first available indication of the quality of the imagery that has been recorded from the satellite. The imagery so obtained is useful for many purposes requiring rapid response time.

More advanced versions of the system are built around a mini-computer and provide such features as date, time and geographic annotation, radiometric correction for sensor imbalance, and automatic frame control, as well as offering the facility for the application of elementary image correction and enhancement techniques.

Earth Survey Problems — Akademie-Verlag, Berlin 1974

# REPORT OF THE FIRST PAN AMERICAN SYMPOSIUM ON REMOTE SENSING

L. Goldstein

Defênse Mapping Agency, Inter American Geodetic Survey,
Fort Clayton, Panama Canal Zone

This paper describes the major scientific and technical results of the First Pan American Symposium on Remote Sensing which was held in Panama, Republic of Panama, in conjunction with the X General Assembly of the Pan American Institute of Geography and History (PAIGH).

The First Pan American Symposium on Remote Sensing was held in Panama from 27 April to 2 May 1973 in conjunction with the X General Assembly of the Pan American Institute of Geography and History (PAIGH), a specialized component of the Organization of American States (OAS). The symposium was jointly sponsored by the PAIGH, the National Geographic Institute "Tommy Guardia" of Panama, the United States Geological Survey (USGS), the Defense Mapping Agency Inter American Geodetic Survey (DMA IAGS) and the OAS.

Approximately 300 delegates from Canada, the United States, Mexico, Central America, Panama, South America and Europe attended the Symposium. Forty-seven papers were presented relating to remote sensing technology and its application to investigation of earth resources through satellite and airborne platforms.

The papers were given in four technical sessions: The investigation of satellite imagery for earth resources, Basic research instrumentation and automatic data processing, Pan American remote sensing programs and Remote sensing applications in geography and cartography.

In the wide spectrum of remote sensing technology, papers were presented to encompass data acquisition, processing and presentation. M. Beilock of the Bendix Corporation spoke on the Brazil ERTS Data Receiving and Processing Station which is now in the final stages of construction. Data acquisition and processing were discussed by M. A. Jager of the Canada Center for Remote Sensing and A. Higer of the USGS. Analog experiments in multi-spectral scanner data collected in Panama were discussed by T. R. Ory of Daedalus Enterprises. Distribution of ERTS imagery as well as publications relating to the field of remote sensing was presented by P. Truxton of IAGS who has traveled throughout Latin America setting up distribution centers. The many interesting reports on geology included the application of ERTS imagery to geology by C. Brockmann of the Geological Service of Bolivia and W. Hemphill of USGS who spoke on ERTS experiments of the EROS Program.

In the field of engineering geology, T. West of Purdue University explained automatic data processing to obtain soil characteristics for highway planning.

A study of the seismic damage patterns during the recent earthquake in Managua, Nicaragua, using photo interpretation techniques was presented by G. Navarro of Nicaragua. D. Brunnschweiler of Michigan State University investigated geomorphic and ecologic aspects of the Andes of Peru comparing existing maps with imagery from the Gemini spacecraft. Utilizing infrared aerial photography, L. LeSchack of Maryland gave illustrations in predicting and reducing costs in jungle land clearing projects. These techniques were applied in studying a road construction project in the Darien of Panama. Tropical land use discrimination was explained by R. Mower using multi-spectral imagery obtained from aircraft and satellite platforms over Puerto Rico.

In the field of agriculture, numerous presentations were made which have enabled scientists and technicians to study changing crops patterns as well as to make inventories of crops and to effect the control of crop diseases and blights. The utilization of ERTS RBV imagery in the Atacama Desert of Chile, presented by A. Cortinez, University of Chile, covered fundamental aspects of geology, drainage and irrigation. The use of aircraft and spatial imagery in the inventory of agricultural and forest natural resources was given by F. and S. Lopez of the Escuela Tecnica Superior, Madrid, Spain. Control of coffee blight and inventory of crops were presented in a paper by M. H. Velloso of Brazil. A. Cortinez presented a paper on the detection and control of plagues in the cultivation of sugar beets. F. Canney of USGS discussed the search for mineral deposits by the use of thermal scanner imagery and the study of advanced geochemical and geophysical processes.

In the fields of geography and geomorphology, satellite imagery was utilized in the interpretation of the Llanos of Colombia by W. Elbersen. R. Holz and D. Conway of the University of Texas examined the morphology of alluvial floodplains in northwest Argentina using Ektachrome infrared aerial film. Resource inventories, utilizing radar and satellite imagery, to manage an ever-increasing world population were presented by J. Everett of Earthsat Corporation. He stated that a carefully planned and integrated use of ERTS and SLAR imagery can provide a powerful system for the economic and timely acquisition of geographic information of Latin America. SLAR imagery of Venezuela was discussed by R. Lairet of Caracas for natural resource inventories in the Amazon Basin. C. Molina of CIAF (Centro Interamericano de Fotointerpretacion) discussed mapping of the Colombia Tropical Forest. Land use studies utilizing remote sensing data from aircraft and satellite systems utilizing computer processing was discussed by R. Nash of North America Rockwell. Evaluation of natural resources on the frontier of Mexico and the United States was presented by M. Aceves, University of Mexico. He presented remote sensing techniques and methods of delineating area borders of water, vegetation, fauna and soil use.

In the field of archaeology exploratory techniques utilizing color and false color imagery from aircraft platforms were described by H. Svehlak of General Electric.

In oceanography and hydrology, A. Higer discussed a prototype multi-parameter data acquisition network, installed and operated by USGS Water Resources Division. This is a viable approach to the hydrologic problems confronting nearly 2.5 million residents of central and south Florida. Monitoring this vast environment is now possible from ground stations, via radio telemetry communication with the ERTS satellite. J. Uliberrena of Argentina explained the application of multiband imagery for the study of coastal processes in the South Atlantic for the determination of suitable deep water port facilities. B. Van Tries from the United States Bureau of Sport Fisheries and Wildlife explained the ERTS Alaska experiment in which migratory habits of fish as well as upwelling of sea nutrients are being studied utilizing different bands of MSS imagery from the ERTS 1 satellite.

In the field of cartography, ERTS imagery was discussed by C. Brockmann. Mapping errors, using SLAR imagery of Australia were discussed by H. Jensen of Aero Service Corporation. R. Joosten of NASA discussed one of the cartographic investigations of Skylab. It will involve 14 Latin American nations in association with IAGS. These Latin American investigators will join with other scientists to determine the usefulness and potential of space acquired data for direct application to individual mapping objectives. The cartographic experiments of Skylab are expected to obtain photographic source data suitable for the preparation of photomaps, thematic maps, land use maps of medium and large scales and planimetric compilations for map revision activities.

There were also four discussion panel meetings held during the symposium: Future earth resources satellite needs, Tropical environmental study area, Educational needs, and Standardization of remote sensing terms. These panels were composed of experts in many disciplines related to earth resources as well as scientists and professors of universities throughout the Western Hemisphere. Significant recommendations made by the panel of Future Earth Resources Satellite Needs were: (i) That there should be improved communications related to earth resources satellites between members of PAIGH and established scientific and technical organizations. (ii) In regard to the concept of multinational funding of operational earth satellites, it was the consensus of the panel that such proposals are premature at this experimental phase of the program. It was suggested that nations begin careful consideration of the benefits to be derived from operational systems so that meaningful governmental recommendations can be developed for future financial participation. (iii) It was the consensus of the panel that a sun-synchronous satellite, containing an infrared scanner would be of great value. (iv) Also, it was suggested that it could be desirable for an operational satellite to contain a radar system which would survey floods which often take place at times when clouds prevent observations with visible or infrared systems.

The panel on Tropical Environmental Studies proposed that there should be: (i) regional and multidisciplinary research; (ii) technological exchanges and training; (iii) selection of multinational regional research areas.

The panel of Standardization of Remote Sensing Terms recommended that a work group be formed to study the feasibility of publishing a trilingual glossary on remote sensing terms for use in the western hemisphere. The languages would be English, Spanish and Portuguese.

The panel on Educational Needs recommended that representatives in PAIGH member nations obtain complete information concerning all facilities offering training in remote sensing techniques and the applications of these techniques. This information should include length of course, levels of instruction, content of the program, student capacity, instructor staff, cost, and the availability of scholarships for students.

Albert Einstein once said, "Concern for man and his fate must always form the chief interest of all technical endeavors". This is especially true in Latin America, where despite a decade of dynamic, economic and social progress, the problems of population explosion, efficient utilization of natural and human resources, industrialization and the social well-being of the individual continue to present major obstacles that must be surmounted in order to achieve the desired rate of development.

This symposium has illustrated that the developing nations of Central and South America are accelerating their efforts, and are drawing abreast of the more developed nations of North America and Europe. Brazil and Mexico have strong programs in research and are implementing programs to solve their natural resources and environmental needs through the use of remote sensing technology. Other countries' scientific and technical organizations are fast approaching these goals. Countries actively fostering the applications of remote sensing technology are Argentina, Bolivia, Peru, Colombia, Venezuela, Guatemala, Nicaragua and Panama. This technology will add to the information already available in the more developed areas where Latin America is beginning to experience environmental problems that accompany a country's growth. Hopefully, this technology will be used to avoid many of the mistakes made by other nations in their development process. However, it was pointed out that although this imagery contains a wealth of information, it will serve no useful purpose unless people are trained to extract this information and put it into a medium that can be used as a tool for the planning and implementation of development programs.

This First Pan American Symposium on Remote Sensing served to bring together dedicated men and women of science for an interchange of ideas and applications of remote sensing technology to the immediate and future needs of the nations of the western hemisphere.

# National Reports on Activities in Earth Surveys

Earth Survey Problems — Akademie-Verlag, Berlin 1974

# NATIONAL REPORT OF SWITZERLAND ON EARTH RESOURCES OBSERVATION FROM SATELLITE IMAGERY

H. Haefner and K. Itten

Department of Geography, University of Zürich, Zürich, Switzerland

The following activities of earth resources observation from space have been undertaken or are under progress by Swiss researchers: geographical application of Gemini photography; earth resources studies with ESSA-APT and Nimbus-HRIR pictures; cloud pattern, circulation of air masses, and fog distribution with meteorological satellite pictures; the Swiss ERTS 1 research program; future earth resources satellite projects.

## 1. Introduction

Earth resources studies including research on weather elements from satellite imagery in Switzerland date back to 1965. Various institutions are actively engaged in this research, in particular the Swiss Institute of Meteorology, Zürich, the Department of Geography, University of Berne, and the Department of Geography, University of Zürich.

## 2. Geographical Application of Gemini Photography

Gemini photography for the first time enabled an overview over broad areas on high quality photography. It reveals the structure of large landforms, their broad natural and cultural pattern, their distribution, combination, composition, synthesis in geographical regions and even their arrangement and interrelationship. Studies were undertaken mainly in Northern Africa [1, 2] which lead to a much better understanding of the formal, functional and genetic aspects and to a regionalization of these landscapes.

## 3. Earth Resources Studies with ESSA-APT and Nimbus-HRIR Pictures

Capabilities and limitations of meteorological satellite images for earth resources studies were examined systematically at the Department of Geography, University of Zürich [3, 4]. In addition various specific applications [5] were studied and processing techniques and instruments tested for an improvement and facilitation of the interpretation. The great advantage of this imagery is the daily repetitive coverage of the same area which allows an elaborate observation and mapping of fast changing phenomena such as snow, ice, water, flood, rainfall pattern etc. and the seasonal fluctuation of the vegetation growth, the snow cover, the soil moisture etc. Its most serious handicap is the rather poor ground resolution. In particular the following problems were studied:

*Geographical interpretability.* For earth resources studies as above-mentioned it is necessary to select a time period of relatively long and continuous good weather, or "geographical interpretability". Therefore twelve monthly maps, showing the number of days of good and fair interpretability of geographical features were composed on the basis of a spheric grid of 5 degrees for Europe and Northern Africa from ESSA-satellite pictures [3].

*Thematic applications.* A selection of the regional and thematic examples studied in combination with different techniques and instruments, used as an interpretation aid is given in Table 1. Methods and results are summarized best in a paper by Itten et al. [6].

Table 1

Methods and Examples of Earth Resources Studies from Meteorological Satellite Images ([6] p. 365).

| Technique | Interpretation aids | Expenditure | | Selected regional and thematic examples |
|---|---|---|---|---|
| | | Finances | Time | |
| Direct optical comparison | Projector, enlarger, stereoscope Interference projector | small | large | Pack ice between Eastern Greenland and Iceland |
| Direct optical-mechanical comparison | Enlarger Constructed transparent interpretation templates | small to medium | large | Snow lines in Swiss Alps |
| Optical-electronic processing + direct optical comparison | Isodensitracer | large | small to medium | Humid and arid regions of the Iberian Penninsula |
| Photographical processing + direct optical comparison | Agfacontour film | small | small to medium | Desert areas of Arabian Penninsula |

## 4. Cloud Pattern, Circulation of Air Masses, and Fog Distribution with Meteorological Satellite Pictures

### 4.1. Cloud Coverage and Geomorphology

Studies of atmospheric circulation and the cloud coverage from satellite imagery not only provide a much more accurate picture of the dynamic processes in the atmosphere but allow additional conclusions regarding wind, precipitation etc. In particular in remote areas with very few stations for meteorological ground observation this information may be very useful for hydrological, geomorphological and regional research. The relations between weather elements as determined on satellite photography and geomorphological processes in its regional differentiation in Northern Africa, especially of the Hoggar and Tibesti Mtns, are the subject of extensive research at the Department of Geography, University of Berne [7, 8], producing new and interesting results.

### 4.2. Fog Distribution

Another weather element which is of great importance for geographical studies and for planning purposes is the distribution of fog which again can be mapped best in its areal extent and periodicity from satellite pictures. A local study was undertaken for the Cantone of Berne and neighbouring areas [9], combining the information from satellite pictures with those of ground observation and aerial photography.

A more comprehensive and comparative investigation on a smaller scale reviewing the situation around the Alps from the Po Valley to the Swiss Plateau, the Rhine Valley and Burgundy is under progress [10]. It already takes advantage of the occasional high resolution ERTS images for an improvement of the accuracy as reached on the daily available ESSA-satellite pictures.

## 5. The Swiss ERTS 1 Research Program

Switzerland has an ERTS 1 project accepted by NASA on "Snow Survey and Vegetation Growth in the Swiss Alps". Several research institutions are combining their efforts and rendering their facilities and equipment for an extensive and thorough interdisciplinary research of the 25 different ERTS 1 image series received between August 1972 and February 1973. The following institutions and investigators are involved:

| | |
|---|---|
| Geography: | Dept. of Geography, Univ. of Zürich (Prof. H. Haefner, Principal Investigator; K. Itten)<br>Dept. of Geography, Univ. of Berne (Prof. B. Messerli) |
| Snow Research: | Swiss Federal Institute of Snow and Avalanche Research, Weissfluhjoch-Davos (Prof. M. de Quervain) |
| Meteorology: | Swiss Institute of Meteorology, Zürich (Dr. A. Piaget) |
| Cartography: | Dept. of Cartography, Swiss Fed. Inst. of Technology, Zürich (Prof. E. Spiess) |
| Photography: | Dept. of Photography, Swiss Fed. Inst. of Technology, Zürich (Prof. W. F. Berg) |
| Photogrammetry: | Dept. of Photogrammetry, Swiss Fed. Inst. of Technology, Zürich (Prof. H. Kasper) |
| Geophysics: | Dept. of Geophysics, Univ. of Milano (Prof. R. Cassinis) |

The investigations are under progress in different regions and on different subjects. The most important aspects will be summarized briefly.

1

2

3

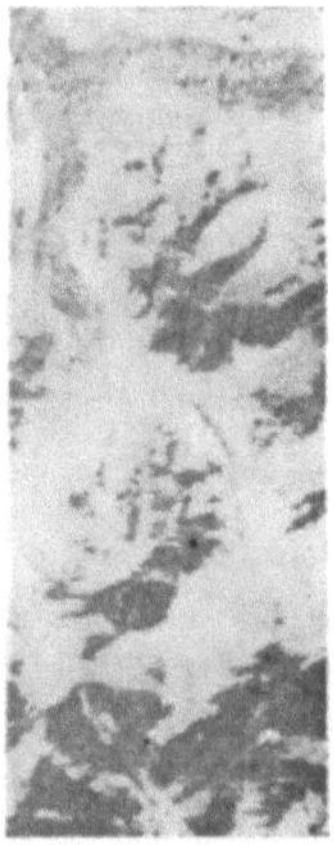

4

5

6

### 5.1. Snow Survey (Fig. 7)

The Department of Geography, University of Berne, again concentrates on more local and detailed studies in the Cantone of Berne to map the temporary snow lines and to examine their variation in different small alpine valleys, in different exposures etc. [11].

A team of investigators from the Department of Geography, University of Zürich and the Department of Photography, Swiss Federal Institute of Technology, Zürich, concentrates on the development of an operational mapping system first from the photographs and later directly from magnetic tape.

So far a method has been developed which combines the possibilities of a Quantimet QTM 720 to discriminate the exact density level of the snow cover for each individual image by visual control with the higher resolution of photographic techniques [12]. The density level determined is used as index for the exposure time of a high resolution lith-film. In addition a careful comparison of the different bands showed that the registered extent of an old snow cover in band 7 is smaller than in band 4. Photos taken immediately after a new snowfall do not show this phenomenon. It has to be concluded that with a combination of the different bands possibilities for a separation of different snow-types exist [13].

### 5.2. Meteorology (Fig. 7)

A combination of meteorological satellite pictures (ESSA/NOAA), ERTS images, weather maps and vertical temperature and humidity profiles allows a synoptic interpretation of the evolution of mesoscale, even microscale meteorological systems and their influence on different ecosystems. First results from the investigations by Piaget give a description of the organization of cumulus clouds

⟵

Figs. 1—6. Cartographic aspects of ERTS imagery (after Hermann [16]).

Fig. 1. A traditional form of oblique illuminated shading (so called Swiss manner), Art. Institute Orell Füssli AG, 1971. The test site shows the area of the Swiss Alps from the Lake of Walenstadt in the North to the Vorderrhein-Valley in the South.

Fig. 2. Topographic map of Switzerland 1:500000 with traditional form of plastic shading.

Fig. 3. Negative of ERTS 1 image of 7 October 1972, taken at 1055 with a sun-angle of 33° and an azimuth of 154°. Negative from color composite MSS bands 4, 5, 7.

Fig. 4. Mosaic of small scale aerial photographies reduced to 1:500000 of 18 December 1972 (Courtesy Luftaufklärungsdienst, Dübendorf, Switzerland).

Fig. 5. Oblique illuminated shading from ERTS image. A negative copy (Fig. 3) was exposed in such a way that on the illuminated slopes no grey tones were visible any longer and with additional corrections, drawn with a fine brush and black ink. The low sun-angle shows the relief in a way similar to Fig. 1 but with more details.

Fig. 6. Combination of traditional topo-map 1:500000 with ERTS image and small scale aerial photography. The linear elements are from the traditional topographic map (Fig. 2). The plastic shading is based on ERTS image (Fig. 5) and the forested areas are interpreted from ERTS image and aerial photo mosaic (Figs. 3 and 4).

over mountain terrain, generation and decay of cloud pattern, aspects of cloud layers just before dissipation in or over mountain valleys, the development of fading of convections, night inversions, the determination of cumulus tops and partly of the bottoms etc. [14]. The investigations of which the first results are presented in [15] will also be continued.

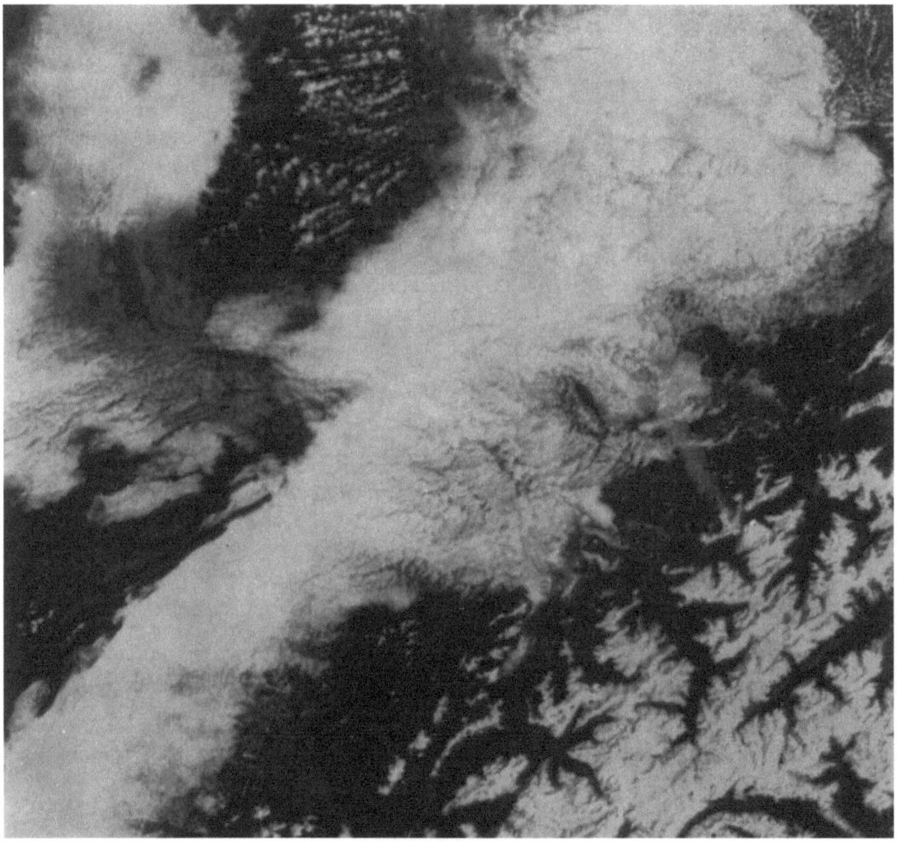

Fig. 7. The ERTS image of 20 September 1972 shows in the lower right, part of the snow covered Swiss Alps (taken by NASA Goddard Space Flight Center, Greenbelt, Maryland) with its typical dendritic contour-line appearance. The large white diagonal band represents a fog layer over the Swiss Plateau, a quite common feature in spring and autumn, caused by an inversion layer. Most of the large cities of Switzerland are covered by this fog layer.

### 5.3. Cartographic Aspects (Figs. 1—6)

ERTS images are an excellent tool for cartography, especially for the presentation of the relief of mountain terrain. A careful visual comparison between drainage and woodland features on the photos and the National Map 1:500000 were made of a test area of about 17000 km². As a result it was found that ERTS photos are precise enough as compilation basis for maps smaller than 1:250000. A black and white negative copy 1:500000 of an enlarged ERTS 1 color composite of a test area (Lake Walenstadt and vicinity) was used as model for plastic shading. With a relative minimum amount of time the oblique illuminated shading could be completed and printed on a traditional topographic map 1:500000 [16]. The method could well improve and rationalize the shading technique for the graphical presentation of the relief in small-scale mapping (Figs. 1—6).

## 6. Future Earth Resources Satellite Projects

At the moment of writing this report the countdown for the Skylab has just started and we are looking forward to a successful launch. The same Swiss group of investigators will study similar problems of snow distribution and vegetation growth in the high mountains, to improve and supplement the ERTS 1 investigations. More detailed information regarding ground resolution, differentiation of snow and vegetation types etc. is expected.

Finally it is hoped to continue and extend the ERTS 1 and EREP projects regionally, thematically and methodologically towards a natural resources inventory and land evaluation and classification, as suggested to NASA in our ERTS B project. In combination with the daily receivable imagery from the NOAA it could contribute tremendously to improve and accelerate the data acquisition for monitoring and inventorying our environment.

## 7. References

[1] H. Haefner, in: Earth Resource Surveys from Spacecraft, NASA, Washington, D.C. 1966 (D-5), partly reprinted in: Ecological Surveys from Space, NASA, Washington, D.C. SP-230 (p. 6).
[2] B. Messerli, in: Jahresber. Geogr. Ges. Bern, Bd. 49, 1970 (p. 139).
[3] K. Itten, M. A. Thesis, Dept. of Geography, University of Zürich (1969).
[4] R. Gfeller, M. A. Thesis, Dept. of Geography, University of Zürich (1972).
[5] K. Itten, in: Berichte des III. Int. Symp. für Photointerpretation, Dresden 1970 (p. 455).
[6] K. Itten, R. Gfeller and H. Haefner, in: XIX Rassegna Internazionale Elettronice Nucleare ed Aerospaziale, Roma 1972 (p. 365).
[7] M. Winiger, M. A. Thesis, Dept. of Geography, University of Berne (1971).
[8] M. Winiger, Hochgebirgsforschung 2, 87 (1973).
[9] H. Wanner, Mitteilungen, Geogr. Institut, Universität Bern (1971).
[10] F. Schacher, M. A. Thesis, Dept. of Geography, University of Zürich (1974) (under progress).
[11] M. Winiger, in: Mitteilungen, Geogr. Institut, Universität Bern (1972).
[12] R. Gfeller and K. Seidel, in: Proc. Coll. Int. Satellites Méteorologiques, Centre Nat. d'Etudes Spatiales, Paris (1973).
[13] H. Haefner, R. Gfeller and K. Seidel, this volume, p. 351.
[14] A. Piaget, Arbeitsber. Schweiz. Meteorolog. Zentralanstalt, No. 37 Zürich 1973.
[15] A. Piaget, paper presented at this Symposium, not published.
[16] Ch. Herrmann, Kartographische Nachrichten 4, 148 (1973).

Earth Survey Problems — Akademie-Verlag, Berlin 1974

# SOUTH AFRICAN ACTIVITIES IN EARTH SURVEYS

O. G. Malan

National Physical Research Laboratory, Council for Scientific and Industrial Research, Pretoria, Republic of South Africa

During 1972 the Republic of South Africa has become actively engaged in the application of space techniques for earth surveys: two South African proposals for participation in the ERTS 1 programme were accepted by NASA. The first of these is a multi-disciplinary proposal for assessment of satellite imagery in national resource surveys and the second an attempt to monitor vegetation growth on mine dumps. In order to stimulate interest and assess the present state-of-the-art, a Symposium on Remote Sensing was held in Pretoria in May 1972, attended by over 150 delegates including several from the neighbouring states Botswana, Lesotho, Rhodesia and Swaziland.

The preliminary analyses of ERTS 1 imagery of South Africa have indicated that these data could play a vital role in accelerating and lowering the cost of national surveys of soil and terrain type, natural vegetation, geology and regional land use.

## 1. Background

The South African Trigonometric Survey, the official mapping agency, has been using aerial photography since 1932 as the basis of topographic mapping of the country's 1.22 million km$^2$. Between 1932 and 1964 every part of the country has been photographed officially at least once, some areas as many as 9 times [1].

The Trigonometric Survey, through the interdepartmental National Advisory Survey Council, also procures the aerial photography required by other government departments. Under this arrangement about 4% of the country has also been covered by black and white infrared photography since 1964. During recent years colour and colour infrared photography has also been acquired experimentally and this type of photography is being introduced into the routine photographic programme. Since 1966 the Geologic Survey had about 180000 km$^2$ of aeromagnetic and radiometric survey flown for the purpose of geological mapping and mineral exploration, and a number of private companies are also engaged in aerial photographic surveys including multi-spectral photography.

Since the Second World War aerial photography has been utilized increasingly in surveys of land use, vegetation type, soil erosion, etc. [2]. Interest in resource surveys by what has become known as "remote sensing" was greatly stimulated by the invitation from NASA to participate in the programme of the Earth Resources Technology Satellite as well as the invitation from the United States

Government to attend the International Workshop on Earth Resources Survey Systems in Ann Arbor in May 1972, which was accepted by a South African delegation of six persons.

In order to stimulate interest in the application of modern remote sensing technology to earth survey problems and to assess the local activities and capabilities in this respect, the Council for Scientific and Industrial Research (CSIR) organized a Symposium on Remote Sensing which was held in Pretoria during May 1972. This symposium was attended by about 150 delegates representing a multitude of disciplines and included delegates from neighbouring countries. Thirty papers were read, including five by overseas visitors.

A paper which deserves special mention was one by the Division of Sea Fisheries which described a semi-operational system which employs image intensification of the phosphorescence of plankton disturbed by fish shoals, as well as additional information, to calculate the shoal tonnage [3, 4].

## 2. Small Scale Colour Infrared Photography

In order to bridge the gap between conventional aerial photography and small-scale multi-spectral photography such as acquired by ERTS, a programme of multiscale, multiband photography, including small-scale colour infrared, was initiated in cooperation with the South African Air Force. A test strip about 14 km $\times$ 250 km from Ventersdorp to Vaalwater, including representative geologic, vegetational and agricultural features, was repetitively photographed at different dates on black and white infrared, colour and colour infrared at scales ranging from 1:20000 to 1:120000.

For the small scale colour infrared photography (from 13 km altitude) a semi-quantitative technique was developed to obtain the optimum colour balance by means of Kodak Wratten filters. The method is based on a triangular colour space diagram, Fig. 1, (calculated for Kodak Ektachrome Infrared film) which can be utilized to select the appropriate filter-exposure combination to adjust the colour balance of the film in a predictable manner in order to compensate for variations in film or development characteristics, atmosphere effects or seasonal effects in the scene. A full description of the method will be published elsewhere [5].

## 3. ERTS Programme

Two South African proposals for participation in the ERTS A project were accepted by NASA: The first "To assess the value of satellite imagery in resource inventorization on a national scale" is a multi-disciplinary effort comprising a soil and agricultural land capability survey, plant ecological survey as well as geological and land use survey undertaken by a team of scientists of the CSIR and government departments. The second, "Monitoring growth or decline of vegetation on mine dumps", is by a private company.

The first images of the Republic of South Africa (acquired 3 August 1972 and received on 6 October 1972) were RBV products. Since then almost 500 scenes were photographed (almost exclusively by MSS), giving an average of four complete coverages of the South African test site (south of 21° S). Every part within the borders of the Republic including Lesotho and Swaziland was covered at least once, about 97% of this area with zero cloud cover. These MSS images were largely of excellent quality, providing a wealth of information which mainly exceeded expectations.

The interpretation of these images has reached the stage where preliminary significant findings can be reported. Interpretation of vegetational, agricultural

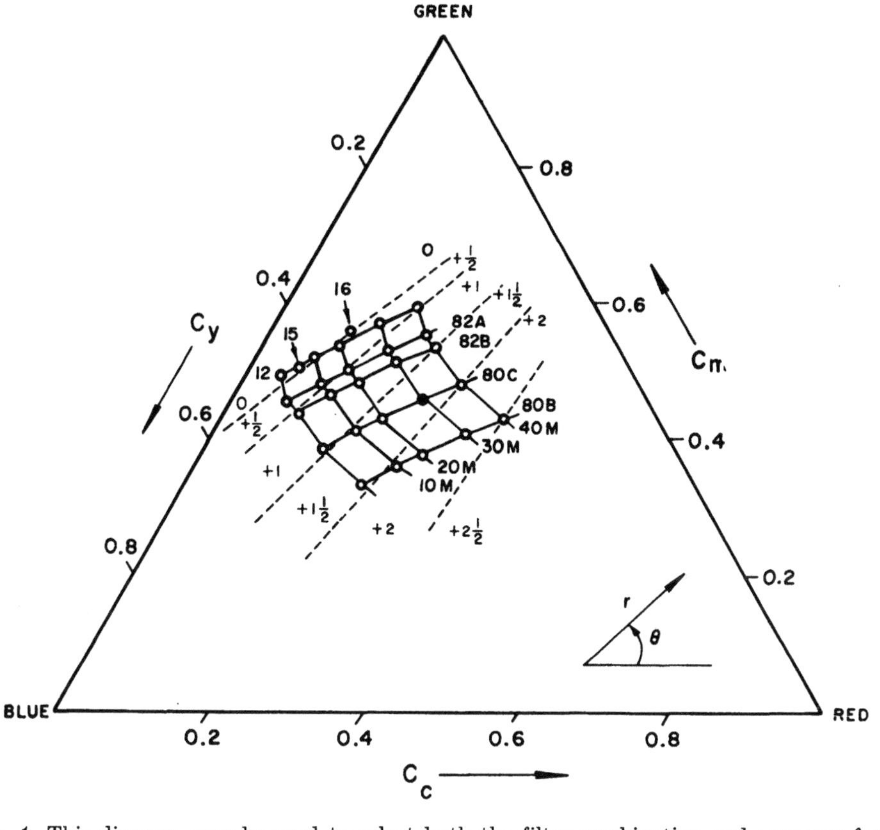


Fig. 1. This diagram may be used to select both the filter combination and exposure for a desired shift in colour balance of Kodak Ektachrome Infrared Film.

and land use information was initially severely hampered by the fact that for the major part of the country the winter and early spring landscape is almost featureless. This condition was prolonged by the very severe drought experienced during the early summer of 1972—1973. Imagery obtained later in summer shows much more pronounced vegetational features and is currently being analysed.

For geologic interpretation the opposite is true. Early spring photography was acquired under low sun elevation (30°—40°) which accentuated topographic features whereas these are less pronounced in summer photography with sun elevation up to over 50°. It is thus very unfortunate that ERTS photography during June—July 1973, when sun elevation of between 20° and 30° will prevail, will probably not be available due to failure of the tape recorder.

## 4. Results from Interpretation of ERTS Imagery [6]

### 4.1. Geology

In view of the relatively advanced state of geologic mapping in South Africa, no significant new information with respect to lithology was revealed by the ERTS images studied thus far. On the other hand the potential application of ERTS imagery to regional tectonic analyses shows great promise; lineaments, folds and structural contours extend further and are much better revealed than on conventional photography or in the field.

The Tugela fault can be traced for more than 150 km and is characterized by lineaments to the north and the south which indicate that this is a fault zone rather than a single fault. Furthermore, the western part of the fault does not coincide with the position on the $1:10^6$ geological map. These features are currently being investigated in the field. The north-south lineaments characteristics of the Lebombo monocline seem to be cut off by the Tugela fault. This fault may constitute an important hinge zone. The inward dipping Transvaal quartzites around the Bushveld complex and folds in the Cape System are more easily distinguished on ERTS imagery.

Analysis of all available imagery could form the basis of a $1:10^6$ scale tectonic map of South Africa which would have wide application in geology and the identification of areas of potential economic mineralization.

### 4.2. Soil and Terrain Mapping

ERTS imagery is currently being used in the final preparation of the first detailed 1:1500000 terrain-morphological map of South Africa. This type of imagery is ideally suited for this purpose because it eliminates the cost and drudgery of analysing conventional information bases (aerial photography, large-scale topographic maps) in which the data are too detailed, too voluminous and non-synoptic.

ERTS imagery is also being introduced into current pedosystem mapping operations at scales of 1:250000 and smaller.

### 4.3. Plant Ecological Mapping

In spite of problems associated with the unfavourable seasonal conditions of the first imagery analysed, tests of the usefulness of ERTS imagery in small scale vegetation mapping have shown great promise.

The facility with which burnt areas are recognized has been outstanding. Burning of pasture is an accepted pasture management operation and has been the subject of considerable research in South Africa during the past forty years. Satellite photography for the first time provides a practicable method of monitoring the time, frequency and extent of burning and enables the pastoralist to determine its relationship with socio-economic and climatic factors and its effect on plant ecology and pasture quality on a national scale.

In the more densely vegetated areas all large evergreen forests are readily identified, as are forest plantations. Difficulties have, however, been experienced in the separation of smaller linear indigenous evergreen forests from scrub and thicket which are usually distinguishable. Despite difficulties arising from the high diversity of vegetation in these areas of rugged topography, many of the physiognomic boundaries drawn from ERTS imagery have coincided exactly with those on existing maps of the area.

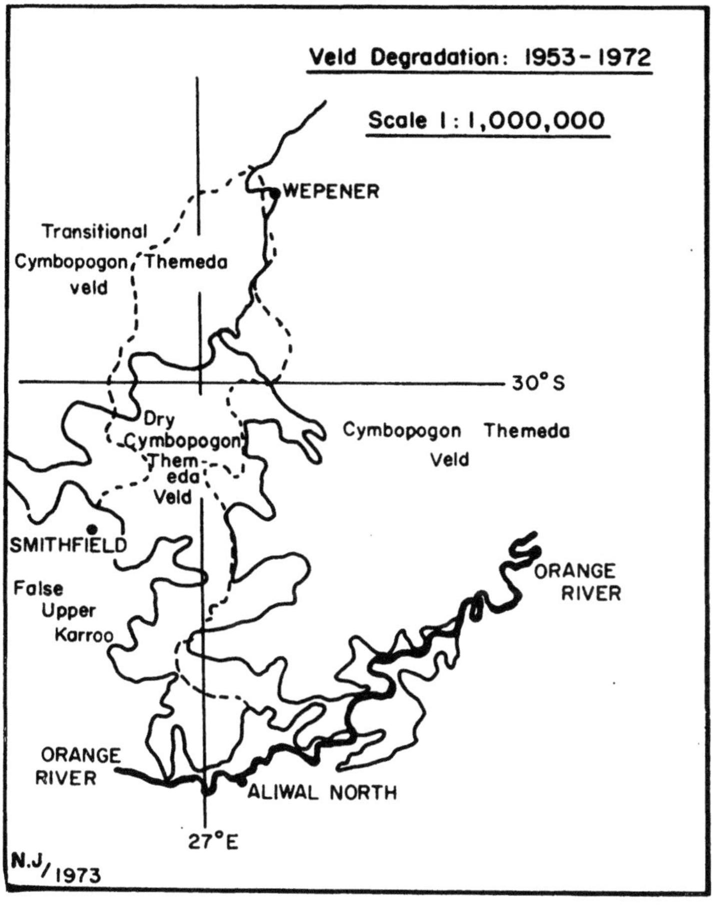


Fig. 2. Veld degradation. The dotted area delimits the grassed area which has been invaded by dwarf scrub communities during 1953—1972.

The detection of evergreen forest is currently of major interest in view of the recommendation of the International Biological Programme for its conservation. Furthermore the detection and conservation of this vegetation type has become an urgent task in view of the accelerated economic and political development of the Bantu Homelands. These are mainly situated over an area of 15 million hectares of southern and eastern South Africa where the rugged topography and patchy distribution of the forests make their detection by conventional means a time-consuming and expensive operation.

Differentiation of the various savannah and grassland types has been less successful on early spring ERTS imagery. Imagery acquired later in the season, however, shows great promise of being easier to interpret. Nevertheless two very significant examples of degraded vegetation could be detected: Firstly, invasion of *Acacia karroo* savannah vegetation into degraded grassveld around Kroonstad could be identified. Secondly, in an area in the south-eastern Free State invasion of the False Upper Karroo, a dwarf scrub type of vegetation, into grassland could be detected. A map of the newly degraded vegetation drawn from ERTS imagery (Fig. 2) indicated that the boundary appeared to have moved about 70 km compared with the boundary mapped by Acocks in 1953. This movement has since been confirmed and its exact location is being field checked. This invasion of scrub into valuable grassland is a problem of major national concern. Whereas 30 million ha (26% of the country) is covered by grassland, less valuable savannahs and dwarf scrub Karroo cover, respectively, 31% and 35% of the country. Acocks in 1953 estimated that of the latter 12 million ha or 10% of the country was grassland which degraded into scrub. Updating of information concerning savannah or Karroo invasion of grassland is therefore of major urgency.

### 4.4. Regional Land Use Mapping

The scale of ERTS imagery has proven to be eminently suitable for land use mapping on a regional scale. This was demonstrated in the Pretoria—Johannesburg region which includes highly urbanized as well as cultivated, uncultivated and trampled land. This type of imagery could contribute significantly in accelerating and lowering the cost of such surveys on a regional and national scale.

## Acknowledgments

The author is indebted to the co-investigators of the South African ERTS programme: Dr. J. M. de Villiers and Dr. D. Edwards of the Department of Agricultural Technical Services, Dr. W. L. van Wyk of the Geological Survey and Mr. J. J. la Grange of the Department of Planning as well as their assistants, for the interpretation of the ERTS imagery.

## References

[1] W. C. WATSON, Symp. on Remote Sensing, Pretoria, May 1972.
[2] C. BOARD, Photogrammetria **20**, 163 (1965).
[3] D. L. CRAM, Symp. on Remote Sensing, Pretoria, May 1972.
[4] D. L. CRAM, Fish. News Intl. **12**, 65 (1973).
[5] O. G. MALAN, Photogrammetric Engng, in press.
[6] O. G. MALAN, ERTS Type II Report NASA-CR-130012, CSIR-FIS-38 (February 1973).

Earth Survey Problems — Akademie-Verlag, Berlin 1974

# EARTH SURVEYS BY REMOTE SENSING IN ISRAEL

J. Otterman and A. Ginzburg

Tel-Aviv University, Department of Environmental Sciences, Ramat-Aviv, Israel

Recent earth surveying activities in Israel by the new techniques of remote sensing are described. The described activities span the period of the last $3^1/_2$ years, 1970—1973. In 1970 the thermal saline springs of Lake Kinneret (Sea of Galilee) were surveyed, using the technique of thermal infrared mapping from the cliffs by the Lake. Starting in 1971, a programme was undertaken to map the temperatures of the Mediterranean by an airborne radiometer for the purpose of locating schools of sardines. In the fall of 1972, the imagery from the ERTS 1 Satellite started to arrive to Israel. The arriving imagery has been subject of analysis in a multi-disciplinary research programme. The objectives of the programme are mapping of agricultural crops, mainly wheat and orange groves, mapping of natural vegetation and studies for the Israeli hydrology, arid regions research, geology and oceanography. The preliminary results include studies of jet contrails over the Eastern Mediterranean, mapping the flow of sedimentary material along the coast and discovery of a seasonal giant eddy in the Gulf of Suez. Cases of water, air and pollution observed on the images are under study.

## 1. Introduction

Israel is deficient in natural resources, including water, and study of the available resources and search for new ones has been vigorously pursued. Because Israel is a very small country, the methods used have been by and large *in situ* studies, rather than remote sensing techniques. During the last few years, modern remote sensing techniques became increasingly applied and these activities are summarized in this paper. Ground-to-ground observations, measurement from airborne instruments, and mapping by satellite techniques are described, and some recent results in pollution studies included.

## 2. Remote Sensing Activities, 1970—1973

The first earth survey in Israel using a remote sensing techniques was an experimental investigation of saline springs in the Sea of Galilee (Lake Kinneret). The Lake is used as a storage and regulating reservoir of fresh water in the Israeli national water carrier scheme, in which fresh water is pumped from northern Israel to the central and southern parts of the country for irrigation purposes.

In and around the lake there are some thermal saline springs, whose effect is to raise the salinity of the lake, especially in the summer months, when the inflow of fresh water is low and pumping rate high. In order to control the salinity, it is necessary to study the flow rate and pattern of the saline springs. The higher temperature of the saline water made it possible to apply thermal mapping to the problem. An infrared television system, AGA Thermovision, was used to scan selected parts of the lake [1]. The imagery obtained showed the irregularity of the flow, and the fact that sources of thermal saline flow could be detected on the surface of the lake.

Another use of the same instrumentation in ground to ground measurements of emitted infrared radiation was in searching for thermal activity over land. The area investigated was a fault block near the Sea of Galilee near which there are a number of hot springs [1]. Thermal foci in this area could be indicative of thermal activity at a shallow depth. Hot springs near the Sea of Galilee and in other places in the Jordan Rift Valley such as the Dead Sea indicate that sources of thermal energy of economic importance may exist in the Rift Valley. Thermal mapping by remote sensing is an obvious reconnaissance tool and in conjunction with plans for such a survey, laboratory measurements of the emissivities of the dominant rock types of Israel were carried out [2].

Aircraft-borne remote sensing surveys were conducted in conjunction with fisheries studies. It is known that in the Eastern Mediterranean sardines are available to inshore fishing where surface water temperature exceeds 18 °C. It is therefore necessary to map water temperatures in order to known when and where fishing can take place. As dynamic changes in the structure of coastal waters occur in the spring, monitoring of sea surface temperatures is necessary. In an investigation carried out in the spring of 1972 [3] several survey flights covering the coastal waters of Israel from Nahariya in the north to Gaza in the south were made. The aircraft was equipped with a Barnes PRT 5 radiometer which recorded surface temperatures. The airborne measurements were compared with water temperature measurements conducted by a research boat. The comparison indicates that the radiometric data express well the relative temperatures. Correlation of the temperature recorded with sardinella catch shows that the fish concentrate in locations where the temperature gradient is highest. Although the experiments must be continued, it appears that the approach used shows promise of turning into a routine tool in commercial fishing in Israel. A typical thermal map of the off-shore areas that results from such flight is shown in Fig. 1.

Israel participates in the ERTS 1 programme. The programme, in which scientists in such diverse disciplines as agriculture, geology, oceanography and meteorology participate, has been described in the ERTS proposal submitted to NASA, and the data analysis for the program described by Otterman [4].

The ERTS 1 images of Israel started to arrive in the fall of 1972. The standard products that arrive are in form of black-and-white transparencies, on a scale 1:1000000. The images are in four spectral bands of the multi-spectral scanner, ranging from green-yellow (MSS 4) to the infrared (MSS 7). The imaging swath is 180 km (100 n mi), and orbital geometry is such that one day ERTS 1 passes over the Rift Valley, from the sources of the Jordan to the Tiran Straits, and next day it provides us with a coverage along a parallel 180 km swath (with some 60 km overlap) from the Sea of Galilee to the Gulf of Suez. Preliminary results were presented in March 1973 [5]; the application areas include oceanography, meteorology, vegetation studies, geology and hydrology [6—8].

For oceanographic uses, the green-yellow band (MSS 4) of the ERTS imagery proved extremely useful. In this spectral band the subaqueous northward movement of sediments is clearly seen from the Damietta branch of the Nile. The area in which sediments are dispersed is a broad strip which is some 50 km offshore in the south and becomes narrower and comes closer to the shore in the Gaza strip.

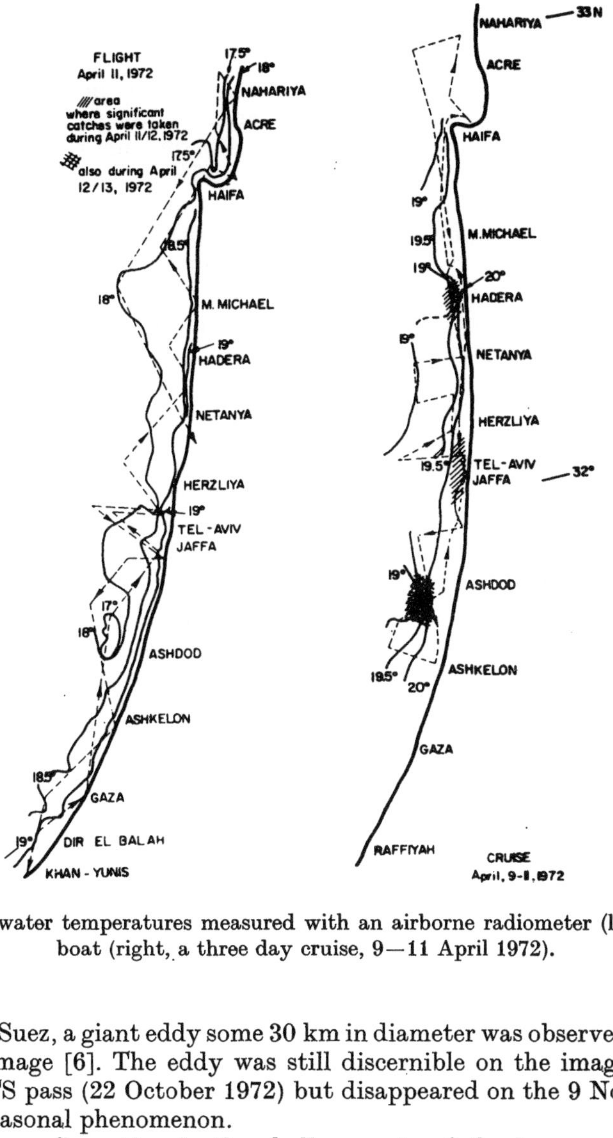


Fig. 1. Surface water temperatures measured with an airborne radiometer (left) and from a boat (right, a three day cruise, 9—11 April 1972).

In the Gulf of Suez, a giant eddy some 30 km in diameter was observed in 4 October 1972 MSS 4 image [6]. The eddy was still discernible on the image obtained in the next ERTS pass (22 October 1972) but disappeared on the 9 November pass. It is thus a seasonal phenomenon.

Sea bottom configuration in the shallow parts of the lagoon of Bardawil is clearly seen on the MSS 4 image, as are the coral reefs of the Gulf of Suez.

The high ground resolution of the ERTS images makes them particularly useful for studies of small-scale atmospheric cloud phenomena. An interesting example is a study of man-made clouds, jet contrails which appeared in a pair of ERTS frames taken near Cyprus [7]. In this case an attempt was made to estimate the average size of ice crystals in the contrails by using their relative reflectivity.

The efficacy of ERTS images for vegetation studies is demonstrated in the Negev area [8]. A striking difference can be observed along the old armistice line separating the Israeli and the Egyptian sides of the Negev. The areas on the Israeli side appear dark while the Egyptian side appears light coloured, because of the lack of vegetation cover. The dividing line can be attributed to overgrazing by sheep and goats on the Egyptian side.

The main advantage of ERTS images for geology is in offering a look at a large area under equal lighting conditions. In addition, the separation into spectral bands magnifies small variation in colour thus permitting the mapping of some formation boundaries with ease. A study of ERTS image of Israel enables the mapping of the main rock types. The dark areas of the crystalline rocks in the red and infrared images contrast very sharply with the light coloured areas of sedimentary rocks [8]. The main structural elements such as folds and faults can be mapped easily as well as linear features such as dykes in both the sedimentary and igneous areas. Morpho-tectonic features such as the cliffs of Gebel e-Tih, which are the impression of the huge northern Sinai syncline, can be readily distinguished.

In hydrology the synoptic view of the ERTS images is utilized for studies of drainage patterns such as that of Nahal Faran in the Negev, while the green yellow and infrared bands clearly show water depths and shorelines respectively.

Agricultural survey from ERTS 1 imagery has been aimed at two crops: wheat and citrus. The poor resolution of the multi-spectral scanner of some 80 m makes it entirely impossible to determine photometric characteristics of fields smaller than some 7 dunams (1 dunam is 1000 $m^2$). But even in the case of fields larger by a factor of 4 or even 10, accurate multi-spectral analysis poses severe problems [9]. Israeli fields are mostly small in area, but relatively large fields of wheat and groves of citrus exist. The agricultural mapping has lagged behind expectation, since it is based on the use of the Spectral Data Corporation projector/viewer, which arrived in Israel late and was put in operation only in May 1973.

Pollution is not a natural resource, but in the framework of remote sensing, pollution detection and monitoring is quite often included in natural resource activities. The possibilities of detecting and measuring pollution from space has been studied by a group at the Tel-Aviv University [10]. Cases of water, air and soil pollution have been observed in ERTS imagery [11]. Oil slicks, an important form of sea pollution, show up, in all the MSS bands, as spots darker than the surrounding, unpolluted ocean.

The oceans are dark in all bands, since the coefficient of reflection at the surface is low; the brighter areas that show up, are caused by light that penetrated the surface and is reflected either from a shallow bottom or from particulate matter suspended in the water. The intensity of such reflection from depth depends on transmission through the water, and thus is greatest in green yellow, MSS 4, band where transmission is high. Transmission, and therefore the intensity of such reflection, decreases at longer wavelengths. Thus even quite shallow water or waters laden with particulate matter are quite uniformly black in the infrared MSS 7 band, as long as the surface is not polluted by the slicks. This band is most appropriate for locating oil slicks. However, in the areas where the reflection from bottom or particulate matter does not make the surface non-uniformly dark, any band can be used for this purpose. In such areas the band MSS 5 is especially recommended, since the contrast due to oil slicks is greatest. Oil slicks have been observed in the Gulf of Suez, in the 22 October 1972 imagery. Smoke plumes, nearly 100 km long, are visible in the same image. Soil pollution from the Timne copper mines, north of Eilath, can be readily seen as a dark spot on the ERTS image. Photometric space signatures of those cases of pollution have been determined by an iris photometer and will be included in paper [11].

## 3. Conclusions

Earth surveys by remote sensing in Israel had a vigorous start in this decade, owing largely to the valuable ERTS 1 imagery of the area. Israel participates in the Skylab-EREP experiments, and it is hoped that valuable data, even though not in repetitive coverage, will be obtained. Later the remote sensing group in Israel looks forward to the ERTS B, to be launched in 1976.

## References

[1] J. Otterman, A. Kahanowitz and R. Shilo, Israel J. Earth Sci. **20**, 80 (1971).
[2] R. Givati, J. Otterman and E. Nathaniel, Tel-Aviv University, Department of Environmental Sciences, Rep. No. ES-73010 (1973).
[3] M. Ben-Yami, S. Pisanty, J. Otterman and R. Zach, Proc. 4th Scient. Conf. of Israel Ecological Society, Tel-Aviv, Israel I 70 (1973).
[4] J. Otterman, Rassegna Internazionale Elettronica Nucleare ed Aerospaziale XIX, 269 (1972).
[5] J. Otterman, Rassegna Internazionale Elettronica Nucleare ed Aerospaziale XX, 285 (1973).
[6] J. Otterman and M. Dayan, Proc. 4th Scient. Conf. of Israel Ecological Society, Tel-Aviv, Israel I 1 (1973).
[7] G. Ohring, J. Otterman, J. H. Joseph and Z. Levin, Proc. 4th Scient. Conf. of Israel Ecological Society, Tel-Aviv, Israel I 15 (1973).
[8] A. Flexer, A. Ginzburg, A. Kahanowitz and J. Otterman, Proc. 4th Scient. Conf. of Israel Ecological Society, Tel-Aviv, Israel I 37 (1973).
[9] J. Otterman, J. Brit. Int. Soc. **23**, 349 (1970).
[10] G. Ohring, J. Otterman and J. H. Joseph, Water, Air and Soil Pollution **1**, 89 (1973).
[11] J. Otterman, A. Ginzburg, G. Ohring and Yu. Mekler, Water, Air and Soil Pollution, Submitted.

Earth Survey Problems — Akademie-Verlag, Berlin 1974

# NATURAL RESOURCES RESEARCH AND DEVELOPMENT IN LESOTHO USING ERTS IMAGERY

A. A. Jackson[a] and P. H. Nixon[b]

[a]Department of Biology, University of Botswana, Lesothe and Swaziland, Roma, Lesotho, Africa
[b]Department of Mines and Geology, Maseru, Lesotho, Africa

Lesotho is a small mountainous country in Southern Africa. Due to the geographical nature of the country, conventional ground-based surveys are extremely difficult and expensive. In order to identify areas which are worth ground-based investigation, remote sensing techniques are essential. For these reasons, Lesotho is participating in the ERTS 1 project in order to identify areas which are possible locations of mineral resources and to monitor agricultural production as regards both crops and range management.

Early black and white imagery received from ERTS 1 indicates that, although it may be possible to identify mineral resource areas and monitor grazing potential, assessment of crop production is more difficult due to the small field size typical of under-developed countries. Experiments are therefore being carried out with photographic techniques, particularly density separation and colour enhancement, in order to simplify the interpretation of the imagery.

## 1. Introduction

Lesotho is a small mountainous country of 30350 km² surrounded entirely by the Republic of South Africa. It is the southernmost of the Black Independent African States.

The country consists of two distinct geographical regions. The smaller of the two regions, comprising roughly one quarter of the area, lies in the west of the country at altitudes between 1500 and 1850 m. These lowlands constitute 75% of the arable land and home for 75% of the total population of just over 1000000. The eastern three quarters of the country consist of a mountainous region rising to heights of over 3300 m (Fig. 1). This is not permanently inhabited, except by herdsmen, and is used mainly for the grazing of livestock. Communications in the mountainous region are difficult, and rely heavily on horses, and light aircraft operating from rough landing strips.

Rainfall is highly erratic, varying from 45 to 90 cm annually in the lowlands, with the highest rainfall occurring between October and April. The lowland temperature varies between $-7$ °C in the winter and 33° in the summer. In the mountains, the summers are cooler and the winters are extreme, with temperatures less than $-20°$ being common.

Due to the nature of the geography of the country, conventional ground-based surveys are extremely difficult and expensive. As a result, the country is heavily dependent on aerial photographic and remote sensing techniques to identify areas which are worth the expense of conventional surveys. Also, due to the communications difficulties, it has not been possible, in the past, to monitor agricultural production, on which the country depends for 75% of its exports.

For the reasons mentioned above, Lesotho is participating in the ERTS 1 project (and has submitted a proposal for ERTS B) in order to obtain data pertaining to natural resources in both the agricultural and geological fields.

## 2. Agriculture

Agriculturally, Lesotho is hampered by a rainfall pattern which is erratic in terms of both time and space. During the summer (wet season) months, much of the livestock is moved from the valley regions into the mountains where better grazing is normally available. However, due mainly to communications difficul-

Fig. 1. Composite of four ERTS images to show the whole of Lesotho. Band 5 images of 20—21 November 1972.

ties, some areas are underutilized whereas others are overgrazed. The overgrazed areas suffer a large loss of topsoil due to sheet erosion on the steep mountain slopes. Data on range potential, derived from ERTS imagery, are beginning to be used to direct the movement of livestock into areas which are capable of supporting the animals adequately. If this programme can be continued, it seems that it may be possible to halt, or at least reduce, the extent of soil erosion in the mountain areas, while, at the same time, increasing the economic usefulness of the region.

Due to differences in agricultural techniques between relatively developed and under-developed countries (Republic of South Africa and Lesotho, respectively), political boundaries are easily discernible in the ERTS imagery of Lesotho. (See, for example, the western border of Lesotho in Fig. 1.) This is mainly due to the small field size typical of under-developed countries. Another factor which contributes to the clear delineation of the border in Fig. 1 is the drought conditions which were prevalent over Southern Africa at the time when the image was gathered.

Although individual fields are not, in general, resolvable in the ERTS images of Lesotho, the situation is alleviated by the common practice of blocks of fields being planted with the same crop. As a result, it is possible, under favourable circumstances, to identify the crop pattern (Fig. 2). We are therefore using the images to try to construct a crop inventory for the country. The drought conditions mentioned above have had the effect of simplifying the identification of different soil types and areas of irrigation. One of the major projects connected with the ERTS imagery analysis is the construction of an accurate soils map. Ground

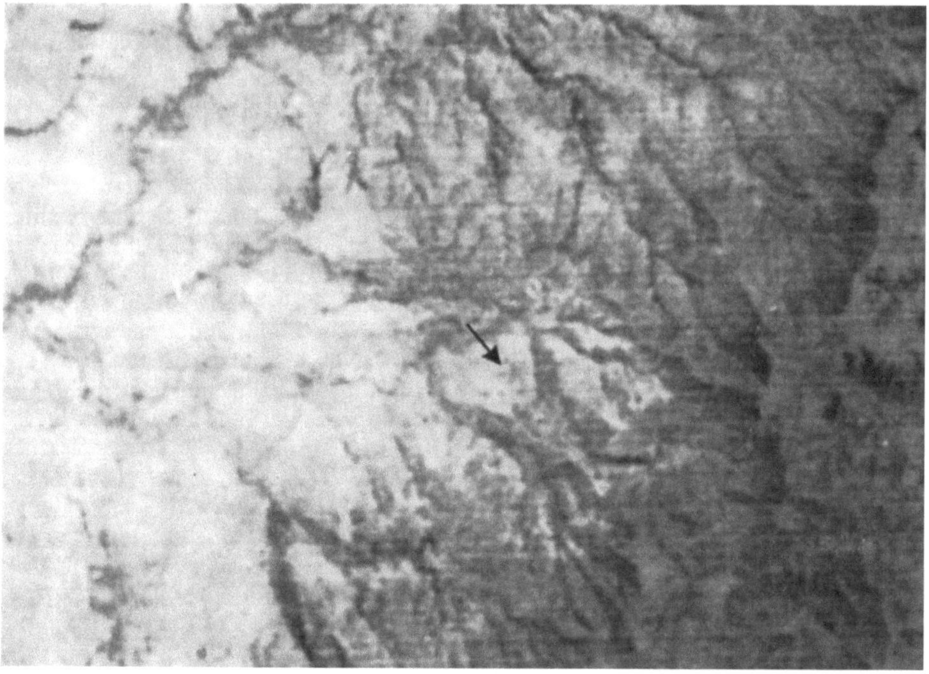

Fig. 2. Greatly enlarged portion of ERTS image E-1121-07315-5 of 21 November 1972, to show crop patterns in the Roma Valley region.

truth data for this project are already available in the form of a preliminary soils map for Lesotho [1]. Whilst giving an outline of the major soil types of the country, the preliminary map is not sufficiently accurate for planning purposes. Linked to this project is an attempt to relate the growth characteristics of various crops to soil type in the different geographical regions. This is aimed at forming the basis for a more rational distribution of cropping patterns, and identifying areas which would be suitable for the introduction of new crops and programmes of reafforestation.

## 3. Geology

Geologically, Lesotho lies at the centre of the Karroo basin. The country consists of a simple stratification of shales and sandstones which form the western lowland strip, but these are overlain by a sequence of basaltic lavas in the eastern mountain region which is up to two km thick. The structure is essentially horizontal, but recent mapping of the lowlands has indicated the presence of numerous gentle basins and domes about 15 km in width. In addition, there is a series of dolerite dykes of late Karroo age (Triassic/Jurassic) which cut the sediments and lavas. These are traversed by well marked joint directions, one of the most conspicuous being the WNW—ESE trend which typifies the later kimberlite intrusions of Cretaceous age.

ERTS imagery is being used mainly to identify dyke swarms which are worth concentrated ground survey for diamond prospecting. Diamonds form the only known mineral resource of Lesotho, although there is the possibility of petroleum existing in the eastern part of the country. Ground truth is fairly well catered for, geologically, in that UNDP have financed an aerial photographic survey and a multi-spectral survey of the known diamond bearing areas of northern Lesotho. We are at present working on the correlation of ERTS imagery characteristics with the aerial photographs, with a view to identifying other diamondiferous areas. Some of the work has been hampered by the drought, in that lack of plant growth has made it impossible to obtain detailed data on the geobotanical relationships of kimberlite intrusions. It is hoped that data from the ERTS B project will enable us to identify the botanical patterns which characterize kimberlite outcrops, and thus identify outcrops which occur in other parts of the country. Some progress has been made, however, in that kimberlitie intrusions are often indicated by sponge areas, especially when associated with dyke swarms and other lineaments. We have been able to identify sponges in the southern part of the country which seem to be associated with linear formations, but, as yet, we have not been able to carry out ground surveys of the features in order to confirm or deny the existence of kimberlite.

The major geomorphological units which exist in the country are easily identified on the imagery. Work is being carried out on the relationships between the various types of erosion visible on the orbital imagery and the geomorphological units on which it occurs. Soil erosion is a feature of special importance in Lesotho, and it is hoped that, if the relationships can be determined, a rational programme of soil conservation can be established.

As mentioned in § 2, the major soil types of the country are relatively easily distinguished. These major types seem to correlate intimately with specific geomorphological units. Also related in this field are the lithological boundaries between the basalts and cave sandstones and between the cave sandstones and the lower Stormberg series. These boundaries are, in the main, easily distinguishable on the ERTS imagery.

## 4. Interpretation Techniques

Most of the interpretation of the ERTS imagery undertaken so far has been accomplished using conventional aerial photographic techniques on black and white positive transparencies. Colour composites have only recently arrived from NASA and there has been little opportunity for work to be done on them.

As there is no advanced interpretative equipment, such as a multi-spectral viewer, available in Lesotho, recourse has been made to various photographic techniques to aid interpretation. Colour enhancement following image slicing on Agfacontour film is being used as a routine interpretative technique. As has been found by Sayn-Wittgenstein [2], this is extremely valuable for the differentiation of various soil and vegetation types. By taking image slices from the different MSS bands and superimposing them in black and white or in colour, use can be made of the spectral signature of various features to produce maps of their distribution.

Fig. 3 shows an image slice composite of northwestern Lesotho taken from a scene of 21 November 1972. This figure shows the distribution of plant growth.

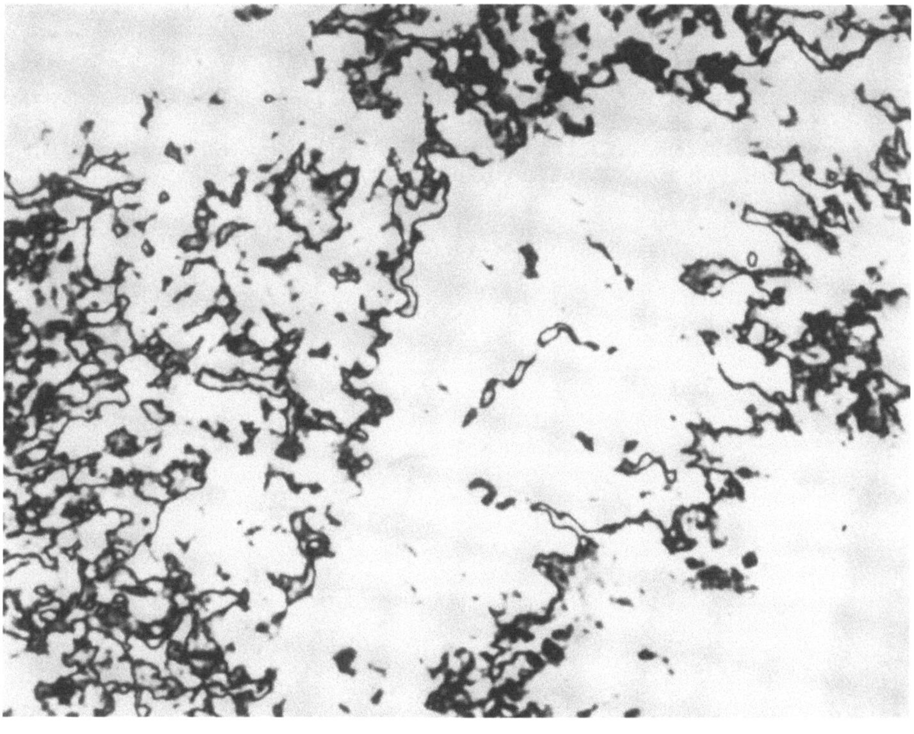

Fig. 3. Image slice composite of northwestern Lesotho and the eastern Orange Free State mapping areas of plant growth. Derived from a scene gathered on 21 November 1972.

The growth regions on the left of the figure are cultivated areas in the eastern Orange Free State, Republic of South Africa. On the right are areas in the foothills of the mountains where basaltic soils overlie sandstone. These areas in the foothills tend to retain groundwater through the winter months, thus enabling vegetation growth to proceed in the spring, even during periods of relative drought. The central region of the photograph shows growth only in the river valleys, mainly on alluvial deposits. This central area is typical of the major part of the arable land in Lesotho.

## 5. Future Work

The experience gained from the ERTS 1 imagery received so far indicates that satellite reconnaissance of Lesotho can be of extreme value in the development of natural resources. Although the detail is not as great as can be obtained from aircraft carrying similar sensor systems, great benefit is obtained, especially in the agricultural field, from the repetitive coverage and general overview available. We are therefore hoping to participate in the ERTS B programme to extend our knowledge of what natural resources may be available in the country and to optimize the utilization of those which are already known. We shall also continue development of the photographic enhancement techniques, as these seem to form an inexpensive alternative to the more sophisticated equipment available to more highly developed countries.

## References

[1] M. G. BAWDEN and D. M. CARROLL, Land Resources of Lesotho, Directorate of Overseas Surveys, Tolworth, England (1968).
[2] L. SAYN-WITTGENSTEIN, in: ERTS 1 Symp. Proc., Goddard Space Flight Center, Greenbelt, Maryland, Pub. No. X-650-73-10, p. 48 (1973).

Earth Survey Problems — Akademie-Verlag, Berlin 1974

# THE PREPARATORY PHASE OF THE EARTH RESOURCES PROGRAMME IN THE FRG

A. E. LÖFFLER

Gesellschaft für Weltraumforschung mbH, Porz-Wahn, FRG

The concept of the preparatory phase of the earth resources programme of the FRG is reviewed and the different activities are described within this framework. The results of industrial studies sponsored by the Gesellschaft für Weltraumforschung are discussed with respect to the basic research and technology required for the implementation of a successful earth resources programme.

## 1. Introduction

In July 1972 the Gesellschaft für Weltraumforschung mbH (GfW) was charged by the Federal Ministry for Research and Technology (BMFT) with the analysis of the conditions and possibilities for the implementation of a successful earth resources programme in the FRG. This analysis was necessary as it was not quite obvious that a small country with a highly developed infrastructure and without remote areas would derive benefits from remote sensing activities.

The following questions had to be studied: Which of the known remote sensing applications correspond to the needs of the FRG? What benefits could be reasonably expected?; in which areas is it necessary to develop research and technology in order to provide the basis for a successful programme?

## 2. Study Activities

Five studies were undertaken to obtain initial answers to the questions raised. First, GfW asked for two concurrent studies to investigate the possible benefits of remote sensing using space technology [1, 2]. The objectives were: review of all known remote sensing applications with respect to FRG needs; evaluation of all relevant applications to determine the possible economical, scientific, technical and political impacts and benefits; identification of potential users and their requirements. These studies now define the baseline of all activities and it is intended to update them as required.

The main objective of the second study was to analyse the problem areas of the data system and to relate them to user requirements [3]. Four data systems were defined: (i) ground-based data collecting platforms and buoys; (ii) aircraft;

(iii) orbiting satellite; (iv) Spacelab. For every platform a typical sensor package was taken and the data stream was analysed from the sensor output to the final user and the associated problems investigated to provide guidance for technical development in the area of data processing and evaluation. The main objective of the DFVLR study [4] on the aircraft programme was to analyse the scientific requirements for such a platform. Scientists of various disciplines proposed a number of test areas from which were selected the most appropriate ones for interdisciplinary research. As a result an aircraft programme providing systematically repetitive coverage of four interdisciplinary test areas was proposed.

Finally we looked at the possibilities for integrating our activities into the progressing ERTS-programme: one study "Description of the current ERTS system and projections of future earth resources sensors, experiments and processing techniques" [5] was performed by the main contractor of the ERTS system and a second one dealt with the requirements and the alternatives for an experimental ERTS-data receiving and processing station [6].

## 3. Other National Activities

### 3.1. Programme of the German Research Council

(Deutsche Forschungsgemeinschaft, DFG)

In 1970 the DFG established a working group on remote sensing to co-ordinate ongoing research projects and to plan new projects. To enforce interdisciplinary research the group consists of members from many scientific and technical disciplines. Up to now false color, multi-spectral aerial photographs and IR-thermal scanner pictures have been taken over various test areas to investigate the following subjects: surface water control, detectability of intakes of lakes and rivers, mapping of subsurface fresh water intakes into salty sea water, sedimentation processes in coastal areas, water pollution mapping, hydrobiology of rivers, mapping of minerals and soils, localization of faults and geothermal anomalies, localization of subsurface cavities (mining damage). To increase the effectivity of these activities the DFG established in 1972 a central facility for "Geophotogrammetry and Remote Sensing" at the University of Munich. This serves as a pool of smaller instruments for use by scientists and provides availability of major equipment.

### 3.2. ERTS Experimenters

Most of the scientists who are members of the DFG-working group are also taking part in a multi-disciplinary geoscientific experiment in Central Europe. The first attempt of evaluating ERTS imagery of Central Europe has largely confirmed the anticipated value of geoscientific studies. Many known features were identified on the imagery. But in spite of the advanced state of knowledge on Central Europe a number of new observations could be made. These were mainly in the field of linear tectonics. In addition studies of dynamic processes (coastal sedimentation, air pollution) are showing promise. Yet, a limiting factor, especially concerning investigations of the biosphere, was the poor repetitive coverage of the test sites and consequently it has not been possible up to now to use the best features of the ERTS system.

### 3.3. Commercial Activities

There are essentially two aerial photography companies who have proceeded beyond the use of photography: Aero Exploration and Hansa-Luftbild. Both have used thermal-IR scanners for various applications in the FRG. In addition, Aero Exploration used laser profilers in Chile for projecting highways.

### 3.4. Application of Remote Sensing Information

Up to now there have been essentially three agencies using advanced remote sensing techniques for their purposes: the Federal Institute of Regional Geography and Planning (BfLR), the Regional Planning Association Untermain (RPU) and the Ruhr Regional Planning Authority (SVR). The BFLR was one of the first institutions in the FRG to use aerial photography for environmental control. This investigation was focused on water pollution problems of the Saar river. The data were analysed by the Geographical Institute of the University of Saarbrücken, the State Institute for Hygiene and Infectious Disease and the Federal Agency for Vegetation, Natural Protection (BfLR) [7]. The RPU is engaged in an interdisciplinary research programme on air hygiene and meteorological investigations of the Frankfurt area using IR-scanner data. It is the objective of this programme to evaluate the flow and spread of fresh air and its effect on the air hygiene conditions. The examination of air movement in this large area of 1400 $km^2$ by means of conventional methods created great difficulties even for a dense measurement network. One special subject within this research programme is the investigation of the effect of various plants on the production of the desired cold air flow.

"Keeping air clean" usually brings to mind emissions of the most substantial air polluters: industry, traffic, domestic heat. On the other hand little is known of the interdependences of the structural conditions of an area and their effect on air hygiene. The investigations show that many pollutant problems in our cities could be less if settlements had not developed in undifferentiated agglomerations. In addition, a computerized data basis is developed from remote sensed data which will supplement existing economic and sociological statistics [8].

The SVR is systematically mapping the whole area by means of an IR-scanner and aerial photography. The data are expected to yield good results in the following fields [9]: information on the amount and extent of water pollution; information on water pollution through industrial and other waste material; quantitative data on the urban atmosphere and climatic changes; heat insulation used by the building industry; information on heating systems in residential areas; detection of dumped burning waste and decomposing refuse; location of open-cast mining operations; fissures and depressions resulting from old mining operations; air pollution.

## 4. Study Results and Conclusions

The two "benefit studies" mentioned before offer about thirty different applications of remote sensing information for the needs of the FRG. From this number those applications likely to yield the most important benefits were selected by a group of experts [10]. However, the discussion held clearly indicated that the best

Table 1

Summary of Most Important Applications [10]

| Application | Area of interest | Benefit | Optimum platform | Remarks |
|---|---|---|---|---|
| Land use mapping in urban areas | FRG, Europe | Actuality of information | Aircraft<br>Spacelab | |
| Traffic control | FRG, Europe | Non-interacting observation | Future satellite systems | Aircraft test of SVR not successful up to now |
| Air pollution<br>Measurement of global air pollution | Global | Basis for global agreements | Satellite | Remote sensing has to be combined with local measurements |
| Regional planning | FRG<br>Europe | Understanding of interactions | Aircraft<br>Spacelab | |
| Continuous pollution Control | FRG<br>Europe | Basic data for decisions | Satellites plus local measurements | Aircraft already used for regional tasks |
| Information for structural planning in agriculture | Europe, FRG<br>dev. countries | Reduction of effort | Aircraft<br>Spacelab (satellite) | Aerial photography already used of next higher performance class |
| Land use mapping for agricultural planning and statistics | Europe<br>dev. countries | Actual information for planning of food-production and statistics | | |
| Characterization of optimal land use | Europe<br>dev. countries | Increase of effectiveness<br>Cost reduction through better basic information | Aircraft<br>Spacelab (satellite) | Satellites for measurement of climate parameter, aerial photography in use<br>ERTS-experiment |
| Forestry detection of damage | FRG, Europe<br>dev. countries | Damage minimization by early warning and countermeasures | Aircraft<br>Spacelab (high resolution satellite) | Important for FRG also as damages can be detected earlier |
| Fishery | Global | Shortening of search time leads to smaller fleets with constant catch | Satellite | Basic information for international agreements on catch quota |
| Water control | FRG, Europe | Planning of actions<br>Early warning | Aircraft<br>Spacelab (satellite) | For FRG resolution of satellites has be increased |

features of remote sensing—*giving one concise data set of all interacting parameters of an observed site*—could not be evaluated by studying specific applications in various disciplines. For the five most important zones the problems which should be analysed by means of remote sensing systems are:

*Urban areas and their environment.* Special interest is in the interaction of climate and urban structure and development, organization of the landscape, land use in agriculture and forestry, detection of dumped waste, desolation of agricultural area, traffic control and air pollution control.

*Agricultural areas.* For these the most interest is in the improvement of the effectiveness of agricultural projects in developing countries by providing basic data on soil quality, water supply and climatic parameters.

*Arid and semi-arid zones.* Research on the hydrologic cycle aiming at more efficient use of available water resources and prospecting of new water supply. Correlation with other climatic parameters is essential for the planning of agricultural projects.

*Oceans and coastal zones.* Mapping of various parameters and analysis of dynamic processes to provide the basic knowledge for a more efficient use of the oceans.

*Atmosphere.* Measurement of climatic parameters and pollutants is a task which can be solved only by means of integrated observatory systems. For all of these applications satellites and aircraft will have to be used to provide synoptic data while additional ground stations will provide supporting data.

The analysis of user needs showed that for most applications the present state of technology and knowledge is not sufficient to provide immediately the answers required. Improvements, however, should not be attempted by looking at the

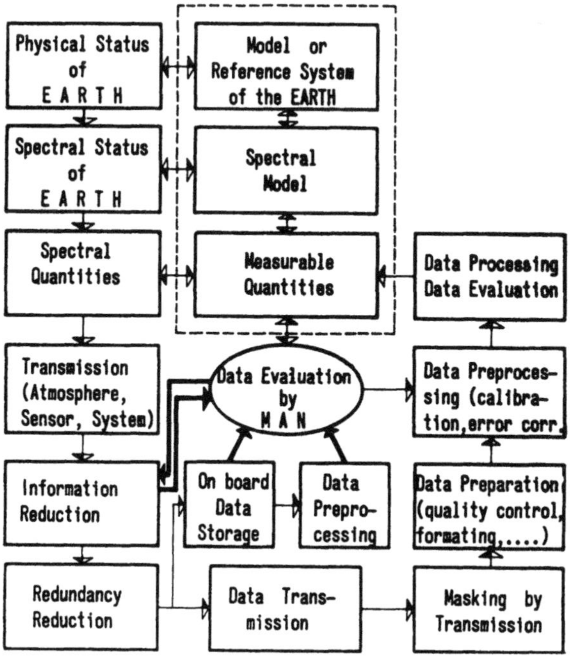


Fig. 1. Earth observation system elements [2].

individual elements of a remote sensing system, as for example the sensor resolution, but by looking at the weak elements of the information system model as given by Fig. 1. The space surrounded by the dotted line indicates our knowledge of the earth and the processes acting on it. As only very few parameters can be measured directly the knowledge of the interaction of the various parameters will be the key for success or failure of an earth resources programme. Following the information flow the next element of an earth observatory system is the sensor. Sensor technology is considered to be sufficiently advanced at the present time and future developments should be guided by user requirements after the data products at present provided by available sensors have been thoroughly understood and evaluated. The possible and necessary interaction of man with the data stream is another major item requiring our special attention in order to use fully the capabilities of Spacelab for earth observatory missions; hence this will be one of our major study areas in the future. Although great technical problems exist with respect to collection, storage, transmission and distribution of large quantities of data as inherent in earth observatory missions the major problem is seen in interpretation techniques and how they are linked with user requirements.

Study of an aircraft programme showed that an experimental aircraft facility would provide the means for establishing the required models and developing the procedures and techniques needed. Four test areas have been selected where most of the applications as defined by the benefit studies can be examined: coastal zone, urban area, agricultural and forestry test site, geoscientific test site. In addition, test areas in other European countries will be necessary for scientific and application reasons.

As long as ERTS is gathering data over European test sites these provide the unique chance to combine spacecraft data with the data from the aircraft programme. As the same type of equipment and procedures is used the advantage of the various platforms and instruments can easily be evaluated.

## 5. Summary

The results of our present study phase show clearly that an experimental earth resources programme is necessary to evaluate fully the capabilities of an earth resources application programme in the FRG. Such an experimental effort is especially needed for the establishment of proper models of the areas to be observed later during this decade using the Spacelab as a sensor platform. The key features of such a programme will be the following: (a) systematic collection of aircraft data on four interdisciplinary test sites; (b) development of an experimental data processing facility to handle ERTS data as well as aircraft products; (c) research work in the areas of automatic data processing and evaluation, future sensor systems especially in the area of active systems and microwaves, physics of remote sensing. Although only rough estimates on the potential benefits can be given at this time there is evidence that our highly populated urban areas and the coastal zones will be the parts of our country for which remote sensing information is most valuable. For the urban areas this is concluded from the fact that repetitive coverage is mandatory for following the rapidly developing demands and impacts of man on his environment. The oceans and the coastal zones are remote and inaccessible. In addition, the space platform for the first time makes it possible to observe interacting phenomena as it maps their distribution in space and time, a feature which can be hardly expressed in returns for an investment.

Hence an experimental programme oriented by needs of the known users will be proposed by GfW. If this programme is approved it can begin within one year and will certainly help to improve our knowledge of our environment.

## References

[1] Dornier System mbH, GfW-Contractor Rep. RV II 1, 9 May 1972 (1973).
[2] Messerschmitt-Bölkow-Blohm, GfW-Contractor Rep. RV II 1, 10 May 1972 (1973).
[3] Dornier System mbH, GfW-Contractor Rep. RV II 1, 17 May 1972 (1973).
[4] DFVLR, Studie für ein Erdwissenschaftliches Flugzeugmeßprogramm, Oberpfaffenhofen (1973).
[5] General Electric Space Division, GfW-Contractor Rep. RV II 1, 16 May 1972 (1973).
[6] Messerschmitt-Bölkow-Blohm, GfW-Contractor Report RV I 3, 4 May 1973 (1973).
[7] S. Schneider et al., in: Rundbrief des Instituts für Landeskunde, Bonn-Bad Godesberg **4** (1972).
[8] A. v. Heseler et al., Infrarot-Thermographie 3. Arbeitsbericht Regionale Planungsgemeinschaft Untermain (1972).
[9] F. Hirt et al., Schriftenreihe Siedlungsverband Ruhrkohlenbezirk 44 (1972).
[10] A. Löffler et al., Gesellschaft für Weltraumforschung TN-73/4 (1973).

Earth Survey Problems — Akademie-Verlag, Berlin 1974

# NEED FOR AND ASPECTS OF A COOPERATIVE EUROPEAN EARTH RESOURCES PROGRAMME

J. Bodechtel[a] and D. Davidts[b]

[a]Zentralstelle für Geo-Photogrammetrie und Fernerkundung, Institut für allgemeine und angewandte Geologie der Universität München, München, FRG

[b]Messerschmidt-Bölkow-Blohm GmbH, Unternehmensbereich Raumfahrt, München, FRG

**Based on national and European studies and investigations on earth resources survey from space, the demands from European aspects are derived. The special environmental conditions and variety of the European continent and areas of European interest call for modified mission programmes and the adaptation of data evaluation to our requirements.**

## 1. Goals for Space-borne Monitoring of the Earth

Our environmental problems more and more assume global proportions, and space technology as a tool for handling these problems is at long distance superior to conventional techniques. Earth-observation satellites will therefore play a basic yet extraordinary role in the future of space technology. The realization that the direct, large-scale interaction between man and his environment can be precisely measured by earth-observation satellites has placed the utility of space technology in an entirely new light. As a consequence of this realization, the control and conservation of our environment and the detection and future management of natural resources should become a task of the highest priority.

Imagery from satellites in relatively low earth orbits and its subsequent interpretation will be part of the efforts aimed at accomplishing these permanent future tasks in an efficient manner. This applies particularly to the investigation and control of environmental pollution, a problem of the first order in view of worldwide industrialization and growing population pressure.

For the definition of tasks concerning global earth observation by space-borne remote-sensing techniques, the following main advantages may be mentioned: a synoptic view and analysis of regional conditions and interrelationships of large areas; continuous observations of transient as well as permanent targets on the earth's surface; observation under conditions of equal illumination for quantitative comparison of the spectral behaviour of different targets of large area; observation at the same time of different parameters for the detection of interrelationships. The last two points are particularly important when different regions are considered; the structure and variety of Europe demand modified and different tasks.

Within the many-fold applications, the observation of changing phenomena and their regional relationships are of great importance. The thematic mapping of those phenomena for Europe shows important tasks within geosciences, hydrology, oceanography including fishery, regional planning, agriculture, forestry and climatology. For the continuous observations of permanent and transient phenomena over Europe satellites are considered to be the best means of guaranteeing synoptic and repetitive coverage.

Tables 1—6 show some typical European remote monitoring requirements, based on industrial studies carried out by Messerschmitt-Bölkow-Blohm GmbH, Munich, and Battelle Institute, Frankfurt/M.

Besides selected fields of application for earth observation present used methods of investigation and monitoring are mentioned. Indicators relevant for remote observation and the requirements under special European aspects are listed. The different applications and requirements are adjusted to physical quantitative spectral bands and remote sensing instrumentation, emphasizing European needs.

## 2. Earth Observation and Quantitative Measurement

Future planning of earth observation systems will have to consider the necessity of omitting the actual qualitative interpretation of images of the earth surface in order to establish a quantitative measurement system. This procedure will be a main requirement to keep the data output on a level which is compatible with the data evaluation capacity and capability of the user community. In order to achieve clarity in the related areas of information and communication engineering, an earth observation system has been defined, whose elements are shown in Fig. 1.

The major components and related research work of this system are:

(1) *A mathematical model of the spectral behaviour of the earth surface:* derivation of model elements, which can be correlated to the physical state and to the physical quantities which can be measured by remote sensing. This model should enable the user to identify the minimum information needed (information reduction).

(2) *The sensor system,* covering the whole electromagnetic spectrum as far as is required by the user for reduced information needs: sensor units must be modified or developed to fit quantitative measurement requirements with respect to information reduction.

(3) *Data processing related to information needs:* software has to be developed for the evaluation of the spectral data with respect to the elements of the mathematical model.

## 3. European Aspects for Earth Resources Satellite Programme

Remote earth observations are necessary similar to meteorological observations as they are carried out within the global atmospheric research programme operation systems. European requirements must be considered in relation to special instrumentation, data evaluation and experiments. The scientific and economic advantages of orbital remote sensing calls for a European research programme in this field.

Compared with the USA, Europe with its varying climatic conditions, its inhomogeneous topography and its natural landscape which has been camouflaged by continuous human activity, requires different detecting and evaluation techniques. The development of sensors, and, in particular, of optimally applicable spectral ranges, with variation of flying altitude according to ground resolution, must be adapted to these conditions.

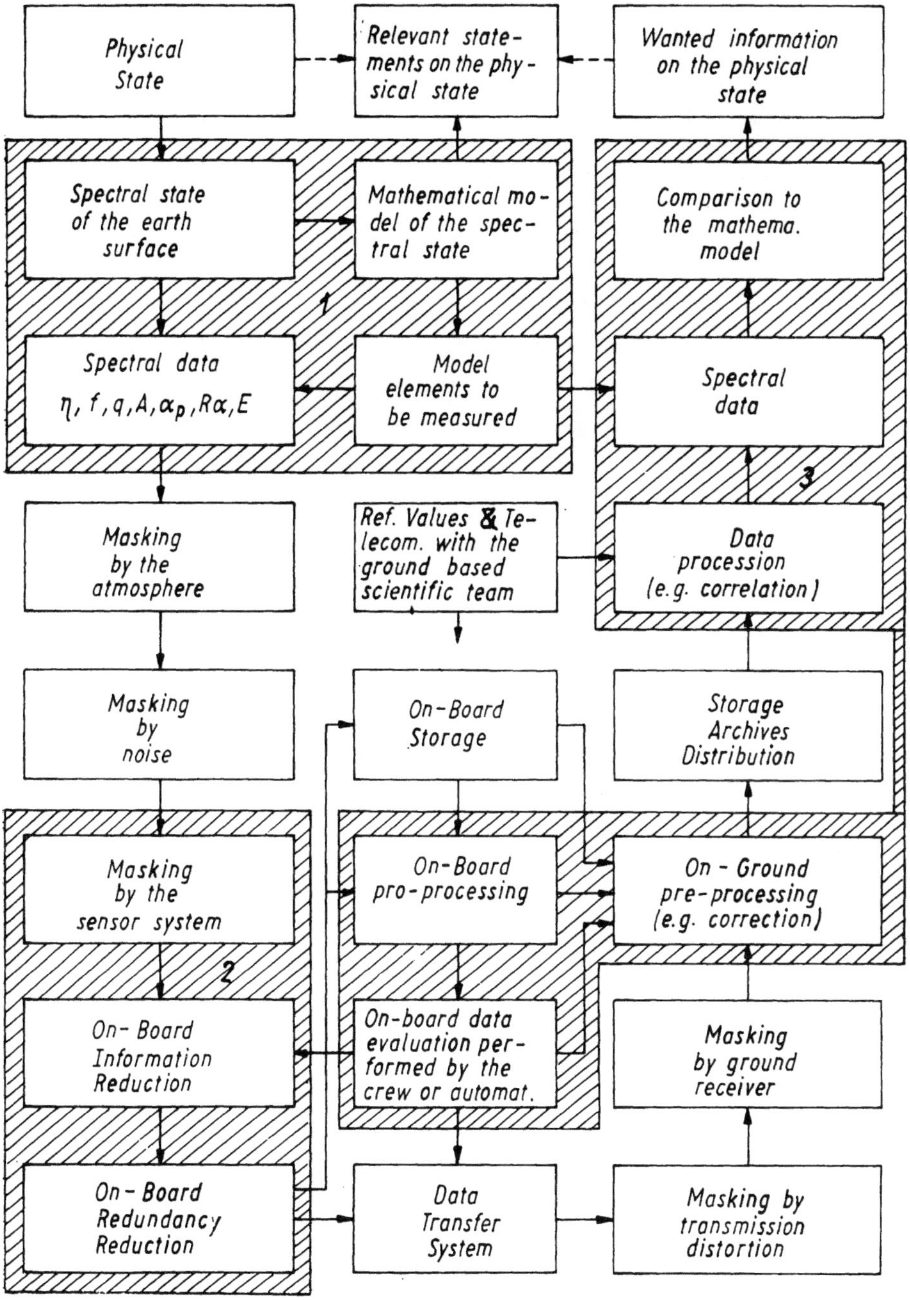


Fig. 1. Earth observation system elements (reflecting manned and automated space platforms).

A mere participation in current US research programmes offers no satisfactory equivalent. The equipment, the repetition-rate dependent on inclination, the orbital altitude and the time of coverage of US satellites (for example weather satellites, noon orbit), must also be varied for European interests. Typical demands derived from European remote sensing activities and studies can be summarized with higher special resolution, and weather independence which requires the application of active and passive microwave techniques.

Besides weather satellites with a high repetition-rate which are needed for the detection of large-scale phenomena the first full-time application of a spacecraft

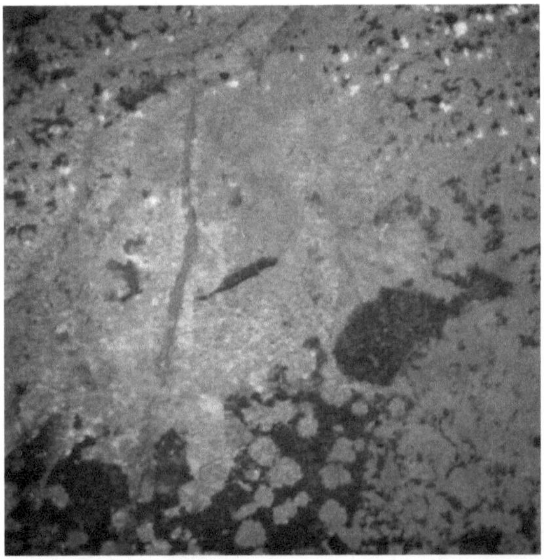

**Fig. 2. Colour composite (ERTS-MSS 4, 5 and 7, August 1972) of the Munich Area, West Germany.**

for earth surveillance has been realized in NASA's ERTS 1. Because of the relatively low repetition rate of 18 days and the lack of near real-time data due to no existing receiving station in Europe, the application of ERTS 1 for European investigations is feasible only for some objectives (Fig. 2). Furthermore the climatic conditions in Europe only allow a limited use of the visible bands.

The manned space programme with the earth resources experimental package EREP in Skylab enlarges the experimental possibilities for the use of satellite remote sensing. The concept of Skylab shows the significant importance of manned space missions of predominantly experimental character and the restricted use for operational monitoring. Manned missions seemed to be indispensable and of utmost importance in preparing optimal concepts and definitions for operational monitoring, because only by a most complete and variable sensor package and the possibility of a permanent interaction of man can such missions be of value.

Worldwide operational space-borne monitoring should show the following characteristics for study of European conditions:

(i) Sun-synchronous orbit of 570 km altitude. The optimum transit time for quantitative investigation of the spectral properties would be noon (for observation in the visible part of the spectrum), in order to eliminate shadowing effects.
(ii) A special resolution of sensors in the visible and near infrared between 20 and 30 m regarding the small scale pattern of European landscape.
(iii) A net of two or three satellites within an orbital distance of 120° latitude to observe fast changing environmental targets and their regional interrelationship. This could be performed by the Spacelab or shuttle system making use of the satellite deployment-capabilities.

European climatic conditions with a cloud cover of often more than 50% emphasize the need for active and passive microwave sensors.

In the manned space program the European contribution of the Spacelab within the Post Apollo Program will be of great importance for earth observation missions. The mission duration of initially 7 and later a maximum of 30 days defines the platform Spacelab to mainly experimental missions. But within those experimental missions the Spacelab will serve as an optimum sensor platform to define operational models.

Typical tasks are the testing and correlation of complex sensor equipments for defined user requirements, investigations and development of on-board data selection and data processing, taking advantage of the role of man within the experiment

Typical European missions of an earth observation Spacelab with 7 days duration could be the repetitive observation of one optimal selected overpass, covering a strip of selected typical test-sites for different tasks. The ideal sensor combination, transit time, and data reduction possibilities would be selected for operational missions.

## References

J. Bodechtel, Int. Symp. on Space and Communciation, Paris 1971 (p. 529).

J. Bodechtel, XX Rassegna Int. Elettron. Nucl. ed Aerospaziale, Roma 1973 (p. 103).

Definition of the Technical Requirements for an Earth Resources Payload for the Spacelab, ESRO Contr. No. SC/3/73/HQ, Messerschmitt-Bölkow-Blohm GmbH, Ottobrunn/München 1973.

Sortie-Laboratory Preliminary Definition Study, Earth Observation Payload Definition and Accommodation Report, Messerschmitt-Bölkow-Blohm GmbH, ESTEC Contr. No. 1644/72 EL, 1972.

Untersuchung des Nutzens der Fernerkundung der Erde mit den Mitteln der Weltraumtechnik I, GFW Vertrag No. RV II 1 V 10/72 KA 10-00-00, Messerschmitt-Bölkow-Blohm — Battelle Institut, 1973.

Views of the ESRO Part Group on Earth Resources Missions in the Post-Apollo Program, Paris 1973.

Abbreviations used in Tables

*Present methods:*

| | |
|---|---|
| field | f |
| measuring stations | st |
| photointerpretation | pi |
| remote sensing | RS |
| sonar | So |
| laboratory | L |
| radar | RA |

*Requirements:*

Columns:

(1) size of target
(2) Repetition rate
(3) Time
(4) Frequency of observations
(5) Coverage
(6) Data output.

| | |
|---|---|
| necessary | + |
| not important | 0 |
| not required | — |
| as required | ar |
| constant surveillance required | const. |
| immediately | im |
| single pass required | 1× |
| hour | h |
| day | d |
| week | w |
| month | m |
| year | a |
| spring | sp |
| summer | su |
| autumn | au |
| winter | wi |
| image | I |
| map | Mp |

*Physical quantities:*

Columns:

(7) Indirect
(8) Direct

| | |
|---|---|
| Emission | E |
| Reflection | R |
| Absorption factor | $\alpha$ |
| Transmission time | L |
| Polarization | P |
| Phase | Ph |
| Amplitude | A |

*Instruments:*

Columns:

(9) UV-IR
(10) Microwave

| | |
|---|---|
| Photographic camera | P |
| Multi-spectral camera | M |
| Multi-spectral scanner | S |
| Infrared scanner | IRS |
| Infrared radiometer | IRR |
| UV absorption spectrometer | UVAS |
| Synthetic aperture radar | SAR |
| Radar altimeter | RH |
| Microwave radiometer | MWR |
| Microwave scan radiometer | MWSR |
| Microwave scatterometer | MWSC |
| Laser altimeter | LH |

Table 1

Regional Planning

| Application | Present methods | Indicator | Requirements under European Aspects | | | | | | Remote Sensing Methods: Physical Quantities | | Remote Sensing Methods: Instruments and Spectral Bands | |
|---|---|---|---|---|---|---|---|---|---|---|---|---|
| | | | (1) | (2) | (3) | (4) | (5) | (6) | (7) | (8) | (9) | (10) |
| *Thematic mapping:* | | | | | | | | | | | | |
| Vegetation maps | F, PI | vegetation communities | 50 m | 2—4/a | sp—au | m | + | I, Mp | E, R | A, L | P, M, S | SAR |
| Maps of local climate | F, St | vegetation temperature | 100 m | d, a | const | m | + | Mp | E, $\alpha$ | A | S | IRS |
| Maps of urban development | F, PI | contours, tonal | 10—50 m | 1/a | 0 | m | + | I, Mp | E, R | A, L PH | P, M, S | SAR |
| traffic flow | | difference | 5 m | const | | d | | Mp, diagr | | A, L, PH | | SAR |

Table 2

Environmental Monitoring

| Application | Present methods | Indicator | Requirements under European Aspects (1) | (2) | (3) | (4) | (5) | (6) | Remote Sensing Methods: Physical Quantities (7) | (8) | Instruments and Spectral Bands (9) | (10) |
|---|---|---|---|---|---|---|---|---|---|---|---|---|
| *Water supply* | | | | | | | | | | | | |
| Ground water level (ground water supply etc.) | F, RS | soil moisture thermal emission, tonal differences vegetation | 10—100 m | 2/a — 4/a | wi, su | 0 | + | I, Mp | E, R, α, S | A, P | M, S, IRS | MWR |
| Snow coverage and predicted melting rate for energy floods | ST | grey tone temperature contours | 100 m | w—m | wi | m | + | I, Mp | E, R, α, S | A, PH, P | P, S, IRS IRR | SAR, MWSR, MWR |
| Water level in rivers, lakes etc. | ST | contours | 10—50 m | m | ar | + | + | I, Mp | E, R, α | A, PH, P, L | P, S, LH | RH |
| *Water Pollution* | | | | | | | | | | | | |
| Discharge detection status | ST | temperature tonal differences | 10—50 | d—m | ar | + | + | I, Mp | E, α, R, S | A, P | M, S, IRS, IRR | MWSR, MWR |
| Monitoring river Lake, sea | | | 50—100 m | d—m | ar | + | + | I, Mp | E, R | A, PH, LP | M, S IRS | MWSR, MWR (SAR) |

Table 2 (continuation)

| Application | Present methods | Indicator | Requirements under European Aspects (1) | (2) | (3) | (4) | (5) | (6) | Remote Sensing Methods: Physical Quantities (7) | (8) | Instruments and Spectral Bands (9) | (10) |
|---|---|---|---|---|---|---|---|---|---|---|---|---|
| *Air Pollution* | | | | | | | | | | | | |
| Emission sources | F, ST | tonal differences spectral behaviour | 10 | d—m | ar | + | + | I, Mp | E, R, $\alpha$, S | A, P | M, S UVAS | IRS, MWSC |
| movement and distribution | F, ST | tonal differences spectral behaviour | 100 | d—m | ar | + | + | I, Mp | E, R, $\alpha$, S | A, P | M, S, UVAS | IRS, MWSC |
| *Degradation and Desertisation in land use* | F, PI | tonal differences spectral emission | 50—100 m | m | ar | mo | + | I, Mp | E, R, $\alpha$ | A, P | M, S, IRS, IRR | SAR |
| *Disaster Assessments* | | | | | | | | | | | | |
| Floods | F, PI | contours reflectance of vegetation | 10—100 m | 1× | im | im | + | I, Mp | E, R | A | P, M, S, IRS | |
| Sea ice | PI, ST | contours temperature | 10—100 m | w—m | im | im | + | I, Mp | E, R, $\alpha$, S | A, P, PH, L | P, M, S, IRS, IRR | SAR, MWR |
| Surveillance of volcanic activity | ST, RS | thermal emission, heat flow | 10—100 m | ar d—m | const | + | + | I, Mp | E, R, $\alpha$ | A | IRS | |

Table 3

Agriculture — Land Use

| Application | Present methods | Indicator | Requirements under European Aspect (1) | (2) | (3) | (4) | (5) | (6) | Remote Sensing Methods: Physical Quantities (7) | (8) | Instruments and Spectral Bands (9) | (10) |
|---|---|---|---|---|---|---|---|---|---|---|---|---|
| Detection of soil quality | F | plant communities | field width | 2/a | sp, su | mo | + | I | E, R, S | A | P, M, S | MWR |
| Detection soil moisture | F | IR-reflection of vegetation | 20 m | 4/a | | d, mo | + | I | R, $\alpha$, S | A, P | M, S, IRS | MWR, MWSR |
| Immediate disease assessment | F | spectral characteristics in emission, reflectance and absorption | 20 m | w | | w | + | I | R, E, $\alpha$ | A, P | M, S IRS IRR | |
| Land use inventory classification harvest forecast | F, PI | spectral characteristics in emission, reflectance and absorption | 20 m | mo | sp, su | w | + | Mp | R, E, $\alpha$ | A, PH, L | P, M, S | SAR |

Table 4

Forestry

| Application | Present methods | Indicator | Requirements under European Aspects | | | | | | Remote Sensing Methods: Physical Quantities | | Remote Sensing Methods: Instruments and Spectral Bands | |
|---|---|---|---|---|---|---|---|---|---|---|---|---|
| | | | (1) | (2) | (3) | (4) | (5) | (6) | (7) | (8) | (9) | (10) |
| Production, correction of thematic maps | F, PI | tonal differences colour, signature, contours | 10—50 m | ar | ar | 0 | + | I, Mp | E, R | A | P, M, S | SAR |
| Forest inventory, improvement and range assessment, range management | F, PI | spectral characteristics in emission, reflection and absorption | 10—50 m | ar | ar | 0 | + | I, Mp | E, R, $\alpha$, S | A, PH, L | P, M, S | SAR |
| Insect and disease survey classification of disease (biotic, abiotic) | F | spectral characteristics in emission, reflection and absorption | 1—50 m | 1 × | su | + | + | I, Mp | E, R, $\alpha$ S | A, P | P, M, S, IRS | |

Table 5

Applied Geology

| Application | Present methods | Indicator | Requirements under European Aspects (1) | (2) | (3) | (4) | (5) | (6) | Remote Sensing Methods: Physical Quantities (7) | (8) | Instruments and Spectral Bands (9) | (10) |
|---|---|---|---|---|---|---|---|---|---|---|---|---|
| Dynamic geology (sedimentation in coastal and transport areas etc.) | F, PI | tonal differences, colour, contours | 10—100 m | d—m | const | + | + | I, Mp | E, R, $\alpha$ | A, PH | P, M, S | SAR |
| Hydrogeology | F, RS | soil moisture, thermal emission, tonal differences, vegetation | 10—100 | 4/a | su | 0 | + | I, Mp | E, R, $\alpha$, S | A, P | P, M, S, IRS | SAR MWSR, MWR |
| Prospecting (mining, oil) | F, PI | especially secondary, geological indicators, thermal anomalies tonal differences colour, plant communities | 10—100 m | ar | ar | 0 | + | I, Mp | E, R, $\alpha$, S | A, P, PH, L | P, M, S, IRS | MWR, SAR |
| Detection of thermal anomalies | RS | thermal emission | 100 m | w—m | const | + | + | I, Mp | E, R, S | A | IRS, IRR | |

Table 6

Oceanography

| Application | Present methods | Indicator | Requirements under European Aspects (1) | (2) | (3) | (4) | (5) | (6) | Remote Sensing Methods: Physical Quantities (7) | (8) | Instruments and Spectral Bands (9) | (10) |
|---|---|---|---|---|---|---|---|---|---|---|---|---|
| Fishery location and movement of fishing grounds | So | bioluminescence (fish, oil-slicks), temperature, colour (plankton) | 30—1000 m | d—m | const | h | + | I, Mp | E, R, $\alpha$, S | A | PMS, IRS, IRR | MWSR, MWR |
| Pollution location, movement and classification of pollutants | L | temperature, colour salinity, surface, films, turbidity | 100 50 m | d—m | ar | + | + | I, Mp | E, R, $\alpha$, S | A | UVAS, M, S, IRS | MWSR, MWR |
| Ocean circulation (currents), tidal phenomena | ST, RS | temperature, colour wave pattern, wave-height | 50—1000 m | d—w | ar | 0 | + | I, Mp | E, R, S | A,P | M, S IRS, IRR | MWSR, MWR |
| Sea state; condition of sea surface | ST | contours wave pattern | 10—50 m | d | const | h | + | I, Mp | E, R, S | A | P, LH, IRS, IRR | MWSR, MWR |
| Detection of sea ice | RS, PI | temperature contours | 50 m | d | const | jm | + | Mp | E, R, $\alpha$, S | A, U | P, IRS | MWR, SAR |
| Maritime activities (ship routing) | RA | colour temperature | 10 —50 m | d—w | const | h | + | Mp | E, R, E, R, $\alpha$ | A A, U | P, MS IRS | RH, MWR, SAR |

# Latest Results in Meteorology and Earth Surveys

Earth Survey Problems — Akademie-Verlag, Berlin 1974

# AEROLOGICAL SOUNDINGS OF THE ATMOSPHERE FROM NOAA 2 DATA FOR OPERATIONAL SYSTEMS

D. Q. WARK, A. WERBOWETZKI, M. P. WEINREB, J. A. PRITCHARD, L. M. MCMILLIN, R. C. LUEBBE, L. A. LAURITSON, H. E. FLEMING, D. S. CROSBY and F. E. BITTNER

NOAA, National Environmental Satellite Service, Washington, D.C., USA

**NOAA 2 carries the Vertical Temperature Profile Radiometer (VTPR), which measures spectral radiance in eight intervals between 12 μm and 19 μm. Step-scanning the 55-km field of view builds an array of observations. By a combination of linear extrapolation and statistical filtering, the radiances of clear areas are deduced, after which temperature profiles to 30 km and mean humidities to 5 km are deduced. Results are transmitted to meteorological services through existing weather communications systems. Central to this system of aerological observations are a reliable quality control procedure and the timely processing and dissemination of data. Portions of the planned system were made operational in December 1972. Examples are shown of temperature, thickness, and height contour charts, and comparisons are made with conventional data. These results indicate several areas of development in which improved products may be anticipated; these include the elimination of certain biases in deduced radiances of clear areas, and the extension of analyses to land areas.**

## 1. Introduction

An active program has been carried out during more than a decade in preparation for operational sounding of the atmosphere for temperature and humidity from meteorological satellites. The Nimbus satellites have carried a series of experiments to measure infrared spectral radiances in multiple spectral intervals. Instruments in this class have included Satellite Infrared Spectrometers (SIRS) on Nimbus 3 and 4; Infrared Interferometer Spectrometers (IRIS) on the same satellites; Selective Chopper Radiometers (CSR) on Nimbus 4 and 5; and the Infrared Temperature Profile Radiometer (ITPR) on Nimbus 5. Measurements from these instruments have demonstrated the feasibility for deriving temperature profiles, and for the operational applications of these results to meteorological analysis.

Anticipating the success of those experiments, the NASA and the NOAA collaborated in planning an advanced technique for deriving soundings on a fully operational basis. The instrument developed under NASA sponsorship was designated the Vertical Temperature Profile Radiometer (VTPR). Two nearly identical instruments of this design are carried on the NOAA 2 satellite, launched 15 October 1972.

## 2. Observations and Data Reduction

The VTPR instrument [1] consists of a rotating filter wheel which is interposed before a single detector; a Cassegrain fore-optics telescope; and a scan mirror which directs the earth beam through the filter to the detector. Calibration is provided from views of space and an internal black body.

Eight narrow-pass filters mounted on the wheel give a set of observations of the earth each half second. Six spectral intervals are located in the 15 μm band of carbon dioxide, and observations there are used for the retrieval of temperature profiles. Spectral intervals at 12 μm and 19 μm are in a window, used as a boundary control, and in the water vapor rotation band for the retrieval of a one-parameter humidity estimate in the lower troposphere. Table 1 lists the nominal central spectral frequency, the nominal bandwidth at half-maximum, and the function for each filter.

Table 1

Nominal (Design) Characteristics of VTPR Filters

| No. | Center ($cm^{-1}$) | Half-width ($cm^{-1}$) | Function |
|---|---|---|---|
| 1 | 668.5 | 3.5 | Temperature profile |
| 2 | 677.5 | 10 | Temperature profile |
| 3 | 695 | 10 | Temperature profile |
| 4 | 708 | 10 | Temperature profile |
| 5 | 725 | 10 | Temperature profile |
| 6 | 747 | 10 | Temperature profile |
| 7 | 535 | 18 | Water vapor |
| 8 | 833 | 10 | Control (window) |

Fig. 1 shows the weighting functions of the six channels for a typical atmosphere in the sub-tropics (Lihue, Hawaii, at 0000 GMT, 31 January 1973). These functions show the levels from which contributions are made to each spectral radiance. The independent variable is $x = p^{2/7}$, where $p$ is pressure; abscissa values refer to the dashed lines, while values for the solid lines must be divided by 2. It will be noted that the spectral frequencies given in this figure for NOAA 2, Instrument 2, are slightly different from the nominal design values of Table 1.

Scan of the VTPR from the 1465 km orbit is $\pm 30.5°$ transverse to the orbital path. The 2.5° field-of-view provides approximately contiguous views of the earth in 23 steps. At each step the filter wheel is rotated once, enabling the detector to provide eight spectral radiances. Average dimensions of the earth projection of the field of view is about $75 \times 75$ km$^2$.

Fig. 2 depicts the scan pattern projected on the earth. From left to right the dimension is about 2000 km, which provides 2/3 areal coverage at the equator and full coverage poleward of 50° latitude. Analysis for "clear radiances", which are radiances deduced statistically from considerations of "clear radiances" in the window and the observed radiances in individual spectral channels, is accomplished by comparison of contiguous readings in matrices of $8 \times 8$ or $8 \times 7$ observed spots in Fig. 2. In each array a set of seven "clear radiances" is deduced for the nominal location of the centers noted by crosses (adjustment to a centroid of clear areas is achieved by examination of the window radiances).

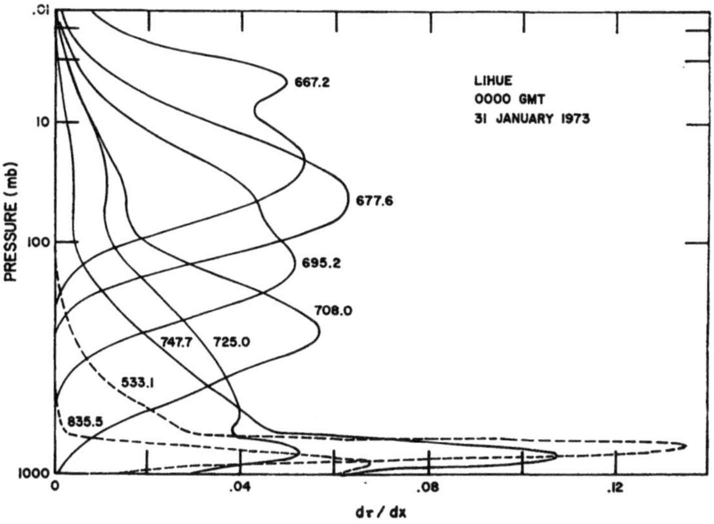


Fig. 1. VTPR weighting functions, $d\tau/d(p^{2/7})$, for Lihue, Hawaii, 0000 GMT, 31 January 1973.

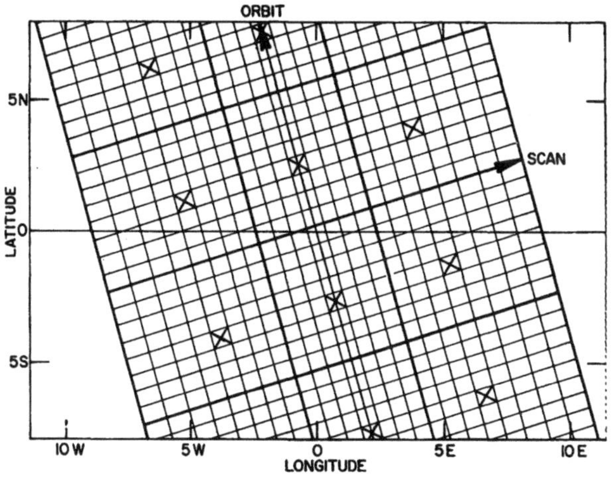


Fig. 2. VTPR scan pattern projected on the earth.

## 3. Temperature and Humidity Retrieval

Results from the preceding analysis provide six "clear radiances" for the deduction of temperature profiles and one for the estimate of lower-troposphere humidity.

Temperature profiles are deduced from the "minimum information" solution [2]. This is a matrix-inversion solution for the difference from a first-guess profile. The first guesses are the sea-surface temperature, which is used for the boundary term, and the forecast height (and temperature) fields generated by the National

Meteorological Center for the standard meteorological levels and the tropopause. The most recent analysis or a fixed first-guess is used in the southern hemisphere or the tropics until a global forecast can be substituted.

Humidity is deduced by a two-step procedure. Regression upon the retrieved temperature profile provides a first estimate of the humidity profile. The solution for the humidity profile is constructed as a linear combination of the first estimate and an empirical orthogonal function [3]. The unknown coefficient in this combination is determined through the requirement that the calculated radiance equals the measured radiance in the 535 $cm^{-1}$ channel.

From retrieved profiles the heights are calculated at each of the 15 standard levels from 1000 mb to 10 mb. Absolute heights are determined from a forecast reference surface; relative heights are used south of 18° N pending implementation of a global forecast. Before transmitting the aerological data to users, they are subjected to several objective quality checks to eliminate gross errors in the troposphere (1000—100 mb). Tests are made for superadiabatic lapse rates and for inconsistency between adjacent soundings. An entire sounding is rejected if it fails a test at any single standard level.

Profiles have been transmitted to users since December 1972 by means of the SIRS telecommunications code, which is a USA national practice. VTPR data are processed immediately after readout of the satellite each orbit; processing time is normally about 30 minutes, after which they are transmitted immediately to domestic and foreign weather services. A somewhat more flexible code for transmission of data will be initiated soon.

For weather services which may wish to make independent analyses of the observational data, "clear radiances" will be made available. A separate code will be employed in the transmission of these data. Radiances which have not been reduced to this form are available to special users (i.e. to those who possess adequate communication facilities); the volume of these data is 70 times greater than for "clear radiances".

## 4. Discussion

An example of meterological analysis based upon satellite-derived temperature profiles is shown in Fig. 3, which is the thickness pattern for the 1000—300 mb levels in the southern hemisphere at 0000 GMT 12 January 1973, $\pm 6$ hours. Dashed lines show major cloud patterns, which may usually be identified with active or degraded fronts. Warm and cold regions are related to these frontal patterns. In this analysis there were 500 VTPR soundings and 35 radiosondes; three of the former and one of the latter were rejected.

Beginning about July 1973, new satellites will have direct readout capability; the VTPR data will be transmitted through the spacecraft beacon. Receiving stations will be able to acquire VTPR data on a real-time basis for local evaluation and analysis.

Soundings now are produced only over ocean areas because surface temperatures over land vary greatly with time and one must therefore rely upon measurements rather than persistence. An approach which makes use of the 747 $cm^{-1}$ measurements to deduce surface temperatures rather than atmospheric temperatures will be attempted soon. An alternate scheme, which would use the much higher resolution of the scanning radiometer, is being considered; the

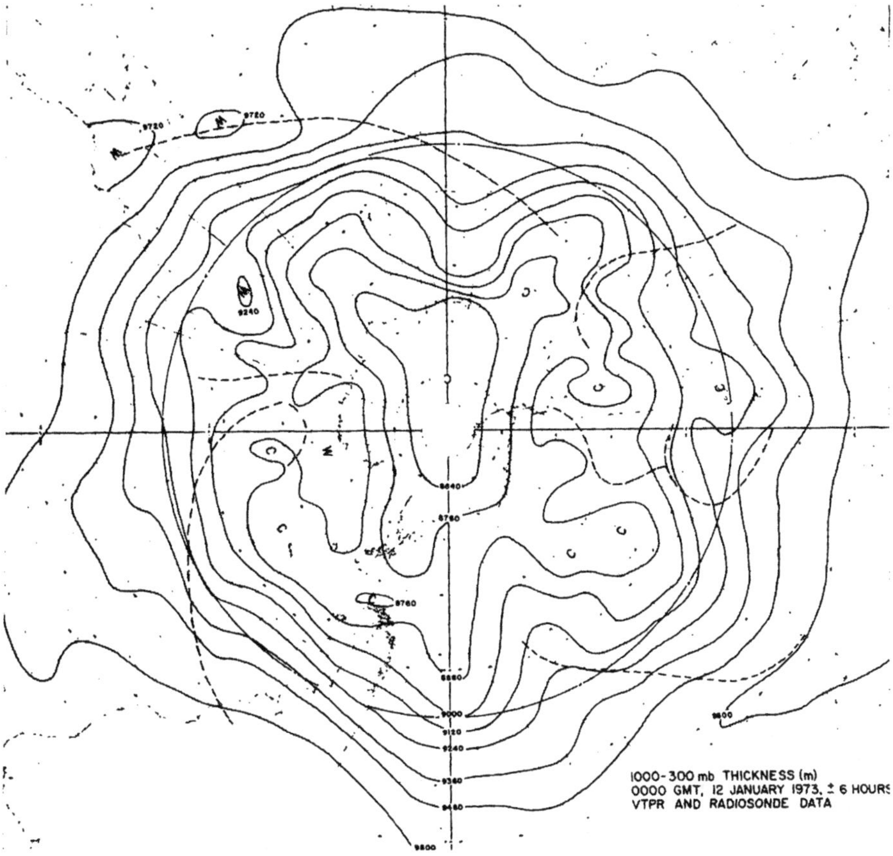


Fig. 3. Contour chart at 300 mb over the southern hemisphere, 0000 GMT, 12 January 1973, drawn from VTPR and radiosonde reports.

resulting delay in processing of VTPR data makes this approach less desirable from an operational standpoint.

All VTPR data will be filed with the Environmental Data Service, Ashville, N.C. Radiances (the complete volume of earth-located and calibrated values) will be in an archive for a limited time yet to be determined; "clear radiances" will be in a second and longer-lived archive; and the temperature, height, and humidity profiles will be contained in a permanent archive.

## References

[1] G. Falbel and D. Zink, Proc. IERE Conf. on Infrared Techniques, University of Reading, UK, 1971 (p. 231).
[2] M. R. Foster, J. SIAM, **9**, 387 (1961).
[3] J. C. Alishouse, L. J. Crone, H. E. Fleming, F. L. VanCleef and D. Q. Wark, Tellus, **19**, 477 (1967).

Earth Survey Problems — Akademie-Verlag, Berlin 1974

# INFRARED TRANSMITTANCES FOR INDIRECT SOUNDINGS OF THE ATMOSPHERE FROM SATELLITE-BASED MEASUREMENTS

D. Q. Wark

NOAA, National Environmental Satellite Service, Washington, D.C., USA

Deductions of atmospheric temperature and humidity soundings from spectral radiance measurements by satellite-borne instruments depend upon a good knowledge of transmittances by carbon dioxide, water vapor and ozone. These are calculated in detail, both spectrally and through the stratified atmosphere, and are applied to instrument spectral responses. Comparisons are then made between radiances calculated from known temperature profiles and measurements from satellites. Differences may be ascribed to errors in transmittances, atmospheric profiles from radiosondes and rocketsondes, and calibrations and other instrumental influences. Data from Satellite Infrared Spectrometers on Nimbus 3 and 4, and from Vertical Temperature Profile Radiometers on NOAA 2 are used as radiometric references. These indicate that some systematic upward or downward shifts are needed in the transmittances to bring calculated radiances into agreement with measurements. Variability in these results is ascribed to causes within the instruments. The need for correction of the calculated transmittances is supported by direct measurements from balloons. Standards of spectral transmittances by the three gases for application to sounding systems to be employed during the GARP period are proposed.

## 1. Introduction

Inference of atmospheric parameters such as temperature, humidity, and ozone structures can be made from satellite measurements of infrared spectral radiances. These are related through the integral form of the radiative transfer equation, which relates observed spectral radiances to local atmospheric temperature, through the Planck relation, and to the local rates of change of optical transmittances, or weighting functions. Solution for one of these two quantities requires that the other must be known.

Verification of observed radiances is usually accomplished from values which coincide in time and location with radiosonde or rocketsonde measurements of atmospheric temperature, humidity, and ozone concentration. Spectral radiances are calculated from the integral equation, using the atmospheric soundings and transmittances evaluated from established parameters. Systematic differences from the measured values are attributed to one of three elements: measurements, assumed profiles, or transmittances.

The present study is part of a continuing program to evaluate spectral transmittances by carbon dioxide, water vapor, and ozone in the infrared with accuracies which will largely eliminate errors from this source.

## 2. Satellite and Balloon Measurements

From coincident satellite and atmospheric measurements during 1969—1973, using SIRS A and SIRS B [1] and the VTPR's [2], some preliminary estimates have been made of the transmittances by carbon dioxide in the range 668 to 750 cm$^{-1}$. Detailed spectral transmittances were calculated point-by-point [3] and convoluted with each instrument's spectral responses. In order to bring radiances calculated from these results into agreement with measured values, it has been necessary to multiply each transmittance by a continuum value of the form $\tau_c(\nu, p) = \exp[-k(\nu)p^2/p_1^2(\nu)]$, where $p_1$ is the pressure level of unit optical depth, used as a normalization of $k$. The rationale for this is that the wings of lines were under- or over-estimated.

In Fig. 1 the values of $k(\nu)$ are given at the discrete locations of each instrument's spectral channels. Also given are values deduced from direct measurements in the atmosphere by a scanning radiometer for a balloon [4].

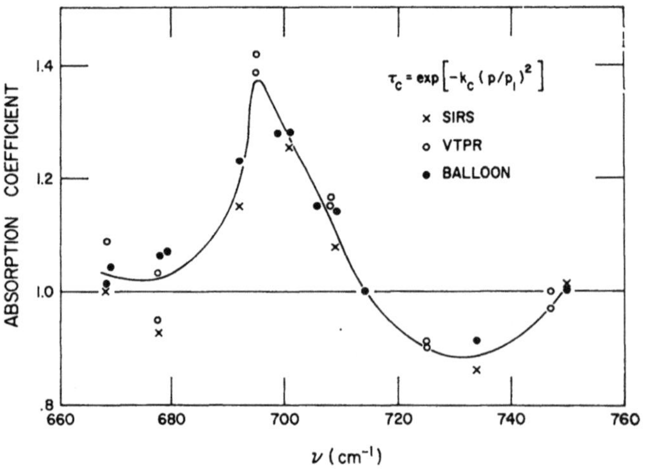


Fig. 1. Continuum absorption coefficients for correcting NESS carbon dioxide transmittances.

The balloon program has been continued, to measure transmittances directly in the stratosphere. On 16 October 1972, two Selective Chopper Radiometers [5], as well as the spectrometer in [4], were carried to 40 km. Deductions of transmittances are in an intermediate state. Fig. 2 shows preliminary values in four SCR channels. Carbon dioxide gas in three cells leads to apparently greater transparencies of the atmosphere where they are interposed in the beam. Because these are direct measurements, they contain the variable solar (source) angle with pressure altitude. The solar angle, $\theta$, is shown on the right. From these results and from those of other planned balloon flights, aircraft flights, and laboratory measurements, it is expected that carbon dioxide transmittances in the 15 $\mu$m band can be established firmly.

Water vapor and ozone transmittances require a different approach because of their variable mixtures in the atmosphere. Point-by-point transmittances may

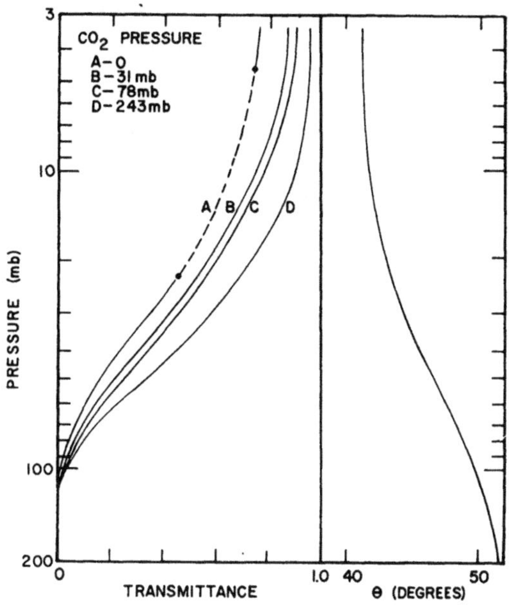


Fig. 2. Observed transmittances for four SCR channels (left) and corresponding solar angles (right).

be calculated for homogeneous paths and applied to the atmosphere by quadrature of increments [6]. Water vapor requires, in addition, the inclusion of the "e-type" continuum [7].

## 3. Standard Transmittances

With the expansion of observing capabilities, and the prospect of additional instruments being employed for the WWW and during the first GARP Global Experiment (FGGE), the need for standard values of transmittances becomes more necessary. Observations from separate observing systems can be coordinated through such standards. Four proposals are set forth to establish standard transmittances:

(i) Line parameters of strength (intensity), pressure-broadened half-width, line position, and the lower-state energy level have been evaluated during the past several years by a group of experts, basing their evaluations upon laboratory measurements and theoretical models. The resulting compilation by the Air Force Cambridge Research Laboratory [8] is proposed as the starting point from which all calculations of transmittance should proceed.

(ii) Numerous models of line shape have been developed from theory and from pragmatic considerations. Each has failed in some degree to predict transmittances at all spectral frequencies and under all conditions of temperature, pressure, and optical path. Because the difficulties arise mainly in the distant wings of lines, a reduction in the quasi-continuous absorption may be achieved by terminating a Lorentzian line at a fixed distance from the line center. In order to account for lines of different intensity and half-width, the distance should be proportional to $S\alpha$ or to $S\alpha u$.

(iii) Lorentzian profiles sometimes are inadequate far from line centers. If only the line shape is the cause, this condition may be corrected by superposing a pressure- and temperature-dependent continuum. For water vapor a second, and as yet not adequately explained, continuum which is dependent upon the partial pressure of water vapor exerts an influence from the inconsequential in the band centers to overwhelming dominance in the windows.

(iv) Calculation of spectral transmittances should be carried out point-by-point for standard conditions of atmospheric temperature and constituent mixture. These standard values, given for about 1—2 km intervals through the atmosphere, should be averaged over small spectral intervals of about 1 $cm^{-1}$. This proposal is for the establishment of such a standard set of transmittances. In order to allow for generalization of results, several standards for coverage and extreme conditions of each parameter should be included.

## Acknowledgments

Data presented in Fig. 1 were produced in cooperation with H. E. Fleming and F. L. VanCleef. The curves in Fig. 2 were generated by P. G. Abel and L. D. Johnson.

## References

[1] D. Q. WARK, D. T. HILLEARY, S. P. ANDERSON and J. C. FISCHER, IEEE Trans. **GE-8**, 264 (1970).
[2] D. Q. WARK et al., this volume, p. 469.
[3] S. R. DRAYSON, Appl. Opt. **5**, 385 (1966).
[4] T. G. KYLE, J. N. BROOKS, D. G. MURCRAY and W. J. WILLIAMS, Appl. Opt. **8**, 1926 (1969).
[5] J. T. HOUGHTON and S. D. SMITH, The Nimbus 5 User's Guide, Goddard Space Flight Center, Greenbelt, Md 1972.
[6] M. P. WEINREB and A. C. NEUENDORFFER, J. Atmos. Sci. **30**, (in press) (1973).
[7] K. J. BIGNELL, Quart. J. Roy. Met. Soc. **96**, 390 (1970).
[8] R. A. MCCLATCHEY, W. S. BENEDICT, S. A. CLOUGH, D. E. BURCH, R. F. CALFEE, K. FOX, L. S. ROTHMAN and J. S. GARING, Tech. Rep. AFCRL-TR-730096, ERP 434, January 1973.

Earth Survey Problems — Akademie-Verlag, Berlin 1974

# ON THE RELATIONSHIP BETWEEN THE EARTH-ATMOSPHERE SYSTEM ALBEDO AND THE EARTH'S SURFACE ALBEDO

K. YA. KONDRATYEV, L. N. DYACHENKO and N. P. PIATOVSKAYA

Department of Atmospheric Physics, University of Leningrad, Leningrad, USSR

On the basis of investigation of measurement data from the Soviet and American meteorological satellites, methods for determination of the underlying surface albedo are suggested, under cloudless sky conditions. The results obtained are compared with the surface climatological data.

The relationship between the earth-atmosphere system albedo and the underlying surface albedo is calculated, and its global distribution discussed.

## 1. Introduction

During the last few years there has been increased attention to the problem of the global net radiation (the earth-atmosphere system radiation balance). The long-term operation of meteorological satellites has made it possible to investigate regularities of variability of the net radiation and its components using direct measurements. Specifically, increased interest in the problem of the earth-atmosphere system net radiation arises from the fact that regularities of the atmospheric general circulation are closely connected with specific features of the geographical distribution of the net radiation on the planetary scale. Investigations of the earth-atmosphere system net radiation are of great interest from the point of view of understanding the basic laws of the planet's energetics as the major factor of the thermal regime and of the dynamics of the atmosphere. These investigations have also become very important in connection with the problem of the present climatic changes.

Satellite radiation data obtained during the past five years have made it possible to draw annual and seasonal maps of the planetary distribution of the earth-atmosphere system net radiation and of its components [1—3].

These data may be used to find relationships between the net radiation components of the system and of the earth's surface, and also to determine the net radiation of the atmosphere (radiative flux divergence and its components for the whole atmosphere). On the other hand, we can use the relationship between the components of the system net radiation and the surface net radiation to develop empirical methods for determining the net radiation components of the earth's surface and the atmosphere from satellite measurements of outgoing radiation (this makes it possible to solve adequately the problem of spatial smoothing as well).

The object of this paper is to compare measured values of the earth's surface albedo $A_0$ with the albedo of the earth-atmosphere system $A_s$. Such a comparison is of interest because it may also be useful when assessing man's possible impact on climate arising from anthropogenic changes in atmospheric optical properties.

The surface actinometric observations characterizing the specific features of the geographical distribution of the surface albedo are not always representative from the point of view of albedo for large territories. The problem of finding the correlation between the system albedo and the surface albedo with the aim of determining the surface albedo from the satellite measurement data, therefore, becomes urgent.

Naturally, such a problem can be considered only in the absence of cloudiness, the transforming effect of the cloudless atmosphere is more or less stable and can be treated using the empirical correlation as the basis or by calculating the transfer function of the atmosphere for albedo.

## 2. Determination of Albedo from the Meteorological Satellite Measurement Data

The earth-atmosphere system albedo is computed from the formula

$$A_s = \frac{R}{S_0 \cos i}, \tag{1}$$

where $R$ is the outgoing shortwave radiation flux determined from the satellite (OSR); $S_0$ is the solar constant and $i$ the solar zenith angle.

Satellite measurement of the OSR values for calculating the albedo with formula (1) is a complex problem because satellite measurements generally provide brightness (intensity) values for small sites under various sighting angles. For determining OSR fluxes there was elaborated a special calculation scheme [4, 5]. In the case of the cloudless atmosphere, this scheme enables calculation of the atmospheric short-wave net radiation and its components.

## 3. Determination of the Surface Albedo from the Albedo of the Earth-Atmosphere System measured from the Satellites

The analysis of the theoretical calculations for the optical model of the cloudless atmosphere, performed in [4], has shown that the albedo of the earth-atmosphere system is practically a linear function of the surface albedo. The following simple approximating formula was obtained to determine the albedo of the earth-atmosphere system depending on the surface albedo and the solar zenith angle

$$A_s = \frac{0.07}{S_0} \sec i + \frac{1.31}{S_0} (1 - 0.07 \sec i) A_0. \tag{2}$$

Paper [4] shows that formula (2) holds for the standard radiation model of the atmosphere considered, and in this case the values of $A_s$ for all $0 \leqq i \leqq 80°$ can be determined to an accuracy of about 5%. The standard radiation model of the atmosphere is characterized by the following parameters: optical thickness $\tau_0 = 0.3$; horizontal visibility near the earth's surface $S_m = 20$ km; amounts of precipitable water and $CO_2$ respectively, $W_{H_2O} = 2.1$ g cm$^{-2}$, $W_{CO_2} = 264$ atm cm.

Under other conditions of atmospheric transparency, the coefficients in formula (2) are, of course, different. The examination of formula (2) with the help of measurement data of the satellites Cosmos and Meteor has shown [6] that in 260 of 314 cases considered deviation of the values of $A_s$ calculated with formula (2) from the measured $A_s$ values computed from formula (1) does not exceed $\pm 10\%$. Hence, to the same accuracy, it is possible to determine the surface albedo from the satellite data $A_s$ under cloudless conditions. Based on Eqs. (1) and (2), we obtain

$$A_0 = \frac{A_s - 0.037 \sec i}{0.697(1 - 0.070 \sec i)}. \tag{3}$$

Here the solar constant value $S_0$ for the $0.3 \leq \lambda \leq 3\ \mu$m spectral region is 1.88 cal cm$^{-2}$ min$^{-1}$ (measurements of the OSR intensities from the Soviet meteorological satellites are taken over this spectral region).

Some results of the surface albedo calculations with formula (3) are shown in Fig. 1. Here the straight lines are drawn according to theoretical calculations carried out in [4]. The points correspond to the $A_0$ values obtained with formula (3) and from the Cosmos and Meteor measurement data for various surfaces under cloudless sky conditions.

It is evident from Fig. 1 that the calculated albedo values differ slightly from those measured for real atmospheric conditions. This difference is quite natural, since calculations are carried out for the standard atmospheric radiation model. The essential difference was observed only in the region of the Antarctic. Apparently, the reason is that it is very difficult to distinguish clouds from snow cover on the video images received from satellites, and thus choosing of the cloudless sky cases was not reliable enough. Also, our standard atmospheric model differs most from the real one under conditions of high transparency and low humidity, characteristic of the Antarctic.

When using formula (3) for satellite data processing, the smoothing was performed within 1—3° intervals for the sun's height and over 4—15 scan lines.

This made it possible to pick out the characteristic types of the surface, homogeneous as to their reflecting properties. In the case of cloudless sky, this data selection practically excludes systematic errors in determining the values of $A_s$ and $A_0$. It has been shown in [7] that the systematic errors in determining $A_s$ and $A_0$ are close to zero.

When elaborating the techniques of determining the surface albedo $A_0$ from the system albedo $A_s$ measured values, the satellite data analysed by Vonder Haar were also used [3, 9].

These data cover the period from July 1963 to November 1965 and represent sufficiently correct climatological characteristics of the whole globe. Because of the absence of tabulated data on the mean seasonal distribution of the system albedo in [3], the $A_s$ values were taken from the maps of the mean seasonal distribution of these values in 10° grid intervals.

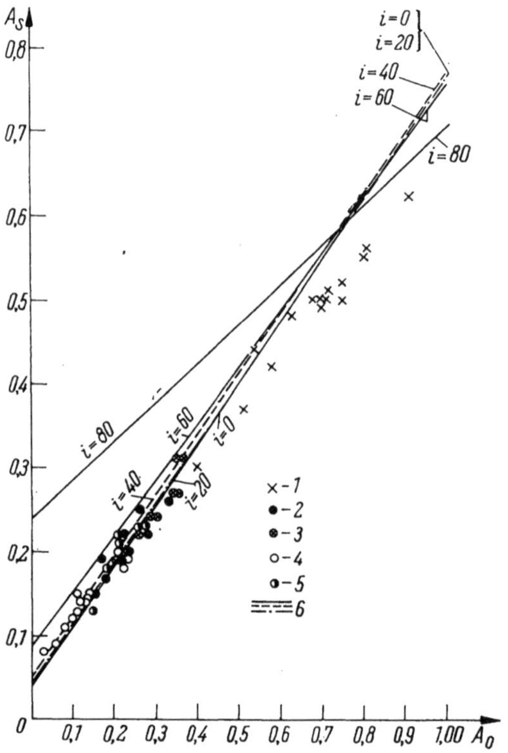


Fig. 1. The earth-atmosphere system albedo against the underlying surface albedo from Cosmos and Meteor observation data. 1, Antarctica (ice, snow); 2, Central Africa (tropical forest, tropical savannahs); 3, The Pacific, Atlantic and Indian oceans (water); 4, Africa (Sahara desert); 5, Hindustan (tropical forest, tropical savannahs); 6, theoretical calculations [4].

For evaluation of the atmospheric contribution to the system albedo it is necessary to have data on the surface albedo for large territories. With this in mind, the authors have plotted the mean seasonal maps of the surface albedo on the basis of the monthly maps of geographical distribution of albedo for large territories all over the globe [8].

The results of intercomparison between $A_0$ values calculated with the use of satellite measurement data (Eq. (3)), and climatological values of the large territories albedo $A_0'$ for some regions of the globe, are listed in Table 1. For Table 1 were selected the cases of cloudless sky over the deserts of Africa, some oceanic regions and other typical surfaces.

The discrepancy between $A_0$ and $A_0'$ values is small in most cases. The existing discrepancy can be explained: (i) by the effect of thin clouds which are difficult to identify when selecting the meteorological satellite measurement data, and (ii) by the low values of the climatological ocean albedo (this albedo is assumed to be 6% everywhere).

Table 1

| Region of globe lat. $\Phi°$ | long. $\lambda°$ | Mean sun's heigt at non $h_{\odot}°$ | Season | $A_s$ | $A_0$ | $A_0'$ | Underlying surface |
|---|---|---|---|---|---|---|---|
| 30N | 0 | 51.4 | Sept.—Nov. | 0.25 | 0.28 | 0.28 | Sahara |
| 20N | 10W | 60.7 | | 0.21 | 0.26 | 0.28 | |
| 20N | 0 | 60.7 | | 0.23 | 0.29 | 0.28 | |
| 20N | 10E | 60.7 | | 0.22 | 0.28 | 0.28 | |
| 20N | 10W | 50.2 | Dec.—Feb. | 0.22 | 0.27 | 0.28 | |
| 20N | 0 | 50.2 | | 0.25 | 0.28 | 0.28 | |
| 10N | 10W | 59.8 | | 0.20 | 0.24 | 0.20 | |
| 20N | 10W | 75.5 | Mar.—May | 0.21 | 0.26 | 0.28 | |
| 20N | 20E | 75.5 | | 0.23 | 0.30 | 0.28 | |
| 20N | 30E | 75.5 | | 0.22 | 0.28 | 0.28 | |
| 10N | 10W | 78.6 | | 0.19 | 0.21 | 0.21 | |
| 10N | 30E | 78.6 | | 0.20 | 0.22 | 0.23 | |
| 30N | 10W | 77.5 | Jun.—Aug. | 0.22 | 0.23 | 0.24 | |
| 30N | 0 | 77.5 | | 0.21 | 0.26 | 0.28 | |
| 30N | 30E | 77.5 | | 0.23 | 0.29 | 0.25 | |
| 20N | 30W | 82.5 | | 0.19 | 0.24 | 0.28 | |
| 20N | 0 | 82.5 | | 0.20 | 0.25 | 0.28 | |
| 30N | 40E | 51.4 | Sept.—Nov. | 0.25 | 0.32 | 0.28 | Arabian penins. |
| 20N | 50E | 60.7 | | 0.24 | 0.28 | 0.28 | |
| 10N | 50E | 70.6 | | 0.22 | 0.24 | 0.23 | East Africa seaboard |
| 0 | 70W | 80.0 | | 0.20 | 0.21 | 0.18 | South America |
| 0 | 60W | 80.0 | | 0.18 | 0.19 | 0.19 | same |
| 0 | 50W | 80.0 | | 0.17 | 0.18 | 0.18 | same |
| 20S | 40W | 75.5 | | 0.18 | 0.19 | 0.18 | Atlant. ocean |
| 60N | 110W | 38.8 | Mar.—May | 0.45 | 0.56 | 0.58 | North America |
| 60N | 70W | 38.8 | | 0.50 | 0.71 | 0.67 | same |
| 60N | 60W | 38.8 | | 0.43 | 0.60 | 0.62 | Atlant. ocean |
| 0 | 160W | 79.8 | | 0.13 | 0.10 | 0.06 | Pacif. ocean |
| 0 | 150W | 79.8 | | 0.14 | 0.10 | 0.06 | same |
| 0 | 50W | 79.8 | | 0.12 | 0.10 | 0.10 | South America |
| 10S | 100E | 70.6 | | 0.13 | 0.11 | 0.06 | Indian ocean |
| 10S | 120E | 70.6 | | 0.17 | 0.17 | 0.13 | same |
| 30S | 20E | 51.4 | | 0.19 | 0.19 | 0.22 | South Africa |
| 70N | 60W | 39.7 | Jun.—Aug. | 0.34 | 0.45 | 0.43 | Atlant. ocean |
| 30N | 40E | 77.5 | | 0.25 | 0.28 | 0.28 | Arabian peninsula |
| 0 | 40W | 70.3 | | 0.15 | 0.12 | 0.06 | Atlant. ocean |
| 10S | 160W | 59.8 | | 0.18 | 0.17 | 0.06 | Pacif. ocean |
| 10S | 150W | 59.8 | | 0.19 | 0.19 | 0.06 | same |
| 10S | 20E | 59.8 | | 0.22 | 0.23 | 0.24 | South Africa |
| 30S | 20E | 40.7 | | 0.24 | 0.24 | 0.24 | same |
| 30S | 30E | 40.7 | | 0.23 | 0.23 | 0.24 | same |
| 10S | 60E | 59.8 | | 0.15 | 0.12 | 0.06 | Indian ocean |

## 4. The Ratio of the Earth-Atmosphere System Albedo and the Surface Albedo

With the mean seasonal $A_s$ and $A_0$ values available, one can determine the ratio of these two values and study the peculiarities of its distribution all over the globe. It is obvious that generally the ratio $A_s/A_0 \lesseqgtr 1$ depending on the albedo of the earth's surface and clouds, the sun's height and the optical properties (turbidity) of the atmosphere. The clouds can increase the system's brightness (if the surface albedo is small) as well as decrease it as is the case in polar regions, where the surface albedo may exceed the cloud albedo.

Under cloudless sky conditions, the turbid layer of the atmosphere, when covering a low reflection surface, increases the amount of scattered light, and therefore the system albedo is greater than the surface albedo. In the case of high reflection surfaces, the effect of atmospheric haze is to decrease the system albedo. With a certain ratio of $A_0$ and the optical properties of the atmospheric layer, the effect of the atmospheric layer will be zero, i.e. $A_s = A_0$.

The comparison of measured $A_0$ (surface measurement data) and $A_s$ (satellite measurement data) for sufficiently homogeneous large territories [6] is of great interest from the point of view of the greenhouse effect of the atmosphere. Table 2 is an example of such a comparison making use of the calculated values of $A_s$. It shows that only over the oceans with their low albedo is $A_s/A_0$ greater than 1.

Table 2

Ratio of Mean Values of the Surface Albedo $A_0$ and the Earth-Atmosphere System Albedo $A_s$ under Cloudless Sky Conditions for Typical Underlying Surfaces

| Region of the globe | Underlying surface | $A_s$ | $A_0$ | $A_s/A_0$ |
|---|---|---|---|---|
| The Antarctic | ice, snow | 0.49 | 0.69 | 0.71 |
| Central Africa | tropical forests, savannahs | 0.21 | 0.23 | 0.91 |
| The Pacific, Atlantic and Indian oceans | water | 0.13 | 0.10 | 1.30 |
| Sahara desert | desert | 0.25 | 0.30 | 0.83 |

In all other cases $A_s/A_0 < 1$, i.e. the positive greenhouse effect of the atmosphere exists in the short-wave region of spectrum. If one takes into consideration the real cloudiness, then the effect of high cloud albedo almost always leads to $A_s/A_0 < 1$.

A more detailed intercomparison of geographical distributions of $A_0$ and $A_s$ obtained from the data of surface and satellite radiation measurements for four seasons, has shown that in summer $A_s/A_0 > 1$ over the oceans, as a rule.

In the Arctic, the minimum ratio $A_s/A_0 \approx 0.5$ is observed. Over the continents, in deserts, $A_s \leq A_0$. The maximum value $A_s/A_0 = 2$ over the continents is

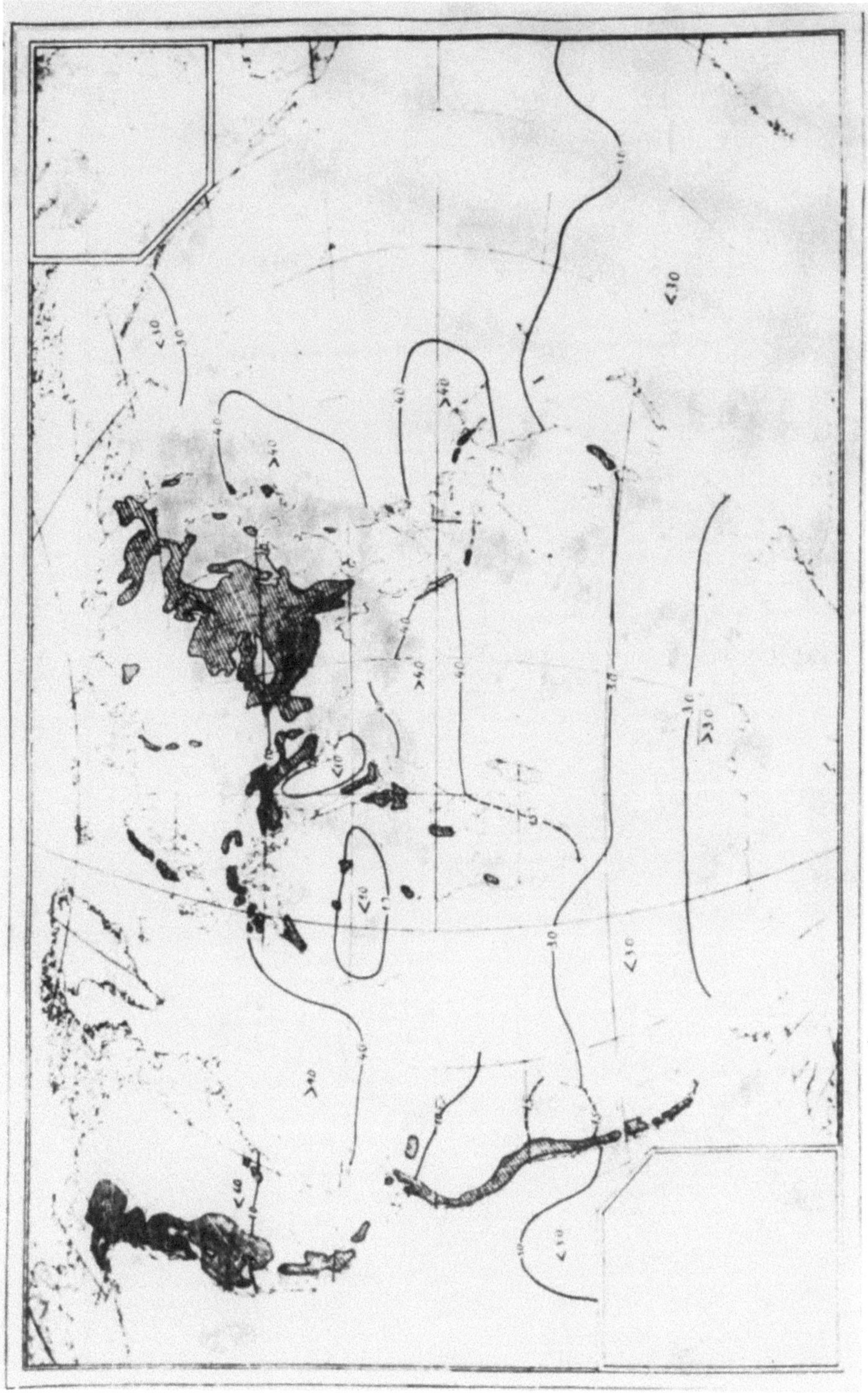

Fig. 2. Global distribution of the ratio $A_S/A_O$, winter.

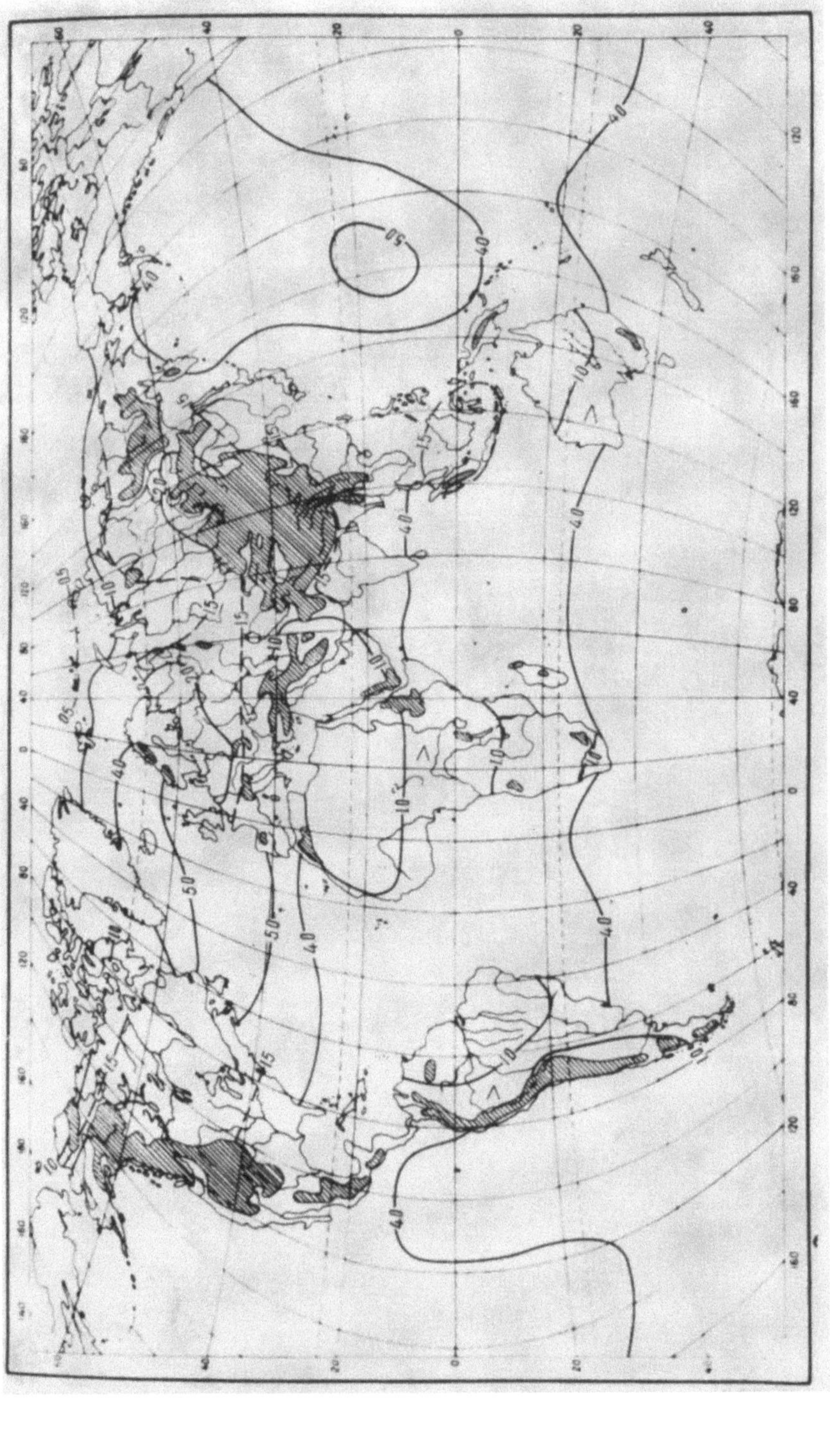

Fig. 3. Global distribution of the ratio $A_s/A_0$, summer.

observed in the regions of Siberia, in north-western Europe and in the north-western part of North America. Naturally, the field of $A_s/A_0$ values changes substantially depending on seasons. Investigation of the smoothed dependence of $(A_s/A_0)(A_s)$ has shown that "critical" values of $A_0$ corresponding to $A_s/A_0 = 1$ vary within 0.20—0.35.

To study the character of the geographical distribution of the ratio $A_s/A_0$ the globe was subdivided into grid squares of 10° latitude by 10° longitude, for which the values of $A_s/A_0$ were calculated.

From these calculated data, the maps of geographical distribution of $A_s/A_0$ for the four seasons were drawn. The maps of global distribution of the ratio $A_s/A_0$ in winter and summer are shown in Figs. 2 and 3. In winter, the same qualitative regularities of $A_s/A_0$ global distribution as in summer are observed, although a definite specificity of the fields of $A_s/A_0$ takes place depending on the season. Fig. 2 shows that the maximum values of the ratio are observed both in winter and in summer over the oceans, but their absolute maximum is of lower value and is somewhat shifted southward from the equator. $A_s$ is approximately 2—4 times $A_0$ everywhere over the oceans. Over the continents, both in summer and in winter, the system albedo in desert regions is either lower than or comparable with the surface albedo.

Figs. 4 and 5 show the ratio $A_s/A_0$ against the underlying surface albedo for the total globe (winter and summer) from satellite data for $A_s$ and from the mean climatological values of the surface albedo $A_0$ [8]. As the surface albedo increases, the scattering of $A_s/A_0$ decreases since $A_s$ is influenced by the cloudiness to a greater extent when $A_0$ is small.

The lower points on the left of Fig. 5 correspond to the cases of cloudless sky. When $A_0$ is high (on the right of Fig. 5), the "cloudless" curve must be drawn

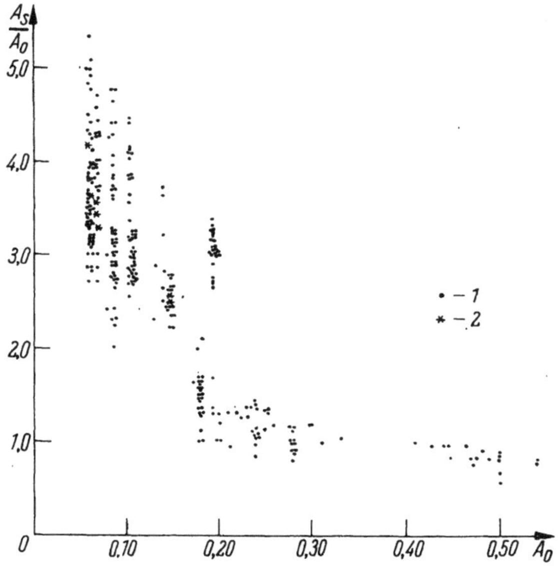


Fig. 4. The ratio $A_s/A_0$ against the surface albedo, winter. 1, single values; 2, points containing from 10 to 30 similar values.

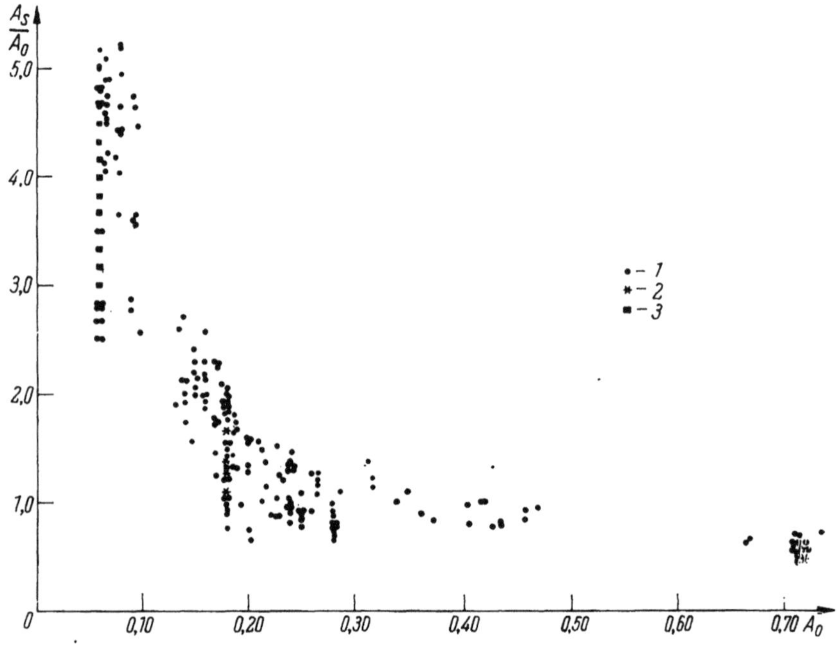


Fig. 5. The ratio $A_s/A_0$ against the surface albedo, summer. 1, single values; 2, points containing from 4 to 10 similar values; 3, points containing from 11 to 34 similar values.

across the upper points. This conclusion is verified also by the examination of Fig. 3.

In summary it may be said that further accumulation of meteorological satellite radiation data will make it possible to perform quantitative assessment of reliability of the above relationship between the earth-atmosphere system albedo as well as to establish a more universal dependence of the system albedo on the sun's height and on the atmospheric transparency than has been done, for example, in [6].

## References

[1] K. YA. KONDRATYEV and L. N. DYACHENKO, Trudy GGO, No. 235 (1970), p. 62.

[2] K. YA. KONDRATYEV, L. N. DYACHENKO and K. YA. VINNIKOV, in: Problemy phiziki atmosphery, LSU Publ., No. 7 (1968), p. 21.

[3] T. H. VONDER HAAR, Dept. Meteorology, Univ. of Wisconsin, NASW-65, February 1968.

[4] K. S. SHIFRIN, V. YU. KOLOMIYTZOV and N. P. PIATOVSKAYA, Trudy GGO, No. 166 (1964), p. 2.

[5] N. P. PIATOVSKAYA, Trudy GGO, No. 275 (1972), p. 3.

[6] N. P. PIATOVSKAYA, Trudy GGO, No. 275 (1972), p. 16.

[7] K. YA. KONDRATYEV, L. N. DYACHENKO, V. V. MUKHENBERG and N. P. PIATOVSKAYA, Proc. Int. Radiation Symp., Sendai, Japan 1972, p. 419.

[8] V. V. MUKHENBERG, Trudy GGO, No. 139 (1963), p. 43.

[9] T. H. VONDER HAAR and V. E. SUOMI, J. Atmos. Sci. 28, No. 3 (1971).

Earth Survey Problems — Akademie-Verlag, Berlin 1974

# REMOTE SENSING WITH VHRR SATELLITE IMAGERY

H. Kaminski

Institute for Space Research, Observatory Bochum, Bochum, FRG

The "new generation" of weather satellites is equipped with high-resolution scanner systems, which operate simultaneously in the ranges of visible light and near and middle infrared. For investigations of static and ephemeral phenomena of the earth's surface, the use of VHRR imagery (Very High Resolution Radiometer) from NOAA 2 and NOAA 3 with a ground resolution of 0.8 km and providing daily coverage of the entire globe both day and night, provides basic material for such geo-scientific fields as geography, geology, hydrology, cartography and environmental control. Particular examples are given for an overview of selected areas in Europe, the Mediterranean and North Africa, illustrating the possibilities of thematic mapping in both visible and infrared ranges.

## 1. Introduction

The institute of Space Research, Observatory Bochum, has received the VHRR transmissions from NOAA 2 beginning in October 1972. Originally, it was planned to receive the real-time transmissions from Nimbus 5 Surface Composition Mapping Radiometer, which operated in the near and medium infrared range, scanning simultaneously in two ranges: 0.8—1.1 μm and 8.3—9.3 or 10.2—11.2 μm during the day and in the two latter ranges during the night. However, the SCMR was shut off after four weeks in orbit due to technical difficulties. The next scanner of this type is planned for Nimbus 6, to start in 1974.

## 2. Examples of VHRR Imagery

For the purposes of this paper, only a few selected examples have been chosen to illustrate possible uses of synoptic satellite photography. Since for many of the areas chosen there are neither accurate maps nor ground-truth data available, it may be that there are errors in interpretation. Only recently have programs been realized which use these VHRR data in combination with underflights and ground-checks at the same time as the satellite images certain test sites [1—4].

The same consideration applies to the results of this preliminary photointerpretation for the two wavelengths of the scanner: the range of visible light, 0.5—0.7 μm and the medium infrared range, 10.5—12.5 μm. Too little experience is available for determining the "best" infrared ranges for determination of geothermic phenomena, or for determining the "best" time of day or night when imagery should be made, and in which geographical areas and disciplines the "best" combination of factors apply. However, the use of synoptic imagery will provide some answers to these problems in the future.

### 2.1. Arctic Regions

The energy balance of the earth is a matter of extreme importance for many fields, beginning with agriculture and down to theoretical physics. An important factor in determining the energy balance of the earth, are measurements made in the two main "heat sinks" of the world, i.e. the Arctic and the Antarctic. There are continuous measuring programs, both on a local basis and on international levels for selected areas, but one main problem has been the inaccessibility of test sites in polar regions. One chief source of information is synoptic "weather" satellite photography, which was previously used only for cloud distribution measurements in these areas [5]. Now with VHRR imagery, actual running measurements of snow and ice can be made on a routine day and night basis.

A good illustration of this possibility of measuring snow and ice is Iceland. Fig. 1 shows the contours of the island on 2 May 1973, at 1149—1208 GMT. The process of snow melt is much further advanced than usual for this time of year, due to the very warm winter of 1972/1973, and areal measurements of bare rock as against still snow-covered highlands are possible with an accuracy of $\pm 2$ miles.

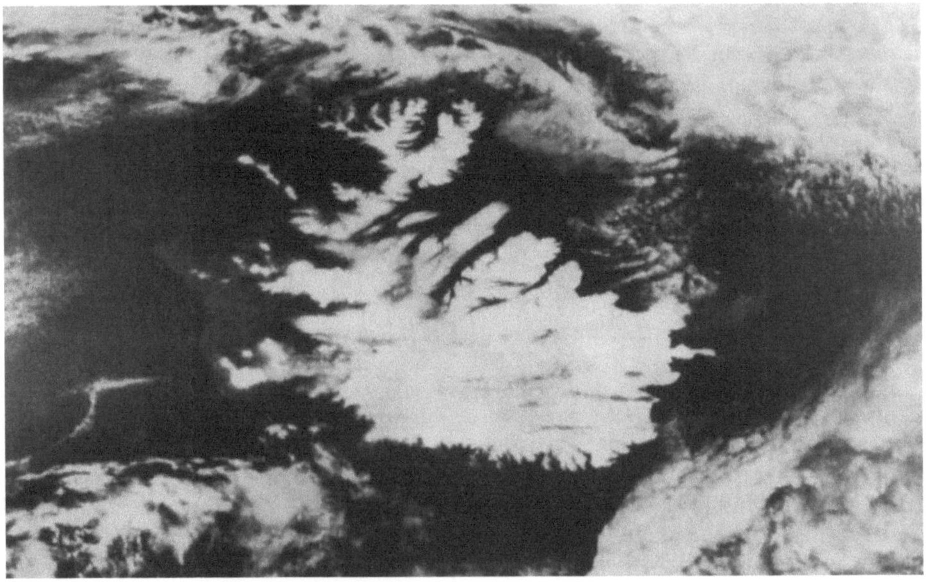

Fig. 1. Overview of Iceland. VHRR, NOAA 2, 2 May 1973, 1149—1208 GMT, 0.5—0.7 μm.

## 2.2. Northern Continental Regions

Two examples illustrate the difficulties attending the interpretation of infrared imagery, even in areas which are well mapped. Fig. 2 is an infrared picture of southern Scandinavia and northern Denmark. On this type of imagery, warm phenomena are shown in shades of darker grey; cold phenomena are lighter grey in tone. It is immediately obvious that on this imagery, taken on 3 May 1973 at 0857 to 0919 GMT, there is definite change in the relative "warmth" of land and water masses from north to south and from east to west. In this case, some of the darker spots on land (Sweden and Denmark) can definitely be correlated to cities having a higher thermal potential than the surrounding countryside. However, many other darker spots are forested coasts or valleys (Sweden) or perhaps geothermal differences in surface morphology (Denmark).

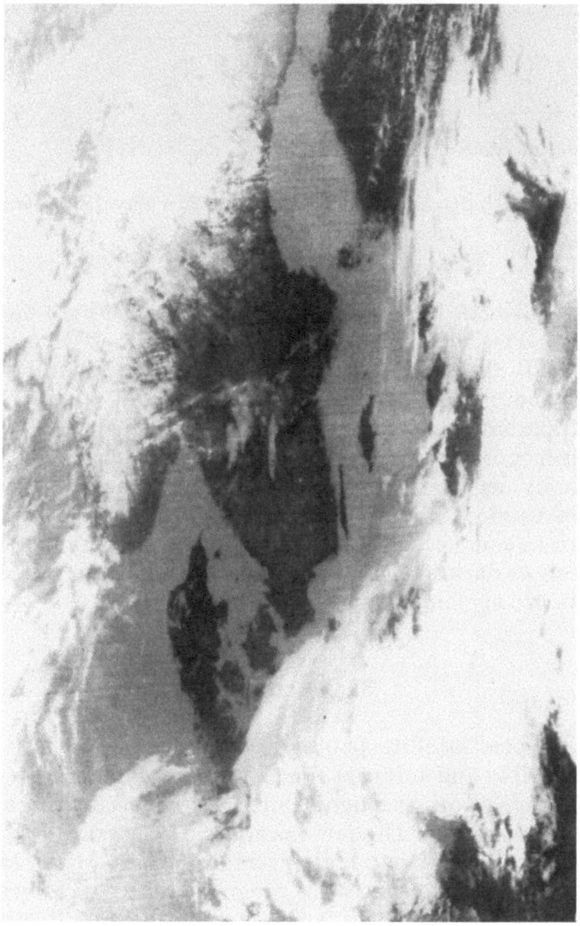

Fig. 2. Southern Scandinavia and Northern Denmark. VHRR, NOAA 2, 3 May 1973, 0857–0919 GMT, 10.5–12.5 μm.

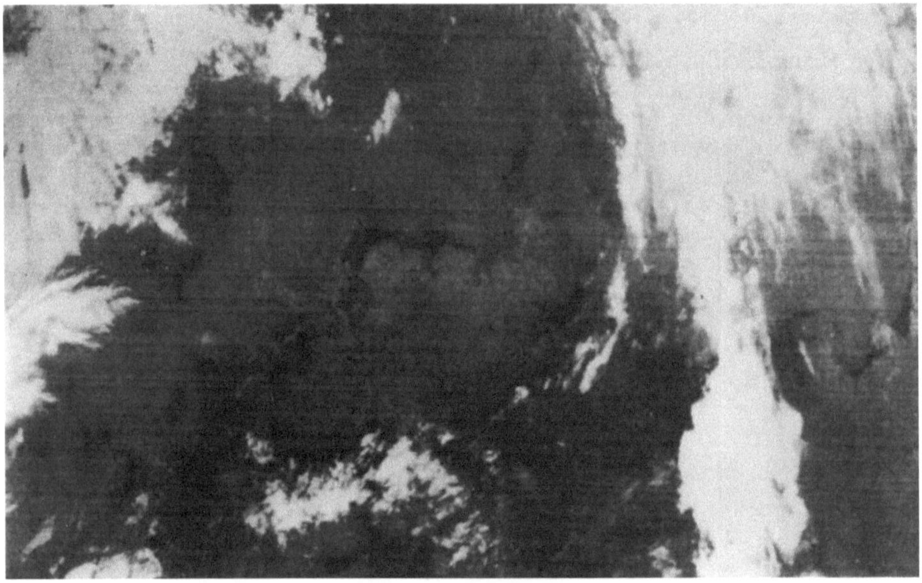

Fig. 3. European North Sea Coast. VHRR, NOAA 2, 4 May 1973, 1903—1926 GMT, 10.5—12.5 μm.

Fig. 3 shows the North Sea Coast — Denmark, Northern Germany and the Benelux countries. This infrared imagery was sent from NOAA 2 on 4 May 1973 at 1903—1926 GMT. The difficulties of comparative photointerpretation are immediately obvious, when related to the preceding photography. Thermal differences are apparent, but an exact correlation with specific objects or phenomena is no longer possible. Additionally, the land-water interface cannot be measured accurately, as the different surfaces have almost the same temperature in the 10—12 μm window at this time of day, on this particular date. The dark spots on land can in some cases be identified as urban agglomerations and many rivers may be seen as dark lines on the imagery. But all dark lines are *not* rivers, *not* all dark spots are agglomerations.

## 3. Conclusions

The uses of synoptic satellite photography, now available as high-resolution imagery in both visible and infrared ranges, extend from studies on atmospheric circulation and composition, through the range of geosciences to earth resources exploration and mapping [6]. The few examples discussed in this paper illustrate how necessary daily (i.e. every 12 hours) remote sensing of the earth's surface is, in order to carry out running research programmes in an effective manner. At the same time, it is also possible to study the parameters affecting the visibility and recognition of objects and phenomena in different wavelengths. For many parts of the world, this is the only type of satellite imagery now available for overview and detailed studies of earth resources.

## References

[1] Program of Ice and Snow Exploration in the Polar Regions, US Navy Fleet Weather Service.
[2] Program: Measurements of Cloud Cover in North Africa, Geographical Institute, University of Bern (Switzerland) and Observatory Bochum.
[3] Program: Measurements of Fresh Water Springs on the Coasts of Northern Africa and Southern Greece, Polytechnic Institute, University of Karlsruhe and Observatory Bochum.
[4] Program: Observation and Measurements of Thermal Phenomena in the Ruhr Industrial Area, Ruhr Siedlungsverband and Observatory Bochum.
[5] G. Weller and E. W. Bierly, Bull. Amer. Met. Soc. **54**/3, 212 (1973).
[6] H. Kaminski, Bild der Wissenschaft, **10**/4, 362 (1973).

Earth Survey Problems — Akademie-Verlag, Berlin 1974

# OPTICAL INVESTIGATIONS FROM THE MANNED SPACECRAFT VOSKHOD 2

A. I. Lazarev, A. A. Leonov and M. M. Miroshnikov

USSR

A number of optical phenomena observed from Voskhod 2 manned spacecraft at the earth's night and twilight horizon on 18—19 March 1965 and one phenomenon photographed on the earth's surface illuminated by the sun are described and interpreted. A supposition is made of observing the sun's corona near the earth's visible horizon, and the possibility of such an observation from space is analysed. The peculiarities of observing colour layers near the earth's twilight horizon are discussed. The earth's specular reflection at reflection angles close to 90° is described. The possibility of detecting noctilucent clouds from space on the earth's night side is discussed.

During the flight of Voskhod 2 manned spacecraft on 18—19 March 1965 one of the authors observed a number of interesting optical phenomena at the earth's night and twilight horizon. Using these observations pictures [1—3] have been painted, sketches for some of which were made in flight on the pages of a log-book. One of the interesting optical phenomana was photographed on 18 March 1965 on the day side of the earth during an exit into open space. The paper contains a description and a possible interpretation of these phenomena. It should be noted that no other Soviet and American cosmonauts observed similar phenomena during flights although more than eight years have elapsed since the flight of Voskhod 2.

In the preface to an album of pictures [1] produced after the flight of Voskhod 2, Yu. A. Gagarin, the first cosmonaut, wrote: "The pictures published in the album tell us, as if through a prism of artistic perception, about things that are already known to science, and about things the scientists do not yet know. "Unexpected" results and "unexpected" hypotheses are often to be met with in research. This is sometimes what happens to the authors of the pictures". This prophecy of Yu. A. Gagarin is splendidly confirmed by the analysis of the optical phenomena depicted in the pictures produced after the flight of Voskhod 2.

Let us begin by analysing the optical phenomena shown in the picture "Sunrise in Space" [1—3] the sketch of which was made with coloured pencils on the pages of a log-book during the flight of Voskhod 2. In the picture, the sun is imaged at the earth's horizon; above the sun are seen the rays of the corona at a distance of up to $2R$. None of the other Soviet and American cosmonauts observed the sun's corona after sunrise or before sunset during a flight along an orbit of an AES*.

* AES = artificial earth satellite.

The possibility of observing the sun's corona from space has already been discussed by various authors. This problem is most completely considered in [4] which shows that the sun's corona cannot be observed from a spacecraft in daytime at high solar elevation. The main obscuration of the corona is background created by scattering of sunlight by spacecraft debris and the elements of a spacecraft structure (mainly, by a window). To provide for the possibility of observing the corona from space (in the absence of the sun's eclipse), it is necessary to reduce considerably the background brightness, which is usually effected by coronagraphs having an external occulting disk [5, 6].

In connection with the observations from Voskhod 2, it is of great interest to analyse the possibility of investigating the corona from space with the sun just above the visible horizon. In this case, the brightness of an artificial background (a debris environment, a window and other elements of the spacecraft structure) is considerably reduced because the sunlight is attenuated in the dense layers of the atmosphere, and the natural background is reduced to the earth's atmospheric emission. Therefore the possibility of studying the corona from space near the earth's horizon is determined primarily by the transparency of the atmosphere in the spacecraft—sun direction and by the relationship between the coronal brightness and that of the natural atmospheric background. The transmittance and the brightness of the atmosphere in the visible spectral region change sharply, in observation from space, at the sighting line elevation $H_p$ of 20—30 km. In observation from space with $H_p \leq 20$ km, the transmittance of the atmosphere is in the range $10^{-6}$—$10^{-2}$; and above 40 km it is very close to unity. The atmospheric brightness from the edge of the earth's limb to the height $H_p = 30$ km depends weakly on $H_p$, and above 40 km quickly diminishes as $H_p$ increases.

The conditions that are necessary for carrying out investigations of the sun's corona from space in the selected spectral range are determined by the sun's elevation above the horizon and the height of the spacecraft. The selection of the sun's elevation above the horizon is related primarily to the degree of decrease in brightness of an artificial background (window, debris environment) and determined by the dependence of atmosphere transmittance on the height of the sighting line perigee (with allowance for atmospheric refraction). The selection of the spacecraft flight height depends primarily on the relationship between the angular dependence of the sun's coronal brightness and that of the earth's atmospheric brightness in the vertical direction at the selected elevation of the sun above the horizon. The important factor is that as the height of the spacecraft increases, the angular size of the glowing layer of the earth's atmosphere decreases, whereas the angular size of the sun's corona remains practically constant.

The analysis of the dependence of atmospheric transparency on the height of the sighting line perigee shows that for a substantial decrease in the artificial background brightness within the visible spectral region, it is necessary that the upper edge of the visible sun's limb be in the direction of the sighting line perigee, $H_p = 20$—$30$ km. Under these circumstances, the illuminations due to the sun and the artificial background brightness, respectively, are 4 to 5 orders of magnitude lower than those at an elevated position of the sun.

The natural atmospheric background brightness at dawn or twilight is governed by light scattering in the dense layers of the atmosphere and by emission of the upper atmosphere. The relationship between the contributions of the scattered radiation and the emission within the selected spectral region depends primarily on height. Within the visible spectral region, at altitudes of up to 90—100 km, the main contribution is due to scattered atmospheric radiation, whereas above 100 km atmospheric brightness is determined primarily by emission [8]. Calculations of the scattered atmosphere radiation brightness up to the altitude $H_p = 100$ km for observations from space in the case of low positive solar elevation were carried out in [9]. The results of these calculations are shown in Fig. 1. The estimates of the upper atmosphere dayglow have been made according to the results of rocket probing of the upper atmosphere in separate emission lines [8].

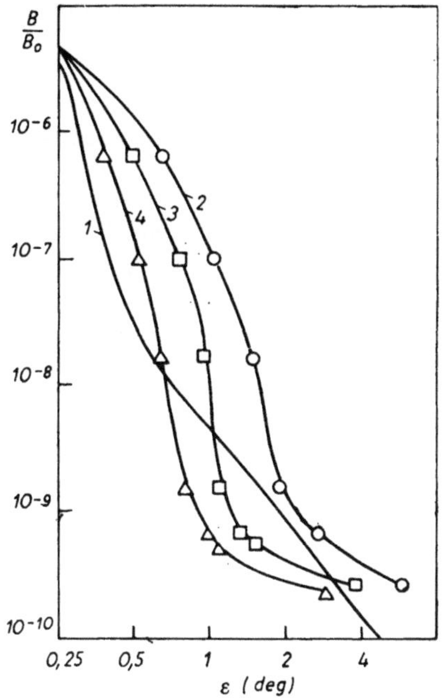


Fig. 1. Angular dependence of the corona brightness (1) and the earth's atmospheric brightness in observations from space at altitudes of 200 km (2), 500 km (3) and 1000 km (4), when the upper edge of the sun's disk is located at the height of the sighting line perigee equal to 20 km.

The angular dependences of the sun's coronal brightness and the atmospheric brightness in observations from space were analysed for altitudes of 200, 500 and 1000 km. According to the analysis, observations of the sun's corona at the height of 200 km are practically impossible, because the brightness of the natural background (twilight atmosphere emission) is considerably greater than that of the sun's corona. At the height of 500 km the corona brightness, in case of elongations $\varepsilon < 0.5°$, is comparable with that of the natural background, and, consequently, it is possible to carry out visual, photographic and other investigations of the corona without taking special measures for shielding the sun's direct radiation. This conclusion is confirmed by the visual observations from Voskhod 2, whose orbit apogee was at a height of about 500 km [7]. The visual observations of the sun's corona from other Soviet and American spacecraft which orbited as AES at altitudes of 160 to 300 km, could not be carried out. At 1000 km direct investigations of the sun's corona within the visible spectral region are possible for elongations $\varepsilon < 3°$.

Selection of the sun's elevation above the horizon and the height of the spacecraft for studying the corona within other regions of the spectrum can also be made according to the appropriate dependences of the transparency and the radiance of the earth's atmosphere and the sun's corona. For instance, in the range of 200 to 300 n mi, the upper edge of the sun's limb should, in all appearances, be in the direction of the sighting line perigee $H_p = 45-50$ km, and the spacecraft height should be in excess of 400 km. Coronal investigations from space at dawn or twilight are possible by means of visual observations and photographic, television, thermovision and spectrophotometric equipment depending on the spectral region selected and the experimental aim. Space observations should be especially promising in the ultraviolet, which is inaccessible to investigations from the earth and where the relative and absolute brightness of the corona is appreciably higher than in the visible or the infra-red.

During a sunrise, three sharply outlined coloured layers were observed at the twilight terrestrial horizon from Voskhod 2: red, yellow and blue, which are shown in the picture entitled "Sunrise in Space". The angular size of each layer amounted to about 10′, which corresponds, from the height of 500 km, to a vertical layer depth of about 8 km. The appearance of sharply outlined coloured layers during a sunrise can apparently be accounted for, first, by dispersion of the direct and scattered radiation of the sun in refraction within the atmospheric dense layers, and, second, by variation in the composition of the atmospheric scattered radiation with height under twilight conditions. Paper [9] contains the results of theoretical and experimental investigations of the coloured pictures of a twilight halo in observation from space at the height of 252 km, the theoretical studies being carried out without allowing for the dispersion in refraction by the atmospheric dense layers. Separate coloured layers, with an indistinct borderline, begin already to show during observations from a height of 252 km. In observations from the height of about 500 km (Voskhod 2) the separation of coloured layers grows sharper in association with a substantial decrease in the angular size of the layers and the transitional zones between them.

One of the observations from Voskhod 2 testifies to the ability of the earth's atmosphere to produce a specular reflection at reflection angles of about 90°. This observation is shown in the picture entitled "Orbiting in a big space station" [3]. During a sunrise, when half the sun's disk was above the horizon of the earth, the same half of the sun's disk was seen in the form of a specular reflection from the earth's atmosphere at a reflection angle of about 90°. In this connection, the photographs of the sun's corona taken by the crew of Apollo 12 space ship on a moon—earth track are also of interest. In [10, 11] are produced photographs, and in [10], besides the photographs, there is a brief description of this phenomenon: "Coasting back to earth in their command module Yankee Clipper, the three astronauts beheld a sight never befor glimpsed by man — an eclipse of the sun by earth. The atmosphere of earth glows white along the limb of the earth behind which the sun has just disappeared. Still stranger sights and new challenges lie ahead as man continues to pursue his greatest adventure to date, the conquest of outer space".

A. Bean, C. Conrad and R. Gordon, the cosmonauts, have photographed a part of the sun's corona and the earth's twilight halo with the sun located behind the earth's limb at a small angle of depression. In the photograph the same part of the sun's corona is seen in the form of a specular reflection on the earth's atmosphere at a reflection angle of about 90°. Attention should be paid to the fact that in association with a considerable decrease in the angular size of the coloured layers, in observations from Apollo 12, the earth's twilight halo was seen as a light bandlet of white.

To analyse the results of observing the reflected radiation, the coefficient of the specular reflection of the earth's atmosphere for reflection angles close to 90° was estimated using some simplifying assumptions. The atmosphere is considered as a flat homogeneous layer having a refractive index $n = 1.000292$. The estimates made according to Fresnel's formula are given in Table 1.

Table 1

Earth's Atmosphere Specular Reflection Coefficient observed from Space

| | | | | | |
|---|---|---|---|---|---|
| Angle of incidence | 89°54′ | 89°48′ | 89°42′ | 89°36′ | 89°30′ |
| Coefficient of reflection | 0.75 | 0.56 | 0.42 | 0.32 | 0.24 |
| Angle of incidence | 89°24′ | 89°18′ | 89°12′ | 89°06′ | 89°00′ |
| Coefficient of reflection | 0.19 | 0.14 | 0.11 | 0.09 | 0.07 |

Naturally, these estimates provide only an approximate idea of the value of the coefficients of Fresnel reflection on the earth's real atmosphere. In forming these estimates, the specular reflection from the conducting layer of the ionosphere was not taken in consideration. Nevertheless, it is clear from Table 1 that at angles of incidence differing from 90° by less than 16′ (half the sun's disk), the specular reflection coefficient exceeds 50%. The ability for specular reflection is, apparently, chaarcteristic of other planets featuring a dense atmosphere as well. Therefore it can be used for studying the physical properties of planetary atmospheres. Ample opportunities for such investigations will come into existence in studying the specular reflection from the atmospheres of planets at large reflection angles over various portions of the electromagnetic spectrum from ultraviolet to UHF radio waves.

In the photographs taken from Voskhod 2 on 18 March 1965 during a flight in the day-time over the territory of the USSR, a very bright spot is seen in gaps of terrestrial cloudiness which is concluded to be the sun's reflection on spring snow with a frozen crust in a direction close to a specular reflection. According to Krynov's investigation [12], the brightness coefficient of spring snow with an icy crust can amount to 7—8 in the visible spectral region in a direction close to specular reflection.

The results of one of the interesting observations of the earth's night horizon are depicted in the picture "Blue Belt of Earth" [1—3], where a sharply outlined greyish-blue layer is shown, in observation through which the colour and blaze of stars and planets changes sharply. The angular size of the layer is approximately 1.5° (from an altitude of 500 km) and the height is 80 km. During an observation of stars and planets on the earth's night side it was found that a change in sighting from the night sky direction to the greyish-blue layer direction results in variation of the colour of stars and planets from white to red and in the diminution of their blaze. The phenomenon depicted in "Blue Belt of Earth" was observed in the flight of Voskhod 2 during one turn only (on the earth's night side).

It may be supposed that a sharp change in the colour and blaze of stars and planets over the earth's night horizon at the sighting line perigee altitude of about 80 km testifies to the presence of noctilucent clouds in the direction of sighting. The change in the colour of stars and planets is related to a substantial selectivity of noctilucent cloud transparency in observation from space in the direction of the earth's horizon. The calculations carried out in [13] show that the transparency of noctilucent clouds in the red portion of the visible spectral range is considerably better than in the bluish-green portion. Therefore in sighting stars or planets from space through a layer of noctilucent clouds a considerable change in their colour ("reddening") is observed. "The earth's blue belt phenomenon" can apparently be used for detection of noctilucent clouds from space on the earth's night side.

The results of the optical investigations carried out on board Voskhod 2 permit a considerable extension of opportunities for studying the natural environment from space. Observations of a twilight horizon (after a sunset or before a sunrise) are promising from the viewpoint of studying the sun's corona and the specular reflection from the earth's atmosphere at large reflection angles, and also extend the opportunities of studying the optical and colorimetric characteristics of the atmosphere; observations at earth's night provide an additional opportunity of detecting and studying noctilucent clouds.

## References

[1] A. A. LEONOV and A. A. SOKOLOV, The Stars are awaiting us, Molodaya Gvardiya Moscow 1967.
[2] A. A. LEONOV and A. A. SOKOLOV, To the stars, Avrora, Leningrad 1970.
[3] A. A. LEONOV and A. A. SOKOLOV, Starry Paths, Molodaya Gvardiya, Moscow 1971.
[4] N. S. KOVAR, R. P. KOVAR and G. R. BONNER, Planet. Space Sci. **17**, 143 (1969).
[5] M. J. KOOMEN, J. D. PARCELL and R. TOUSEY, Nature **226**, 1135 (1970).
[6] R. TOUSEY, Sky and Telesc. 1971, **42**, 346 (1971).
[7] Mastering of Space, TASS Official Communications and Materials of Central Press of 1957—1967, Nauka, Moscow 1971 (p. 457).
[8] J. F. NOXON, Space Sci. Revs **8**, 92 (1968).
[9] O. I. SMOKTII, Proc. Acad. Sci. USSR, Physics of Atmosphere and Ocean **5**, 782 (1969).
[10] S. BEDINI, W. VON BRAUN and F. L. WHIPPLE, The Moon, Man's Gratest Adventure, Harry N. Abrams Inc. Publrs; New York 1970.
[11] Soviet Encyclopaedia, 3rd Edn, **9**, 384 (1972).
[12] E. L. KRYNOV, Spectral Reflectivity of Natural Formations, USSR Acad. Sci. Press, Moscow, Leningrad 1947.
[13] CH. J. WILLMAN and V. N. SERGEYEVICH, Proc. Acad. Sci. USSR, Physics and Mathematics **18**, 200 (1969).

Earth Survey Problems — Akademie-Verlag, Berlin 1974

# PRELIMINARY RESULTS FROM SKYLARK EARTH RESOURCES ROCKET EXPERIMENT IN ARGENTINA

D. D. Clark[a], J. R. Hardy[b], A. J. Parsons[b], R. B. Ridgway[b], R. A. G. Savigear[b], and J. R. G. Townshend[b]

[a]Department of Trade and Industry, London, UK
[b]Department of Geography, University of Reading, UK

In March 1973 an earth resources Skylark rocket was used to obtain imagery of central Argentina in a joint Anglo—Argentine programme for the purpose of preparing a crop and land use inventory. Both colour and false colour infrared films were used in the bank of cameras (3 Hasselblads and one F24) to obtain many photographs from heights above 200 km covering a total area of about 300000 km². Ground observations were made for purposes of identification. One typical photograph in false colour infrared is reproduced. A preliminary visual inspection showed that it was possible to distinguish between healthy green crops with high chlorophyll content, ripe mature crops, scrubland and natural pasture, stubble and fallow land, and bare soil. Some cultural features and other natural resources are also indicated. A detailed analysis of the data is being made and results will be reported at a later date.

## 1. Introduction

In March 1973 an area of central Argentina of approximately 300000 km² was photographed from above 200 km altitude by means of cameras carried aloft in a double firing of a British Aircraft Corporation Skylark rocket. This firing formed part of a collaborative programme between Argentina and the UK, the immediate purpose of which was to obtain an inventory of crops and land use and to carry out a resources survey of the central region of Argentina.

The payload section of the Earth Resources Skylark stabilizes itself automatically in the horizontal plane after separation from the rocket motor and is

Table 1

| Camera | Focal Length (mm) | Format (mm) | Film | Filter |
|---|---|---|---|---|
| F24 aerial survey | 127 | — | Kodak aerochrome false colour | Wratten 12 |
| Hasselblad 500 EL | 150 | 70 | IR type 2443 | |
| Hasselblad 500 EL | 150 | 70 | Kodak aerial colour type | None |
| Hasselblad 500 EL | 250 | 70 | SO-242 | |

Table 2

| Camera | Angle from horizontal | Lens Focal Length | Contact Scale Near edge | Contact Scale Far edge | 10× Enlargement Scale Near edge | 10× Enlargement Scale Far edge |
|---|---|---|---|---|---|---|
| Hasselblad A | 45° | 250 mm | 1/1.4 m | 1/1.7 m | 1/140000 | 1/170000 |
| Hasselblad B/C | 45° | 150 mm | 1/2.1 m | 1/3.2 m | 1/210000 | 1/320000 |
| F. 24 | 25° | 127 mm | 1/2 m | 1/3.4 m | 1/200000 | 1/340000 |

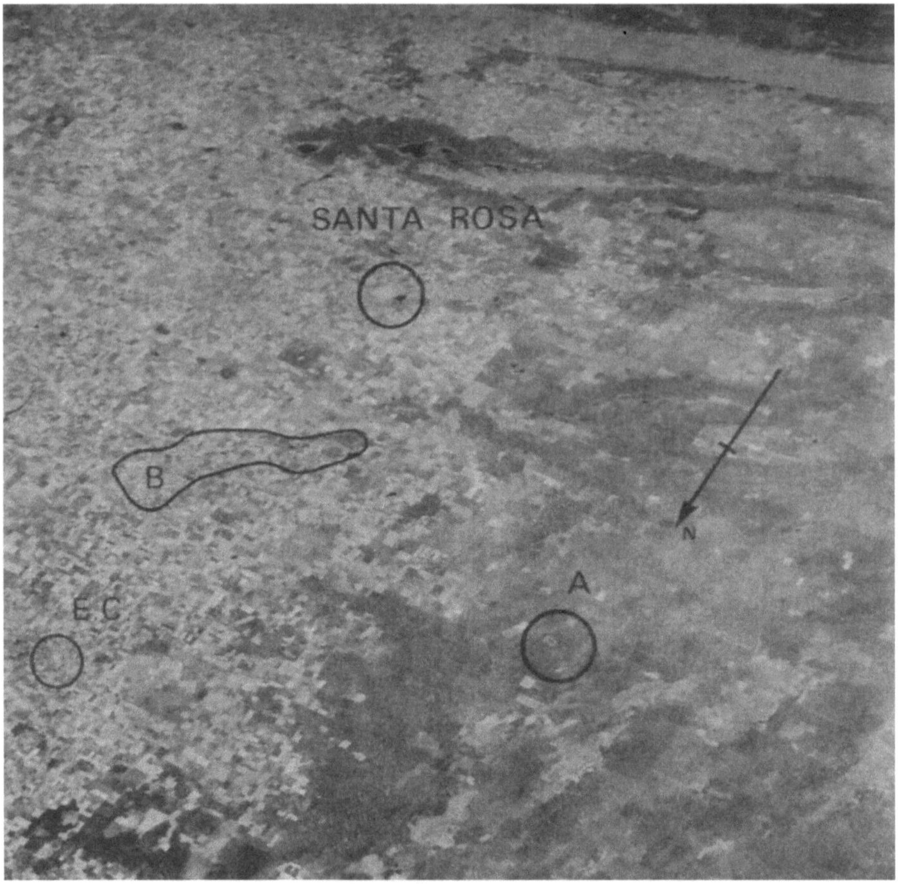


Fig. 1. A print prepared from false colour infrared (Kodak 2443) exposed at an altitude of approximately 200 km. The camera was a Hasselblad, with a 150 mm focal length lens and 70 mm film. The camera was tilted at 45° to the horizontal, and the near (lower) edge shows approximately 100 km of country, and the far (upper) edge about 160 km. The direction from bottom to top of the picture is approximately south south-east.

pre-programmed to perform successive rotations about the vertical axis. The cameras are inclined obliquely to the vertical so as to cover a sector of the sub-apogee circle. At each successive position in the rotation cycle photographs are taken thus providing the required coverage of the whole area. The payload contained four cameras of which particulars are given in Table 1. Camera angles,

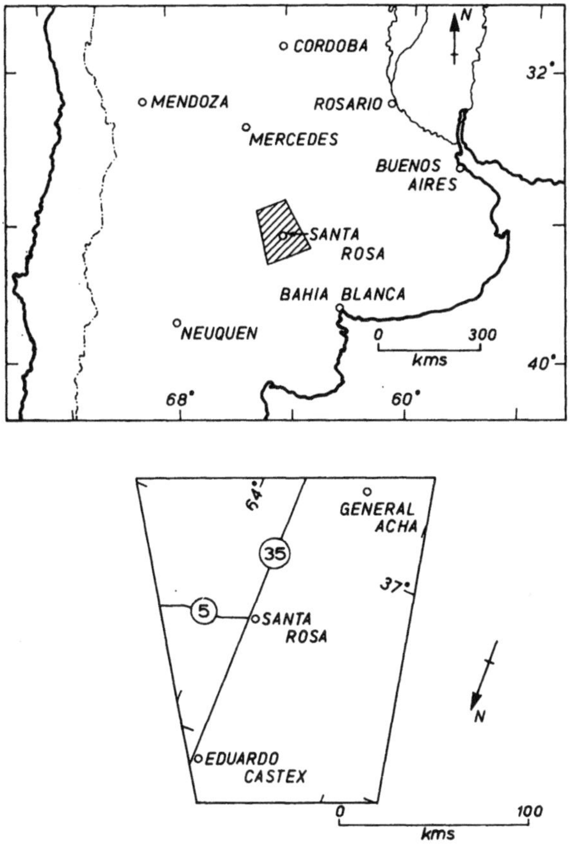


Fig. 2. Top: A map to show the general location of the print in Fig. 1. The area is shown hatched. Bottom: The more detailed location of the print shown in Fig. 1. Note the different north directions and scales on the upper and lower maps on this figure.

focal lengths, contact and ten times enlargement scales are listed in Table 2. The 10 × enlargement is approximately the most suitable for visual work. Satisfactory trial rectifications have been carried out to near cartographic accuracy and provide excellent images. Laid as a mosaic, some twenty of these rectified prints are required to cover the whole area photographed. One of the photographs is reproduced in Fig. 1, with a location map in Fig. 2.

The ground survey programme related to field based crop identification (in some 1000 fields) and to sampling of terrain was carried out at the time of firing by Instituto Nacional de Tecnologia Agropecuaria (Argentina) (INTA) and Reading University. Aerial photography was performed from an Argentinian Air Force aircraft which flew more than 2000 kilometres at 3000 m altitude making 33 transects of approximately 80 km each. Spectroradiometric observations were obtained in some 130 of the surveyed fields, together with low level helicopter photography of the same areas, the helicopter also being provided by the Argentine Air Force.

## 2. The Interpretation of Land Use

Visual inspection of the photograph (Fig. 1) enables a primary distinction to be made between the cropped area of the eastern (left-hand) side of the imaged area and the more uniformly toned low caldena woodland with isolated fields or groups of fields to the western (right-hand) side. In this latter area, the density of the blue tone provides an indication of the density of the woodland cover. Where the tone is lighter, this indicates a less dense woodland cover, or grassland, due either to poorer soil and moisture conditions, or to clearance. Ground survey has enabled these categories to be recognized.

In the cropped areas, initial distinctions can be made between the following land use types: (i) healthy green crops with high chlorophyll content, (ii) ripe mature crops, (iii) scrubland and natural pasture of tall grasses and low shrubs, (iv) low weeds, stubble and fallow land, (v) bare soil. These categories approximately correspond to reddish purple, violet, dark blue, light blue and bluish white respectively on the photograph. The sharp contrast between high stubble and freshly ploughed land is particularly well illustrated in field A, marked on Fig. 1, where only the outer edge is ploughed. Fields as small as 20 metres across can be recognized, where the contrast with adjacent fields is high.

## 3. Interpretation of Cultural Features

Many cultural features (apart from the field pattern) can be clearly seen in Fig. 1. The towns of Santa Rosa and Eduardo Castex appear as indistinctly defined white areas on the print, although on the original transparency the grid pattern of the streets is visible. These streets are about 15 metres wide and are separated by blocks about 100 m across. On the true colour photography taken with the 45° oblique 250 mm focal length lens Hasselblad camera (Table 2) individual streets within towns are distinguished. The extent to which roads and railways are apparent varies with their linearity and contrast vis-à-vis adjoining land use. Narrow dirt roads and tracks with light brown sand and gravel surfaces in the scrubland area stand out clearly from their background and can be seen easily. Even roads of only 3 metres width have been distinguished on photographs taken with the 150 mm lens in favourable conditions. On the other hand, major roads (10 metres wide) may not stand out clearly. Usually, where the road itself cannot be seen, its alignment appears as a distinct linear break in the pattern of the fields.

## 4. Interpretation of Natural Resources

In the cultivated area, the linear alignment of zones of greener crops and of scrubland and bare soil, which are associated with field sizes of different dimensions, may indicate the presence of different terrain types which relate to an eastern extension of the rock, soil, water, landform and plant relationships clearly seen in the uncultivated west (for example, area B, Fig. 1). Some illustrations of the potential of the imagery may be noted. These are derived from field traverse observations. For example, the mottled pattern in the north-west (bottom right) of the area imaged is interpreted as an area characterized by linear zones of vegetated sand dunes and sand drifts within which a considerable amount of detail can be discerned. The east-west lineation of cultivation and caldena zones in the south-west (top right) of the image relates to the known occurrence of ridges of tosca (a type of calcrete) and intervening vales of saline sandy soils, lakes and lacustrine deposits. The lakes, showing as the darkest blue-black areas, within any given belt are not connected by surface drainage lines, but each group lies within a separate basin of internal drainage. Former higher water levels around the lakes can also be distinguished.

## Acknowledgments

The authors gratefully acknowledge the assistance provided in Argentina by Comision Nacional de Investigaciones Espaciales (CNIE), particularly Comodoro J. J. Tasso, by Fuerza Aerea Argentina, and by the Director and staff of Instituto Nacional de Tecnologia Agropecuaria (INTA) particularly Dr. O. Dominguez and Dr. D. Cappannini, and the staff of the INTA experimental stations at Anguil, General Pico, General Villegas, Manfredi, Marcos Juarez, Mendoza, Mercedes and San Martin. Acknowledgment is also made of the technical assistance provided by the British Aircraft Corporation and the Royal Aircraft Establishment and also to Dr. D. S. H. Drennan and Mr H. Walkland of Reading University who formed part of the field team and to Dr. E. S. Owen-Jones of Bedford College, London, who made the spectroradiometric observations. The UK participation in this project was financed by the Department of Trade and Industry, and managed for them by the UK's Procurement Executive.

# INDEX OF AUTHORS